Czechoslovak Academy of Sciences

Geology of Recent Sediments

Scientific Editor
Doc. Dr. Jan Petránek

Translated by
Helena Zárubová

Geology of Recent Sediments

Dr. ZDENĚK KUKAL, CSc.

Central Geological Survey, Prague

Academia
Publishing House of the
Czechoslovak Academy of Sciences
Prague

Academic Press
London
and New York

Prague 1971

Distribution throughout the world with the exception of Czechoslovakia:
Academic Press Inc. (London) Ltd., Berkeley Square House, Berkeley Square, London
Library of Congress Catalog Card No. 73-145667 ISBN 0-12-428750-6

Printed in Czechoslovakia

Contents

Page

Introduction 8

1. Sedimentary environments and their classification 11
2. Weathering 17
3. Rate of denudation of continents 25
4. The types of contemporary tectonic movements in relation to sedimentation 41
5. The rate of sedimentation 43
6. Cosmic material in Recent sediments 54
7. Biological components of sediments 56
8. River sediments 72
9. Sediments of alluvial fans 110
10. Eolian sediments 116
11. Glacial sediments 138
12. Lacustrine sediments 157
13. Deltaic sediments 183
14. General characteristics of the marine environment 195
15. Beach sediments 209
16. Shallow-marine sediments 225
17. Tidal flats (waddens, wades) 264
18. Sediments of inland seas 280
19. Deep-sea sediments 291
20. Shallow-water carbonate sediments 343
21. Changes in sedimentation in the Sub — Recent and their causes 376
22. Changes in sediments following their deposition — transition of Recent sediments into ancient deposits 389
23. The application of knowledge of Recent sediments to ancient sediments 398

Conclusions 463

Index 465

Preface

The present book on Recent sediments is intended for all geologists studying sedimentary rocks and all scientific workers who are interested in contemporaneous geological processes on the Earth's surface. The book has two main tasks: to give a concise survey of the composition and genesis of Recent sediments and the laws controlling their deposition, and to point out such sedimentological data that could be applied with advantage to the study of sediments from earlier geological periods. In writing this book I have intended to make up, at least partially, for the lack of synthetizing works that would sum up all the problems bearing on Recent sedimentary rocks.

This subject is, however, so comprehensive that it is impossible to become acquainted directly with all varieties of Recent sediments. Therefore, I had to gather the necessary information from literature and discussions with respective specialists.

My foremost thanks are due to those who, by profound analytical studies, have promoted the science of Recent sediments to its present-day level. Without their great efforts this book could never have been written. Further, I am much indebted to all those who in stimulating discussions have helped me to solve many difficult problems. No work of this kind can be flawless and I shall be grateful to those who will point out any deficiencies of this book or offer their constructive comments and criticisms.

Dr. Zdeněk Kukal, CSc.

Central Geological Survey, Prague

Introduction

The rapid progress of many scientific branches leads to their increasing specialization; comprehensive papers and monographs are today published in fields which not long ago were regarded as highly specialized. This is also true for the science of Recent sediments: instead of brief chapters in textbooks on geology and petrography, several monographs and dozens of short papers appear each year dealing with special aspects of this problem. The interest in Recent sediments is steadily increasing because a perfect knowledge of their deposition contributes to an understanding of ancient sedimentation and makes it possible to reconstruct sedimentary processes throughout geological history. On this basis many palaeogeographical problems of past periods can be solved.

The sedimentology gathers information mostly from a direct observation of contemporary processes, and only subordinately from laboratory experiments and theoretical considerations. In spite of the attention paid to Recent sediments in some countries, their recognition on a world-scale is still insufficient, and the study of them is often underrated. But with regard to the modern requirement of exactness in geological sciences, it is clear that one approach to the fulfilment of this demand is a perfect knowledge of the regularities of Recent sedimentation, which in many cases can be expressed by mathematical formulae and precise figures. Thanks to the study of Recent sediments, sedimentology is becoming an exact geological science.

Table 1

Year	Number of papers	Number of pages	Year	Number of papers	Number of pages
1947	11	272	1955	38	1244
1948	11	333	1956	40	1096
1949	18	408	1957	33	1012
1950	17	524	1958	42	1364
1951	21	680	1959	44	1297
1952	23	635	1960	46	1331
1953	22	724	1961	51	1486
1954	35	906			

The extremely comprehensive literature scattered through many journals, bulletins and monographs on the one hand, and a lack of synthetizing works on the other, make the full usage of information on Recent sediments almost impossible. To illustrate the disproportion between analytic and synthetic papers, works on Recent sediments published during the years 1947—1961 in wel-known journals are listed in Table 1. Their number is roughly doubled by the reports published in special, less significant journals. On the other hand, there are very few synthetizing works; they include books by J. WALTHER (1893), D. V. NALIVKIN (1956), N. M. STRACHOV (1962) and H. and G. TERMIER (1960), and some books dealing only with marine sediments (K. ANDRÉE 1920, H. U. SVERDRUP *et al.* 1942, F. P. SHEPARD 1963, PH. H. KUENEN 1950). Description of Recent carbonate sediments is given in the symposium „Carbonate Rocks“ 1967 (G. V. Chilingar et al., editors). A concise survey of Recent sediments is given, for instance, in textbooks on sedimentary petrography (A. LOMBARD 1956), or oceanography (K. KALLE 1943, E. BRUNS 1958). Of importance are some collections of papers (Recent Marine Sediments 1939, Origin of the Sediments in Recent Basins 1954, Recent Sediments of Seas and Oceans 1961, The Sea 1963), Special issues of Fortschritte d. Geol. Rheinland und Westphalen – 1963, Marine Geology – 1967 and SEPM Symposium – 1969. Although several authors have endeavoured to apply the knowledge of Recent sediments to the research of ancient deposits at least in part (mainly V. P. BATURIN 1944, D. V. NALIVKIN 1956, and F. P. SHEPARD 1963), there is so far no book available that has been written directly with the aim of using the principles established in Recent sedimentary processes for the reconstruction of depositional conditions of ancient sediments. The submitted work is intended to fill this gap. In the detailed description of Recent sediments preference is given to those properties which are of importance also for the diagnosis of ancient sediments. The many-sided and varied data had to be generalized to some extent, so that they could serve as a clue for the study of fossil deposits and their depositional environment. Limited space did not allow all the problems concerned to be treated in full; for this reason some topics (e.g. fundamentals of petrography) discussed in modern textbooks and standard reference books have been omitted. The author is aware of the shortcomings of the submitted publication, which result partly from the divergence of approach to the problems dealt with, but hopes that is will contribute to the knowledge of Recent and ancient sediments and the conditions and environment of their deposition.

References

ANDRÉE K. (1920): Geologie des Meeresbodens. P. 1—689, Leipzig.

BATURIN V. P. (1944): Petrographic analysis of geological past by means of terrigenous components (in Russian). P. 1—338, Leningrad—Moscow.

BRUNS E. (1958): Ozeanologie. Bd I., p. 1—420, Berlin.

Fortschritte d. Geol. Rheindl. u. Westphalen (1963): Special issue, vol. 10: 1—390, Krefeld.

CHILINGAR G. V. - BISSELL H. J. - FAIRBRIDGE R. W. (Editors) (1967): Carbonate rocks. Development in Sedimentology, vol. 9a; 1—471, Amsterdam.

Gilbert G. K. (1914): Transportation of debris by running water. US Geol. Surv. Prof. Pap. 86, p. 1—263, Washington.

Hill M. N. (Editor) (1963): The sea, ideas and observations on progress in the study of the seas (Symposium). Vol. 3:1—963, New York, London.

Kalle K. (1943): Stoffhaushalt des Meeres. P. 263, Leipzig.

Klenova M. V. (1948): Marine geology (in Russian). P. 1—495, Moscow.

Kuenen Ph. H. (1950): Marine geology. P. 1—568, New York.

Laporte L. F. (1969): Ancient environments. P. 1—162, New Jersey.

Lombard A. (1956): Géologie sédimentaire. Les séries marines. P. 1—792, Paris.

Marine Geology (1967): Special issue — Depth indicators in marine sedimentary environments. Vol. 5: 329—562, Amsterdam.

Nalivkin D. V. (1956): Learning about facies (in Russian). Vol. 1: 1—534, vol. 2: 1—393, Moscow—Leningrad.

SEPM Symposium (1969): Sedimentary environments. Bull. Am. Assoc. Petrol. Geol., vol. 53: 703—751, Tulsa.

Shepard F. P. (1963): Submarine geology, 2nd edition. P. 1—557, New York.

Strachov N. M. (1962): Fundaments of the theory of lithogenesis (in Russian). Vol. 2: 1—573, Moscow.

— (Editor) (1954): Origin of sediments in recent basins (Symposium, in Russian). P. 1—791, Moscow.

— (Editor) (1961): Recent sediments of seas and oceans (Symposium, in Russian). P. 1—644. Moscow.

Sverdrup H. U. - Johnson M. W. - Fleming R. H. (1942): The oceans, their physics, chemistry and general biology. P. 1—1060, New York.

Termier H. - Termier G. (1960): Érosion et sédimentation. P. 1—412, Paris.

Trask P. D. (1932): Origin and environment of source sediments of petroleum. P. 1—364, Tulsa.

— (Editor) (1939): Recent marine sediments (Symposium). P. 1—736, Tulsa.

Walther J. (1893): Einleitung in die Geologie als historische Wissenschaft. P. 1—1055, Jena

1. *Sedimentary environments and their classification*

The composition and distribution of Recent sediments result from a number of very diverse processes — weathering of the parent rock, transport and depositional conditions; which impress characteristic features on the rocks. The mode of weathering and transport plays an important role especially in the deposition of continental sediments; the differences in sedimentary conditions come to the fore in marine sedimentation and their significance increases with depth. On the whole, the influence of the conditions at the site of deposition is most important. The composition of Recent sediments deposited under the same conditions is mostly similar, even if sedimentation takes place in different climatic zones under different conditions of rock weathering and material transport. On the other hand, widely different sediments originate under identical weathering and transport conditions but in different sedimentary environment.

The origin and composition of sediments are controlled by the following factors: 1. mechanical, 2. physico-chemical, 3. biogenic.

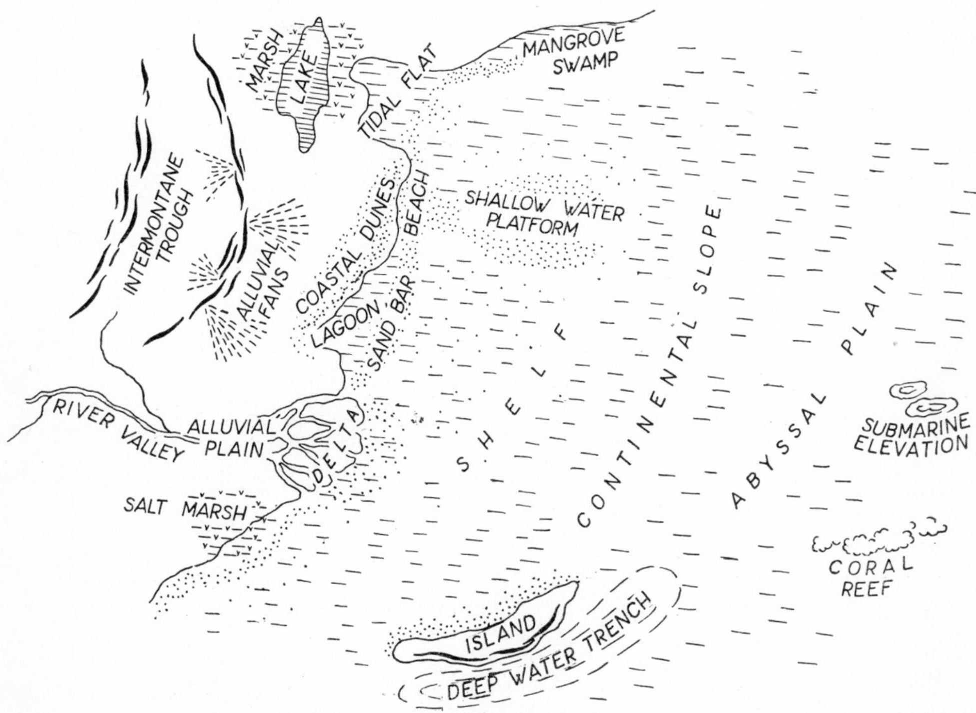

Fig. 1. Schematic distribution of Recent sedimentary environments.

The mechanical factor can be measured as the energy of the environment, i. e. by the current velocity or by the movement energy of water particles. The comparison of (for instance) wave energy and velocity of currents in various environments reveals great differences (Table 2) which are reflected in the composition of sediments.

Table 2

Environments	Current velocity cm/sec	Wave energy kg/m^2
Great rivers	200 and more	max. 10
Tidal flats	160	usually very small
Pelagic ocean	up to 30	max. 8450
Inland seas	up to 20	about 100

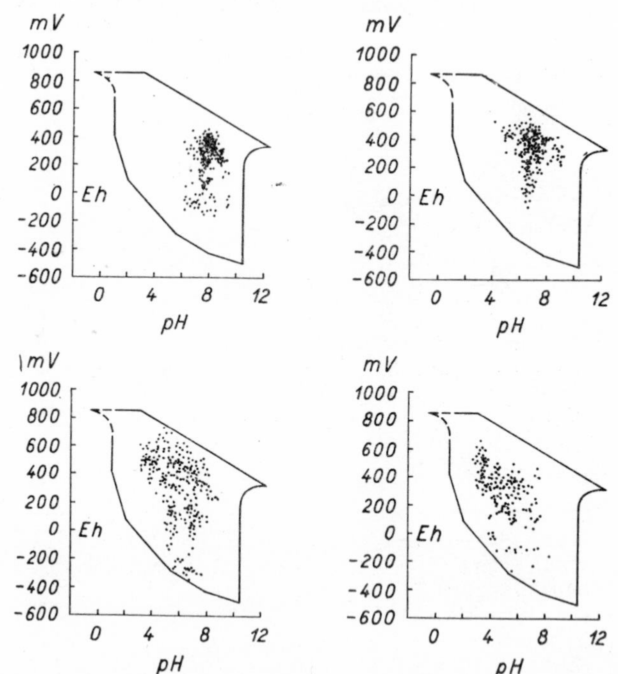

Fig. 2. Eh-pH characteristics of sea water (upper left), fresh water (upper right), soils (lower left) and marshes (lower right). After L. G. M. Baas Becking *et al.* (1960).

Physico-chemical conditions likewise vary within broad limits. Changes of pH (H˙ion concentration) and Eh (oxidation-reduction potential) are the most sensitive indicator. Examples of various physico-chemical conditions are shown in Fig. 2. In addition to pH and Eh, it is mainly the salinity, temperature and degree of $CaCO_3$ saturation which influence sedimentation.

Of the biological factors, quantitative and qualitative differences in the production of organisms and organic matter must be primarily considered. The range of differences can be instanced by the differences in the intensity of the formation of organic carbon under diverse continental conditions:

	Annual production of organic carbon
Forests	$8{\cdot}8 \times 10^9$ t
Cultivated land	$4{\cdot}3 \times 10^9$ t
Steppe land	$1{\cdot}9 \times 10^9$ t
Deserts	$0{\cdot}1 \times 10^9$ t

In addition to the formation of organic matter, its preservation in sedimentary rock is likewise of importance.

From these three fundamental factors controlling the composition of sediments the existence of the main components of Recent and ancient rocks can be inferred:

1. Mechanical components,
2. Chemical components,
3. Biological components.

They participate in the composition of individual sediments in various proportions, and their interrelationship is often indicative of the conditions of deposition. In Recent sediments these conditions can be investigated and checked against the composition of the rock; in ancient sediments the composition can readily be determined, but the sedimentary conditions must be derived from analogy with the relations established for Recent sediments. The establishment of the conditions and environment of deposition is the most important problem of sedimentology, whose solution will help in unravelling palaeogeographical and geological relations in the individual epochs.

Sedimentary conditions on the Earth's surface are most varied and their classification is not easy. The complex of sedimentary conditions (mechanical, physico-chemical and biological) represents the environment of sedimentation, which is the basic unit in the study of Recent sediments. The prerequisite of such a study is, as far as possible, a natural classification of the environments of deposition. Owing to the heterogeneity of geographical and other conditions, a uniform classification valid for the whole of the Earth's surface has not yet been compiled, although a number of excellent classifications already exist for certain parts of the surface. It is obvious that all classification schemes are not of the same order: the more detailed and thorough is the research, the more detailed the classification based on it. The differentiation of marine, continental, and transitional environments may be quite sufficient for reconnaissance survey purposes, but in very detailed investigation, even the minor lagoons, for instance, are differentiated into lagoons in the proximity of river mouths, sand bars, in the vicinity of oyster bioherms or inlets, etc.

Table 3

Marine environments	Transitional environments	Continental environments
Open shelves Sheltered shelves Inland seas Continental slopes Pelagic oceans Deep water trenches Coral reefs	Lagoons and bays Deltas	Mountain ranges Intermountainous troughs Deserts River valleys Lakes Alluvial plains Coastal plains

First order	Second order	Third order	Fourth order
Continental shelf	Subaerial delta	1. Upper flood plain	
		2. Lower flood plain	a) Channel b) Point-bar c) Levee d) Backswamp e) Cut-off channel
		3. Mangrove swamp	a) Main channel b) Gullies c) Inter-creek flat d) Point-bar e) Intra-swamp delta
		4. Beach	a) Active b) Beach ridge c) Transverse channel
	Marginal barrier island-lagoon	1. Beach 2. Channel and creek 3. Lagoon 4. Lagoon delta 5. Marginal swamp	
	Marginal estuary	1. Marginal swamp 2. Open water	
	Continental shelf under delta influence	1. River mouth bar 2. Delta-front platform 3. Pro-delta slope 4. Open shelf	

The basic classification of sedimentary environments compiled mainly after D. V. NALIVKIN (1956) and W. C. KRUMBEIN- L. L. SLOSS (1951), as given in Table 3, seems to comply best with natural conditions.

Each of the environments can be subdivided into innumerable sub-environments, depending on the thoroughness of investigation. Examples of a detailed division of some shallow-water environments are given above (p. 14).

Not all kinds of Recent sediments have been investigated to the same extent. Whereas some parts of the Earth's surface have been explored most thoroughly, a great part of the continents and the sea bottom still awaits to be studied in detail. Present-day knowledge of the principles governing Recent sedimentation is based on detailed investigation of these environments:

Rivers:	Mississippi, Colorado, Volga, Rhine, Fyris, Tessin, etc.
Eolian sediments:	Dune sands from coastal plains of North Sea, Atlantic coastal plains of North America; loesses of Ukraine, Germany, North America, Argentine and China
Alluvial fans:	Semi-arid regions of North America
Glacial sediments:	Russian Table, North America, Scandinavia
Alluvial plains:	Louisiana, Eastern China
Small lakes:	North Germany, North Poland, Russian Table, North America, the Alps
Great lakes:	Aral Sea, Balkhash, Great Lakes of North America
Intercontinental seas:	Black Sea, Caspian Sea, Mediterranean Sea, Baltic Sea, Adriatic Sea, Gulf of Mexico
Beaches:	Gulf of Mexico, North and Baltic Seas, Egyptian coast of Mediterranean Sea, etc.
Shelf:	Atlantic shelf of North America, Ochotsk Sea, Scandinavian shelf, Atlantic shelf of South America and Central Africa
Bays and lagoons:	Texas Bay, Naples Bay, Scandinavian fiords, Finland Bay, Bays of North American Atlantic coast
Deltas:	Mississippi, Volga, Amu-Darya, Rhine, Rhône, Fraser, Niger, Colorado, etc.
Continental slope:	Atlantic of North America, Scandinavian, etc.
Abyssal ocean:	All oceans
Deepwater trenches:	Aleutian, Kurile, Mariana, Peru-Chile
Coral reefs:	Bikini, Eniwetok, Saipan, Kapingaramangi, Florida Reefs, Great Barrier Reef

The data presented in this book should serve mainly as a clue to the recognition of sedimentary environments of ancient sediments. As the above-mentioned detailed classification is not usable for ancient deposits, a basic classification has been drawn up, based on Recent environments but respecting the possibilities of the sedimentological methods in the investigation of ancient sedimentary deposits.

1. Continental environments:
   - River valleys and alluvial plains
   - Deserts and regions of eolian sedimentation
   - Mountain ranges and intermountainous troughs
   - Lakes

2. Marine environments:
   - Deltas a) subaerial parts
     b) submarine parts
   - Shallow plains a) sheltered
     b) open
   - Shallow basins
   - Regions of coral reefs and bioherms
   - Environments analogous to recent bathyal
   - Environments analogous to recent abyssal

References

ANDEL TJ. VAN - POSTMA H. (1954): Recent sediments of the Gulf of Paria. Verhandel. Koninkl. Nederl. Akad. Wetenschap., afd Natuurk., Erste Reeks, D. 20, No 5: 1—244, Amsterdam.

BAAS BECKING L. G. M. - KAPLAN I. R. - MOORE D. (1960): Limits of the natural environments in terms of pH and oxidation-reduction potentials. Journal of Geology, vol. 68: 243—284, Chicago.

CROSBY E. J. (1969): Classification of sedimentary environments. Bull. Am. Assoc. Petrol. Geol., vol. 53: 714, Tulsa.

HEDGPETH J. W. (1957): Classification of marine environments. Mem. Geol. Soc. Am., vol. 67, P. I: 93—100, Baltimore.

KRUMBEIN W. C. - SLOSS L. L. (1951): Stratigraphy and sedimentation. P. 1—409, San Francisco.

LAPORTE L. F. (1969): Ancient environments. P. 1—162, New Jersey.

NALIVKIN D. V. (1956): Learning about facies (in Russian). Vol. I: 1—534, vol. 2: 1—193, Moscow—Leningrad.

SHEPARD F. P. - MOORE D. G. (1955): Central Texas coast sedimentation; characteristics of sedimentary environment, recent history, and diagenesis. Bull. Am. Soc. Petrol. Geol., vol. 39: 1463—1593, Tulsa.

2. *Weathering*

This chapter will deal with the rate of weathering processes and the relation of weathering to Recent sedimentation. A detailed description of weathering processes and their products is beyond the scope of this publication.

The rate of weathering can be measured according to the rate of removal of material and soil formation and according to the development of talus cones in the case of mechanical weathering.

The rate of soil formation depends mainly on the climate and subordinately on the parent rock. The differences in weathering rates of the same rock under different climatic conditions are much greater than those established in weathering of different rocks under identical climate conditions. The data given in Table 4. clearly show that neither the climate-weathering nor parent rock-weathering relations are linear functions.

During the last half century, a considerably quantity of field and experimental data bearing on weathering rates have accumulated, much of which was summarized by H. JENNY (1941). Values vary from several centimetres in a year to less than 2 cm in 5,000 years. The definition of the degree of weathering is in itself variable, but a review of the data suggest that changes produced by weathering including the increase in depth of weathering zone, are an exponential rather than a linear function of time.

In humid tropical climatic conditions the rate of soil formation is recorded as roughly 10—30 cm in 1,000 years. The weathering rate decreases, naturally, with the increasing thickness of the soil layer. Under all climatic conditions the weathering of granite and gneiss is approximately twice as rapid as the weathering of clastic terrigenous sediments.

Under a temperate humid climate the weathering rate was studied on the basis of the rate of decalcification of marine sediments in the reclaimed polders of the Netherlands. During 300 years the original 10% of $CaCO_3$ has been removed by the activity of subaerial agencies. According to other data, a 30—40 cm thick soil layer developed on clays of the dried-out Ragunda lake in Sweden over 1,000 to 1,500 years; the thickness of the part including the A_2 horizon was 10 cm, that of B horizon attained 10—20 cm. The development of soil can also be observed on the increase of organic matter, and of organic carbon and nitrogen in particular. A layer of volcanic ash in the near-surface 30 cm contained 1 – 2% organic matter (i. e. 0·22—0·035% org. N) 14 years after deposition and 2·1% (i. e. 0·1% org. N) after 30 years.

Data on the weathering of lava in the Krakatoa island are well known from literature. During 45 years of tropical weathering the chemical composition of lava

Table 4

Depth of weathering (After L. B. Leopold *et al.* 1964)

Location	Rock type	Depth (cm)
Wakefield, New Hampshire	Granite	300—450
Keene, New Hampshire	Granite	300—600
Gorham, New Hampshire	Granite	300—450
North Conway area, New Hampshire	Granite	up to 600
Washington, D.C.	Granite	640
Pikes Peak, Colorado	Granite	600—900
Atlanta, Georgia	Granite	2,800—9,000
South California	Granodiorite	up to 6,000
Transvaal, South Africa	Granite	up to 6,000
Hongkong, China	Granite	6,000
Transvaal, Natal	Granite	1,900
Kwantung, China	Granite	1,000
Cameroons	Granite-gneiss	1,900
Madagascar	Mica schist	3,000
British Guiana	Granite	480
Compos Cerrados, Brazil	Granite	1,000
Sandwich, New Hampshire	Syenite	60—300
Rowan County, North Carolina	Diorite	600
Juneaz, Alaska	Diorite	600
Medford, Massachusetts	Diabase	300
Minnehala County, South Dakota	Diabase	750
Rowan County, North Carolina	Meta-gabbro	120
Saipan	Andesite	1,500
Cameroons	Andesite	300—1,950
Puerto Rico	Andesite tuff	1,200
Bintan	Liparite breccia	4,800
Cuba	Serpentine	1,200—1,500
Saipan	Limestone	180
Rio Grande do Sul	Shale	1,170

underwent the following changes (after E. C. J. Mohr - P. A. Van Baren 1954) (Table 5). The rate of weathering of bedrock-till in New Hampshire is according to N. M. Johnson *et. al.* (1968) 800 kg/ha/year.

The main features of lateritic weathering were apparent already after this short period, particularly the formation of free hydrated Al and Fe oxides and the decrease in SiO_2/Al_2O_3 ratio.

The values recorded for the weathering rate of limestones are somewhat different. According to V. V. Akimcev (1932) a 10—40 cm thick layer of soil with 5% $CaCO_3$ and 50—56% clay developed on limestones under a humid temperate climate during

Table 5a

Chemical composition of fresh and weathered lava from Krakatoa
After E. C. J. Mohr - P. A. Van Baren 1954

	SiO_2 %	Al_2O_3 %	Fe_2O_3 %	FeO %	CaO %	Humus %	pH	SiO_2/Al_2O_3
Weathered surface of lava	61·13	17·24	2·59	3·60	0·45	0·45	6·0	6·03
Transitional layer	65·87	16·37	1·74	2·05	3·07	0	5·8	6·86
Fresh rock	67·55	16·19	1·52	2·15	2·89	0	5·3	7·56

a period of 231 years. Other writers, however, note that under the same climatic conditions only a few centimetres of soil is formed on limestones in 250 years. From the surface of Egyptian pyramids (built of nummulitic limestones) 0·2 mm of material are removed yearly, i. e. a 5 cm thick bed in 250 years.

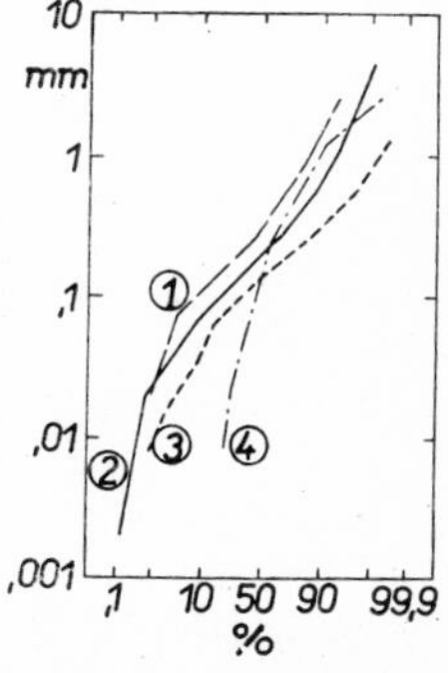

Fig. 3. Particle-size distribution of soil material without profile development, caused principally by freeze and thaw, compared with decomposed granite from Malaya. (Schist and syenite sample from Norway, shale from Alaska.) 1. Hornblende schist, 2. Slate, 3. Mangerite syenite, 4. Decomposed granite. After L. B. Leopold *et al.* (1964).

On the major part of the Earth's surface, the process of soil formation is not always completed, the partly weathered material being removed and occassionally transported into the basin of deposition. On the basis of the above-mentioned values, the mean rate of soil formation over the whole of the Earth's surface seems to be about 10 cm in 1000 years. The total rate of the denudation of continents, however, is sure to be higher than the rate of soil formation, because the material removed from the surface consists mainly of products of imperfect, mostly only physical, weathering.

Although the influence of climatic conditions on the weathering rate is decisive, the nature of the parent rocks must also be taken into account. The weathering products of the most widespread rocks the intrusives of granitoid character have been petrographically studied in fairly great detail (e. g. M. C. McEwen *et al.* 1959, E. P. Ruxton 1957). Granular parameters of the residual granite eluvium are as follows:

Md — range of values 0·28 mm—3·70 mm, average value 1·47 mm.

So (coefficient of sorting according Trask) — range of values 1·66—4·44. Sorting is therefore from very bad to good.

The coefficient of skewness is always positive, indicating that in the weathering products the sorting of fraction coarser than the Median is better than that of finer fractions. The slight redeposition of material induces considerable changes in the character of the weathering products. The median moderately decreases (mean value 0·52 mm) and the coefficient of sorting declines (mean value So 1·49), which is indicative of the rising degree of sorting. The coefficient of skewness is not greatly affected by redeposition.

The amounts of grain-size fractions originating by the physical weathering of granite show the following values (Table 5b).

Table 5b

Amounts of grain-size fractions originating from the physical weathering of granite

	Average amount %	Maximal amount %	Minimal amount %
Gravel	12	36	1
Sand	48	84	36
Silt	28	40	1
Clay	12	34	0

From the above data it can be inferred that the weathering of plutonic igneous rocks gives rise mainly to medium grained sand. Unfortunately, detailed data on the weathering products of basic plutonic and effusive rocks are not available. The sporadic data reveal that the main products of their weathering are fine gravel, sand and silt. The amount of coarse and medium grained sand is lesser than that in the weathering products of granitoids.

---

* Note to presentation of grain-size analyses and parameters

It was desirable to chose a method of presentation of the grain-size analysis which would be suitable to illustrate the genesis of sediments. It was also necessary to chose a method which had been widely used so there were thus many analyses of a wide variety of sediments already expressed in this way in the literature.

The parameters proposed by D. L. INMAN (1952) comply best with these requirements. Arithmetic and logarithmic grade scales have been used. The value of logarithmic scale (Φ) can be converted into the arithmetic ones according to the following table:

Φ	mm	Φ	mm
−3·00	8·00	2·25	0·21
−2·75	6·73	2·50	0·18
−2·50	5·66	2·75	0·15
−2·00	4·00	3·00	0·125
−1·75	3·36	3·25	0·105
−1·50	2·82	3·50	0·088
−1·25	2·38	3·75	0·074
−1·00	2·00	4·00	0·0625
−0·75	1·68	4·25	0·0526
−0·50	1·41	4·50	0·0442
−0·25	1·19	4·75	0·0372
0·00	1·00	5·00	0·0313
0·25	0·84	5·25	0·0263
0·50	0·71	5·50	0·0221
0·75	0·53	5·75	0·0186
1·00	0·50	6·00	0·0156
1·25	0·42	6·25	0·0131
1·50	0·35	6·50	0·0110
1·75	0·29	6·75	0·0092
2·00	0·25	7·00	0·0078

The relation between the 2 values can be expressed by the equation:

$$\Phi = -\log_2 \xi$$

where ξ is diameter of particles in millimetres.

The following parameters have been used:

Median (Md) is a point of crossing of the cumulative curve and the 50% line.

Coefficient of sorting is used either in the sense of Trask, as $S_0 = \sqrt{(Q_3/Q_1)}$; $Q_3 = 75$ percentile, $Q_1 = 25$ percentile, or after D. L. INMAN as Phi Deviation measure:

$$\sigma_\Phi = 1/2(\Phi_{84} - \Phi_{16})$$

Φ_{84} and Φ_{16} designate 84 and 16 percentiles (i. e. point of intersection of cumulative curve with 84% and 16% lines).

The Phi Skewness Measure indicates skewness of the grain-size curves. It is expressed by an equation:

$$\alpha_\Phi = \frac{M_\Phi - Md_\Phi}{\sigma_\Phi}.$$

If the value of the coefficient of α_Φ is positive, the mean grain-size (M) is finer than median (Md) and the sorting of fractions finer than median is poorer (grain-size curve is flatter within this range). The negative coefficient α_Φ indicates mean (M) coarser than median (Md) and a poorer sorting of fractions coarser than Md.

M is mean grain-size expressed by the formula:

$$M = 1/2(\Phi_{16} + \Phi_{84}) .$$

In spite of certain shortcomings, these parameters are widely used today for the expression of the analyses of Recent sediments. For detailed information on grain analyses and the presentation of their results I refer particularly to the works by D. L. INMAN (1952), and G. MÜLLER (1964).

The materials developing during the formation of a soil profile differ widely. The grain-size distribution of some tropical and arctic soils is given in the diagrams below (Figs 4, 5). The median of tropical soils is roughly on the silt-clay boundary (about 0·01 mm). A number of analyses shows even much lower values of median.

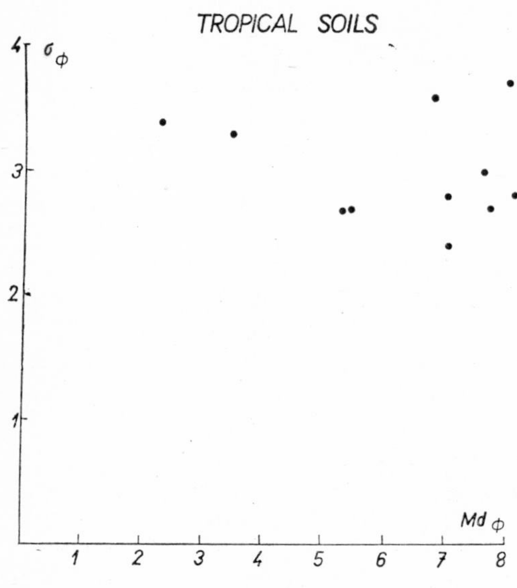

Fig. 4a. Grain-size parameters of tropical soils. Relation between Phi Median Diameter and Phi Deviation Measure. Data compiled from E. C. J. Mohr-P. A. Van Baren (1954), G. W. Robinson (1932) and L. Smolík (1957).

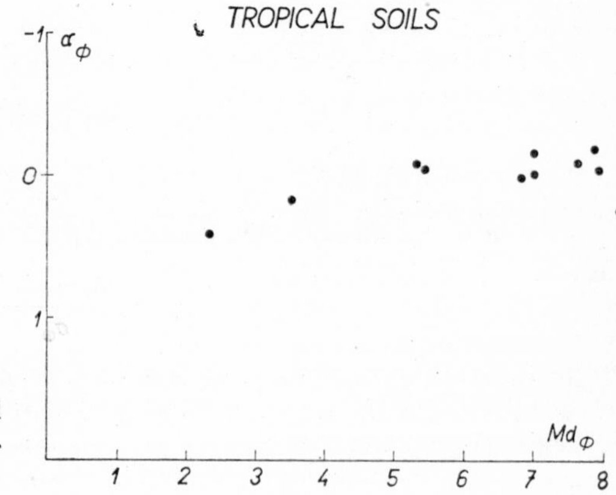

Fig. 4b. Grain-size parameters of tropical soils. Relation between Phi Median Diameter and Phi Skewness Measure.

Contrary to tropical soils, which are rich in clay and silt, the soils of polar regions have the median on the silt — sand boundary; most values fall in fine sand grades. Soils of temperate humid climate are, as far as grain-size is considered, intermediate between the 2 above types; in the major part, they are richest in silt and clay, occasionally in fine-grained sand. In German soils, for instance, the mean content of silt exceeds 30%.

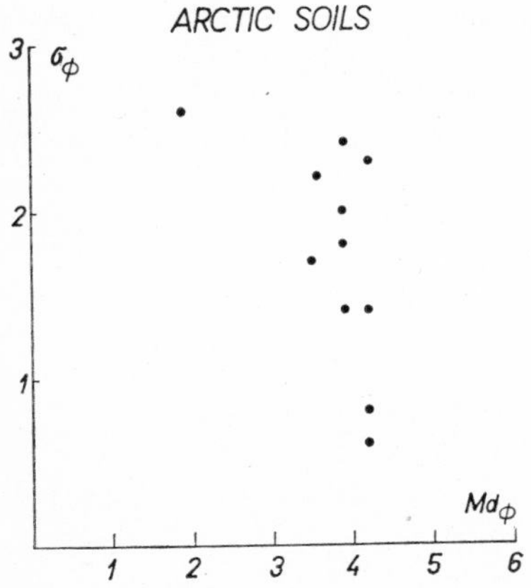

Fig. 5a. Grain-size parameters of arctic soils. Relation between Phi Median Diameter and Phi Deviation Measure. Data compiled from G. W. Robinson (1932) and L. Smolík (1957).

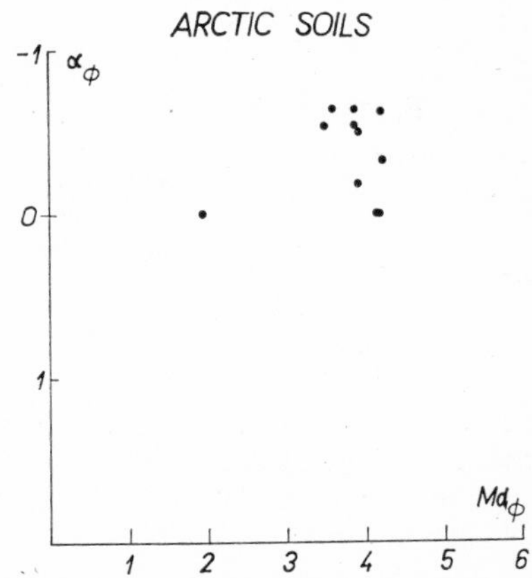

Fig. 5b. Grain-size parameters of arctic soils. Relation between Phi Median Diameter and Phi Skewness Measure.

The soils seem to be one of the main source of silt and clay transported into the Recent sedimentary basins. Indeed it is very likely that the majority of these fine fractions — which are components of sediments at present — originate during the weathering processes, the percentage produced by abrasion being negligible. Thus, the high content of silt and clay in Recent and ancient sediments can be explained as being derived from the supplied weathering material, which is already primarily rich in these fine fractions.

The diverse influence of climate on the production of various grain-size fractions during weathering can be inferred from the different grain-size of soils in individual climatic zones. The humid tropical climate produces the greatest amount of clayey material with a high percentage of silt. The tropical semi-arid and arid climate produces mostly coarser clay-poor material. Silt with a large amount of clay and fine-grained sand originates under a humid temperate climate. The polar climate gives rise mainly to fine sandy material.

References

AKIMCEV V. V. (1932): Historical soils of the Kamenetz-Podolsk fortress. Proc. II. Internat. Cong. Soil, vol. 5: 130—132.

INMAN D. L. (1952): Measures for describing the size distribution of sediments. Jour. Sedim. Petrology, vol. 22: 125—145, Menasha.

JENNY H. (1941): Factors of soil formation. P. 1—281, New York.

JOHNSON N. M. - LIKENS G. E. - BOHRMANN F. H. - PIERCE R. S. (1968): Rate of chemical weathering of silicate minerals in New Hamsphire. Geochim. Cosmochim. Acta, vol. 32: 511—545, London.

KELLOG C. E. (1941): The soils that support us. P. 1—370, New York.

KELLY W. C. - ZUMBERGE J. H. (1961): Weathering of a quartz diorite at Marble Point, McMurdo Sound, Antarctica. Journal of Geology, vol. 69: 433—446, Chicago.

LEOPOLD L. B. - WOLMAN M. G. - MILLER J. P. (1964): Fluvial processes in geomorphology. P. 1—522, San Francisco.

MCEWEN M. C. - FESSENDEN F. W. - ROGERS J. J. W. (1959): Texture and composition of some weathered granites and transported arkosic sands. Jour. Sedim. Petrology, vol. 29: 477—492, Menasha.

MOHR E. C. J. - VAN BAREN P. A. (1954): Tropical soils. P. 1—498, Bandung.

MÜLLER G. (1964): Methoden der Sediment-Untersuchung P. 1—303, Stuttgart.

MÜNCHENHAUSEN L. (1959): Die wichtigste Bodentypen des deutschen Bundesrepublik. P. 1—212, Bonn.

ROBINSON G. W. (1932): Soils. P. 1—409, London.

RUXTON E. P. - BERRY L. (1957): Weathering of granite and associated erosional features in Hong Kong. Bull. Geol. Soc. Am., vol. 68: 1253—1292, Baltimore.

SMOLÍK L. (1957): Pedology (in Czech). P. 1—329, Prague.

3. *The rate of denudation of continents*

The rate of denudation is closely related to the rate of weathering. Rocks liable to weathering are removed at a greater rate than rocks more resistant to this form of disintegration. The difference in the rate of erosion of 2 rock types which show various resistance to weathering is never so great as the difference in the rate of weathering of the same rock, in the fresh or weathered state.

The main erosional factor is the run-off, which is most active on vegetation-free slopes, i. e. a condition which undoubtedly existed frequently during geological history. The rate of erosion depends on: 1. the climate, 2. the configuration of the relief, 3. the presence of vegetation, and 4. the degree of cultivation of the area.

Erosion proceeds at the greatest rate in the mountainous areas of the arid and semi-arid climate. W. M. DAVIS (1938) noticed that the sheetfloods and streamfloods are the main tools of the erosional process. Sheetfloods occur after heavy downpours, recurring repeatedly within the periods of several tens or hundreds of years. Material-laden water flows down minute channels (joining in fan-shaped sheets) into perennial lakes or piedmont depressions. The erosion and transportation of material accomplished by sheetfloods during several minutes are greater that the erosion and transportation produced by rivers over a few years. Sheetfloods were mostly effective during those geological periods when the slopes were devoid of vegetation cover. We speak about streamfloods when the mass of rock material and water is confined to channels; their activity proceeds at a slower rate.

The rate of erosion has been investigated mainly by:

1. The determination of the amount and character of material transported by rivers into the sea during one definite time unit;

Table 6

Different kinds of denudation in Alps
(After H. Jäckli 1958)

Type of denudation	Vertical movement of material by different processes in 10^6 t/year
Solifluction	0·75
Running water	5,930
Earth slides	274
Avalanches	45
Glaciers	27

Table 7

Intensity of mechanical denudation in the basins of major streams of the world

River	Amount of suspension load transported in 10^6 t/year	Mechanical denudation in drainage basin in t/km^2/year
Mekong	1,000	1,200
Ganges	1,800	1,040
Euphrates and Tigris	725–1,000	690–1·000
Irawadi	350	850
Rioni	8·5	633
Terek	26	600
Indus	400	420
Rhône	31·5	320
Colorado	160	271
Po	18	240
Yangtze	275	234
Mississippi	500–750	154–230
Yukon	88	103
Danube	83	101
Zambezi	100	75
Amazon	1,000	60
Columbia	26	47
Orinoco	45	47
La Plata	96·5	32
Niger	67	32
Nile	88	31
Rhine	4·5	20
Volga	25·7	19
Ob	14·2	6
St. Lawrence	4	4
Yenisei	10·5	4

2. The measurement of the rate of sedimentation in lakes and reservoirs.

The rate of denudation is given either in millimetres per 1,000 years, or in square metres per square kilometres yearly.

The denudation is accomplished not only by sheet or streamfloods and rivers, but also by many other agents; their activity, however, is mostly of a local character or is restricted in another way. The correlation of the intensity of different kinds of erosion active in the Alps is given in Table 6.

The Alps, as are most areas on the Earth's surface, are eroded most rapidly by the effects of running water (including sheetfloods and streamfloods). The denudation is mechanical (studied according to the transportation and deposition of solid

Table 8

Dissolved and suspended load in selected rivers in different climatic regions in the United States (After L. B. Leopold *et al.* 1964)

River and Location	Elevation m	Average suspended load	Average dissolved load	Dissolved load as % of total load
		millions of tons/year		
Little Colorado Woodruff, Arizona	1,900	1·6	0·02	1·2
Canadian River near Amarillo, Texas	1,360	6·41	0·124	1·9
Colorado River Near San Saba, Texas	350	3·02	0·208	6·4
Bighorn River at Kane, Wyoming	1,200	1·60	0·217	12
Green River at Green River, Utah	1,300	19·0	2·5	12
Colorado River near Cisco, Utah	1,350	15·0	4·4	23
Iowa River at Iowa City, Iowa	305	1·184	0·485	29
Mississippi River at Red River Landing, Louisiana		284	101·8	26
Sacramento River at Sacramento, California	0	2·85	2·29	44
Flint River near Montezuma, Georgia	85	1·4	0·132	25
Junista River near New Port, Pennsylvania	120	0·322	0·566	64
Delaware River at Trenton, New Jersey	2·8	1·003	0·830	43

particles) and chemical (determined according to the amount of dissolved material removed). Much data is available in the literature on the intensity of mechanical

Table 9

Rate of denudation and its relation to topography and climate
(After J. Corbel 1959)

Lowland	Rate of denudation mm/1,000 years	Rate of chemical denudation	
		% of total	mm/1,000 years
Climate with cold winter	15	87	13
Intermediate maritime climate (Rhine)	27	83	22
Continental climate (Mississippi, Missouri)	58	18	10
Hot dry climate (Mexico)	22	10	1
Tropical desert climate (Sahara)	1	—	—
Equatorial climate with dry season	22	70	15
Mountains			
Periglacial climate	604	34	205
Oceanic climate of mountain chains	217	51	110
Climate of Mediterranean high mountain chains	449	18	78
Semi-arid climate	100	40	40
Arid climate	177	4	7

denudation in the basins of major streams of the world (the values given in Table 7 are mainly from G. V. Lopatin 1952 and I. V. Samoilov 1952).

The overall denudation of large basins requires more time than that of small basins because sediment may be deposited in the valley several times before it reaches the basin mouth.

The study of the dependence of the denudation rate on topographic and climatic factors revealed the following facts (Table 9).

The differences in the denudation rates are substantially greater between the lowland and mountain areas of the same climatic zone than between the individual climatic zones. This suggests that the topographical character of the area is the main controlling agent.

The maximum rate of stream erosion (excluding the glacial streams) has been found in the basin of the Himalayan river Kosi: 1,144 mm in 1000 years. The glacial streams, whose activity due to the small extent of glaciated regions is rather limited, show a far greater erosion rate than other rivers, as evidenced by studies in the glaciated areas of the Alps, Greenland and Scandinavia. The maximum erosion rate — 30,000 mm in 1,000 years — has been established for the stream issuing from the Hidden Glacier in Alaska. The yearly erosion value of 3 cm is indeed astonishing. Values recorded for other glacial rivers are listed in Table 10.

Table 10

Rate of river denudation in glaciated areas
(After J. Corbel 1959)

Drainage basin of river	Rate of denudation mm/1,000 years
Bosson (Chamonix)	1,800
Blanc (French Alps)	1,600
Heilstuga (Norway)	1,400
Memurelven (Norway)	1,600
Auserfjötur (Norway)	2,200
Jokullsá (Finland)	2,200
Hoffeksjökul (Iceland)	3,200
L-Isortok (Greenland)	2,500
Saskatchewan (Canada)	2,000
Muir (Alaska)	5,000

Under similar topographical conditions the erosion rate of glacial streams is about four times greater than that of the non-glacial ones. In glacial drainage basins mechanical denudation always dominates over chemical denudation, comprising more than 90 % of the total value.

It is very difficult to calculate the average rate of denudation of all continents. Calculations based on the best available data after S. A. Schumm (1963) indicate that denudation will occur at the average maximum rate of 90 cm per 1,000 years in the early stage of the cycle of erosion and that the average rate of denudation will be about 8 cm/1,000 years. G. V. Lopatin (1952) determined the average rates of denudation for the individual continents (Table 11); the differences found are due mainly to the differences in the relief. The mountainous tracts of Asia succumb to denudation at a maximum rate, the lowland stretches of Europa and Australia are denuded at the slowest pace.

Table 11

Average rate of denudation of continents (in 10^6 t/year)
(After G. V. Lopatin 1952)

Continents	Area 10^6 km^2	Mechanical denudation	Chemical denudation
Europe	9·67	420	305
Asia	44·89	7,445	1,916
Africa	29·81	1,395	757
North and Central America	20·44	1,503	809
South America	17·98	1,676	993
Australia	7·96	257	88

Table 12

Rate of limestone denudation
(After J. Corbel 1959)

	Rate of limestone denudation mm/1,000 years
Arctic climate	40
More moderate arctic climate	
Svartisen (Norway)	400
Gold Creek (Alaska)	530
Capilano	420
Oceanic cold climate	
Lismore (Ben Nevis)	150
St. Casimir (Quebec)	160
St. Thérése (Quebec)	120
Continental climate	
Jasper (Alberta)	40
Whitehorse (Canada)	32
Fort-Simpson (Mackenzie)	120
Moderate humid climate	
Lesse (Belgium)	27
Mediterranean climate	
Jugoslavia (humid)	60
South Algeria (arid)	6
Tropical humid climate	
Lowland (Yucatan)	16
Mountains (Guatemala)	45

The proportion of mechanical to chemical denudation changes with the increase of the total denudation. The presumption that chemical denudation decreases with increasing denudation rate is not correct, but whereas the rate of mechanical denudation can increase as much as 500 times, the increase in chemical denudation is only 70 times. The increase of the former with the increase of topographical differences, the mean annual temperature and amount of precipitation is much more rapid than the increase of the latter.

The composition of individual components of the material in suspension and solution indicates the rate of denudation of rocks and rocks complexes. The rate of denudation of limestone under different climatic conditions has been established on the basis of the $CaCO_3$ content in the rivers (Table 12).

The data of Table 12 suggest that the rate of limestone denudation increases towards the cooler climatic zones, the increased solubility of calcium carbonate being caused by the lower temperature of the water. The influence of other factors, such as lithology and the nature of vegetation, cannot be omitted either. For example M. M. SWEETING (1964) ascertained that the limestone ground in north-western England is lowered very slowly at a rate of 30 mm in 1,000 years, whereas this rate is frequently higher at a higher temperatures.

The rate of erosion of silicites has been determined on the basis of the amount of SiO_2 transported, it proved to increase generally towards the warmer climatic zones (Table 13).

Table 13

The rate of silicites denudation
(After J. Corbel 1959)

	Rate of silicites denudation mm/1,000 years
Cold, humid climate	0·1—4·0
Oceanic moderate climate	0·9
Continental climate	5·0
Tropical humid climate with dry season	0·1—4·0
Equatorial climate (Amazon, Java)	4·0—9·0

The interrelation of mechanical and chemical denudation is also indicated by the amount of elements carried in suspension and solution. Some elements are transported predominantly in suspension, others more frequently in solution. The elements pass from suspension into solution in the following succession:

$$(V, Be, Ga, Zn, Cr, Ni) - (Fe, Mn, P, Pb, Sn)—Ba—Co—Sr .$$

The first group of elements migrates in suspension under all conditions. Elements of the other group are present in compounds mostly in solution at low rates of denudation, migrating mainly in suspension at higher rates. Co and Sr under practically all conditions migrate only in solution.

Organic substances are a good indicator of the denudation rate and of climatic and morphological conditions at denudation times. They can be transported both in solution and suspension. The greater the rate of denudation and the total content of suspended load, the higher is the percentage of organic substances carried in suspension. Some examples are given in Table 14.

Contrary to earlier opinions, it has been recognized that a considerable part of calcium carbonate can also be transported in suspension. The ratio of suspended carbonate to carbonate in solution is also a useful indicator of the rate of denudation. The values taken from N. M. STRACHOV (1960) are listed in Table 15.

Table 14

Amounts of organic carbon transported in suspension load and in solution (After N. M. Strachov 1960)

River	Suspension load g/l	Org. C	
		in suspension %	in solution %
Pripet	0·005	16·8	83·2
Dnieper	0·027	19·6	80·4
Don	0·155	44·5	55·5
Danube	0·179	35·7	64·3
Rioni	0·251	67·1	32·9
Tchoroch	4·219	95·5	4·5

Table 15

Amount of carbonate transported in suspension load and in solution (After N. M. Strachov 1960)

River	Suspension load g/l	$CaCO_3$	
		in suspension %	in solution %
Dnieper	0·027	0·8	99·2
Don	0·155	6·2	93·7
Danube	0·179	12·0	88·0
Rioni	0·251	24·1	75·9
Tchoroch	4·219	91·9	8·1

These facts suggest that in some basins the carbonate can for the most part be of detrital origin.

The selected data on total amount of $CaCO_3$ carried in suspension by various rivers is presented on Table 16.

Table 16

Amount of carbonate in suspension load of various rivers

River	% of $CaCO_3$ in suspension load
Volga	1·98– 8·38
Kura	7·04–16·00
Ural	13·1 –29·81
Don	4·0 – 8·20
Fyris (Sweden)	about 40

The grain-size distribution of suspension load in streams

The available data on the grain-size distribution of the material suspended in streams from regions of varied topography and climatic conditions is plotted in the diagrams below (Fig. 6a, b).

Md ranges from 0·06 to 0·007 mm, i. e. within the range of fine silty or clayey fraction. The amount of sandy and clayey fraction varies widely.

The values of Phi Deviation (σ_{Φ}) are scattered between 1 and 4, thus indicating that the sorting of suspended sediment is usually poor.

Phi Skewness has a value of about zero. Only some samples show a high positive skewness and a lesser amount of samples show a fairly high negative skewness. The values vary between −0·12 and 0·13.

The large amount of suspended silt in some streams is noteworthy. An average silt content in the Mississippi River is 37% of the total load and in Alpine Rhine up to 70%. In the lowland rivers of Siberia and the Russian Plateau 25% silt has been determined. The Colorado River carried 52% and intermitently even 96·7% silt in suspension. The average amount of silt suspended in the river Amu Darya is 33·6%, in the Volga 22·4%, and in the Kura 26·11%.

The grain-size distribution of suspended material display regional changes, as a result of differences in topography and climate.

Generalizations based on the abundant data on the grain-size distribution of the suspended load are extremely difficult because it changes not only regionally, but also in time. Therefore, the conclusions given below are valid only for approximately 75% of the cases considered:

1. Rivers of mountainous areas with an irregular nongraded profile usually carry the largest amount of clay and silt and a subordinate amount of fine-grained sand;

2. Rivers of lowland areas carry much clay and fine-grained sand in suspension and are usually poorer in silt.

There are, of course, many exceptions to this rule: there are mountain rivers with a suspended load rich in sand, and, on the other hand, lowland rivers rich in silt.

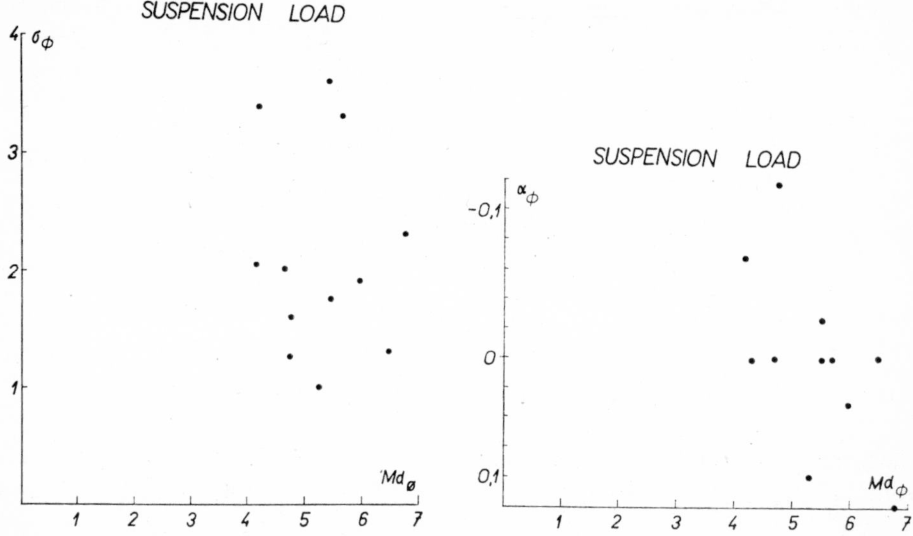

Fig. 6a. Grain-size parameters of river suspension load. Relation between Phi Median Diameter and Phi Deviation Measure. Data compiled from M. V. Klenova (1956), C. S. Howard (1947), N. M. Strachov (1961).

Fig. 6b. Grain-size parameters of river suspension load. Relation between Phi Median Diameter and Phi Skewness Measure.

The periodical changes in the grain-size distribution of suspension load are roughly of the same order of magnitude as the regional changes. Generally, the changes have been observed to follow two trends:

1. The grain-size increases with the increase of the suspended load (the content of sandy fraction particularly rises). This is the case with rivers which do not show any great differences between the normal and flood discharge and which drain basins characterized by a relatively slow rate of denudation.

2. The grain-size decreases with increasing suspended load (the content of clayey fraction increases). This tendency has been observed at rivers whose drainage basins undergo rapid denudation, mainly at the mountain areas.

The Colorado River can serve as an example of streams of the former type. Periodical analyses of the suspended load have shown that the amount of suspended material and concomitantly also of coarser particles in it increases during spring floods. The increase of sandy load during spring floods has long been known to occur in the French rivers; L. W. COLLET (1925) termed it a "sand wave".

As the suspended load of streams is the main component of the terrigenous material in most of the Recent and ancient sedimentary basins the world over, all these changes, both regional and periodical, must be mirrored in the grain-size composition of the sediments. The amount of suspended load greatly exceeds the amount of material transported in other ways (by traction or in solution). The ratio of the material carried by these modes of transport is expressed by the following relation (after G. V. LOPATIN 1952):

The streams annually deliver to the world seas:

in suspended load	12,695 × 10^6 t of material
in bed load	1,000 × 10^6 t of material
in solution load	3,600 × 10^6 t of material

From this it follows that the suspended load: bed load: solution load = 3·5 : 0·35 : 1·0, i. e. the amount of material carried in suspension is 3 times the amount carried in solution and 10 times that transported by traction. As the rivers bring the bulk of terrigenous material into the basin, the suspended load of streams is the main factor determining the nature of detritus in Recent and ancient basins of deposition.

The grain-size distribution of the basinal sediments to a great extent reflects the topographical conditions of the adjacent area. The above-mentioned considerations imply that basins rich in silt were supplied by streams draining areas with great topographical differences, whereas basins filled by clayey and fine sandy sediments obtained material from lowland areas. The frequent vertical alternation of coarser and finer beds (a few cm to dm thick) in Recent and fossil sediments indicates periodical changes

Table 17

Chemical composition of suspension load of Russian rivers in flood

River	Amount of fraction <0·01 mm	Insoluble residue	Fe_2O_3	Al_2O_3	CaO	MgO	Ignition loss
			%				
Terek	48·5	69·52	5·42	3·77	7·49	1·46	11·53
Sulak	39·2	74·45	5·96	3·36	5·41	1·39	8·85
Samur	36·0	76·22	6·36	4·22	3·08	1·69	6·94
Shabran	37·2	75·41	6·24	3·60	3·58	1·55	7·81
Ullu	31·4	77·94	6·32	5·13	0·72	1·57	7·48
Sumgait	35·0	61·11	3·99	5·19	12·81	1·73	14·06
Ak-su	63·7	64·26	3·56	5·58	10·93	1·48	13·09
Kura	30·1	70·60	5·58	5·88	6·56	1·76	9·00

Table 18

Composition of suspension load of rivers from various regions

Region	Insoluble residue %	Al_2O_3	Fe_2O_3	CaO	MgO	Ignit. loss %	Clay fraction %
		% soluble in 10% HCl					
Western slopes of the Caucasus	72·5	5·2	4·9	4·2	2·0	9·2	20·4
Eastern slopes of the Caucasus	70·5	4·7	5·3	6·6	1·8	10·1	40·7
Rivers of moderate forest zone	69·7	6·5	6·5	4·4	2·0	10·8	42·2
Average for lowland rivers	66·7	7·9	5·7	4·9	2·5	11·6	46·3
Rivers of Central Asia	63·8	4·0	3·9	12·1	2·4	13·1	40·5

in the content and grain-size of suspension load transported into the basin. The coarser beds correspond to floods which at present occur yearly in most rivers on the Earth's surface.

The chemical composition of the suspended load can be indicative of the manner of weathering and of the rate of denudation in the drainage basin. M. V. KLENOVA and V. K. NIKOLAIEVA (1961) recorded chemical analyses of the suspended load of some Russian rivers in flood (Table 17).

The same authors also introduced a parameter for the intensity of weathering processes in the drainage basin of a stream. It is expressed by the ratio of the clayey fraction (< 0·01 mm) to the portion of suspended load soluble in 10% HCl (the coefficient of chemical-mechanical weathering). The value of this coefficient ranges from 0·90 to 1·78 in the rivers of the Caucasus and from 1·22 to 1·63 in the lowland rivers in the Russian Plateau.

The composition of that portion of the suspended load which is soluble in 10% HCl can also be of diagnostic significance. The average values (in %) for several regions are presented in Table 18.

The correlation of the chemical composition of suspended load and that of the rocks of the drainage basin aids in locating the source of the suspended load and, consequently, also of the basinal deposits. Thus, for instance, W. H. WONG (1931) proved the origin of the suspended material of the river Hwang-Ho by comparing its composition with the nature of Chinese loesses (Table 19).

Table 19

Comparison of chemical composition of Chinese loesses and suspension load of the Yellow River (After W. Wong 1931)

	Average chemical composition of Chinese loesses %	Chemical composition of suspension load of the Yellow River during flood %	Chemical composition of suspension load of the Yellow River at normal water stage %
SiO_2	64·30	51·27	64·69
Al_2O_3	11·23	13·67	11·44
Fe_2O_3	2·40	4·80	3·55
FeO	1·22		
CaO	9·96	10·60	6·55
MgO	2·50	3·62	2·12
K_2O	2·34		
Na_2O	1·48		
MnO_2		0·10	0·09
CO_2		11·87	7·45
TiO_2		0·53	0·44

The chemical composition of loess corresponds to the composition of the suspended load at a low discharge of the river; during the floods the river erosion obviously attacks other rock material as well. The amount of carbonate in the suspension load of flooded river is usually smaller than during normal water stage (e. g. in the Euphrates River there is 52·1% $CaCO_3$ during flood and 81·3% $CaCO_3$ during normal water stage according to K. H. Al-Habeeb 1969).

The suspended load of the river Fyris in northern Sweden has a strikingly high content of calcium carbonate (according to F. J. Hjulström 1935), as seen from the following analysis.

SiO_2	6·96%	CaO	48·51%
Al_2O_3	2·69%	Alkalies	0·80%
Fe_2O_3	1·32%	MgO	0·12%

CO_2 37·80%

The following conclusions can be drawn from these fragmentary data: the SiO_2/Al_2O_3 ratio, i. e. practically all the content of free quartz increases with decreasing intensity of chemical weathering, i. e. from the tropical through the temperate to the polar climatic zone. There are, however, several departures from this relation. The more intensive the chemical weathering in the river basin, the greater is the content of sesquioxides in suspension load, mainly Fe_2O_3 and Al_2O_3 soluble in

10% HCl. The amount of CaO is variable, depending mainly on the petrographical character of the river basin and the intensity of mechanical weathering.

A comparison of the rate of present-day with the denudation rate during earlier geological periods has given important results. H. W. MENARD (1961) succesfully performed this correlation for some regions of the United States and the Himalayas (Table 20).

Table 20

Comparison of present and past rates of denudation
(After H. W. Menard 1961)

Region	Past rate of denudation $cm/10^3$ years	Present rate of denudation $cm/10^3$ years	Ratio of past to present rate of denudation
Appalachians	6·2	0·8	7·8
Drainage basin of Mississippi	4·6	4·2	1·1
Himalayas	21	100	0·2
Rocky Mountains (Lower Cretaceous)	3		
Rocky Mountains (Upper Cretaceous)	12—20		

The tabulated values show that the rate of denudation within a definite area has undoubtedly fluctuated in the course of geological history. It was closely associated with the orogenic movements, increasing with the increase of topographical differences and decreasing with their levelling. A numerical expression has been developed for this well-known fact.

In this connection, the question of the site of deposition of the suspended load should be mentioned. A prevailing part of it is laid down on the continents or in the transitional environments and only part is deposited in the sea, as is documented by many examples. 50% of the suspended load of the river Nile and Yangtze Kiang is deposited in their deltas. The river Shatt-al-Arab deposits 50% of its suspended load on the alluvial plain, and the Kura laid down almost all the suspended material in its delta between 1907 and 1940. In the delta and estuary of the river Visla nearly 65% of the material in suspension is deposited. About 10% of the suspended load of the Don is deposited on the flood plain and 15% in the delta.

From 1849 to 1913 about 50% of the material carried by the Yuba River and Sacramento River was deposited in river channels and valleys above tidewater. From this it follows that in an environment with a great thickness of sediments such as a flood plain, alluvial plain or delta above tidewater, the material deposited consists almost entirely of the suspension load, which is only in places mixed with the bed load or eolian material. Thus, sediments of all these environments reveal the character of the suspended load and, as a result, the character of the topographical and climatic conditions of the river basins. The following examples (Tab. 21) demonstrate how the composition of the suspended load and thus the topographical and climatic conditions can be deduced from the grain-size and chemical composition of the sediments.

Table 21

Interpretation of composition of suspension load

Example of grain-size	$CaCO_3$ content	Ratio SiO_2/Al_2O_3	Content of free sesquioxides soluble in 10% HCl
1. Mixture of silt and fine-grained sand	minimal	< 3	high, Fe_2O_3 about 5%
2. Suspension rich in silt and fine-grained sand, poor in clay	high - about 30%	high - over 5%	minimal
3. Mixture of silt and clay	about 10%	about 3	content of soluble Fe_2O_3 about 2%

Interpretation:

Example 1. The sediment can be interpreted as material derived from the zone of tropical weathering and dominantly plain character.

Example 2. The suspension load derives from the region of cool, probably polar climate and is the product of mechanical weathering. Moreover, a relatively wide difference in the relief of the source and depositional area can be assumed.

Example 3. As the suspension load shows few characteristic features, the composition is difficult to evaluate genetically. The grain-size distribution and $CaCO_3$ content point most likely to fairly great topographical differences, and other factors to a weathering process in temperate humid, or possibly semi-arid climate.

References

Al-Habeeb K. H. (1969): Sedimentology of the flood plain sediments of the Middle Euphrates River. Thesis, 1—72, Baghdad.

Collet L. W. (1925): Les lacs. P. 1—320, Paris.

Corbel J. (1959): Vitesse de l'erosion. Zeitschr. f. Geomorphologie, N. F., Bd. 3, p. 1—28, Göttingen.

Davis W. M. (1938): Sheetfloods and streamfloods. Bull. Geol. Soc. Am., vol. 49: 1337—1416, Baltimore.

Hjulström F. J. (1935): Studies of the morphological activity of rivers as illustrated by the river Fyris. Bull. Geol. Inst. Uppsala, p. 221—528, Uppsala.

Howard C. S. (1947): Suspended sediment in the Colorado River 1925—1941. Water Supply Paper 998, p. 1—163, Washington.

Jäckli H. (1958): Der rezente Abtrag der Alpen im Spiegel der Vorlandsedimentation. Eclog. Geol. Helv., vol. 51: 354—365, Basel.

Klenova M. V. (1956): Suspended sediments of the river Kura. Recent sediments of the Caspian Sea (in Russian). P. 151—174, Moscow.

Klenova M. V. - Nikolaieva V. K. (1961): Suspended sediments of some Russian rivers. Recent sediments of seas and oceans (in Russian). P. 39—57, Moscow.

Leopold L. B. - Wolman M. G. - Miller J. P. (1964): Fluvial processes in geomorphology. P. 1—522, San Francisco.

Lopatin G. V. (1952): River deposits of USSR (in Russian). P. 1—366, Moscow.

Menard H. W. (1961): Some rates of regional erosion. Jour. Geology, vol. 79: 154—161, Chicago.

Müller G. - Förstner U. (1968): Sedimenttransport in Mündungsgebiet des Alpenrheins. Geol. Rundschau, vol. 58: 229—259, Stuttgart.

Okko V. (1956): Glacial drift in Iceland, its origin and morphology. Bull. comm. Geol. Finland. No 170: 1—117, Helsingfors.

Pardé M. (1947): Fleuves et riviéres. P. 1—224, Paris.

Richards A. F. (1960): Rates of marine erosion of Tephra and Lava at Isla San Benedicto, Mexico. Internat. Geol. Congr., 21. Sess., Norden. p. 10: 59—64.

Samojlov I. V. (1952): River mouths (in Russian). P. 1—524, Moscow.

Schumm S. A. (1963): The disparity between present rates of denudation and orogeny. Geol. Survey Prof. Pap. 454-H: 1—13, Washington.

Strachov N. M. (1962): Fundaments of the theory of lithogenesis (in Russian). Vol. 1: 1—211, Moscow.

— (1961): About some relations between denudation and transport of sedimentary material. In "Recent sediments of seas and oceans" (in Russian). P. 1—27, Moscow.

Sweeting M. M. (1964): Some factors in the absolute denudation of limestone terrain. Erdkunde, vol. 18: 92—95, Stuttgart.

Wolman M. G. - Miller J. (1960): Magnitude and frequency of forces in geomorphic processes. Jour. Geology, vol. 68: 54—74, Chicago.

Wong W. H. (1931): Sediments of the North China rivers and their geological significance. Bull. Geol. Soc. China, vol. 10 (Grabau Anniversary): 247—271, Pejpin.

4. *The types of contemporary tectonic movements in relation to sedimentation*

Tectonics is one of the main factors affecting the formation and composition of Recent and ancient sediments. The effects are, however, only indirect, as the composition of sediments does not express definite tectonic conditions, but only a definite environment of deposition. The tectonic conditions can be determined only from vertical and lateral succession of the sedimentary environments. In other words, all types of sediments can occur under very different tectonic conditions. On the Earth's surface, there are a number of tectonically differing areas with sedimentary rocks of various types, but none of them can be regarded as characteristic for a particular tectonic region. The reconstruction of tectonic conditions requires the consideration of all sedimentary environments together with the size and geological position of the basin. From this it follows that tectonics does not directly control the composition of individual sedimentary rocks, but predetermines the formation of their associations in time and space.

The study of Recent tectonic processes furnished quantitative data which allow their comparison with the absolute rate of other geological processes. The contemporary tectonic processes are divided as follows:

1. Instantaneous and relatively abrupt movements alternating with long periods of rest; they are characteristic of seismic areas;
2. Movements of an oscillatory character and comparatively low speed;
3. Eustatic movements of the Earth's crust.

As is well-known, movements of the first type occur at a rate of 390 to 8,000 cm in 1,000 years. The maximum rate have been described from Japan and Fennoscandia.

Movements of the second type, either of positive or negative character, attain a rate of 50—100 cm in 1,000 years. They are omnipresent and frequently show an oscillatory character. On the shore of the Baltic Sea the amplitudes of 6·5 mm per year have been recorded. Maximum values have been established at some places on the Russian Plateau: 1,200 cm in 1,000 years. The whole of the European continent is rising with an average speed of about 100 cm in 1,000 years.

Rapid tectonic movements can effect the sedimentation in the adjacent basins chiefly by controlling the development of nearshore environments, by the formation of fiords, etc. Slow tectonic movements presumably do not exert influence on the sedimentation.

The comparison with the absolute rate of other contemporary geological processes suggest that the average rate of present-day tectonic processes is undoubtedly of maximum value:

	cm/1,000 years
Rate of weathering (soil development)	average 10—30
Rate of denudation of continents	average 10
Rate of sedimentation	average 5
Rate of recent tectonic movements	average 50—100

The originating tectonic deformations are in most cases not compensated either by denudation, or by sedimentation, so that tectonics is, and was, in the whole of geological history the basic factor in the development of sedimentary environment.

References

GERASIMOV I. P. (Editor) (1963): Recent crustal movements (Symposium, in Russian). P. 1—383, Moscow.

Recent movements, volcanism and earthquakes on continents and ocean bottoms (Symposium for the 8. Congress of INQUA — in Russian, English res.). P. 1 — 276, Moscow.

Symposium über rezente Erdkrustenbewegungen (1962). P. 1—1112, Leipzig.

WEGMANN E. (1955): Lebende Tektonik, eine Übersicht. Geol. Rundschau, Bd. 43, p. 4—34, Stuttgart.

5. *The rate of sedimentation*

The absolute rates of geological processes and particularly the rate of sedimentation have attracted the attention of many workers. The data assembled up to the present have been gained by the following methods:

1. Theoretical calculation based on the amount of suspended load transported by streams into a basin over a definite time interval. In this way the rates of detrital sedimentation in some lakes, inland seas and oceans have been determined.
2. Direct observation, mainly measurement of the depth of the bottom in the given time intervals. This method provides data on the rate of sedimentation in shallow marine environments, such as deltas, tidal flats and bays.
3. Measurement of the thickness of laminae in laminated sediments.
4. Chemical and physico-chemical methods, widely used nowadays. In addition to radiocarbon method, the U: Io : Ra ratio, the ratios of Pb isotopes, Be isotope, or Ti content are measured. Also the isotope ^{32}Si has been used for the determination of the rate of sedimentation.
5. Some special methods, such as determination of the amount of cosmic material or the calculation of the content of suspended load and the rate of its deposition. The latter has recently produced some successful results.

The rate of sedimentation is expressed in several ways. In sediments whose deposition is rather slow and regular (lacustrine, bathyal (etc.) and presumably without major interruptions, it is usually given in cm/1,000 years. When the sedimentation is rapid and irregular, the rate of deposition is expressed in smaller time units, generally in millimetres per year or per day.

Fluviatile sediments

Their deposition is so irregular that absolute values for channel deposits cannot be given. The formation of sand-bars in river channel, as well as their disturbance, occurs at a rate of up to several metres per year. The rate of flood plain deposition is about several mm to several cm per year. Archaeological data have furnished evidence of the average of sedimentation in Mesopotamia and Mohenjo Daro. Detailed data are presented in Table 22.

Eolian sediments

There are few data available on the rate of deposition of eolian sediments. The wind-storms can deposit several millimetres to centimetres of dust during a few hours.

These episodes, however, are separated by long periods of non-sedimentation. According to some estimates, loess was deposited at a rate of 20—100 cm in 1,000 years. The rate of deposition of the loess in central Alaska is 1·5—19·3 cm/1,000 years (T. L. PÉWÉ 1968).

Table 22

River	Time range	Thickness of sediment layer cm
Nile	1 year	0·9
Ohio	January—February 1937	0·240
Connecticut	March 1936	3·42
Connecticut	September 1938	2·10
Kansas	July 1951	2·93
Euphrates	5,000 years	100
Indus	4,500 years	90
Yuba	1 year	10
Sacramento	1 year	7·5

Proluvial sedimentation cannot be expressed in absolute data. During a few minutes several metres of material can be piled up, but a number of years can follow in which no sedimentation or even erosion takes place. This also holds for glacial sediments.

The rate of deposition of organic phosphates (guano) is up to several cm per year in some islands.

Cosmic dust is deposited rather slowly; the rate of sedimentation is expressed by the number of particles falling on the areal unit during time unit. Several data are listed on p. 54—55.

Lacustrine sediments

Data on the rate of deposition of lacustrine sediments are very numerous. Table 23 presents the most important values along with the types of lakes and kind of sediment. The table shows great differences in sedimentation rate, which are accounted for by the differences in the supply of terrigenous material, in the production of biological material and the intensity of chemical sedimentation. The deposition proceeds at a maximum rate in those places where the supply of detrital material is largest. The biological and chemical calcareous sedimentation, however, can also attain a considerable rate, particularly in small lakes. The average rate of deposition in the lakes is about 3 mm yearly, i. e. 300 cm in 1,000 years. This, comparatively considerable speed proves that shallow lakes can be filled up with sediments within quite a short period.

Table 23

Lake	Type of lake and kind of sediments	Rate of sedimentation mm/year
Great Lakes of North America lakes	glacial lakes; varvites	0·15
Lake of Zurich	oligotrophic; clayey and calcareous varvites	0·7
Lake Neuchatel	oligotrophic; clayey and calcareous varvites	0·7
Lake Michigan	oligotrophic; clayey and calcareous varvites	3
Average for fresh-water lakes	diatomites	0·3—1·0
Swedish lakes	eutrophic; gyttja	1—2
Olof Jone Damm (Sweden)	dystrophic; peat	5·3
Maxinkuckee (Canada)	eutrophic; lake marl	0·3
Average for small lakes of North Germany	lake marl	1—3
Lake Vierwaldstätt	alpian oligotrophic; calcareous clays	10·4—31·7
Lake Wallen	alpian oligotrophic; calcareous clays	11·3
Lake of Brienz	alpian oligotrophic; calcareous clays	31·7
Lake Lunz	near mouth of the Rhône	17·9
	average for the whole lake	2·5
Ladoga	great oligotrophic; clays	6·12
Onega	great oligotrophic; clays	7·1

Peat bogs and swamps

Data on the accumulation of peat can be found in a number of papers. Young peat bogs grow at an average rate of 2 mm per year, the older ones approximately at half this rate. E. Firbas (1935) determined the yearly accretion of Swabian high moors at 0·15—0·18 cm/year. For North American peat bogs 0·55 cm/year is estimated as the average value of accretion. The sedimentation in littoral swamps occurs at roughly the same rate. The maximum rate of growth measured in Great Britain is 0·98 cm/year.

Inland seas

Data on the sedimentation rate in the inland seas are not very abundant, but they give a good picture of the sedimentary conditions there. The most important data are listed in Table 24.

Table 24

Sea	Sediment	Rate of sedimentation cm/1,000 years
Mediterranean	calcareous oozes	10
	oozes with eolian admixture	20
Caspian Sea	calcareous clays	10–18
Black Sea	calcareous clays	20
Gulf of Mexico	sandy- and silty-clays on the upper continental slope	7
	silty clays on the lower continental slope	5
	calcareous clays of deep-water parts	4
Baltic Sea	organogenous black clays	30
Tyrrhenian Sea	calcareous and diatomaceous clays	10–50
Barentz Sea	clays with coarser admixtures	0·8–4
Adriatic Sea	shells	1
Milford Sound (New Zealand)	sandy silts	1·25

The rate of deposition in the inland seas is mostly indirectly proportionate to their dimensions. The larger the basin and the smaller the supply of clastic material, the slower is the sedimentation. The decisive factor is the supply of terrigenous material, as the production of biological matter is already smaller in extent. The influence of chemical sedimentation recedes to the background.

Table 25

Bay (lagoon)	Environment and kind of sediments	Rate of sedimentation cm/1,000 years
Gulf of California	deep bay; clays, diatomites	60–100
Texas Bay	shallow lagoons; clays with coarse admixtures, rich in eolian material	max. 383
Kiel Bay	silts and sands	150–200
Drammens fjord (Norway)	deep fjord; black clays	150
Gulf of Paria	great gulf of middle depth; clays prevail	0–100
Clyde Bay	small, shallow bay; clays	240–300
Kara-Bougas-Gol (Caspian Sea)	shallow bay; clays and salts	50– 70

Bays, lagoons, and similar environments

The rate of deposition varies considerably with the type of bay. It depends mainly on whether streams empty into the bay or lagoon or not (Table 25). The Gulf of Paria between Trinidad and South America provides a typical example of the relationship between the rate of deposition and the supply of material by rivers. In the vicinity of the mouths of rivers the rate is 100 cm/1,000 years, in the central parts of the gulf 20—50 cm/1,000 years on the average, and near the straights the rate drops to zero.

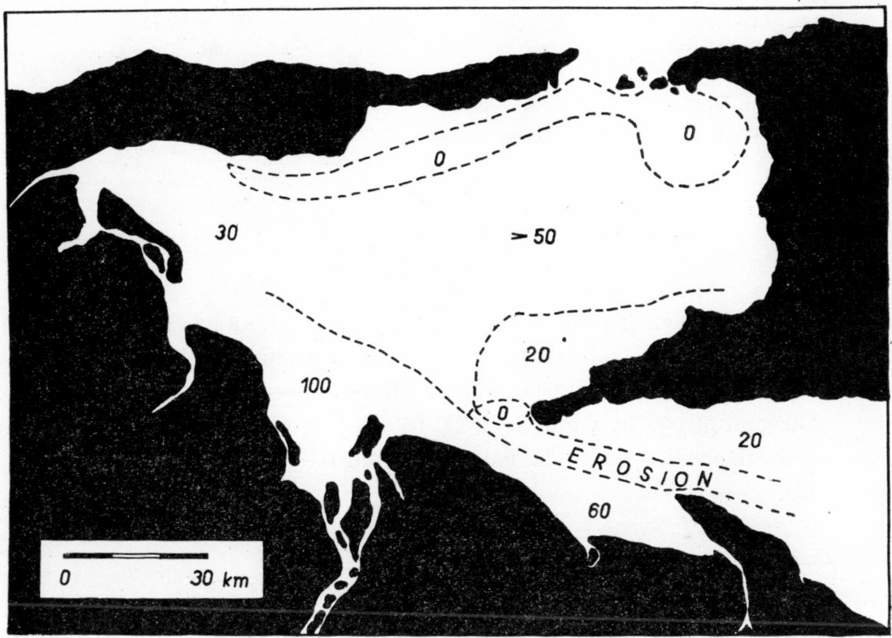

Fig. 7. Rates of deposition in the Gulf of Paria for the period covering approximately the last 700 years (in centimetres per century). After Tj. van Andel-H. Postma (1954).

Deltas

Contrary to other shallow-water environments, deltas of major streams are sites of long-term sedimentation, which, moreover, proceeds at a more rapid pace than anywhere else on the Earth's surface. Some of the values expressed in centimetres per time unit are given in Table 26.

Short-term measurements furnished several astonishing values: for instance, a rate of 3 cm of silt in 4 days has been determined in some sections of the Mississippi delta. In the front parts of the Mississippi delta 30—45 cm of material are deposited during one year. In the prodeltal parts 6—30 cm/year are laid down and in the shelf area under delta influence 4·5 cm/year. In the delta of the Orinoco River 1 cm and in that of the Rhône up to 40 cm of silt are deposited annually, the above-water part of the Nile delta grows by 1 cm and the Don delta by 1·22 mm of clay a year.

Table 26

Delta of river	Rate of sedimentation cm/1,000 years
Fraser (Canada)	5,000–30,000
Hwang-Ho (China)	150
Volga	500–7,000
Amu Darya	2,500
Average of top sets of deltas	1,500–2,000

Tidal flats

Sedimentation in this environment is so irregular that it cannot be expressed in longer time units. In the Netherlands the average rate of deposition in tidal flats has been estimated at 1—2 cm/year. Short-term observations, however, in some places showed a rate of several metres in a few days; G. Evans (1965) records that in the Wash (England) the deposition on a tidal flat varies from 1—2 cm, and up to a maximum of 6 cm/year.

Table 27

Kind of sediment	Region	Rate of sedimentation cm/1,000 years
Blue mud	average of all continental slopes	1·78
Diatomaceous mud	isolated deep basins on Californian coast	88
Diatomaceous clay	Yellow River, China Sea and Sea of Japan	5–20
Calcareous clay	East India Sea	85
Average for hemipelagic sediments of Atlantic Ocean		5–10
Blue mud	NW Pacific Ocean	10
Grey mud	Bering Straits	8–450
Grey silty clay	Antarctic slope	1–16

Oceanic realms

Many data on the sedimentation rate of bathyal deposits have recently appeared in the literature. Since they have been obtained by all the above-mentioned procedures, they are readily comparable (Table 27). Any possible difference of major importance can be interpreted in terms of deep current activity or differences in biological production.

Table 28

Kind of sediment	Region	Rate of sedimentation cm/1,000 years
Globigerina ooze	average for all oceans	1– 8
Other calcareous oozes	Atlantic Ocean, Mediterranean Sea	2–10
Brown clays	average for all oceans	0,2– 1
Radiolarian ooze	average for all oceans	0,5
Diatom ooze	Pacific Ocean	0·5–5
Non-carbonate fraction of calcareous oozes	Northern Pacific	1,0
	Southern Pacific	0·4
	Northern Atlantic	1·2
	Southern Atlantic	2·0
	Indian Ocean	0·6
	Caribbean Sea	1·2

The differences in the deposition of hemipelagic and eupelagic sediments are mainly due to the differences in the intensity of the supply of terrigenous material. The admixture of the marine plankton component induces a more rapid deposition of calcareous and siliceous oozes compared with that of brown clay (Table 28).

Biological sediments

Massive corals grow at an average rate of up to 4 mm/year, whereas the branching corals grow about 1 mm/year. In the Florida Bay the average rate of growth has been determined at 30 cm in 528 years, i. e. less than 1 mm/year. Some coral species, however, such as *Montastrea annularis,* attain a rate of 10·7 mm/year. Recent stromatolites grow up to 4 mm in 24 hours. The average rate of coral growth in the upper parts of atolls is about 14 mm/year, on the submerged outer reef it ranges from 0·33 to 0·91 mm/year, and in the lagoon attains 3·8 mm/year. Thus, the atolls afford an environment of exceptionally rapid sedimentation, the rate of which is more than 1,000 times the deposition of the contiguous oceanic sediments. Calcare-

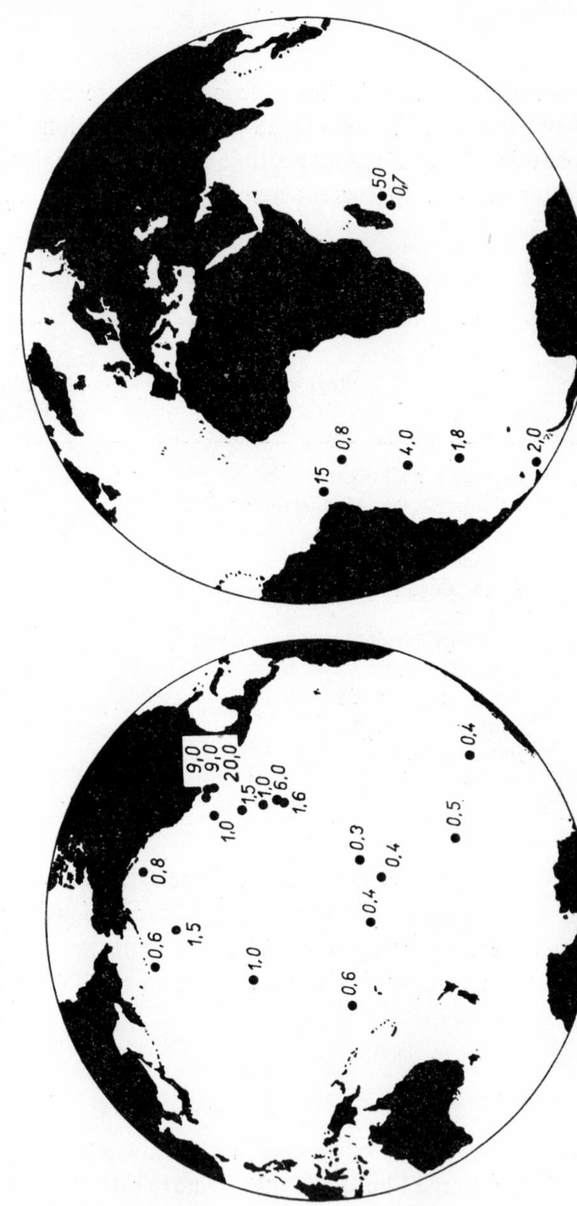

Fig. 8. Rates of deposition of deep-water sediments in the oceans (in milimetres per 1,000 years). After E. D. Goldberg *et al.* (1962).

ous algae, *Lithophyllum incrustans* and *Lithothamnion lenormandi* grow at a rate of 2—7 mm/year; the growth is most intensive in summer months (up to 1 mm/month).

*

As mentioned above, sedimentation keeps a uniform rate for a long period only in the bathyal environment. In all other environments periods of rapid sedimentation alternate with periods of slower deposition or non-sedimentation. One of the important tasks is to determine the ratio of the 2 periods. For this purpose, the rate of sedimentation was measured in the North German tidal flats by means of various

Table 29

Rate of sedimentation	Time of measuring
22 cm/100 years	1,900 years
115 cm/100 years	4 years
$1{\cdot}45 \times 10^4$ cm/100 years	8 days
Rate of deposition of sand layer (stream velocity 100 cm/sec) was $2{\cdot}10 \times 10^8$ cm/100 years	in several seconds (according to experiment)

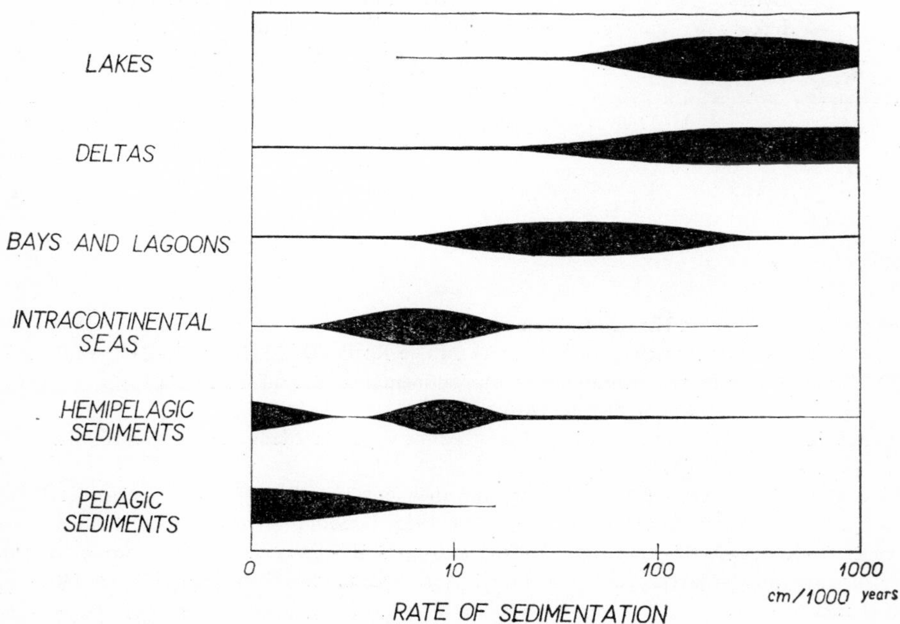

Fig. 9. Rates of sedimentation in individual sedimentary environments.

methods (H. E. Reineck 1960) over different time intervals. When the rate of sedimentation was measured for the short intervals and recalculated to longer intervals, the values obtained were always higher than those resulting from direct measurements performed for the respective longer periods. This conclusion clearly follows from the correlation of values in Table 29.

The comparison of the above values indicates that in the shallow-water sediments of the tidal flats investigated, the period of non-sedimentation was far longer, i.e. 4,060 times longer than the period of deposition. This unusually high value indicates that the period of deposition of shallow-water sediments, either Recent or ancient, represent only a fraction of time in their history. On the basis of the above values the ratio between the periods of sedimentation and non-sedimentation in river plains was computed at 0·018. Table 30 presents a rough estimate of this relationship ascertained in some other types of sediments.

Table 30

Environment	Ratio time of sedimentation/ time of non-sedimentation
Abyssal	0·5—1
Bathyal	0—1
Deltas	variable, most frequently between 0·5—0·75
Lakes	about 0·9
Tidal flats	0·0002 (see above)

References

Andel Tj. van - Postma H.: (1954): Recent sediments of the Gulf of Paria. Verhandel. Koninkl. Nederl. Akad. Wetenschap., afd Natuurk., Erste Reeks, D. 20, No 5: 1—244, Amsterdam.

Carlson O. R. (1968): Sedimentation rate on continental terrace off Columbia River. Bull. Am. Assoc. Petrol. Geol., vol. 52: 560 (Abstr.), Tulsa.

Ericson K. G. (1965): Études de quelques sédiments de la Méditerranée occidentale. Acta Universit. Uppsaliensis, vol. 60: 1—23.

Evans G. (1965): Intertidal flat sediments and their environments of deposition in the Wash. Quart. Jour. Geol. Soc., vol. 121: 209—245, 1965, London.

Fay R. C. - Reuter J. H. - Grady J. R. - Richardson S. H. - Bray E. E. (1963): Biogeochemistry of sediments in experimental Mohole. Jour. Sedimentary Petrology, vol. 33: 140—172, Menasha.

Firbas E. (1935): Die Vegetationsentwicklung des mitteleuropäischen Spätglazials. Bibl. Botanica, Bd. 112: 1—68.

GILES A. W. (1930): Pennsylvanian climates and paleontology. Bull. Geol. Soc. Am., vol. 14: 1279—1299, Baltimore.

GOLDBERG E. D. - KOIDE M. (1962): Geochronological studies of deep sea sediments by ionium-thorium method. Geochim. Cosmochim. Acta, vol. 26: 417—450, London.

HOFFMEISTER I. E. - MULTER H. G. (1964): Growth-rate estimates of pleistocene coral reef of Florida. Bull. Geol. Soc. Am., vol. 75: 353—358, Baltimore.

KHARKAR D. P. - TUREKIAN K. K. - SCOTT M. R. (1969): Comparison of sedimentation rates obtained by ^{32}Si and uranium decay series determinations in some siliceous Antarctic cores. Earth and Planetary Letters, vol. 6: 61—68, Amsterdam.

KUKAL Z. (1957): Rate of sedimentation of Recent and fossil sediments (in Czech). Čas. pro mineralogii a geologii, vol. 2: 155—169, Prague.

LAAGAIJ R. - KOPSTEIN F. P. H. W. (1964): Typical features of a fluviomarine offlap sequence. Proc. 6th Sedimentological Congress, p. 216—226, Amsterdam.

LISICYN A. P. (1963): Sediments of Antarctic shelf. In "Deltaic and shallow-marine sediments" (in Russian with English abstract). P. 82—99, Moscow.

NAYUDU J. R. - ENBYSK B. J. (1964): Bio-lithology of Northeast Pacific surface sediments. Marine Geology, vol. 2: 310—342, Amsterdam.

PÉWÉ T. L. (1968): Loess deposits of Alaska. Int. Geol. Congr., 23. Sess., vol. 8: 297—309, Prague.

REINECK H. E. (1960): Über Zeitlücken in rezenten Flachseesedimenten. Geol. Rundschau, Bd. 49: 149—161, Stuttgart.

WASMUND E. (1938): Lakustrische Unterwasserböden. Handbuch d. Bodenlehre, Bd. 5: 97—189, Berlin.

WERNER F. (1964): Sedimentkerne aus den Rinnen der Kieler Bucht. Meyniana, Bd. 14: 52—65, Kiel.

6. Cosmic material in Recent sediments

The share of cosmic material in Recent sedimentation has not yet been evaluated on a large scale. Particles of presumably cosmic origin were first found in the past century, as metallic magnetic globules (about 0·2 mm across) and silicate globules (0·5—1 mm across) with weathered surfaces, occurring in bathyal sediments. It was estimated that in one litre of sediments 20 minute magnetic globules and 5 larger silicate globules could be found. Subsequent detailed investigations have shown that these extratelluric components of bathyal sediments contain 64·5% Ni, 1·8% Co and 44·4% Fe.

Large numbers of magnetic spherules of unknown origin and composition, ranging in size less than a micron to several microns, are frequently observed in sediments. Analyses of such bodies indicate a heterogeneous origin. Many of the magnetic spherules previously assumed to have been derived from outer space have been shown to consist of volcanic glass with inclusions of magnetite and metallic iron. Others, in size-range of 0·5—5 μ, appear to consist of goethite and might have been secreted by marine bacteria. It therefore appears necessary to define individual cosmic spherules not only on the basis of shape and magnetic properties but also on chemical

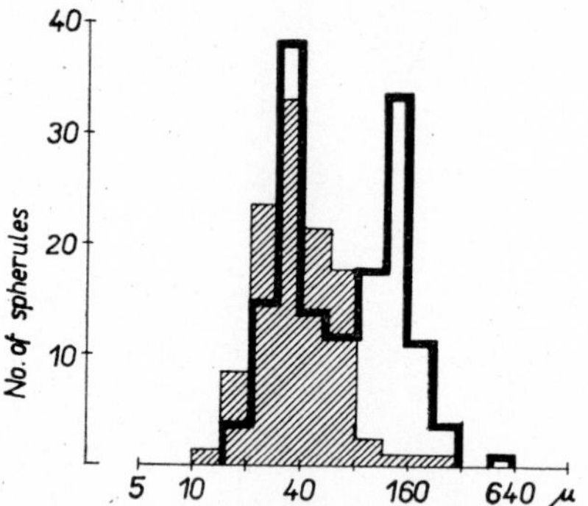

Fig. 10. Size distribution of cosmic spherules from pelagic sediments. Shaded histogram = iron spherules; line-bounded histogram = silicate spherules. After G. Arrhenius in M. N. Hill (1963).

composition. B. P. Glass (1969) described the occurrences of microtectites — small glassy objects — in deep-sea sediments. Their size is from 1 mm down to at least 20 μ in diameter. They were originally accumulated in certain horizon 0·7 to 1 million years old and their origin was connected with the tectite falls in Australia, Asia and the Ivory Coast. Later the tectites were dispersed by the activity of burrowing organisms.

From the content of cosmic spherules in bathyal and abyssal sediments checked against the rate of deposition, R. CASTAING and K. FREDERICKSON (1958) estimated the amount of cosmic material falling annually on the Earth's surface at 1,000 – 5,000 tons.

Cosmic material has been observed directly in eolian suspension, as a multitude of spheric corpuscles, whose average size ranged from 4 to 25 microns. In 1 cm^3 of dust as many as 100 corpuscles of safely cosmic origin have been found. Two hypotheses have been developed bearing on their origin; the particles are either parts of small chondrites moving at a small speed, or fragments of large meteorites detached from them by friction during their passage through the atmosphere. The finding that an enormous number of particles of 3—100 μ in size originates during the flight of meteorites through the atmosphere speaks rather for the latter opinion. The ground round the site of a meteorite fall is generally strewn with such particles.

Cosmic material is most likely an ubiquitous component of Recent sediments. As it presumably accumulates in the most slowly deposited sediments, its amount is indirectly proportional to the rate of sedimentation.

Recently, a number of papers recording the finding of similar particles also in ancient sediments have appeared. Their study obviously deserves more attention than has so far been paid to them. Care must be taken, however, not to mistake similar particles for them, such as, weathered accessory magnetic minerals in fossil sediments or particles, possibly also of anthropogenic origin, in Recent sediments.

References

BAKER J. L. - ANDERS E. (1968): Accretion rate of cosmic matter from Iridium and Osmium contents of deep-sea sediments. Geochim. Cosmochim. Acta, vol. 32: 627–646, London.

CASTAING R. - FREDERICKSON K. (1958): Analyses of cosmic spherules with an X-ray micro-analyse. Geoch. Cosmoch. Acta, vol. 14: 114–117, London.

GLASS B. P. (1969): Reworking of deep-sea sediments as indicated by the vertical dispersion of the Australasian and Ivory Coast microtectite horizons. Earth and Planetary Science Letters, vol. 6: 409–415, Amsterdam.

HODGE P. W. - WILDT R. (1958): A search for airborne particles of meteoritic origin. Geoch. Cosmoch. Acta, vol. 14: 126–133, London.

SKOLNICK H. (1961): Ancient meteoritic dust. Bull. Am. Geol. Soc., vol. 72: 1837–1842, Baltimore.

SACKETT W. M. (1964): Measured depositional rates of marine sediments and implication for accumulation rates of extraterrestrial dust. Annals of the New York Acad. of Sciences, vol. 119, Art. 1: 339–346, New York.

UTECH K. (1961): Über das Vorkommen magnetischer Kügelchen im norddeutschen Buntsandstein, ihren stratigraphischen Wert und die wahrscheinliche Herkunft. N. Jhrb. Geol. Mh, H. 8: 432–436, Stuttgart.

7. Biological components of sediments

One of the fundamental components of sediments are the organogenic substances. The organisms contribute to the formation of sediments in the following ways:

1a) Their solid parts, i. e. tests, valves, exoskeletons, etc. build up sedimentary rocks;
b) Products of the decay of their soft bodies take part in the composition of rocks;
c) Excrements of organisms form a constituent part of sediments.
2a) They affect the sedimentation by mechanical activity;
b) They influence the sedimentation in affecting the physicochemical conditions of environment.

Solid biological components of Recent sediments

The amount of biological components in Recent sediments ranges from 0 to 100%. Recent sediments contain calcite- and aragonite-built shells of molluscs, agglutinated and siliceous tests of protozoans, phosphate bones and teeth of fish, and

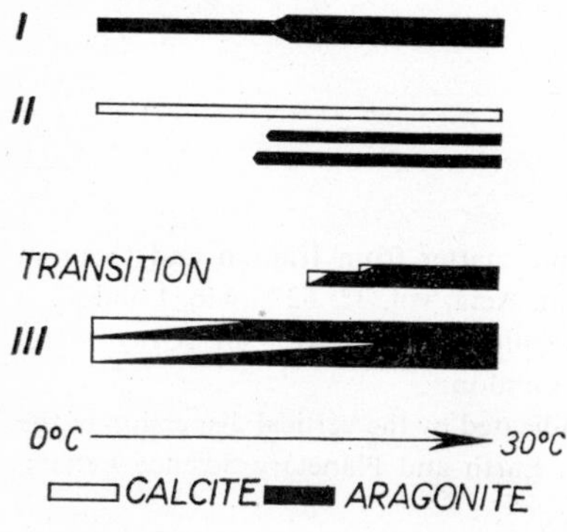

Fig. 11. Schematic presentation of relative stages in complexity of mineralogic responses to environmental temperatures. Category I includes orders composed of species whose skeletons consist entirely of aragonite but in which the number of species is markedly greater in warmer than in colder waters. Category II includes classes or subclasses in which all species of a given order secrete either only calcite or only aragonite but in which the orders with aragonite-depositing species are confined to warmer waters. The transition category contains species which differ from those in category II in secreting traces of calcite in colder climates in their marginal geographic ranges. Category III includes genera with species whose skeletons are composed of mixtures both aragonite and calcite, with the aragonite content increasing with higher temperatures and the calcite with lower temperatures, the temperature effect depending upon the species. After H. A. Lowestam (1954).

chitine carapaces. When these biological components make up a considerable part of the sediment, its composition depends on that of organic remains. As the carbonate components are most widespread in Recent biological sediments, they have been studied most thoroughly.

Table 31

Mineralogy of carbonate skeletal material
(C — common, R — rare, after K. E. Chave 1964)

Organism	Aragonite	Aragonite and calcite	Calcite	Mg-calcite	Mg-calcite and aragonite
Foraminifera					
Benthonic	R			C	
Pelagic			C		
Sponges				C	
Corals					
Madreporian	C				
Alcyonarian	R			C	
Bryozoa	C			R	C
Brachiopods			C	R	
Echinoderms				C	
Molluscs					
Gastropods	C	C			
Pelecypods	C	C	C		
Caphalopods	C			R	
Annelids	C			C	C
Arthropods					
Decapods				C	
Ostracods			C	R	
Cirripeds			C	R	
Algae					
Benthonic	C			C	
Pelagic			C		

Carbonate exoskeletons are composed of the minerals calcite, aragonite and a spectrum of magnesian calcites. The mineral vaterite has been reported as a skeletal mineral, but at best its occurrence is rare. Phase equilibrium studies of the systems $CaCO_3$ and $MgCO_3$ indicate that only calcite, poor in magnesium, is stable under nearsurface pressure-temperature conditions. Aragonite, vaterite and calcite, which contains more than about 4% $MgCO_3$ in solid solution are unstable phases. One of the most important problems is that of the calcite/aragonite ratio. The content of these minerals in the shells is controlled by biological and environmental factors. The amounts of carbonate minerals in the shells of various organisms are tabulated in Table 31.

H. A. Lowestam (1954) differentiated 3 feasible modes of occurrence of calcite and aragonite in tests or shells (Fig. 11). The first group includes those species whose tests consists only of aragonite and whose number increases towards the warm seas. The second group comprises all families in which all species of a definite genus

Table 32

Chemical composition of the tests of some Recent organisms

	Orbitoides marginatis (Foraminifer) %	*Oculina difusa* (Coral) %	*Lithophyllum antilarum* (Ca alga) %	*Discinisca lamellosa* (Phosph. brachiopod) %	*Eupectella speciosa* (Si sponge) %
Ca	34·90	38·50	31·00	26·18	0·16
Mg	2·97	0·11	4·36	1·45	0
CO_3	59·70	58·00	52·50	7·31	0·24
SO_4			0·68	4·43	0
PO_4	traces	traces	traces	34·55	0
SiO_2	0·03	0·07	0·04	0·64	88·56
$(Al, Fe)_2O_3$	0·13	0·05	0·10	0·44	0·32
Org. matter	2·27	3·27	1·32	25·00	10·72

Table 33

Content (p.p.m.) of trace elements in the shells of Recent organisms (after F. Leutwein and H. Waskowiak 1962)

	Calcite Aragonite	Mg	Sr	B	Pb	Mn	Fe	Ni	Cu
Hexacoralla	A	600	8,600	20·5	4·0	1·4	13		1·1
Serpulidae	C	3·2%	1,700	6·6	2·1	33·3	340	19	4·1
Echinoidae	C	1·9%	1,300	2·7	1·8	7·0	26		0·7
Patellidae	AC	1·5%	1,600	2·5	2·7	8·3	40	0·8	0·6
Trochidae	A		2,500	4·5	14·5				6·7
Littorinidae	AC	1,100	2,500	1·6	2·6			14·2	2
Hydrobiidae	A	840	2,300	2·0	1·0	9	80	2·3	1·8
Naticidae	A	310	3,100	1·9	2·1	7	184	3·0	3
Buccinidae	A	290	2,500	5·1	1·3	11	102	1·5	1
Nassidae	AC	660	1,400	4·8	1·4	15	79	2·4	1·4
Arcidae	C	2,500	1,800	3·5	1·0	29	90	1·0	1·3
Mytiladae	C	3,700	1,500	8·1	1·1	224	135	0·4	2·2
Pectinidae	C	4,300	1,400	10·7	1·1	31	127	1·1	1·2
Anomiidae	A	90	1,800	3·7	1·0	3·3	93		1·0
Ostreidae	A	400	3,000	1·9	1·0	0·7	6	0·1	1·7
Cypridinae	A	250	3,100	3·8	1·6	11	169	2·5	1·3
Cardidae									
Veneridae									

secrete aragonite or calcite tests respectively and whose genera with aragonite tests are confined to a warm-water environment. The third group includes genera whose species built up shells of aragonite-calcite mixture, the aragonite content increasing with the rising temperature. In all cases the salinity of the environment produces anomalous relations. With the increasing salinity the aragonite amount increases at the expense of calcite, although the contrary case has also been ascertained, (e. g. in the pelecypod *Mytilus californianus*). J. R. DODD (1963) also reports that the aragonite content rises with the increasing size of tests.

In general, biological carbonate sediments deposited in warm waters are richer in aragonite than those originating in cold waters. The instability of aragonite is well-known, and the decrease in the aragonite amount in favour of calcite has been observed already during diagenesis. In spite of it, aragonite can be preserved partially also in fossil sediments, as shown by the finds of primary aragonite in Triassic limestones.

The chemical composition of the tests of some Recent organisms was studied by F. W. CLARKE - W. C. WHEELER (1917) who published analyses of calcareous shells of a phosphate brachiopod and siliceous sponges (Table 32).

The paper by F. LEUTWEIN - H. WASKOWIAK (1962) completes their data by analyses of the content of trace elements in the shells of Molluscs (Table 33).

Shells of some organisms contain a remarkably large amount of Mg, which formerly was ascribed erroneously to the presence of dolomite. At present it is known, however, that it occurs in the form of solid solution of $MgCO_3$ in $CaCO_3$. The perfect miscibility of these components caused mainly by the physiological factors reaches up to 25% $MgCO_3$. A new mineral — Mg calcite – accedes to calcite and aragonite. In Table 34 the $MgCO_3$ contents in shells of Recent organisms are presented, as given by different authors.

Table 34

$MgCO_3$ contents in shells of Recent organisms (after K. E. Chave 1954)

	$MgCO_3$ %		$MgCO_3$ %
Foraminifera	0·30—12·52	*Lamellibranchiata*	0— 1·28
Spongiae	6·84—14·10	*Gastropoda*	0— 1·78
Hydrozoa	0·26— 8·15	*Echinoidea*	3·24—13·79
Hexacoralla	0·11— 1·11	*Crinoidea*	7·28—13·74
Octocoralla	0·35—16·90	*Asteroidea*	7·79—14·55
Vermes	0— 9·72	*Ophiuroidea*	7·62—14·95
Bryozoa	0—11·08	*Cirripedia*	0·75— 2·49
Brachiopoda	0·45— 8·63	*Decapoda*	0·75— 2·49
Cephalopoda	0— 6·02		

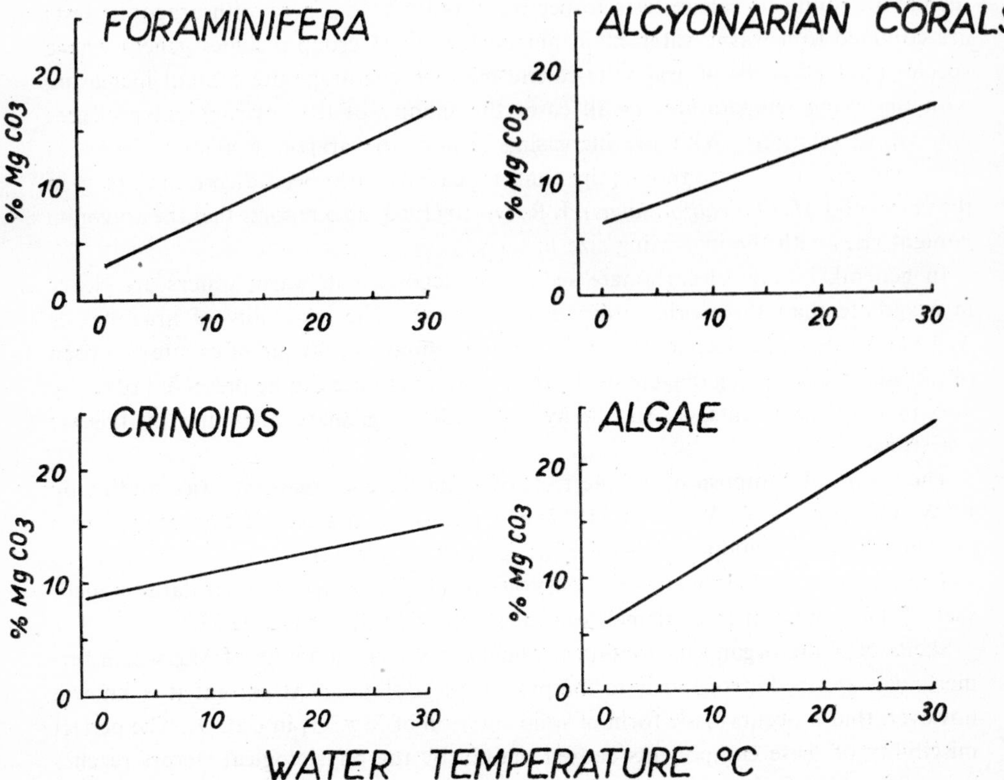

Fig. 12. Relation between water temperature and magnesium content in individual groups of organisms. After K. E. Chave (1954).

Some classes displayed a direct dependance of the $MgCO_3$ content on the temperature of water (Fig. 12). The $MgCO_3$ content is also affected by the composition and structure of shells, salinity and phylogenetic level of the organism. With the exception of a few cold-living forms, the aragonite-built shells are poorer in magnesium than the calcitic ones. Therefore, in deducing the salinity and the temperature of the environment from the Mg-content of shells, we must consider separately those made up of calcite and those of aragonite. The Mg-content and salinity also show a direct relationship, although not so pronounced as the Mg-content and temperature. The relationship between the physiological level of organisms and the Mg-content in shells reveals interesting dependences. It has been ascertained that the gradient of the temperature / Mg-content curve is maximal for organisms standing at the lowest physiological level, decreasing towards the groups at a higher physiological level (Fig. 13). In other words, the Mg-content in shells of the lowest organisms increases with temperature more rapidly than in those of higher organisms. It can be broadly said that the most primitive groups have on the average a higher Mg-content in tests.

The direct relation between the content of magnesium and temperature can serve as a useful tool for ecological and palaeontological investigations. The drawback of this, otherwise fairly sensitive method is its inapplicability to studies of organic remains derived from earlier formations, because a strong demixing of magnesium from shells already occurs in organic remains of Pleistocene age.

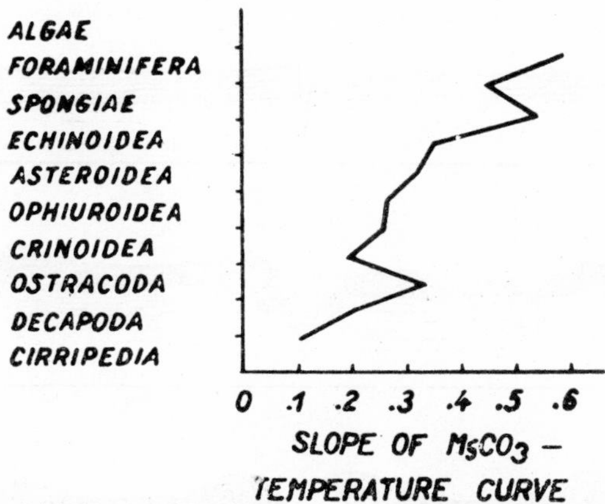

Fig. 13. Relation between slope of temperature-magnesium curve and organic complexity. After K. E. Chave (1954).

A close relation between the Sr-content and the composition of shells has also been recognized. Due to a more open lattice, aragonite entraps strontium more readily than calcite does. A direct proportion of the percentage of aragonite in shells and the Sr-content has been shown on some examples, as, for instance, at the coral genus *Diploporia* from Florida (F. R. Siegel 1960) or at *Patelina corrugata* foraminifer (determined experimentally by W. Knauff 1963).

A large amount of strontium has been found in skeletal parts of tropical reef-building corals and algae formed of aragonite. It is noteworthy that water surrounding these Sr-rich aragonite skeletons is impoverished in this element. The amount of strontium in shells is also controlled by the content of Sr in solution and by the salinity and the temperature of the environment. The interrelationship of salinity and the Sr-content is extremely complicated, being affected by a number of subsidiary factors. Therefore, the Sr-content cannot be used as a reliable indicator. C. H. Pilkey - H. G. Goodell (1963) record that differences in salinity cause greater changes in shell composition than differences in temperature, but this relationship is generally too weak to be used for palaeoecological determinations. Some species secreting aragonite in shells show that there exists a direct relationship between the Sr-content in skeleton and the mean temperature of water. Thus, for instance the Sr-content in serpulid shells varies between 0·45 and 1% depending on water temperature. Moreover, the Sr-content is influenced by the activity of organisms; some of them show a higher affinity to this component (e. g. worms) and contain, therefore, more strontium in shells than aragonite skeletons of other organisms. Strontium is not a very stable component of shells; it is soon lost, and aragonite or strontium-poor calcite develops.

The controversy of the origin of fossil stromatolitic structures excited interest in Recent stromatolites found in some carbonate environments. They are found in shallow-water shelf areas of the Bahamas and Florida, in Great Salt Lake, in some

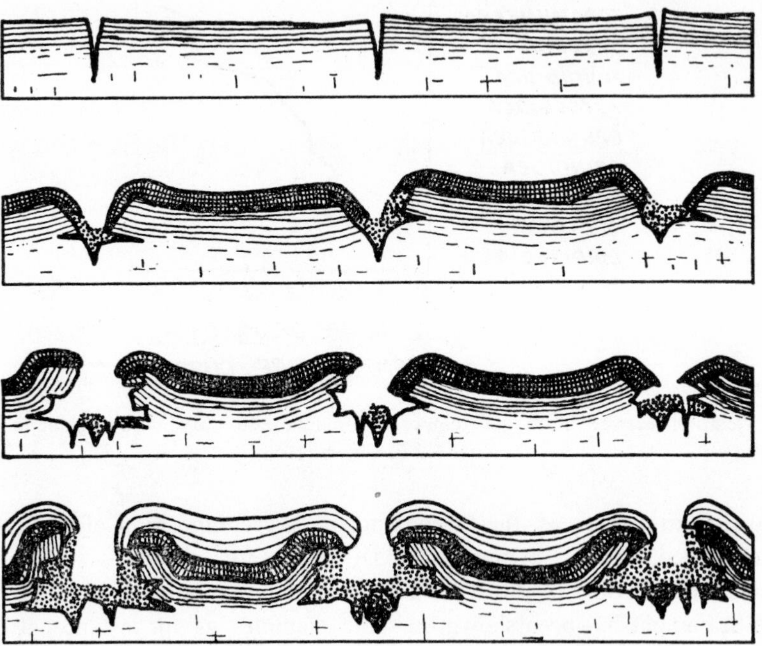

Fig. 14. Sequence of growth of algal deposits and the origin of stromatolitic structures. After M. Black (1933).

littoral lakes of Australia in Canary Islands and near fringing reefs at many places. M. Black was the first to describe them from the Bahamas in 1933. The ecological and palaeoecological significance of stromatolites, however, has not been appraised up to the present. They occur everywhere as separate leaf-like bodies or growing one over the other. In various sedimentary environments their morphology is fairly diverse. Thus, for instance, isolated semispherical forms with a high relief are characteristic of open shallow-platform; they correspond to the fossil genus *Cryptozoon*. The sheltered platforms where the activity of waves and currents is weaker are distinguished by the presence of interconnected not very distinct cupole-like forms corresponding to the fossil genus *Collenium*. There is no doubt today that they are produced by blue-green algae. Their structure is the result of differential growth and precipitation of carbonate on algal filaments. The precipitation is most intensive on the elevations, being usually absent in the depressions. They usually develop even on gentle irregularities of the bottom which increase gradually by the differential growth of algae.

Both Recent and fossil stromatolites show a characteristic lamination in cross-section. In Recent stromatolites, consolidated and unconsolidated or fine- and coarse-grained laminae alternate. The dense white carbonate crust was apparently precipitated directly by the action of the algal colonies under their gelatinous mat. The layers of fragmental material consist of debris of algal crust, oolites and faecal pellets, as well as of silt and clay particles. In the period of accelerated growth an algal filament grows on the detrital lamina.

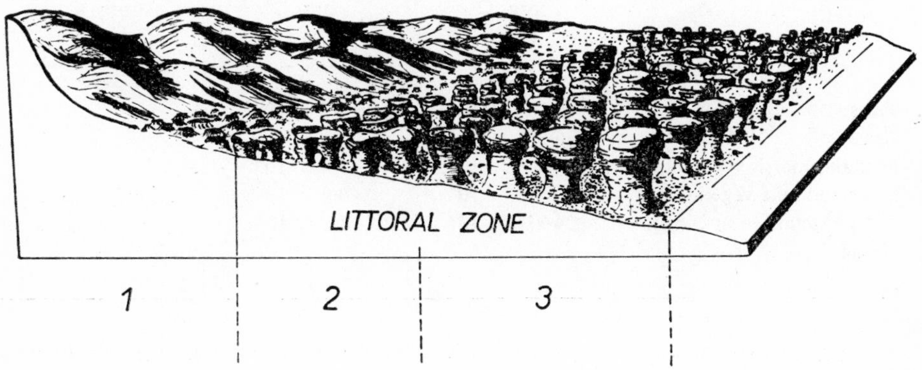

Fig. 15. Recent stromatolitic structures in Shark Bay (Australia). After B. W. Logan (1961).

Thus, stromatolitic structures in ancient sediments are a perfect indicator of sedimentary environment. They originate only in the littoral zone, i. e. in the zone between the average level of the high tide and that of the low tide, extending only sporadically a few metres below the sea level at the low tide. The form of this structures gives a still more exact picture of sedimentary environment. In a sheltered environment, without any intensive water movement, the mound-like structure are indistinct, interconnected, occasionally even overlapping. In a open non-sheltered environment the forms are pronouncedly individualized. At present, stromatolites occur only in very warm waters of normal or increased salinity. In the Shark Bay in Australia, they have been found also in water of 56—65 per mile salinity.

Products of the decay of soft bodies of organisms

Decay products of soft bodies of organisms (organic matter) are an important part of varied sediments or they form separate sedimentary rocks as well. Organic substances are produced both by plant and animal life, but their amount and composition widely differ according to their origin.

The total production of organic matter in the oceans and seas equals $1{\cdot}5 \times 10^{10}$ t organic carbon per year. The production of organic carbon by vegetation under continental environmental conditions is given in Table 35 (after B. A. SKOPINCEV 1961).

Table 35

The production of organic carbon by vegetation (after B. A. Skopincev 1961)

	Autotrophic org. C t/year	Heterotrophic org. C t/year	Dissolved org. C t/year
Trees	$9{\cdot}3 \times 10^9$	$14{\cdot}7 \times 10^9$	
Grasses	$8{\cdot}7 \times 10^9$		
Peat and marsh vegetation	$6{\cdot}2 \times 10^9$	$0{\cdot}4 \times 10^9$	
River and lake vegetation	$0{\cdot}728 \times 10^9$	$0{\cdot}6 \times 10^9$	$0{\cdot}5 \times 10^9$
Total vegetation on continents	$24{\cdot}93 \times 10^9$	$15{\cdot}7 \times 10^9$	$0{\cdot}5 \times 10^9$

The production of organic matter is distributed very unevenly over the Earth's surface. On the continents, it is by far greatest in the tropical zone. The conditions are much more complicated in the oceans, where the amount and the nature of organic matter produced depends on the amount of nutrients, on the temperature and salinity of water, and other subordinate factors. In evaluating the content of organic matter, not only the intensity of primary production, but also the rate of decomposition should be considered, because the differences in production are considerably levelled out by the effects of decay processes, which are most rapid at the sites of maximum productions. According to literature, 0·2 g of organic matter are deposited on one square metre of the sea bottom, which indicates that only 0·12% of total organic production is incorporated into sediments, while 99·88% of organic matter is decomposed in different ways before they become part of sediments. This percentage is still greater on the continents. In the course of diagenesis and weathering, an additional 30—90% of organic matter is lost, so that in ancient sediments only a fragment of 1% of organic matter originated in the seas and on the continents are preserved. The bulk of it is converted into inorganic compounds or serves as source of energy for higher organisms.

Organic matter of marine organisms and marine sediments shows the following composition (Table 36).

There is a fundamental difference between the composition of the organic matter of plant and animal origin. Therefore, organic substances of Recent and ancient sediments are divided into:

1. Humic matter, derived mainly from plant remains. It consists predominantly of cellulose and lignins, the amount of nitrogen and hydrocarbons being subordinate. The org. C/org. N ratio usually far exceeds 10.

2. Bituminous matter derived mainly from animal remains. It is characterized by an increased amount of carbohydrates and nitrogene compounds, such as proteins, fats, etc. The org. C/org. N ratio are low.

Table 36

Composition of organic matter of marine organisms and marine sediments (after P. D. Trask 1936)

	Phytoplankton %	Zooplankton %	Marine sediments %
Ether extract—includes oils, fats, pigments, organic sulphur compounds, and sulphur	2	10	1
Alcohol extract — includes waxes, resins, pigments and alkaloids	9	5	5
Water extract — includes sugars, starches, simple alcohols and simple organic acids and their salts and esters	23	5	3
Hemicellulose	11	0	2
Cellulose	5	2	1
Proteins	7	56	40
Urionic acid	16	0	0
Acid soluble non-nitrogenous compounds including lignin	5	3	40

The organic matter of Recent sediments contains on the average 56% C, 6% N, 8% H, 30% O_2; the mean org. matter/org. C ratio is about 1·8.

The basic components of organic matter have the following composition (Table 37).

The main components of the tissues of present-day plants are cellulose, lignin, tannin, proteins, resins and waxes. Their amount strongly varies depending upon the type of plants. Thus, for instance, the average composition of wood is: 40—60 % cellulose, 20—30 % lignin, 10—30 % hemicellulose and other polysacharides; the remainder of organic matter does not attain 5 %.

The components of organic matter are very comparable with some mineral components of terrigenous sediments:

1. Cellulose and proteins resemble dark Fe and Mg minerals in their low stability.

2. Waxes and resins are resistant, not very abundant components just as are zircon, rutile, tourmaline and some other stable accessory minerals.

3. Lignin converts by chemical and bacterial processes into humus, this displaying an analogy of behaviour to that of feldspars.

Table 37

Composition of basic components of organic matter

Element	Carbohydrates %	Lipoids %	Proteins %
O	49·48	17·90	22·40
C	44·44	69·05	51·30
H	6·18	10·00	6·90
P		2·13	0·17
N		0·61	17·80
S		0·31	0·80
Fe			0·10

Excrements of organisms as components of sediments

The share of excrements in the composition of Recent sediments seems to be underrated. Some authors note that in shallow-water environments more organic matter derive from excrements than from the bodies of organisms. Preserved faecal pellets represent only a minute portion of organic matter of this origin; it has been recognized that usually only faecal pellets of lithophagous animals are preserved due to their high consistency. Carnivores produce pellets without any cohesion, those of herbivores are more consistent.

Faecal pellets are of various size and shape. The simplest are ovoid in shape and several hundredths to a few tenths of a millimetre large. They are composed of organic matter rich in fats and nitrogen compounds, occasionally with an admixture of calcium carbonate and a few per cent P_2O_5. In places they are formed of silty and clayey particles cemented by organic matter called peloglea. The material of pellets is mostly somewhat coarser and better sorted than the sediment in which they are embedded.

Faecal pellets are most abundant in shallow-water environment, in clayey and silty-clayey sediments from tidal flats, lagoons and bays. H. B. Moore (1939) found as many as 3,400 ovoid faecal pellets in 1 cm^3 silty clay of the Clyde Bay. In fine-grained calcareous deposits of the Bimini Island (Bahamas) a sediment consisting of 90% pellets of *Batillaria minima* has been found. They are 0·90 × 0·16 mm in size and consist of minute detrital calcareous particles of silt grain-size which are cemented by organic matter. The amount of faecal pellets found in abyssal sediments is never so large as in shallow-water sediments. Therefore coprolites (fossil faecal pellets) found in a great number point to the deposition of the enclosing sediments in shallow-water environment.

Mechanical effects of organisms on sedimentation

The influence of the environment on the development of faunal assemblages is far reaching. Of unusual importance, however, are also the reverse, both mechanical and physico-chemical effects of faunal assemblages on the evolution of the environment.

The growth of stromatolites, as referred to above, affords one of the most illustrative examples of the mechanical influence of organisms on the environment. Fine-grained material, which in other conditions would be removed by water, is entrapped on the surface of algal filaments, so that the deposition implies finer-grained particles than would correspond to the hydrodynamical conditions. The rate of stromatolitic sedimentation can be faster than 4 mm/24 hours.

Some varieties of sea-grass also exert influences on the environmental conditions. *Thalassia testudinum* forming a dense mat on wide-spread shallow-water areas and near-shore areas creates a currentless environment, in which even the finest material may settle. In addition, they provide a resort to an immense amount of micro-organisms which in turn affect the environment by changing the physico-chemical conditions. The fossil nodules of fine-grained limestones enclosed in coarse-grained limestones may probably correspond to them. In Palaeozoic basins dense covers of crinoids could have played a similar role. Accumulations of pelecypod shells affect the environment in an analogous way, and a current-less space enabling the sedimentation of clayey material arises between them. Fossil sedimentary complexes present a number of analogous cases. The explanation of the "pockets" of dense sediments in coarsely detrital and organogenic limestones would be most difficult without the knowledge of Recent sedimentary conditions.

Coral reefs are the best and most typical examples of the influence of organisms on the sedimentary environment. The great thickness and comparatively high rate of deposition on coral reefs affect the wide environment by changing the direction of current and by the supply of sedimentary material.

Effects of changes in physico-chemical environmental conditions on sedimentation

The activity of organisms does not only modify but also controls the environmental conditions. In the latter case, the main factors are photosynthesis of phytoplankton and the activity of bacteria.

Photosynthesis of phytoplankton consists in the consumption of carbon dioxide and the liberation of free oxygen under considerable changes of pH of the environment. In the open ocean these changes are somewhat suppressed by the buffering effect of calcium carbonate — carbon dioxide — water system, but in a partly closed system lacking a perfect water exchange, as for instance in bays, pH can undergo substantial changes. In the Florida Bay pH of water ranges from 8·9 in midday to 8·0 at night; the values of pH are increased by photosynthesis during the day and decreased by respiration at night. The oscillation of pH exceeds the range of 0·5 also in other tropical bays. The changes of pH value greatly affect the carbonate system. At an intensive photosynthesis, i. e. at the increase of pH value, a decrease of solubility of calcium carbonate and its precipitation take place. On the other hand, dissolution of limestone was observed at the respiration (at night) in the littoral zone.

The number of bacteria in waters and Recent sediments is enormous. The absolute amount of bacteria in bay sediments depending on the grain-size of sediments is given in Table 38 (after C. E. ZoBell 1946).

Table 38

Sediment	Md μ	Organ. N %	H_2O %	Number of bacteria in 1 g of wet sediment
Sand	50—1,000	0·09	33	22,000
Silt	5— 50	0·19	56	78,000
Clay	1— 5	0·37	82	590,000
Colloid	<1	1·00	98	1,510,000

Table 39

Amount of bacteria in sediments of the Pacific Ocean from various depths (after C. E. Zobell 1946)

Bacteria	500 m	780 m	1,322 m
	Number of bacteria in 1 g of wet sediment		
All aerobic	930,000	31,000,000	8,800,000
All anaerobic	190,000	2,600,000	1,070,000
Denitrifying	10,000	1,000,000	100,000
Nitrifying	0	0	0
Sulphate-reducing	0	0	0

In fine-grained sediments bacteria are always more abundant because they contain more organic matter with which they are associated. The individual species of bacteria do not occur in the same amount in all types of sediments (Table 39 after C. E. ZoBell 1946).

The nitrogen cycle, controlled by the activity of bacteria, can be divided into several degrees:

1. Ammonification;
2. Nitrification, including:
   a) Oxidation of ammonia to nitrites;
   b) Oxidation of nitrites to nitrates;
3. Assimilation of nitrogen;
4. Reduction of nitrates and denitrification;
5. Fixation of nitrogen.

Oxidation of ammonia and nitrites are strongly exothermic processes; in sea water they are more frequent than those of reduction character which need a supply of energy from outside.

One of the end products of bacterial decay of organic matter is H_2S. Under anaerobic conditions almost all organic sulphur is altered to H_2S, under aerobic conditions it is dissolved in water, in the form of sulfate or sulfite. Besides bacteria which oxidize hydrogen-monosulfide to sulfides and sulfates there are bacteria inducing a reverse process — reduction of sulfates to sulfites and to hydrogenmonosulfide. The latter reactions occur in an environment rich in organic substances. It has been safely evidenced, for instance, that the major part of sulfides in fine grained sediments was derived from sulfates of marine water reduced by activity of bacteria.

An interesting question arose of whether bacteria can induce precipitation of calcium carbonate. Undoubtedly, im some shallow-water environments calcium carbonate can be precipitated by an indirect effect of bacteria which increase the pH value, but the existence of the so-called calcareous bacteria precipitating directly $CaCO_3$ has not been proved.

References

Baier C. R. (1937): Die Bedeutung der Bakterien für den Kalktransport in den Gewässern. Geol. d. Meere und Binnengew., Bd 1, p. 75—105, Berlin.

Black M. (1933): The algal sediments of Andros Island, Bahamas. Phil. Trans. Royal. Soc. London, Ser. B, vol. 122: 165—192, London.

Bock W. D. (1969): Thalassia testudinum, habitat and means of dispersal for shallow-water benthonic foraminifera. Bull. Am. Assoc. Petrol. Geol., vol. 53: 2033, Tulsa.

Bregger I. A. (1950): The chemical and structural relationship of lignin to humic substances. Conf. on the Origin and Constitut. of coal, Nova Scotia, p. 111—119.

CAROZZI A. V. (1962): Observations on algal biostromes in the Great Salt Lake, Utah. Jour. Geology, vol. 58: 246—252, Chicago.

CHAVE K. E. (1964): Skeletal durability and preservation. In "Approaches to paleoecology". P. 377—387, New York.

— (1954): Aspects of the biogeochemistry of magnesium. A. Calcareous sediments and rocks. Journal of Geology, vol. 62: 587—599, Chicago.

CLARKE F. W. - WHEELER W. C. (1917): The inorganic constituents of marine invertebrates. US Geol. Surv. Prof. Pap. 192: 1—224, Washington.

CLAYTON R. N. - DEGENS E. T. (1959): Use of carbon isotope analyses of carbonates for differentiation fresh-water and marine sediments. Bull. Am. Assoc. Petrol. Geol., vol. 43: 890—897, Tulsa.

CURTIS C. D. - KRINSLEY D. (1965): The detection of minor diagenetic alteration in shell material. Geochim. Cosmochim. Acta, vol. 29, 71—84, London.

DODD J. R. (1963): Paleoecological implications of shell mineralogy in two pelecypod species. Journal of Geology, vol. 71: 1—11, Chicago.

HARRIS R. C. (1965): Disequilibrium precipitation of molluscan skeletal material and its implications regarding the use of trace elements in fossil shells as paleoecologic indicator. Bull. Am. Ass. Petrol. Geol., vol. 49: 343, Abstr., Tulsa.

INGERSON E. (Editor) (1963): Organic geochemistry (Symposium). P. 1—658, Oxford.

KEITH M. L. - EICHLER R. - PARKER R. H. (1960): Carbon and oxygen isotope ratios in marine and fresh-water mollusk shells. Bull. Geol. Soc. Am., vol. 71: 1901—1902, Baltimore.

KINSMAN D. J. J. (1969): Interpretation of Sr concentrations in carbonate minerals and rocks. Jour. Sedimentary Petrology, vol. 39: 486—508, Menasha.

KNAUFF W. (1963): Über Zuchtversuche an *Patellina corrugata* (Foram.) in unterschiedlichen Meerwasserlösungen. Fortschr. Geol. Rheinl. u. Westph., Bd. 10: 363—366, Krefeld.

KORNICKER L. S. - PURDY E. G. (1957): A Bahamian faecal-pellet sediment. Jour. Sedimentary Petrology, vol. 27: 126—128, Menasha.

LALOU C. (1957): Studies on bacterial precipitation of carbonates in sea water. Jour. Sedimentary Petrology, vol. 27: 190—195, Menasha.

LERMAN A. (1965): Paleoecologic problems of Mg and Sr in biogenic calcites in light of recent thermodynamic data. Geochim. Cosmochim. Acta, vol. 29: 977—1002.

LEUTWEIN F. (1963): Spurenelemente an rezenten Cardien verschiedener Fundorte. Fortschr. Geol. Rheinl. Westf., Bd. 10: 283—292, Krefeld.

LEUTWEIN F. - WASKOWIAK H. (1962): Geochemische Untersuchungen an rezenten marinen Molluskenschalen. N. Jhrb. Min., Abh., Bd. 99: 45—78, Stuttgart.

LLOYD R. M. (1960): Shell chemistry of some recent and pleistocene mollusks and its environmental significance. Bull. Geol. Soc. Am., vol. 71: 1917, Abstr., Baltimore.

LOGAN B. W. (1961): Cryptozoon and associate stromatolites from the Recent, Shark Bay, Western Australia. Journal of Geology, vol. 69: 517—533, Chicago.

LOGAN B. W. - REZAK R. - GINSBURG R. N. (1964): Classification and environmental significance of algal stromatolites. Journal of Geology, vol. 72: 67—83, Chicago.

LOWESTAM H. A. (1954): Factors affecting the aragonite-calcite rations in carbonate-secreting marine organisms. Journal of Geology, vol. 62: 284—322, Chicago.

MASLOV V. P. (1959): Stromatolites and facies (in Russian). Doklady AN SSSR, vol. 125: 1085 — 1088, Moscow.

MCMASTER R. L. - CONOVER J. T. (1966): Recent algal stromatolites from the Canary Islands. Jour. Geol., vol. 74: 647—652, Chicago.

MOORE H. B. (1939): Faecal pellets in relation to marine sediments. In "Recent marine sediments". P. 516—525, Tulsa.

Oppenheimer C. H. (1959): Bacterial activity in marine sediments. General Petroleum Geochemistry Symposium, p. 49—55, New York.

Pilkey O. - Hower J. (1960): The effect of environment on the concentration of skeletal magnesium and strontium in Dendraster. Journal of Geology, vol. 68: 203—216, Chicago.

Pilkey O. H. - Goodell H. G. (1963): Trace elements in recent and fossil mollusk shells. Bull. Am. Ass. Petrol. Geol., vol. 47, P. 366, Abstr., Tulsa.

Siegel F. R. (1960): The effect of strontium on the aragonite-calcite ratio of pleistocene corals. Jour. Sedimentary Petrology, vol. 30: 297—304, Menasha.

Skopincev B. A. (1961): Some results of investigation of organic matter in sea waters (in Russian). In "Recent sediments of seas and oceans". P. 285—291, Moscow.

Trask P. D. (1939): Organic content of recent marine sediments. In "Recent marine sediments". P. 428—453, Tulsa.

ZoBell C. E. (1946): Marine microbiology. P. 1—240, Waltham.

8. River sediments

The environment of river valleys affords a suitable opportunity for studying sedimentary conditions, so that the composition of river sediments has been relatively well recognized. The knowledge of the changes in the composition of sediments and of the regularities of their origin is poorer, mainly because the sediments have so far been studied from a static point of view. The deposition in running water has frequently been studied in laboratories, but mainly for technical purposes. The experiments for sedimentological and geological purposes require a somewhat different approach. Basic data on the regularities of transport and deposition of solid particles in running water applied mainly to river sedimentation are contained in the papers by G. K. GILBERT (1914), F. J. HJULSTRÖM (1935), CH. NEVIN (1946), H. W. MENARD (1950a) and L. B. LEOPOLD *et al.* (1964), A. SUNDBORG (1956) and M. MORISAWA (1968).

The description of the processes of sedimentation in rivers has been recently described by J. R. L. ALLEN (1965). A detailed description of the development and geomorphological characteristics of the drainage patterns can be found in all textbooks of geology and geomorphology, in particular in the comprehensive book by L. B. LEOPOLD et al. (1964). Therefore, I think it unnecessary to treat of these basic problems at length and I shall limit myself to a few data bearing on the regularities of the activity of running water and its influence on transport and deposition of solid particles.

Of the two known types of flow of running waters, laminary and turbulent, the latter absolutely prevails in rivers. Standing and horizontal whirls and transversely circulating movement are components of the turbulent flow. On the basis of the absolute velocity the turbulent flow can be subdivided into sheetflood and streamflood. The boundary value between the two types of movement is 10 m/sec. The velocities of flow corresponding to streamfloods occur only in the upper stretches of river valleys. Velocities in the middle reaches range 1 to 5 m/sec, in the lower courses of major streams they drop to 0·8—2 m/sec. The velocity of flow is not uniform through the whole cross-section of a stream, declining with the distance from the streamline and the decrease in depth.

The regularities of the deposition - transport - erosion regime are best shown on F. J. HJULSTRÖM's diagram (1935), compiled on the basis of experiments. The diagram shows that the transport represents an equilibrium state between the velocity of flow and the grain-size of material borne. Provided that the grain-size of material remains unchanged, deposition or erosion takes places depending on whether the velocity of flow rises or decreases. The velocity of flow necessary for erosion is approximately 1·4—1·5 times greater than the lowest velocity of flow at which the sed-

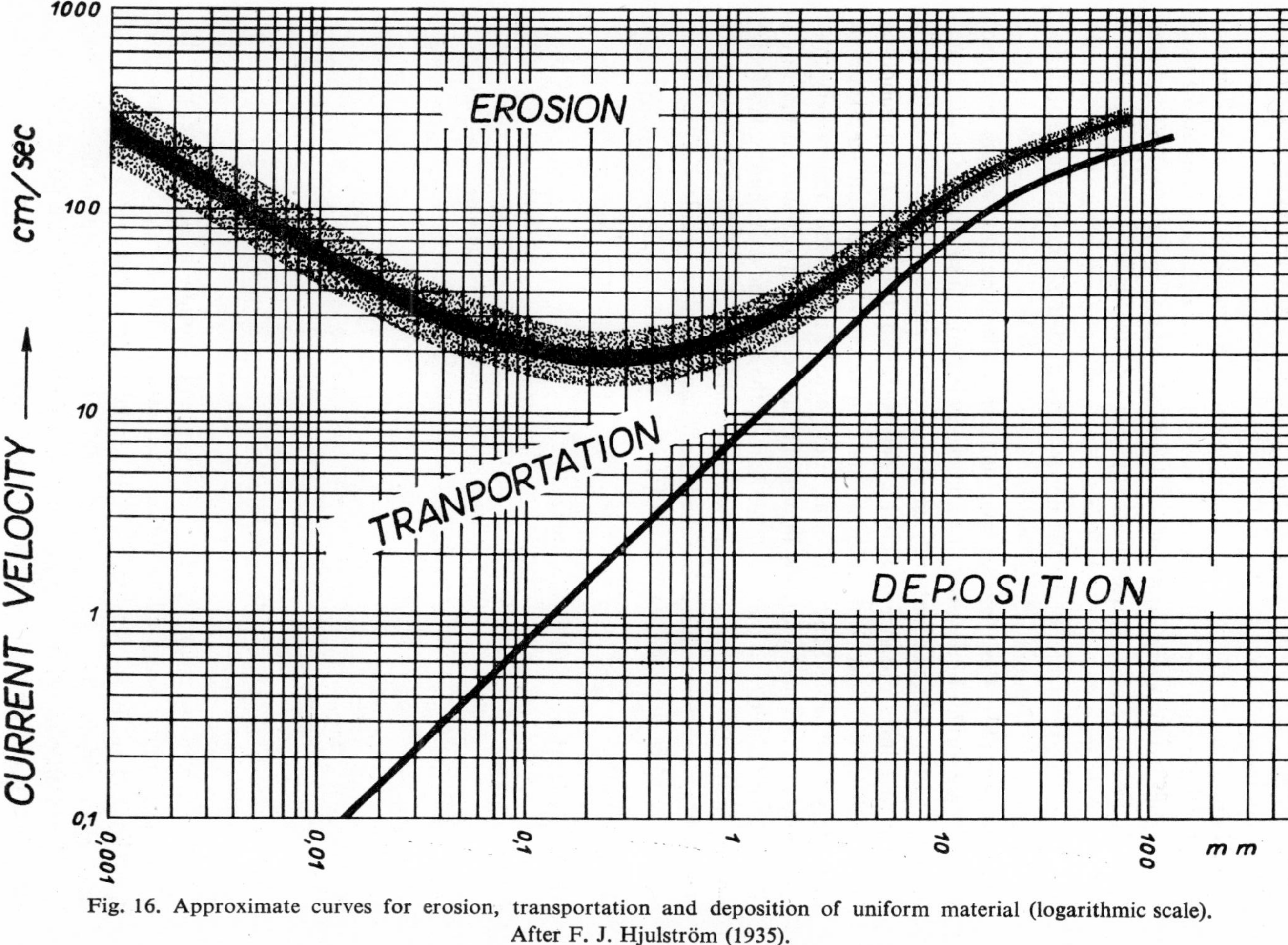

Fig. 16. Approximate curves for erosion, transportation and deposition of uniform material (logarithmic scale). After F. J. Hjulström (1935).

iments can still be transported. This implies that at a given velocity of flow larger grains than those carried by flow are deposited in the stream.

The importance of rivers in the development of Recent and ancient sediments is twofold. On the one hand, they supply the bulk of clastic material into most basins (as mentioned above), and on the other, detain part of the deposits on the continents. Some values of the ratio of sediments supplied into the basin to those deposited in river channels are given on p. 38. These data reveal that many rivers deposit more than 25% of clastic material in the upper parts of deltas, on alluvial plains, river flood-plains, or piedmont alluvial cones.

During transport, the fractions that had not undergone any sorting by former processes, are sorted. Fine clastic silty and clayey material in suspension is transported at a rate corresponding to that of the stream (i. e. at a rate of 1—2 m/sec in the valley reaches); covering a distance of about 2,000—4,000 km in a year. From this it may be inferred that in most rivers this fine material reaches basins with stagnant water after one year's travel. Sandy material, transported as bed load, moves very much more slowly. The velocity of this movement can be reconstructed from the rate of migration of sand bars in the river bed. Rate of the movement of sand waves (height 8—6 m) in the Brahmaputra River rarely exceeds 30 m/day (J. M. Coleman 1969). Some other observations have furnished the following values (Table 40).

Table 40

Velocity of migration of sand bars in river bed

River	Velocity of movement of sand bars m/24 hours
Volga	6.00–7.00
Luga	0·10–0·4
Kemka	0·15–0·25

The approximate mean value of downcurrent motion is one metre per day, so that sand under bed load conditions is transported at a rate 5,000 times lower than clay and silt in suspension. Four thousand kilometres covered by clay in one year, requires in the case of sand about 5,000 years (disregarding the floods). Therefore, river transport brings into the basin clay simultaneously with much older sand, which had been removed from the upper stretches of the stream several hundred or even a few thousand years ago.

As is well-known, the transitional form of transport between those of suspension and bed load is transport by saltation. In addition to these main means of transport, the material can be carried by flotation, by ice, in the roots of trees, etc. Transport in solution, which is of importance for marine and lacustrine deposition, is on the whole of no importance for river sedimentation.

Transport in suspension

The displacement of suspended load to considerable distances is possible only owing to the turbulent movement of water particles. In the turbulent flow ascending components rich in suspended material intertwine with descending components poorer in such particles. The ascending components overcome the gravitational tendency of grains and keep the material in suspension conditions. Moreover, the sinking particles are caught by horizontal and vertical whirls. Part of the material, which is the greater the lower the degree of turbulence, sinks to the bottom and settles down. In a layer of running water, clayey particles are distributed regularly, in contrast to coarser, silty and fine sandy particles, which are randomly scattered, their number increasing towards the bottom. The greater the turbulence of flow, the smaller is the tendency to the accumulation of larger grains at the bottom.

The distribution of grains and their velocity of sinking are also affected by their shape. Angular grains sink more slowly than round grains and, consequently, they are transported to a greater distance. The same holds for disk- and rod-shaped particles. From this it can be deduced that coarser particles of the above forms are carried along with smaller spherical particles, so that they can occur together in the resulting sediment.

The decisive factor of the transport of suspended load and its deposition are flood waters. Most intensive floods occur after short heavy downpours, mainly in vegetation-free areas, where almost all atmospheric water runs off the surface and by stream channels. In the extreme case, the floods can produce mudflows where water is only a passive component brought into movement by solid particles. A comparison of floods of Recent rivers has shown that on most rivers there are annual floods and, in addition, irregular floods recurring in a period of several years or a few decades.

The amount of material transported by floods is immense. Thus, for instance, the river Kish Chai in Central Asia transported, during the flood in 1936, 2,175,000 m^3 material in two hours. Such an amount is displaced during 25 years under normal water conditions. The flood on the river Almatinka (in 1921) carried away 2·5 million cubic metres of material, whose transport at normal water height would take 100 to 125 years. During floods, the suspension load in water is about 30—50%; a maximum of 68·2% has been measured during the above-mentioned flood on the river Kish Chai. Considering the immense amounts of transported material and the several year — or several decade periodicity of floods, the main share of floods in the transport of material in vegetation-free areas is evident. Some mountain rivers carry coarser material in suspension than under normal conditions, as instanced by the grain-size of suspended material in the river Kish Chai in a flood state (Table 41). At the normal water level, the suspension load contains much higher amounts of finer fractions.

Under continental conditions the flood sediments are deposited to a greater extent than normal river sediments. Some mountain streams deposit almost all flood material in the piedmont, the lowland streams deposit on the average one-half of this material on flood plains and alluvial plains.

Table 41

Grain-size of suspended load in the river Kish Chai in a flood stage

Fraction mm	%	Fraction mm	%
1·00—0·25	1·4	0·010—0·005	8·1
0·25—0·05	26·4	0·005—0·001	15·7
0·25—0·01	39·5	<0·001	8·9

Transport by saltation

Material transported by saltation is too coarse to be carried in suspension. As the velocity of streams flow oscillates, grains of a definite size are at one time in suspension, at the other transported by saltation. The leaping movements of particles are separated by intervals when they rest on the stream bottom or are moved by traction. The length of leaps varies between several centimetres and a few metres. This means of transport can be regarded as a transitional form between suspension and traction transport.

Material transported as bed load

The bed load moves along the stream bottom either by slipping or rolling.

According to the ratio of the stream velocity and the grain-size of several phases of transport can be distinguished.

1. Initial movement: Individual grains are picked up by the current, slip and jump, the height of leaps increasing with the increasing velocity of flow.

2. Scattered transport: Whole groups of grains set moving in individual parts of deposits.

3. Smooth transport: At the increased velocity of flow a few millimetres thick bed of sediments with laminar motion is formed.

4. Dune-ripple transport: Grains stop moving at lines perpendicular to the direction of flow and spaced apart at roughly the same intervals. In those places low ridges developing as asymmetrical ripplemarks are formed. The explanation of their origin presumes that under certain conditions the form of ripple marks represents a shape consuming a minimum energy at the friction of two environments. The greater the velocity of flow and the coarser the sediment, the greater is the length of wave and the amplitude of ripple marks. The more material is carried, the flatter the ripple marks and the less definite their ridges. With a lack of material, the ridges become isolated and can change into migrating ridges of various sizes.

5. The next phase, called smooth traction, occurs at considerable velocities of flow. A layer of several centimetres to decimetres of sediments is transported in a mixture with water. This manner of transport is common mainly at floods and longshore drifts.

6. The last phase whose existence is questionable is the so-called antidune transport. It has been observed only experimentally; apart from a few not entirely unequivocal evidenced cases it has not been observed in nature. At greatest velocities of flow, a ripple-shaped surface is presumed to form on the sandy bottom; the asymmetrical ridge of ripples oriented with the steep side upstreams are called antidunes and represent erosion forms of the sandy bottom. Their preservation in ancient sediments is almost impossible, although some similar forms of questionable origin are referred to in literature ("flame structures"). If the velocity of flow continues

Table 42

Computed average monthly scour and fill in Colorado River bed during 1955 water year in the 88-mile reach from Less Ferry to Grand Canyon, Ariz. (after B. R. Colby 1964)

Month	Computed total discharge of sands near Grand Canyon (in 1,000 tons)	Average scour (−) or fill (+) in reach (in ft)
October	370	+0·04
November	22	+0·03
December	9	+0·02
January	4	+0·01
February	6	+0·03
March	790	+0·03
April	1,470	−0·03
May	13,500	−1·11
June	9,340	−0·57
July	500	+0·06
August	660	+0·004
September	33	+0·01

to increase, all sediments are washed away and the erosion of solid rocks in the river bottom sets in. In dependence on the changes in the flow velocity, the stream bottom is a site of alternating scour and fill, as documented in Table 42 (compiled from data on the Colorado River by B. R. COLBY 1963).

These stages of transport are not confined only to the stream environment, but may occur wherever the unidirectional flow is active.

Sediments recurrently transported and lying on the bottom are the stream sediments proper. As long as they are in contact with water environment, they are in a state of permanent changes, and their composition is not constant either. The texture and structure of sediments are stabilized when they are out of contact with water environment, which occurs most frequently when they are covered by other sediments or when the river shifts its channel.

Transport by flotation

Transport by flotation is usually regarded as subordinate, but at certain conditions it can play a fairly important role, when considered from the geological and sedimentological point of view. This mode of transport is achieved either under the influence of the surface tension of the water level, or with the help of air bubbles. In the former case grains can be caught on the water level and carried to a considerable distance before they break through. As it is well-known, grains smaller than 0·1 mm get only with difficulty below the water level at a slight impact and when finally dropping they catch along air, with the help of which they can travel a bit farther. Transport with the help of air bubbles is generally more active than transport by flotation; in stagnant environment it can actually be the only means of transport. Transport by flotation is far more marked in the composition of marine than river sediments, because in the marine environment sandy grains can be transported to areas of clay sedimentation.

There are two main opinions on the geological significance of transport by flotation. One, advocated particularly by C. H. BEHRE (1926), K. H. WOLF (1961) and J. D. HUME (1964) presumes that material transported by flotation can prevail in some deposits. The other (e. g. K. O. EMERY 1950) regards flotation as a frequent mode of transport, but usually masked by other, more effective, ways of transportation. There are some quantitative data on the transport by flotation: The Ohio River carried 130 t material in one hour; the Merimac River in the NE of the United States bore about 1,000 t sand per year. Pure angular quartz grains were transported far into the area of clay sedimentation and laid down in places whose hydrodynamical conditions did not correspond to the deposition of sand. Transport by flotation may be effective also when combined with eolian transportation. Wind-borne grains falling on the surface of water may be transported to a considerable distance under flotation conditions.

The present state of knowledge of this problem, far from being perfect, allows us to assume that under definite conditions the amount of material transported in flotation may even exceed the volume of bed load, but can never attain the quantity of the suspended load.

•

The amount of material carried by streams in whatever manner differs according to the type of rivers. It has been found that the amount of suspension and bed load increases with the increasing discharge and the increasing velocity of flow. The amount of suspended material increases at a far greater rate that that of the bed load. G. V. LOPATIN (1952) reports the following values of the suspended/bed load ratio:

	Ratio bed load/suspension load
Mountain rivers	0·10—0·20
Lowland rivers	0·05—0·10

The results show that in the mountain rivers the suspended load exceeds 5 to 10 times, and in the lowland rivers 10 to 20 times the bed load. The larger the stream, the greater is the difference. Major streams may carry 100—1,000 times more material in suspension that as the bed load. This justifies the above-mentioned assumption that the majority of material brought into the basin by streams had travelled in suspension.

Composition, origin and occurrence of fluviatile sediments

The term fluviatile (river) sediments includes all sediments deposited by rivers, i. e. not only sediments of river channels but those laid within river valleys (levees, marshes). An excellent description of the general character and the development of the stream pattern is given in the book by L. B. LEOPOLD *et al.* (1964). The mountain range supplying the overwhelming part of clastic material to the lowlands is usually cut by large river valleys, through which the bulk of coarse-grained gravel is carried away. On the outer side of the ridge, minor streams or mountain creeks, transport material of smaller grain-size to the lowlands, because they have a lower gradient and drain the outer parts of the mountain range built up generally of sedimentary rocks. In the valleys of major streams cutting the mountain ranges gravels and coarse-grained sands are deposited; finer-grained material is carried beyond the mountain stretches of the streams and is laid down for the most part at the transition from the mountain environment into the lowland tract. In those places the river sedimentation frequently passes into a deposition of alluvial fans, and with the altered character of the surrounding deposits, into eolian sediments.

River sedimentation takes place in the following environments:

1. River channel
   a) in the proximity of the stream-line,
   b) shoals at the margin of the channel;
2. Natural levees (bank deposits) comprise all environments into which the channel passes laterally;

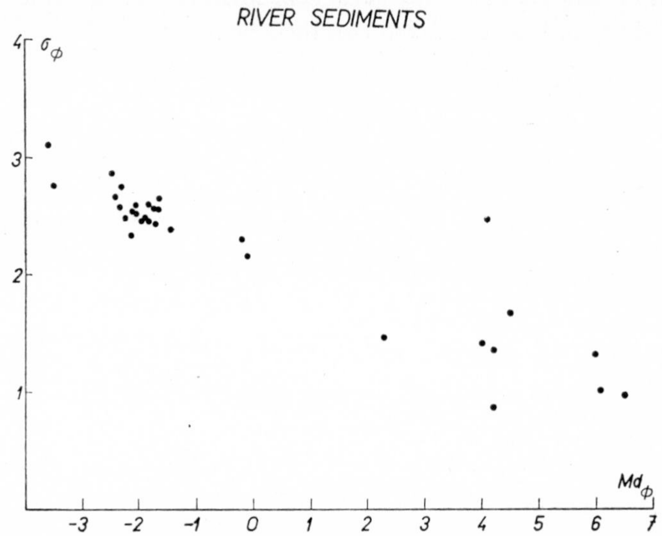

Fig. 17a. Grain-size parameters of various kinds of river sediments. Relation between Phi Median Diameter and Phi Deviation Measure. Data compiled from C. Burri (1930), G. V. Lopatin (1952), C. F. Nordin - J. K. Culbertson (1961).

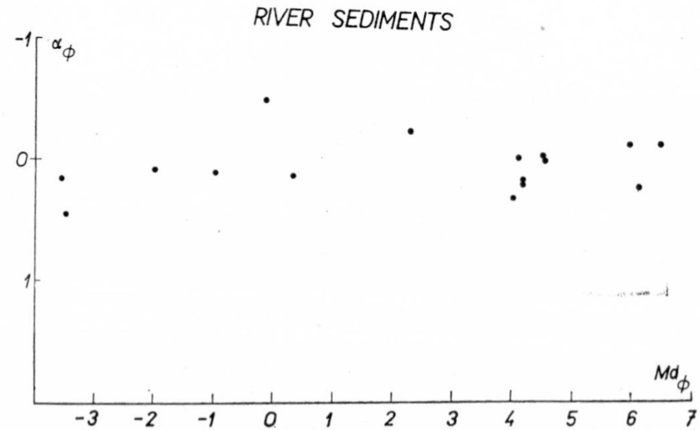

Fig. 17b. Grain size parameters of various kinds of river sediments. Relation between Phi Median Diameter and Phi Skewness Measure.

3. Flood-plains (overbank deposits) including all sediments deposited during the floods above the channel and bank sediments.

The alluvial plains are sometimes defined as a separate, fourth type of stream environments. They are those environments in the lower reaches of streams which appear as a single homogeneous area of stream deposition, in which the individual features of the three above sedimentary areas are already obliterated.

Besides the descriptive classifications, there are those elaborated on the genetic basis, which agree broadly with the classification referred to above. Genetic classifications emphasize the relationship between the processes of erosion and accumulation. Of these, mention should be made at least of that proposed by I. P. KARTASHOV (1961). This author presumes that in the course of the stream development three succesive phases are active. He calls them: 1. Instrative phase, 2. Constrative phase, 3. Perstrative phase. The first phase is active during the downcutting of the stream, when erosion greatly prevails and sedimentation is intermittent. The second phase corresponds to the equilibrium state between erosion and sedimentation; in the last phase the accumulation of sediments is dominant.

Channel deposits

The stream channel appears to be a geological, hydrological, and sedimentological unit and it is not easy to divide it into further, generally valid subenvironments. Although the grain-size often decreases from the central part to the margins of the stream channel, this phenomenon cannot be regarded as a common rule.

The composition of channel deposits is affected by the following factors:

1. The velocity of flow, the amount and character of the material borne.

2. Mixing of the material deposited from suspension with that produced by lateral erosion.

3. Mixing of suspended and bed loads,

4. Scouring of fine-grained sediments, burying of coarse-grained sediments by the finer-grained ones which drop onto the coarse underlying sediments,

5. Arrangement of larger particles by the effects of currents.

On the basis of grain-size distribution two types of channel deposits can be differentiated:

1. Sediments containing more than 5% of gravel (material >2 mm); they are mainly represented by gravels and coarse sands deposited in mountain reaches of streams.

2. Sediments containing less than 5% of gravel, which are deposited particularly in the middle and lower reaches of streams as sandy and finer-grained material.

There are striking petrographical differences between the sediments of mountainous and lowland rivers and those of the upper and lower reaches of the same river. Some

Table 43

Petrographical differences between the sediments of mountainous and lowland rivers (after L. B. Rukhin 1947)

	Mountainous stream	Lowland stream
Sorting	poor, mixture of gravel, sand and clay prevails	good, sand fraction prevails
Changes in Md downstream	great changes, Md rapidly decreases, value ranging between 0·25 and 0·314 mm	Md slightly decreases value ranging between 0·163 and 0·200 mm
Amount of unstable mineral grains	great amount of feldspatnic rock fragment and unstable heavy minerals	quartz grains strongly prevail
Changes in mineral composition downstream	amount of unstable components rapidly decreases	mineral composition remains most frequently unchanged
Grain roundness	angular and subangular grains prevail	about 50% of sand grains are rounded or well rounded

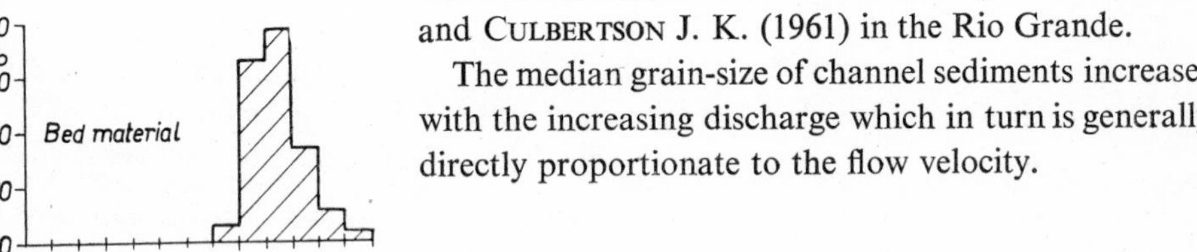

of these, especially qualitative differences, were listed by L. B. Rukhin (1947) (see Table 43).

Table 44 presents an example of how the composition of channel sediments depends on hydrodynamical conditions, chiefly on the velocity and discharge of the stream. The data tabulated were determined by C. F. Nordin and Culbertson J. K. (1961) in the Rio Grande.

The median grain-size of channel sediments increases with the increasing discharge which in turn is generally directly proportionate to the flow velocity.

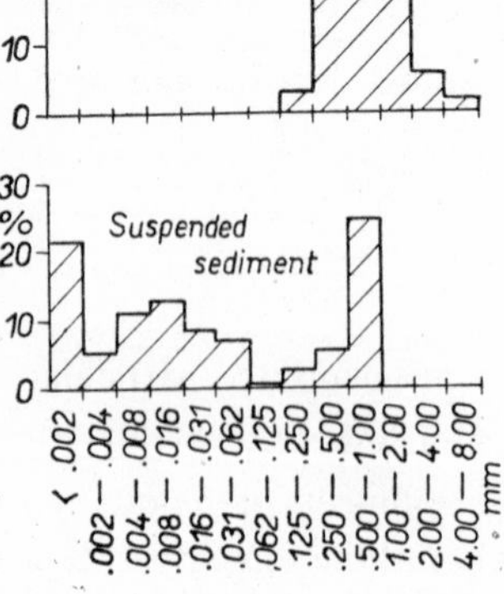

Fig. 18. Histograms of size distribution of bed material and suspended sediment for Rio Grande at Otowi Bridge, July 20, 1961. After C. F. Nordin - J. P. Beverage (1965).

Table 44

Location	Discharge ft^3/sec	Md of bed load		
		minimum	maximum	average
		mm		
Otowi	1,130—10,100	0·20	10·40	1·45
Cochiti	24— 9,810	0·18	3·10	0·44
San Felipe	71— 9,720	0·17	5·10	0·45
Bernalilo	4—10,000	0·11	0·45	0·25
Belen	520— 8,270	0·08	0·37	0·23
Socorro	697— 3,600	0·15	0·25	0·19
San Antonio	368— 8,400	0·15	0·25	0·20
San Marcial	1,990— 8,680	0·13	0·16	0·14

Sediments of the mountain rivers are distinguished by bimodal grain-size composition (Fig. 21). In general, the medium and coarse-sandy fractions are lacking (especially fractions between 0·2 and 0·5 mm). This type of composition is due to long-term reworking and redeposition of sediments. Other opinions, however, have been propounded on the origin of this bimodality, or even polymodality. It seems likely that bimodality results from the mixing of material transported in two different ways i. e. as bed and suspension load. The gravel fraction is moved by traction and the fine sandy and finer fraction in suspension. According to another view, the polymodality of channel sediments is due to abrasion, because it has been experimentally proved that abrasion produces fine sandy and silty fractions, but never the coarse sandy ones. As a result, the gravel and fine sandy + finer-grained fractions (without any transition) develop by abrasion of the initial homogeneous gravel material. Another way of the origin of bimodal sediments, which was observed in the field and experimentally reconstructed, consists in the infiltration of finer-grained material into the coarser material previously deposited. It has been observed that finer-grained material can infiltrate into the freshly deposited material bearing more than 70% of gravel. In spite of all these possibilities the fact remains that in the overwhelming majority of cases bimodal sediments develop in those deposits which were not laid down in one phase, but were many times reworked and redeposited, remaining for a long time in contact with the hydrodynamical environment. The possibility exists, namely, that in the river channels there are also sediments of lognormal distribution and no signs of polymodality. Sediments of this type settled from suspension during one phase and can be found mainly among flood deposits and in alluvial fans.

Channel sediments of mountain streams have the following typical parameters: The Median varies between 0·1 and 0·8 mm. Phi Deviation Measure (σ_Φ) ranges from 0·4 to 2·58 being most frequently around 1·0. Phi Skewness (α_Φ) varies between

—0·68 and 0·53, the negative values roughly of the same frequency as the positive ones. The bimodality, as described above, occurs only when erosion is present in the stream; it has not been ascertained in the aggradation stages of streams. Erosion is most probably the decisive factor for the development of bimodality in sediments, the other factors being of subordinate importance only.

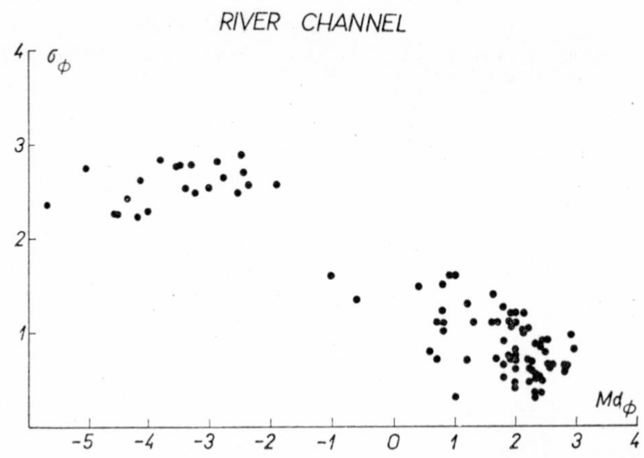

Fig. 19a. Grain-size parameters of river channel sediments. Relation between Phi Median Diameter and Phi Deviation Measure. Data compiled from V. P. Baturin (1947), C. Burri (1930), W. C. Krumbein (1942), J. M. Pollack (1961), V. I. Popov *et al.* (1956), L. B. Rukhin (1947).

Along the mountainous stretches of streams, the pebble material undergoes fairly rapid changes in the grain-size (decline of Md downstream), rounding of grains and in mineral composition (removal of unstable components). Two factors, selective abrasion and selective transport, exert a compensatory influence on the grain-size distribution and mineral composition.

1. Experimentally reconstructed selective abrasion has shown that larger and less stable pebbles succumb more readily to abrasion.

2. According to present knowledge a major importance is postulated for selective transport. Owing to this, the grain-size and mineral composition are brought closer to each other, because under definite hydrodynamic conditions grains of the same size and shape are concentrated, transported and deposited simultaneously.

Channel sediments of lowland rivers have usually less than 5% of gravel; medium or fine-grained sand constitutes the prevailing part of their material. The percentage of silt and clay increases. The bimodal composition is manifested in another form, i. e. by the lack of fraction between 0·25 and 0·1 mm. The bimodality originates most frequently in places where the sediments become coarser as a result of erosion processes. As illustrative example of bimodal sediments in these environments are their occurrence in bank deposits where material produced by erosion is mixed with sediments deposited from suspension.

The grain-size composition of channel sediments, both of mountain and lowland rivers is plotted on the diagram in Fig. 19a,b. Channel sediments of the lowland rivers show the following parameters:

The Median varies between 0·115 and 0·30 mm, occasionally rising above this limit. Phi Deviation Measure (σ_Φ) ranges from 0·5 to 1·5; this value suggests the

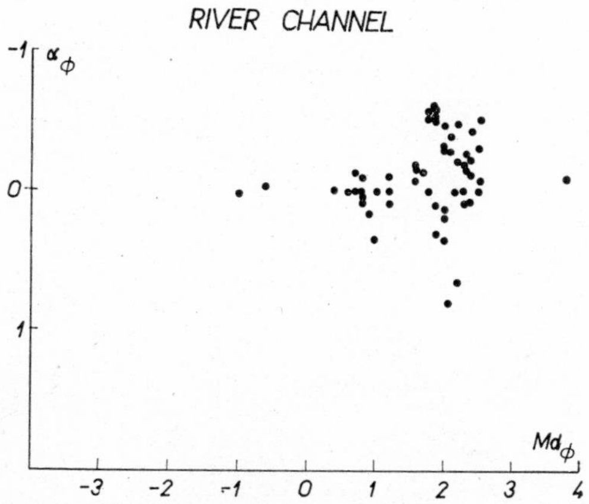

Fig. 19b. Grain-size parameters of river channel sediments. Relation between Phi Median Diameter and Phi Deviation Measure.

possible presence of well-sorted deposits, but the grade of sorting never attains that of beach or dune sands. Phi Skewness (α_Φ) is distributed symmetrically round zero. The minimum value found is —0·6, the maximum value is 0·8.

S. A. Schumm (1961) records the following values of the median grain-size for channel sediments of ephemeral streams (Table 45).

Table 45

Mean values of channel sediments of some ephemeral streams (after S. A. Schumm 1961)

Study area	Type of cross section	Median grain size mm	Trask sorting index	Silt-clay %
Sage Creek	Stable	0·080	5·74	54
	Aggrading	0·067	4·96	65
Arroyo Calabasas	Stable	0·81	1·99	4
Bayou Gulch	Stable	0·61	1·72	4
Medano Creek	Stable	0·24	1·23	0·8

The material deposited in river channel affects its shape. A. S. NAIDU and R. C. BORRESWARA (1965) have found that the present morphology of the Lower Godawari River in India is the manifestation of the bed load material. Higher silt-clay ratios increase the degree of sinuosity and decrease the width/depth rations of the channels. Decrease in the sinuosity of the river course in the last one hundred years is probably a result of the coarsening in the bed load.

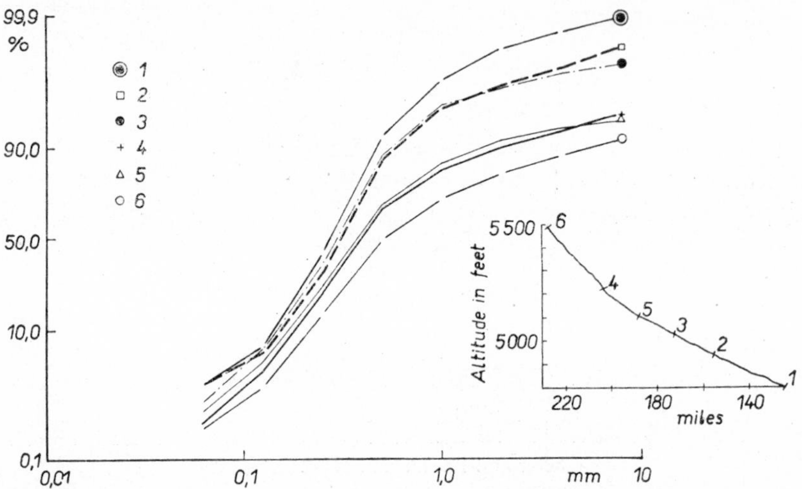

Fig. 20. Channel profile of the Rio Grande from Otowi Bridge to Belem. Relation between channel slope and size distribution of bed material. 1. Belen, 2. Albuquerque, 3. Bernalillo, 4. Cochiti, 5. San Felipe, 6. Otowi. After C. F. Nordin - J. P. Beverage (1965).

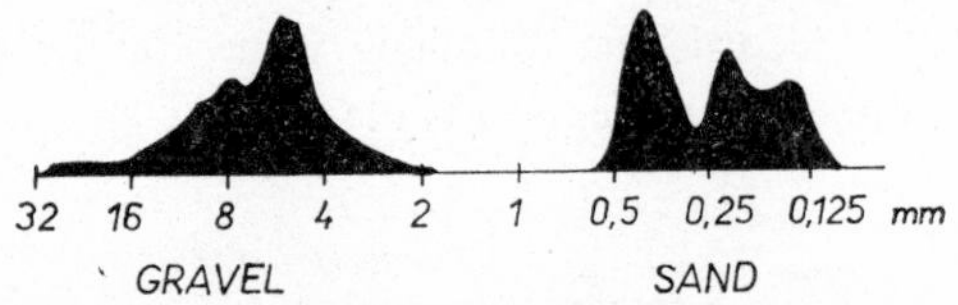

Fig. 21. Grain-size distribution of channel sediments of Brazos River. After R. L. Folk *et al.* (1957).

Sedimentary structures of channel deposits

The fundamental form of channel deposition is the formation of gravel and sand bars and megaripples. Bars develop in places of the decrease in the capacity of streams. They are either separate and most varied in shape or pass into sandy subaqueous dunes and sand waves migrating slowly downstream. The average velocity of their movement is 1 m in 24 hours. The wave length and amplitude of these sand waves changes with the velocity of flow, the depth of the river channel, grain-size and amount of carried material. In minor streams the amplitudes are usually of a few decimetres, the length of wave ranges from several metres to some tens of metres. In major rivers (e. g. the Danube or the Volga) the amplitudes of sand waves are 1·0—1·5 m and the length of the wave amounts to 30—50 m. The largest sand wave (amplitude 4 m and length 143 m) have been described from the Mississippi River channel. Large sand waves are of different forms, they are often barchan shaped, elongated both up- and downstream. Small forms pass into typical asymmetrical ripples. The described forms may also be fossilized, when they are in equilibrium with the hydrodynamical environment, at least for a short time, and are covered afterwards by fine-grained sediments. The knowledge of the internal structure of the megaripples, sand waves and point bars is of special importance for the study of fossil sediments. All these structures are the products of the water current, from which it follows that their main feature will be various forms of the current bedding. S. K. SARKAR - S. BASUMALLICK (1968) described from the Barakar River following vertical sequence of structures (from bottom to top): very large trough cross-stratification – large trough cross-stratification – small trough cross-stratification — ripple drift lamination and deformation structures — small trough cross-stratification — tabular cross-stratification — horizontal stratification. J. COLEMAN (1969) distinguished major bed forms and minor bed forms. By the migration of larger bed forms (sand waves) large scale cross-stratification of trough type develops (with maximum unit thickness up to 6 m). By the migration of minor bed forms ripple drift lamination originates. In the channel just as in other environments sedimentary structures depend primarily on the grain-size; the presence of individual structures can be inferred, to a large extent, from the size distribution of sediments. Small-scale ripple-bedding is frequent in fine-grained sands of the channel (Md = = 0·10—0·25 mm). The sets of cross-strata increase with the increase in grain-size. In coarser-grained sediments planar and trough cross-bedding is common. Sets of cross strata measure up to 30 cm in thickness and are commonly lenticular or wedge-shaped, and therefore are not continuous for very great distances. Maximum dip angles of cross-laminae are 25 to 30 degrees.

The regularity of current bedding, the sharpness of boundaries between current-bedded units in particular, is impaired with the deterioration of sorting. Ripple-bedding passes into streak apparent bedding as the content of clay increases.

The thinning of current-bedded units in the upward direction is a reliable distinctive feature of current bedded channel deposits. In addition, sediments of coarsest grain-size are accumulated at the base of the units in the lower parts of inclined laminae and beds.

Most point bars consist of a sequence of coarse sand and gravel grading upwards to silt and fine sand. A typical section may be subdivided into four generalized zones, each characterized by a particular class of sedimentary structures (from the bottom to the top):

1. Poor bedding, 2. Megaripple cross bedding (giant trough-bedding grading to planar wedge-shaped bedding). 3. Small ripple cross-bedding, 4. Horizontal lamination. Some of these zones, particularly the upper ones, can be absent.

The direction of dip of laminae shows a minor scatter compared to those of beach or eolian deposits (Fig. 131). The coarser and more poorly sorted are the sediments, the lesser the scatter of directions of dip of laminae. Well-sorted quartz sands display a wide scatter of directions and often an indistinct preferred orientation. Gravel deposits of river channels are in places unstratified and of a chaotic structure; in others they are characterized by the so called alluvial pavement, i. e. by beds of tilted pebbles.

Abudant clay pebbles can originate especially in the channels of ephemeral streams. According to I. Karcz (1969) they originate as scree in the lee of the clayey wadi-banks, which collapse during the drought periods. Their initial sorting and sphericity are usually good, because the collapse of the banks is controlled by mudcracks.

Bank deposits (lateral accretion)

Bank deposits are often mistaken for channel sediments or, even more frequently, for overbank deposits (flood-plain sediments-vertical accretion). The indefinite character of their petrography is responsible for this confusion: when coarser, they closely resemble channel deposits (e. g. in the Brazos River — H. A. Bernard - C. F. Major 1963); when of clayey-silty composition, they look like overbank deposits (e. g. in the Amu Darya).

Owing to a rapid lateral migration, some rivers lay down fairly coarse sands (Md up to 3—5 mm) as bank deposits. According to recorded data, their grain-size never exceeds 10 mm. Whereas in stream channels no chemical sedimentation takes

Table 46

Sediment characteristics of bank deposits, Sage Creek, S. Dakota (after S. A. Schumm 1961)

Cross-section	Median grain size mm	Trask sorting index	Silt-clay %
1	0·028	2·20	93
2	0·030	2·24	93
3	0·028	2·00	96
4	0·035	1·83	90
5	0·032	1·68	84
6	0·038	2·00	83
7	0·031	2·68	88
8	0·033	1·91	90
9	0·027	2·13	89
10	0·015	2·00	79

place, bank deposits contain in places considerable amounts of carbonate which is supposed to have settled from suspension, together with clay particles. The content of organic matter is also greater than in channel deposits, increasing with the rise in clay amount. Sediments containing about 75% of silt and clay have 1—3% of organic matter sporadically reaching up to 10%.

S. A. SCHUMM (1961) has recorded parameters of bank deposits in several ephemeral streams of semiarid regions (Tab. 46).

Bank sediments of these ephemeral streams are similar to channel sediments. The presence of fine-grained sediment has been established only in one case.

Owing to their generally homogeneous petrographical composition, bank deposits do not show a great variety of structures. A fairly regular silty-clayey lamination is rather rare. Sporadic seams of silt and sand are usually ripple-bedded. Fine-grained sediments richer in organic matter bear abundant traces of plant roots and occasionally have a mottled structure produced by non-uniform accumulation of organic matter.

Overbank deposits (Flood-plain deposits)

Flood-plain sediments form the bulk of stream valley deposits. They constitute a sedimentary cover both of channel and bank deposits. The shape of the flood plain is controlled by the shape of the stream valley; it attains a great width in the valleys of lowland rivers and is narrow, almost undeveloped in the deep-cut valleys of streams with a steep gradient. The altitude and the thickness of flood-plain deposits are governed by the difference between the stream-level under flood and normal conditions; this difference commonly amounts to a few metres only, but many attain as much as 32 m (the Lena River in Siberia, 420 km from the mouth). Theoretically this flood could have deposited about 30 m of flood-plain deposits. Normal floods,

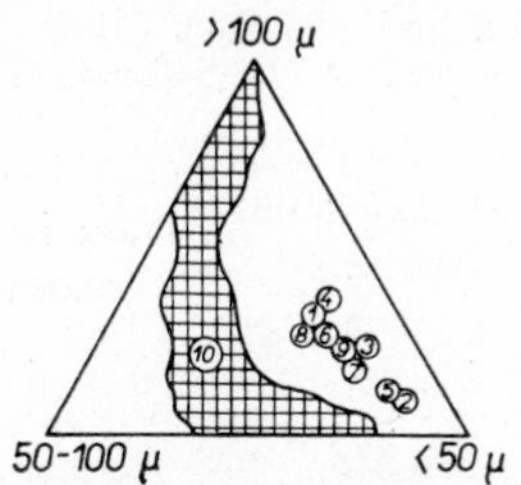

Fig. 22. Grain-size distribution of flood deposits and flood-plain sediments. 1. Ohio Valley flood, 1937, coarser deposits, 2. Ohio Valley flood, 1937, finer deposits, 3. Potomac River deposits, Maryland and Virginia, 4. Brandywine Creek, Pennsylvania, flood-plain materials, 5. Kansas River, 1951, flood deposition on flood-plain, 6. Kansas River flood-plain surface, 7. Seneca Creeks, Maryland, flood-plain, 8 and 9. Watts Branch, Maryland, flood-plain, 10. Connecticut River, flood deposits of September 1938. After M. G. Wolman - L. B. Leopold (1957).

during which the river level rises a few metres only, are repeated annually, but large floods recur only in long intervals, generally of several decades. During one flood the thickness of material deposited is of mm — to cm — order. Thus, for instance, the Nile River deposits during flood 1 mm thin layer of clay on an extensive area, the Amu Darya River in Central Asia laid down 18—20 cm material (additional data are given in the chapter on the rate of sedimentation, p. 44).

The typical course of floods, i. e. an abrupt beginning and a slow dying out, are mirrored in the grain-size distribution of sediments. In the first phase, a great amount of comparatively very coarse material is transported, which settles down immediately upon the slight decrease of the competence of the stream. During the further lowering of the stream level and of the velocity of flow, the grain-size of detritus to be deposited decreases, so that in places perfect graded bedding develops.

The mean grain-size distribution of flood-plain deposits is roughly as follows:

Fne-grained sand	5—10%
Silt	20—40%
Clay	35—60%

In some flood-plain sediments, as in Bijou Creek, Colorado (E. D. McKee *et al.* 1967) or in Euphrates River (K. H. Al-Habeeb 1969) surprisingly great amounts of fine sand have been found (more than 30%).

Md of flood-plain deposits varies between 0·005 and 0·06 mm, i. e. within the range of clay and silt fractions. Phi Deviations is usually between 1 and 3, which indicates a medium or poor sorting. Phi Skewness (α_{Φ}) oscillates moderately about zero.

Table 47

Sediment characteristics of overbank deposits, Sage Creek, S. Dakota (after S. A. Schumm 1961)

Cross-section	Median grain size mm	Trask sorting index	Silt-clay %
1	0·040	3·63	74
7	0·035	1·89	90
13	0·007	5·10	97
14	0·044	1·78	85

Table 47 gives parameters of overbank deposits of some ephemeral streams in semi-arid regions (after S. A. Schumm 1961).

As can be seen, the content of clay and silt is in all cases higher than in channel and bank deposits. This is consistent with a lower value of Md and, on the average, a higher value of So.

The content of carbonates is rather variable, attaining in places a high value; flood-plain deposits of the Amu Darya River contain, for instance, on the average, 22% of carbonate, those of the Volga River 25% and in Euphrates River more than 50%. The mean content of carbonates in flood-plain sediments of other rivers studied is generally about 10%. The amount of organic matter ranges from 1 to 10%; the organic matter, similarly as carbonates, is mostly of detrital origin and settles from suspension, along with clay particles. In addition to the prevalent clay and silt, flood-plain sediments occassionally bear layers of medium and fine-grained sand which are difficult to differentiate from channel deposits on petrographic evidence.

Sedimentary structures of flood-plain deposits are represented mainly by bedding and lamination, which are graded in about one-fifth of cases. The thickness of individual strata corresponds to that of sediment deposited during one flood, varying generally between several mm and a few cm. The dipping of foresets in ripple-drift lamination is either downcurrent but also away from the river channel. Nicely developed convolute structures have been described from Brahmaputra River (J. M. Coleman 1969) and Euphrates River (K. H. Al-Habeeb 1969). Their origin is probably influenced by the asymmetric ripple wandering and deformation. Homogeneous structures, which occasionally occur through the greater part of the profile, correspond to the deposition of homogeneous clays or silt. Coarser beds and laminae are almost

Table 48

Petrographical characteristics of river sediments

	Channel sediments	Bank sediments	Overbank sediments
Grain-size	Md from 0·05 mm up to several metres most frequently 0·1—2 mm	Md from clay size up to 10 mm, most frequently 0·01—0·05 mm	Md from clay size up tp 0·5 mm, most frequently 0·005—0·05 mm
Carbonate contents	most frequently 0—3%	0—10%	often over 10%
Amount of organic matter	up to 1%	1—50%	most frequently 1—10%
Sedimentary structures	tabular-planar bedding with high angles and low angles as well, current ripple bedding	without recognizable structures, sometimes current ripple bedding, mottles, rootlets	regular lamination and bedding, graded bedding, mottles, biogenic structures, streaky bedding, mud cracks

ripple bedded. The deposits bear a great amount of plant debris; non-uniform distribution of organic matter is responsible for mottled structure of the sediments. Mud cracks are often developed on their surface.

The most reliable indicator of individual environments is the grain-size. Structures and chemical composition range as secondary diagnostic features, and the geological position and shape of the body and lateral transitions as auxiliary parameters. All these characteristics are listed in Table 48.

Some stream sediments can be differentiated from the marine ones on the basis of the content of carbonates. Whereas in stream sediments the content of carbonate increases with the decrease in the size of fraction, in marine sediments this relation is inverse. The tendency, referred to for stream sediments, is due to the supply of fine detrital carbonate along with the clay fraction.

Orientation of clastic components

The orientation of pebbles in stream sediments has been thoroughly studied and recently the orientation of sand grains has also been traced.

The particle moving in water or lying on the bottom occupies as stable a position as possible in relation to its environment, i. e. a position which would reduce the friction between the two environments to minimum. The striking imbricate arrangement of pebbles in the stream channel (with long axes dipping upstreams) has long been noted. Theoretically, this position of the upstream inclined pebble can be explained by the tendency of pebbles to turn upcurrent as large a face as possible to increase buoyancy. This orientation is not so pronounced in quite small pebbles that are affected by contact with other grains. The dip of pebbles ranges from 15 to 30°, that of beach pebbles has been statistically established at 2—12°. This difference in the amount of dip can serve as a distinctive indicator. The problem of orientation of the longest axes of blade- and rod-shaped pebbles upstream is far more complicated. In many cases, investigation has brought conflicting results. It is difficult to evaluate exactly the factors responsible for bed surface structure and particle orientation in running water. Among bed types the following surfaces may be distinguished: beds with contact load or noncontact load material; erosion pavements, transport and deposition beds with different fabrics. These are influenced by a great many variables, such as depth and velocity of flow, turbulence conditions, sediment feed and sorting, bed properties, e. g. hardness, roughness, and general configuration, inclination, particle volume, mantle area, shape and density in relation to the transporting medium. In spite of the great number of directional analyses, many questions are still unanswered.

The evaluation of all facts established can be summed up as follows:

In streams where the velocity of flow is fairly high and stable, the pebbles are usually oriented with the longer axis parallel to the flow. Small pebbles show a more pronounced preference for alignment parallel to the direction of flow than large pebbles and cobbles, as was observed by C. E. JOHANSSON (1965). This condition is interpreted as a selective orientation due to particle position during transport and deposition. "Small" (approx. less than 4 cm) particles moved in less contact with the bed and took up a longitudinal orientation more readily as a result of the shearing stress of the flow. They were further turned more easily round obstacles at the deposition than "large" particles, that rolled or slid in more or less permanent contact with the substratum. If the velocity of the current is lower or swift currents alternate

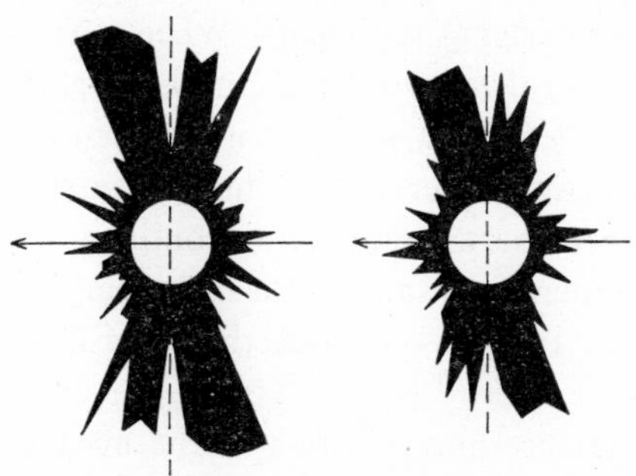

Fig. 23. Example of preferential orientation of channel pebbles. Orientation perpendicular to current direction. Ephemeral stream in South Persia. After M. Kürsten (1960).

with the slow ones, a preferred orientation of longer axes of pebbles perpendicularly to the direction of flow has frequently been observed (Fig. 23). From these conclusions it follows that a safe interpretation of the orientation of long axes of pebbles in river sediments need a perfect evaluation of all conditions. It depends on whether the pebbles were transported in suspension (the resulting orientation of the longer axis would then be theoretically parallel to the flow) or by saltation or rolling (which would support the orientation of the long axes at right angles to the flow. Moreover, the orientation is influenced also after the deposition of the pebbles, by the effects of the current and the surrounding sediments. In places, the long axes are oriented fanwise, which corresponds to the fanwise arrangement of current laminae. In the ephemeral streams — wadis — of Western and Central Iraq only parallel orientation of the longest axes of pebbles with stream direction has been found (Z. KUKAL - A. SAADALLAH 1969). Perpendicular orientation, however, has been found in some other ephemeral streams (M. KÜRSTEN 1960).

The orientation of sand grains in the river deposits is rather of theoretical interest. In all cases studied, a quite expressive preferred orientation of longer axes of sand grains in the downstream direction has been found. As distinct from marine beach deposits, the position of long axes is consistent with the elongation of linear sand body.

Heavy minerals in fluviatile sediments

Fluviatile sands and gravels are in many places enriched in heavy minerals. There are sufficient factual data available on the accumulation of these natural concentrates in stream deposits, but a theoretical generalization of the findings is still lacking.

The accumulation of heavy minerals is based on the differences in the specific weights of light and heavy fractions, which make possible differential transport and deposition of the two fractions. A contributory factor, which in some cases may be of primary importance, is the dropping of grains of heavy minerals into the interstices of much coarser underlying sediments. When a mixture of sandy and silty material is transported along the gravelly bottom, large grains of quartz and smaller grains of heavy minerals, being heavier than is the competency of stream, are laid down; smaller grains of heavy minerals sink into the interstices between the gravels of the bottom, while larger quartz grains often continue in their travel. Besides, the concentration of heavy minerals on sand banks or in marginal shoals occurs similarly as on the beaches: lighter fractions are selectively washed out from the primarily sorted sediment.

The stream concentrates are never so rich as the marine ones in which the percentage in volume and weight of heavy minerals may attain as much as 100%. High concentrations in river sediments can be met with only in point and lateral bars, some parts of sand bank and on the upstream sides of subaqueous dunes.

The study of stream concentrates has shown that the bottom parts, particularly the so-called relict gravel deposits of alluvium, are richest in heavy minerals, although they do not exceed the value of 10—40%. Increased concentration has also been observed in places of a sudden decrease in the velocity of a stream in the upper parts of meanders, near the mouths of brooks, etc. In these places high floods wash out light fractions from the accumulated sands.

The mineral composition of concentrates of minor streams or brooks usually differs from that of concentrates in major streams. The composition of the former is influenced by the character of rocks in the river basin and their content of heavy minerals. The more extensive alluvia of major streams represent, however, a separate unit, and the content of concentrates develops according to the changes in their hydrodynamic character. This difference can be explained by the differences in the composition of valley sediments. In young valleys of minor streams there is still an abundance of non-reworked material of older weathered rocks on the surface, whereas senile valleys of major streams are formed from youngest river sediments or solid older rocks.

Changes in the character of sediments downstream

F. J. Pettijohn (1957) noted that the unknown direction of flow of an ancient stream could be inferred from ancient river sediments: if their medians are connected by isoplethes, the direction of declining isoplethes will show the direction of flow.

Thorough studies of Recent sediments have shown, however, that the above rule can be applied only roughly. It holds good in the sense that there is an overall tendency for the grain-size to decrease downstream, or, in other words, that the channel deposits are coarser in the upper and middle reaches of the stream than in the lower reach. In detail, there are so many exceptions and such a dependence on local factors that it cannot be universally accepted.

Table 49

Coefficient of grain-size changes of several rivers (after L. B. Rukhin 1947)

Mountainous rivers	Coefficient of grain-size changes
Great Laba, upper reaches	3·0
Tessin	1·1
Rio Grande	0·6
Lowland rivers	
Great Laba, lower reaches	0·2
Volga	0·2
Mississippi	0·4

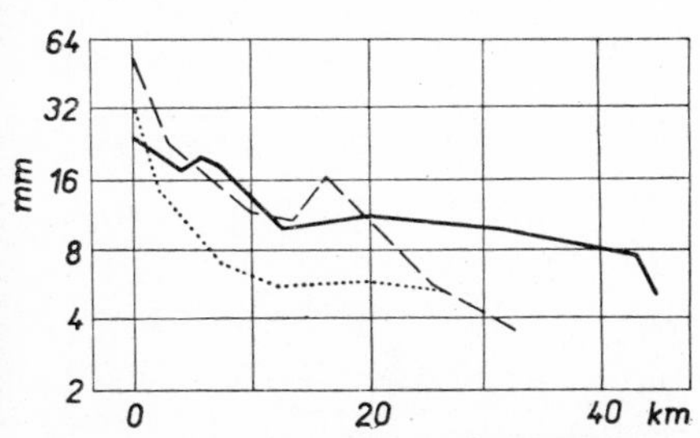

Fig. 24. Relation of mean grain-size to distance of transport. Rapid Creek, Bear Butte Creek and Battle Creek (South Dakota). After W. J. Plumley (1948).

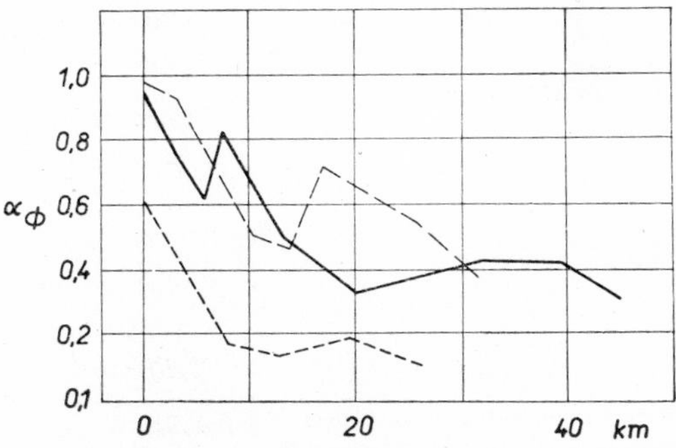

Fig. 25. Relation of Phi Deviation Measure (sorting) to distance of transport. Rapid Creek, Bear Butte Creek and Battle Creek (South Dakota). After W. J. Plumley (1948).

The most detailed data on the changes of grain-size downstream have been gathered for the Mississippi and Colorado Rivers, for the Alpine river Tessin, the Lower Godavari River in India L. B. Rukhin (1947) introduced the so-called coefficient of grain-size change for the characterization of these changes. This coefficient is defined as 100 times the difference between the sizes of 2 samples of grains divided by the distance (in km) between the sites of sampling. The above author records the following values computed (Table 49).

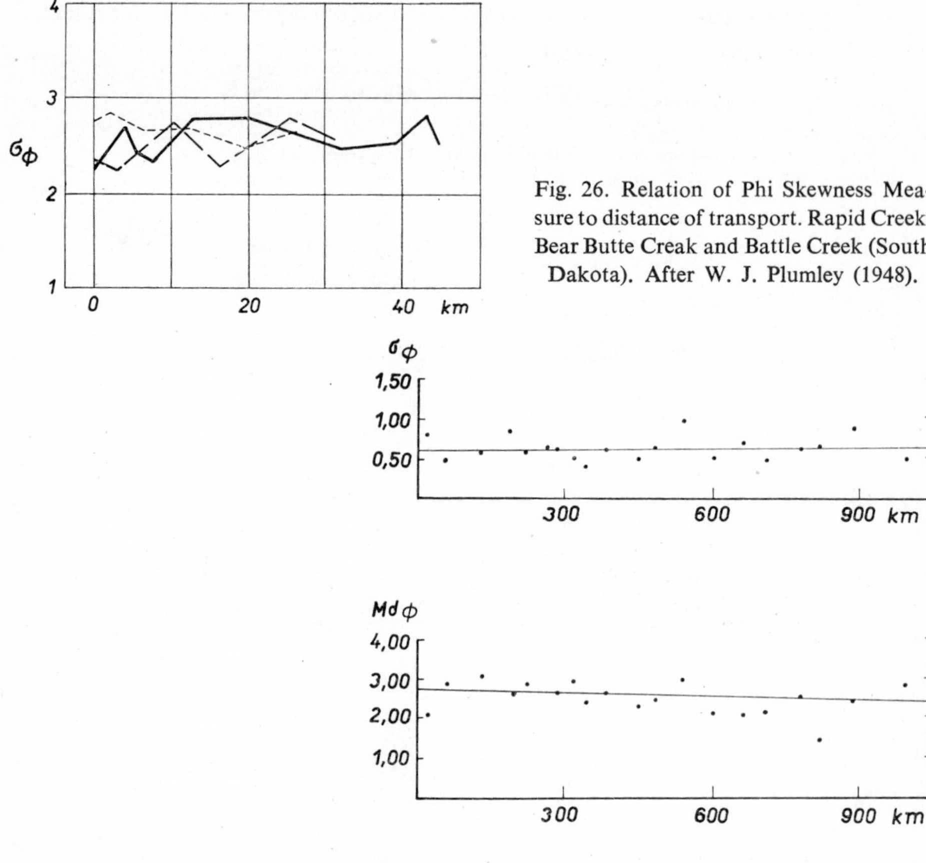

Fig. 26. Relation of Phi Skewness Measure to distance of transport. Rapid Creek, Bear Butte Creak and Battle Creek (South Dakota). After W. J. Plumley (1948).

Fig. 27. Relation between Phi Median Diameter, Phi Deviation Measure and Phi Skewness Measure to distance of transport in South Canadian River channel sands. After J. M. Pollack (1961).

These data confirm numerically the fact mentioned before that the grain-size diminishes more rapidly in sediments of mountain than of lowland rivers. Although in mountain rivers the grain-size considerably diminishes downstream, the coefficient of sorting is only slightly changed. The polymodality, however, is sometimes balanced or converted to bimodality of another type. Phi Skewness (α_{Φ}) undergoes fairly great changes in the direction of flow: the initial positive values drop to zero and even pass to negative ones, e. g. in the Lower Godavari River (A. S. NAIDU - R. C. BORRESWARA 1965).

The changes of parameters Md and So downstream in the Mississippi River are listed in Table 50.

Table 50

Changes of grain-size parameters downstream in the Mississippi River (after R. D. Russell and R. E. Taylor 1937)

Distance downstream from Cairo, Illinois miles	Md			So (Trask's)		
	minim.	maxim.	average	minim.	maxim.	average
	mm			mm		
0—100	0·263	1·616	0·868	1·22	2·54	1·54
100—200	0·318	1·019	0·670	1·18	2·00	1·50
200—300	0·302	0·754	0·510	1·18	1·87	1·35
300—400	0·205	0·885	0·480	1·12	2·79	1·45
400—500	0·239	1·541	0·580	1·10	2·21	1·32
500—600	0·241	0·638	0·391	1·16	1·51	1·25
600—700	0·090	0·591	0·322	1·16	1·44	1·23
700—800	0·173	0·562	0·349	1·14	1·30	1·20
800—900	0·199	0·359	0·272	1·10	1·14	1·12
900—1,000	0·174	0·328	0·224	1·15	1·21	1·28
1,000—mouth	0·115	0·202	0·157	1·14	1·34	1·18

In addition to this general tendency, some local factors are at work which may occasionally be of great importance. They are:

1. Gradient and velocity of the stream. In stretches of greater slope, sediments become coarser and usually bimodal.
2. Width of the stream channel which affects the velocity of flow. In places of narrowing valley, sediments again becomes coarser and bimodal.
3. The presence of tributaries which always bring coarser material.

•

The diminishing of pebbles and grain-size is interpreted in terms of abrasion and selective transport.

1. The influence of abrasion is beyond doubt. It has been experimentally proved that larger particles succumb to abrasion more readily than smaller ones. The

diminishing of particles occurs after the well-known Sternberg law

$$W = W_o e_{kx} ,$$

where W_o = original weight, x = distance, e = nat. log (2.8183), k = constant.

There have been found, however, considerable departures from this principle in natural environment.

For a long time, abrasion was regarded as the single process controlling the diminishing of grain-size. The decisive influence of selective transport has been recognized only over the last twenty years.

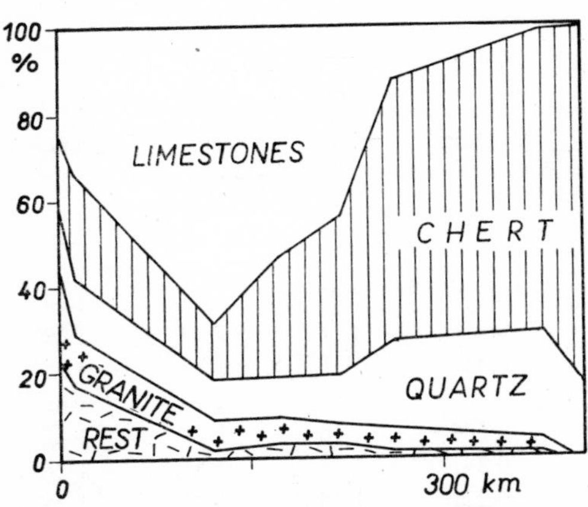

Fig. 28. Relation of river gravel composition to distance of transport in Colorado River, Texas. After E. D. Sneed - R. L. Folk (1958).

2. Selective transport, i. e. selective sorting of material according to the weight, shape and size during transport. In the opinion of some authors, selective transport causes 75—84% of the changes in the grain-size distribution, the remainder falling to the share of abrasion. This process implies a grouping of particles of the same size, weight and shape and their separation from other particles.

The changes in mineral and petrographical composition follow the trend of grain-size changes: they are considerable in the upper reaches of mountain rivers, being not very pronounced in the lower reaches or lowland rivers. Changes in the composition of the gravel fraction have most frequently been a subject of study. The changes in the content of pebbles of discrete rocks depend on the:

1. Diminishing of pebbles by abrasion and their transition to finer fractions,
2. Mixing of the original material with the detritus from other sources.

A. Kodymová (1960) came to notable conclusions in her study of stream gravels in the Bohemian Massif area (Table 51).

From the data in this table it follows that a great loss of pebbles of unstable rocks takes place already in the first kilometres of travel; more stable rocks show the greatest loss in the first tens of kilometres. A rapid decrease of limestone, which is one of the least stable rocks, in pebble material has been evidenced in many streams. Sandstone pebbles (except quartz sandstone) are also very unstable and get lost

Table 51

Changes in composition of stream gravels in the Bohemian Massif area (after A. Kodymová 1960)

Rock	Distance downstream from the last rock outcrop in river valley km	Percentage of the total amount of pebbles
Granulite	20	9·5
	45	1·3
	63	1·1
	98	0·7
	151	0·14
	273	scarce
Aphanitic porphyrites	17	11
	31	9
	56	6
	78	1·5
	107	3
	115	2
	238	0·5
Granites	0	35
	18	24
	58	24
	83	5
Phyllites	14	8·5
	24	2·8
	38	2·8
	125	0

usually after a few kilometres of transport. Some data on the resistence of pebbles of various rocks, compiled from the data of different authors, are given in Table 52.

Generally it can be said that if the changes in the composition of the gravel fraction were not affected by the import of tributaries or by erosion of material in the river itself, the unstable pebbles, i. e. all non-siliceous ones, would decrease and the stable— siliceous material (chert, quartzite, lydite) increase, away from the primary source. The pattern of changes, however, is more complicated, because the material of the original source is mixed with other material sand the amount of unstable components may increase in places.

Sandy sediments show still more intricate changes in the petrographical composition along the course of stream. These changes are studied mainly in terms of the quartz/feldspar ratio. Detailed studies have shown two interdependences. In some streams (e. g. in the Mississippi and Rhine) feldspars decrease and the amount of quartz rises. In the Mississippi River 20% of feldspars is lost over a distance of about 1,800 miles, in the Rhine River the amount of feldspars decreases by 10% from the Swiss frontier to the sea. At present, the opinion that mechanical abrasion is responsible for the decrease of feldspars is no longer held; the increase of quartz at the expense of feldspar is ascribed rather to the supply of quartz from older sediments. In other rivers, the conditions are more complicated, and the quartz/feldspar ratio does not increase downstream (the Colorado River, South Canadian River, Alpine rivers).

Table 52

Resistance of pebbles of various rocks during the river transport

Rock	Length of transport
Gneiss	usually not more than 20 km
Granite	hundreds of kilometres, when the original weight was more than 20 g
Shale	after 40 km transport completely disaggregated
Sandstone	not more than 15—20 km
Greywacke	weight loss after 450 km was 21%
Limestone	abaut 100 times less resistant than granite

Small differences have also been observed in the amount of some heavy minerals (hornblende, pyroxene, etc.) downstream (the Mississippi River, the Lower Godavari River, Alpine rivers). Our knowledge of this problem, however, is not so detailed as to warrant the conclusion whether the decrease is a result of abrasion, selective transport or mixing with other material.

Changes in roundness and sphericity of grains

In psephitic (gravel grade) sediments the roundness of pebbles greatly changes in the course of travel. The degree of roundness is an important diagnostic feature for an estimate of the length of transport. Pebbles of little resistant and abundant rocks (e. g. limestones) afford a suitable material for study of this question, because they are very sensitive to hydrodynamical conditions. It has been found that their rounding is better in reaches with greater velocity of flow than in reaches where velocity is lower. W. J. Plumley (1948) found that the increase in roundness attains its maximum after the first few kilometres of travel, and that after reaching a certain value it does not rise any more. Thus, for instance, most limestone pebbles attain maximum roundness (about 0·73—0·74 for pebbles 16—22 mm across) after a migration of 10—20 km, but already in the first kilometres the increase in rounding is observable. From these data it can be deduced that unrounded limestone pebbles have not covered a longer distance than 10 km. Pebbles of vein quartz show a rounding after a transport of 20—250 km, and chert and lydite pebbles after travelling several hundred kilometres.

The changes in the sphericity of particles depend on the size, type of rock and distance of travel. The following conclusions are gathered from a comprehensive statistical work by E. D. SNEED and R. L. FOLK (1958). It has been found that the development of sphericity of pebbles in fluviatile sediments is most complicated, because every fraction and almost every rock shows a different behaviour. For instance, quartz pebbles 54—70 mm in size, move most frequently along the bottom of the channel and assume a rodlike form, but the value of their sphericity does not change. The rolling movement of quartz pebbles, 30—54 mm in size, is irregular and their sphericity increases owing to a most intense abrasion on both ends. Flat chert pebbles of a size of 38—70 mm are shifted downstream and their sphericity decreases during travel. On the other hand, chert pebbles 30—38 mm across, are moved by rolling as most small pebbles and their sphericity increases. Pebbles of limestone do not show any systematic changes and tendencies in the development of sphericity. The above observations show the relationship between the sphericity and the shape + size of pebbles to be most suitable criterion for the determination of the length of transport. Pebbles of all rock types and sizes display a similar sphericity near the source, but as a result of different modes of transport of individual fractions and individual rocks, the differences in sphericity increase with the distance from the source.

The changes in roundness and sphericity are far less distinct in the sand fraction. Sand grains show a slight increase in roundness downstream, probably rather on account of selective transport than mechanical abrasion. Some experiments indicate that abrasion produces chiefly breaking of grains, so that it rather decreases than increases their roundness. According to present views, the changes in roundness are most affected by the import of material by the tributaries mainly from ancient sediments.

The changes of sphericity of sand grains has received less attention and no general trends have been ascertained.

Alluvial plains

Although the sediments of alluvial plains are sometimes included into river deposits, they occupy a special position. They are deposits of a great thickness and extent, generally in the lower reaches of major streams, not far from their mouths, maybe directly on coastal plains. The great thickness of these deposits attaining in places a few thousand metres, is often caused by tectonic subsidence of the plains. The impossibility of distinguishing the conventional groups of stream environments is another particular feature of alluvial plains. The channel sediments are the only ones which can sometimes be differentiated and even those pass very gently into contiguous sediments. The bulk of sediments of alluvial plains are overbank deposits laid down during the high floods, when the river often innundates extensive areas.

As mentioned above, major lowland rivers bear during high floods 100 to 1,000 times more material than at normal water stage, and most of this is deposited on their alluvial plains. The geology and partly also the lithology of some alluvial plains have been thoroughly studied (the Mississippi, the Euphrates and the Tigris, the Amu Darya or Hwang-Ho Rivers in particular).

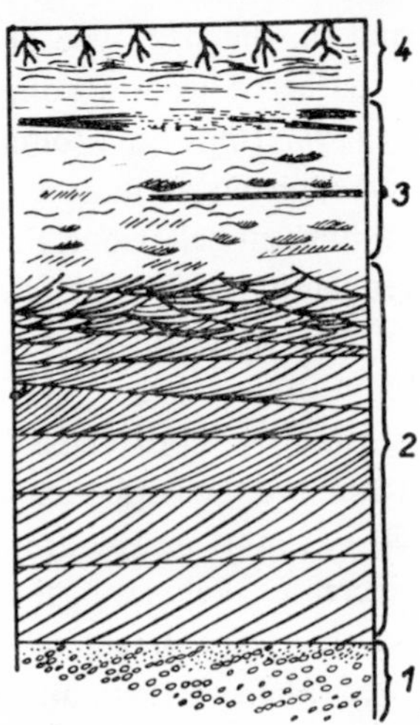

Fig. 29. Ideal vertical section through the set of alluvial sediments. 1. Basal gravel, 2. Channel deposits, 3. Flood-plain deposits, 4. Soil cover. After L. N. Botvinkina (1962).

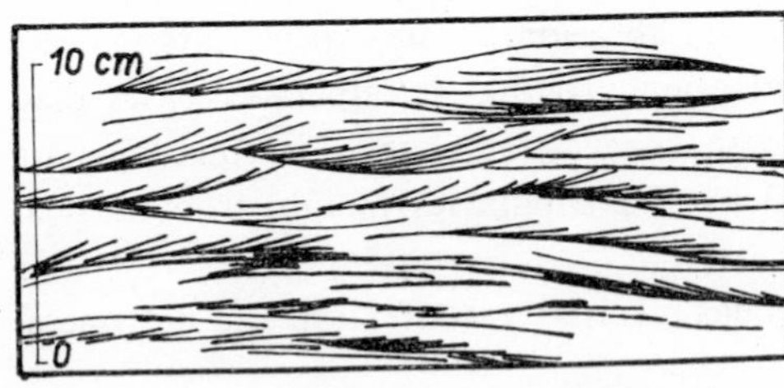

Fig. 30. Type of bedding in flood-plain silts and fine-grained sands. After L. N. Botvinkina (1962).

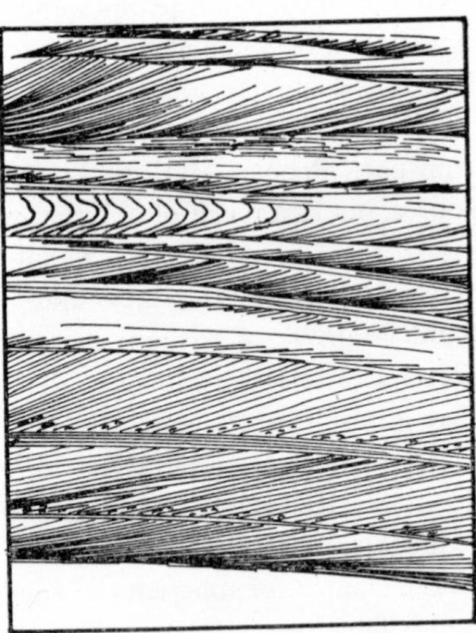

Fig. 31. Current bedding in fine-grained channel sands. After L. N. Botvinkina (1962).

The area round the joint mouth of the Euphrates and the Tigris Rivers and that along the Hwang-Ho River in eastern China are the largest alluvial plains of the world. The former occupies an area of about 100,000 km^2, the latter is approximately half this size. The rate of sedimentation on the Hwang-Ho alluvial plain averages

1·48 mm per year (W. Wong 1931). As has been established by boring the thickness of sediments on this plain attains 2,000 m. The section through the deposits has shown about 70% of clay-silt, 20% of sand, and 10% of gravels, which constitute not very thick but usually regular beds.

Sediments of alluvial plains pass into upper parts of the delta and this transition is very slow. In some cases it is difficult to decide whether the sediments already belong to the delta, i. e. to the transitional area between continental and marine sedimentation, or still to the continental alluvial plain.

Stratigraphy and vertical sequence of fluviatile sediments

The thickness of river deposits in the upper and middle reaches of a stream is usually not very great. Apart from the transitions into alluvial fans and other slope sediments, the thickness of river deposits does not exceed a few metres. However, at the boundary

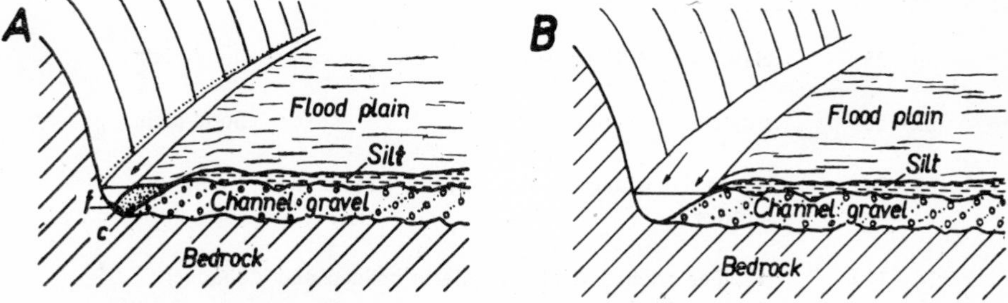

Fig. 32. Formation of flood-plain by lateral swinging of the river, as exemplified by the Shoshoe River, Wyoming. A. Low water stage, f. fine detritus laid down in the stream channel during periods of normal flow, c. coarse detritus laid down in the channel at the end of a period of high water, B. High-water stage. After L. B. Leopold et al. (1964).

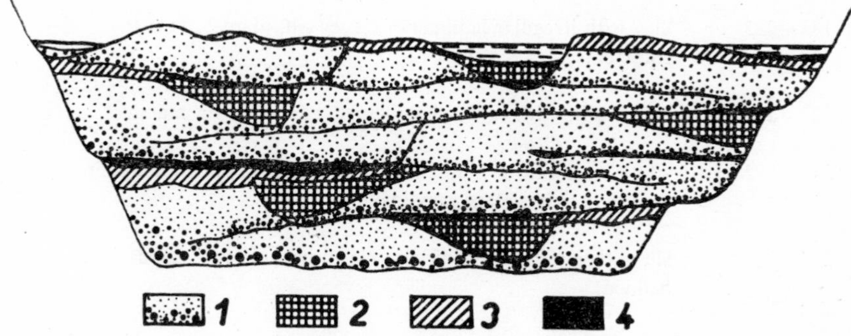

Fig. 33. Schematic illustration of alluvium development during constrative phase (i. e. during equilibrium between erosion and deposition processes). 1. River channel sediments, 2. Lateral accretion, 3. Flood-plain sediments, 4. Sediments of alluvial lakes and swamps. After E. V. Shancer (1951).

between the mountain and lowland area, the slope sediments combined with those of alluvial fans, can be some thousand metres thick.

In the lower reaches, the river valleys may develop in areas which are tectonically stable or subsiding.

In stable areas the river sediments cannot attain an extraordinary thickness. Channel, together with flood-plain deposits of rivers of a medium size are a few dozen metres thick while those of major streams may exceed 100 m in thickness.

Tectonic subsidence of an area causes the deposition of up to several hundred metres thick complexes of sediments.

Table 53

Vertical sequence of river deposits, according to a boring through the flood-plain of the Amu Darya River

Depth cm	Deposits	Environment
0—25	clayey silt, with uniform structure	flood-plain
25—60	silty clay, sometimes mottled	flood-plain
60—74	brown clay with greenish spots, silty streaks and mottles	swamp
74—124	silt, fine-grained sand, irregularly bedded and ripple bedded	channel
124—149	clayey silt, uniform structure	flood-plain
149—183	clayey silt, mottled	flood-plain
183—223	clay with irregular laminae of clayey silt, plant fragments	marsh
223—245	clayey silt, with mottled structure, rootlets	marsh
245—285	silty clay, uniform, sometimes mottled traces of animal activity, plant remains	flood-plain
285—292	silt with intercalation of silty clays, irregularly bedded	flood-plain
292—357	clay with silt and sand mottles (product of activity of benthos); accumulations of plant remains, shells of pelecypods, in lower parts soil profile	marsh-alluvial lake

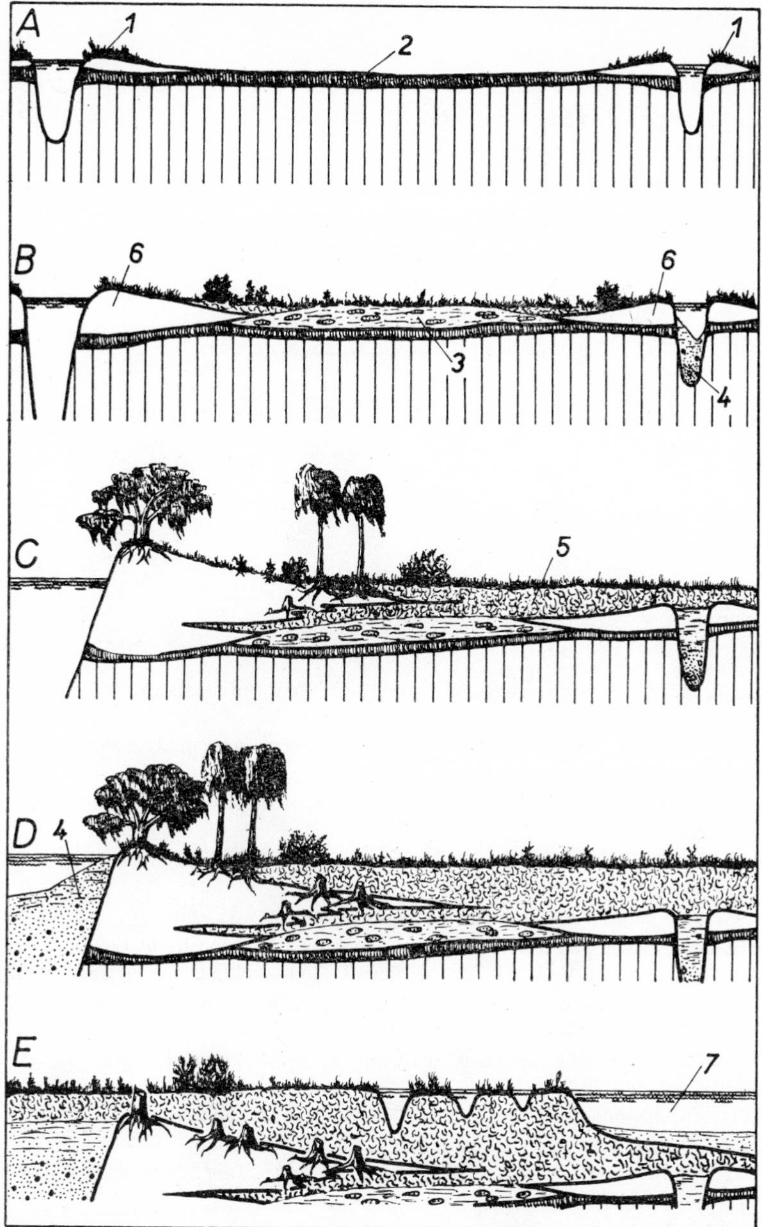

Fig. 34. Progressive stages in peat accumulation acompanying deltaic-plain development.

A. Initial development of distributaries and interdistributary trough.

B. Enlargement of principal distributary and its natural levees — creation of marshes in trough.

C. Maximum development of distributary and its natural levees — creation of swamp as levee subsides.

D. Deterioration of distributary — advance of swamp over subsiding levees.

E. Continued subsidence with partial destruction of marshes. 1. Organic muck, 2. Gulf floor silty clays, 3. Interdistributary trough fill, 4. Channel fill, 5. Peat, 6. Natural levee, 7. Marine deposits of bay and sound. After H. N. Fisk (1961).

The variability of river deposits in the vertical direction is generally considerable. The log of a boring driven through the flood plain of the Amu Darya River provides a typical profile of river sediments (Table 53).

Borings in the alluvial of the Mississippi, Amu Darya, Rhône and some other rivers made it possible to compute the mean percentage of respective deposits of the individual environments in the total volume of sediments:

Flood-plain	68%	Bank deposits	6%
Marsh deposits	14%	Channel deposits	12%

The large amount of flood-plain deposits and a small percentage of channel sediments are noteworthy.

References

AL-HABEEB K. H. (1969): Sedimentology of the flood plain sediments of the Middle Euphrates River. Thesis, 1—72, Baghdad.

ALLEN J. R. L. (1965): A review of the origin and characteristics of Recent alluvial sediments. Sedimentology, vol. 5: 89—191, Amsterdam.

ANDEL TJ. VAN (1950): Provenance, transport and deposition of Rhine sediments. P. 1—129, Wageningen.

AVDUSIN P. P. (1956): Granulometrical composition of alluvial sediments (in Russian). Tr. inst. něfti AN SSSR, vol. 7: 88—107.

BEHRE C. H. (1926): Sand flotation in nature. Science, vol. 63: 405—408.

BERNARD H. A. - MAJOR C. F. (1963): Recent meander belt deposits of the Brazos River. An alluvial "sand" model. Bull. Am. Assoc. Petrol. Geol., vol. 47: p. 350, abstr., Tulsa.

BOTVINKINA L. N. (1962): Stratification of sedimentary rocks (in Russian). Trudy geol. inst. AN SSSR, vd. 59: 1—538, Moscow.

BRUUN P. (1962): Engineering aspects of sediment transport. Engin. Progress. at the University of Florida, vol. 16, No 7.

BURRI C. (1930): Sedimentpetrographische Untersuchungen an alpinen Flusssanden. 1. Die Sande des Tessin. Schweiz. Min. Petr. Mitt., Bd. 9: 205—240.

BYRNE J. V. (1963): Variation in fluvial gravel imbrication. Jour. Sedimentary Petrology, vol. 33: 467—469, Menasha.

COLBY B. R. (1963): Fluvial sediments — a summary of source, transportation, deposition, and measurement of sediment discharge. Geol. Survey Bull. 1181-A: 1—47, Washington.

COLEMAN J. M. (1969): Brahmaputra River: channel processes and sedimentation. Sedim. Geol., vol. 3: 129—239, Amsterdam.

DAPPLES E. C. - ROMINGER J. F. (1945): Orientation analysis of fine grained clastic sediments. Bull. Am. Geol. Soc., vol. 56: 1153—1154, Baltimore.

DAVIS S. N. (1958): Size distribution of rock types in stream gravel. Jour. Sedimentary Petrology, vol. 28: 87—94, Menasha.

EMERY K. O. (1950): Contorted Pleistocene strata at Newport Beach, California. Jour. Sedimentary Petrology, vol. 20: 111—115, Menasha.

ENGELHARDT W. (1939): Untersuchung wasser- und windsortierter Sande auf Grund der Korngrössenverteilung ihrer leichten und schweren Gemengteile. Chemie d. Erde, Bd. 12, p. 451—465, Jena.

FOLK R. L. - WARD W. C. (1957): Brazos river bar: a study in the significance of grain-size parameters. Jour. Sedimentary Petrology, vol. 27: 3—26, Menasha.

FRAZIER D. E. - OSANIK A. (1961): Point bar deposits, Old River Locksite, Louisiana. Trans. of the Gulf Coast Assoc. of Geol., vol. 11: 121—137, Houston.

Genesis and lithology of Anthropogen continental deposits (in Russian). To the VIIth INQUA Congress in USA 1965, p. 1—112, Moscow.

GILBERT G. K. (1914): The transportation of debris by running water. US Geol. Surv. Prof. Pap. 86: 1—263, Washington.

GORECKIJ G. I. (1964): Alluvium of the great ancient rivers of Russian Plateau (in Russian). P. 1—413, Moscow.

GRIFFITHS J. C. - ROSENFELD M. A. (1953): A further test of dimensional orientation of quartz grains in Bedford sand. Am. Jour. Sci., vol. 25: 192—194, New Haven.

HJULSTRÖM F. (1935): Studies of the morphological activity of rivers as illustrated by the river Fyris. Bull. Geol. Inst. Uppsala, vol. 25: 221—528, Uppsala.

HUBBELL D. W. - MATEJKA D. Q. (1959): Investigation of sediment transportation Middle Loup River and Dunning, Nebraska. Water-Supply Paper, No 1476: 1—120, Washington.

HUME J. D. (1964): Floating sand and pebbles near Barrow, Alaska. Jour. Sedimentary Petrology, vol. 34: 532—536, Menasha.

JOHANSSON C. E. (1965): Structural studies of sedimentary deposits. Geol. Fören. Förhandlingar, vol. 87: 3—61, Stockholm.

KALTERHERBERG J. (1956): Über Ablagerungsgefüge in grobklastischen Sedimenten. N. Jhrb. Geol., Abh. 104: 30—57, Stuttgart.

KARCZ I. (1969): Mud pebbles in a flash floods environment. Jour. Sedimentary Petrology, vol. 39: 333—337, Menasha.

KARTASHOV I. P. (1961): Facies, dynamic phases and strata of alluvium (in Russian). Izvestija AN SSSR, ser. geol., p. 77—90, Moscow.

KODYMOVÁ A. (1960): Investigation of heavy minerals in alluvial deposits of Czech rivers. Thesis, p. 1—104, Prague.

KOLDEWIJN B. W. (1941): Provenance, transport and deposition of Rhine sediments. Geol. en Mijnbouw, 17e Jg., p. 37—45, Amsterdam.

KRUMBEIN W. C. (1941): Measurement and geological significance of shape and roundness of sedimentary particles. Jour. Sedimentary Petrology, vol. 11: 48—56, Menasha.

— (1942): Flood deposits of Arroyo Seco, Los Angeles County, California. Bull. Geol. Soc. Am., vol. 53: 1355—1402, Baltimore.

KUKAL Z. - SAADALLAH A. (1969): Paleocurrents in the Mesopotamian geosyncline. Geol. Rundschau, vol. 59: 523—536, Stuttgart.

KÜRSTEN M. (1960): Zur Frage der Geröllorientierung un Flussläufen. Geol. Rundschau, Bd. 49: 498—501, Stuttgart.

LADD G. E. (1898): Geological phenomena resulting from the surface tension of water. Am. Geologist, vol. 22: 267—285.

LANE D. W. (1963): Cross stratification in San Bernard River, Texas. Jour. Sedimentary Petrology, vol. 33: 350—354, Menasha.

LATTMAN L.H. (1960): Cross section of a flood plain in a moist region of moderate relief. Jour. Sedimentary Petrology, vol. 30: 275—282, Menasha.

LEOPOLD L. B. - WOLMAN M. G. - MILLER J. P. (1964): Fluvial processes in geomorphology. P. 1—522, San Francisco.

LUDWIG G. - VOLLBRECHT K. (1957): Die allgemeinen Bildungsbedingungen litoraler Schwermineralkonzentrate und ihre Bedeutung für die Auffindung sedimentärer Lagerstätten. Geologie, Bd. 6: 229—277, Berlin.

McKEE E. D. (1957): Flume experiments on the production of stratification and cross-stratification. Jour. Sedimentary Petrology, vol. 27: 129—134, Menasha.

McKEE E. D. - CROSBY E. J. - HILL N. (1967): Flood deposits, Bijou Creek, Colorado, June 1965. Jour. Sedimentary Petrology, vol. 37: 829—581, Menasha.

McKELVEY V. E. (1941): The flotation of sand in nature. Am. Jour. Sci., vol. 239: 294—307, New Haven.

MENARD H. W. (1950 *a*): Sediment movement in relation to current velocity. Jour. Sedimentary Petrology, vol. 20: 148—160, Menasha.

— (1950 *b*): Transportation of sediment by bubbles. Jour. Sedimentary Petrology, vol. 20: 98—106, Menasha.

MORISAWA M. (1968): Streams, their dynamics and morphology. P. 1—175, New York.

NAIDU A. S. (1969): Texture of modern deltaic sediments of Godavari River (India). Bull. Am. Assoc. Petrol. Geol., vol. 53: 733, Tulsa.

NAIDU A. S. - BORRESWARA R. C. (1965): Some aspects of Lower Godavari River and delta sediments, India. Bull. Am. Ass. Petrol. Geol., vol. 49: 354, Abstr.

NEVIN CH. (1946): Competency of moving water to transport debris. Bull. Geol. Soc. Am., vol. 57: 651—674, Baltimore.

NORDIN C. F. BEVERAGE J. P. (1965): Sediment transport in Rio Grande, New Mexico. Geol. Surv. Prof. Pap. 462-F: 1—35, Washington.

NORDIN C. F. - CULBERTSON J. K. (1961): Particle-size distribution of stream bed material in the Middle Rio Grande basin. Geol. Surv. Prof. Pap. 424-C: 323—326, Washington.

PETTIJOHN F. J. (1957): Sedimentary rocks. 2d ed. P. 1—718, New York.

PLUMLEY W. J. (1948): Black Hills terrace gravels: a study in sediment transport. Jour. Geology, vol. 56: 526—577, Chicago.

POLLACK J. M. (1961): Significance of compositional and textural properties of South Canadian River channel sands, New Mexico, Texas and Oklahoma. Jour. Sedimentary Petrology, vol. 31: 15—37, Menasha.

POTTER P. E. - PETTIJOHN F. J. (1963): Paleocurrents and basin analysis. P. 1—296, Berlin, etc.

RITTENHOUSE G. (1943): Transport and deposition of heavy minerals. Bull. Geol. Soc. Am., vol. 40: 1725—1780, Baltimore.

RUKHIN L. B. (1947): About composition of river sands (in Russian). Vestnik Leningrad. Gos. Inst., vol. 12: 1—41, Leningrad.

RUSNAK G. A. (1956): Sand-grain orientation and geological application. Jour. Paleontology, vol. 30: 996.

RUSSELL R. D. - TAYLOR R. E. (1937): Roundness and shape of Mississippi River sand. Jour. Geology, vol. 45: 1726—1757, Baltimore.

SARKAR S. K. - BASUMALLICK S. (1968): Morphology, structure and evolution of a channel island in the Barakar River, Barakar, West Bengal. Jour. Sedimentary Petrology, vol. 38: 747—754, Menasha.

SARKISJAN S. G. - KLIMOVA L. T. (1955): Orientation of pebbles and its investigation (in Russian). P. 1—162, Moscow.

SCHLEE J. (1957): Upland gravels of southern Maryland. Bull. Geol. Soc. Am., vol. 68: 1371 — 1410, Baltimore.

SCHUMM S. A. (1960): On the effect of sediment type on the shape and stratification of some modern fluvial deposits. Am. Jour. Sci., vol. 258: 177—184, New Haven.

— (1961): Effect of sediment characteristics on erosion and deposition in ephemeral stream channels. Geol. Surv. Prof. Pap. 352-C: 1—70, Washington.

SCHUMM S. A. (1968): Speculations concerning paleohydrologic controls of terrestrial sedimentation. Bull. Geol. Soc. Am. vol. 79: 1573—1588, Baltimore.

SHANCER E. V. (1951): Alluvium of the lowland river of the moderate climatic belt (in Russian). Trudy inst. geol. nauk, vol. 135, geol. ser.: 1—166, Moscow.

SNEED E. D. - FOLK R. L. (1958): Pebbles in the lower Colorado River, Texas. A study in particle morphogenesis. Jour. Geology, vol. 66: 114—150, Chicago.

SUNDBORG Å. (1956): The river Klarälven: a study of fluvial processes. Geograph. Ann., vol. 38: 127—316, Stockholm.

VEJCHER A. A. (1948): Preliminary investigation of river channel sedimentation (in Russian). Litologičeskij sbornik, vol. 2: 7—14, Leningrad.

WADELL H. (1932): Volume, shape and roundness of rock particles. Jour. Geology, vol. 40: 443—451, Chicago.

WILLIAMS P. F. - RUST B. R. (1969): The sedimentology of a braided river. Jour. Sedimentary Petrology, vol. 39: 649—679, Menasha.

WOLF K. H. (1961): The flotation phenomena — a by-passing and sorting process. Jour. Sedimentary Petrology, vol. 31: 476—478, Menasha.

WONG W. (1931): Op. cit. on page 40.

9. Sediments of alluvial fans

Alluvial fans are stream deposits whose surface forms a segment of a cone that radiates downslope from the point where the stream emerges from the mountainous area.

We are justified in grouping of alluvial fans with river deposits, as they are formed, to a large degree, by fluviatile processes. Several particular features, especially their morphology of truncated cones, led some workers to regard them as a separate form of sedimentary deposits. The highest point on an alluvial fan, generally, where the stream emerges from the mountain front, is called the apex.

Table 54

Section of alluvial fan deposits exposed in Arroyo Hondo (Fresno County, California) (after W. B. Bull 1964)

Lithology	Thickness cm	Lithology	Thickness cm
Sand, clayey, pebbly	25	Sand, gravelly, clayey	10
Gravel, diatomaceous shale chips	6	Gravel-clay, oriented shale chips	3
Sand, fine-grained, clayey, silty	3	Sand, well sorted, some clay binder	6
Sand, well-sorted	10	Gravel, shale chips, clay binder	6
Gravel-clay, diatomaceous shale chips	15	Sand, clayey, pebbly, shale chips not oriented	0·14
Clay, sandy, root cavities	10	Sand, well sorted, scattered charcoal fragments	26
Sand, clayey	8	Sand, well-sorted	10
Sand, clayey, silty	3	Silt-clay, sandy	6
Gravel, clayey, diatomaceous shale chips	17	Sand, clay films around grains	6
Clay, pebbly, sandy, root cavities	17	Sand, poorly sorted	3
Sand, clayey	8	Sand, clayey	3
Sand, pebbly, clayey	10	Sand-clay	·6
Sand-clay	3	Sand, clayey	16
Sand, clayey	3	Clay	16
Sand, well-sorted, silty	15	Sand, clayey, some pebbles, root cavities	28
Clay	1·5	Gravel, clayey	17
Sand-clay, pebbly	12	Sand, silty, coarser-grained toward toe top	36
Sand, shale chips	3	Gravel, silty	5
Sand, clayey	13		
Sand, fine-grained, clayey	6		
Sand, pebbly, clayey	6		

Alluvial fans are deposited, essentially, by two processes: fluviatile and mudflow. Fluviatile processes are predominant; the mudflows which also contribute to the formation of alluvial fans are produced by the downslope movement of solid debris. There are a number of transitional processes between the two above-mentioned processes. In the sections through the alluvial fans, layers deposited by fluviatile processes are seen to alternate with mudflow and intermediate sediments (Table 54, Fig. 38).

The deposition on alluvial fans is caused mainly by the decrease in depth and velocity of flow that results from the increase in width as the flow spreads out over the fan. If water from the flow infiltrates into the ground, the decrease in volume of flow also causes deposition.

The classification into mudflow, intermediate, or water-laid deposits was adopted because it is independent of the overall shape of the deposits and place of deposition, which is a definite advantage when materials from core holes are classified. The general properties of each type of deposit, as sampled at the land surface, are as follows (Table 55, according to W. B. BULL 1964 b).

Table 55

The general properties of the deposits of alluvial fans (after W. B. Bull 1964)

Type of deposits	Depositional characteristics	Average parameters
Water-laid sediments	No discernible margins, usually clean sand or silt; crossbedded, laminated or massive	Clay content 5% So 1·8 σ_{Φ} 1·4
Intermediate deposits	No sharply defined margins; clay films around sand grains and lining voids; graded bedding and oriented fragments	clay content 17% So 4·0 σ_{Φ} 3·9
Mudflow deposits	Abrupt well-defined margins, lobate tongues; clay may partly fill intergranular voids. May not have graded bedding or particle orientation	clay content 31% So 9·7 σ_{Φ} 4·7

The part of alluvial fans formed of water-laid sediments has usually an imbricate orientation of pebbles and a conspicuous channel and trough cross-bedding in places. In parts which have originated by a process of mudflows, generally no pronounced orientation of clastic components is developed and the bedding is chaotic and streaky on account of a large content of clay.

The content of individual fractions (gravel, sand, silt, clay) in sediments of alluvial fans is very variable, the deposits are often a mixture of all four fractions (Fig. 36).

Grain-size analyses of sediments of some alluvial fans are given in Fig. 35 a, b.

Some finer-grained parts of alluvial fans contain sediments rich in calcium-carbonate (e. g. in the alluvial fans of California an average content of 10% $CaCO_3$ has been found). Generally, however, it is thought that $CaCO_3$ is precipitated secondarily by the action of ground waters.

The thickness and shape of alluvial fans differ from place to place, depending particularly on the form and tectonic behaviour of the substratum of the fan and its

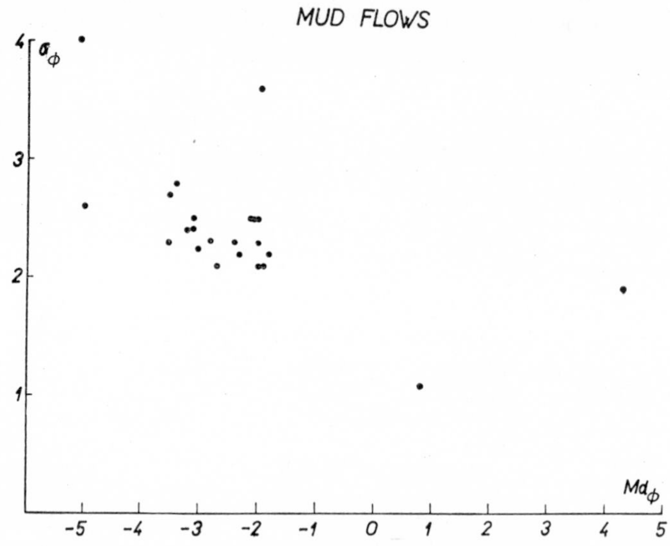

Fig. 35a. Grain-size parameters of mud-flow sediments. Relation between Phi Median Diameter and Phi Deviation Measure. Data compiled from various authors.

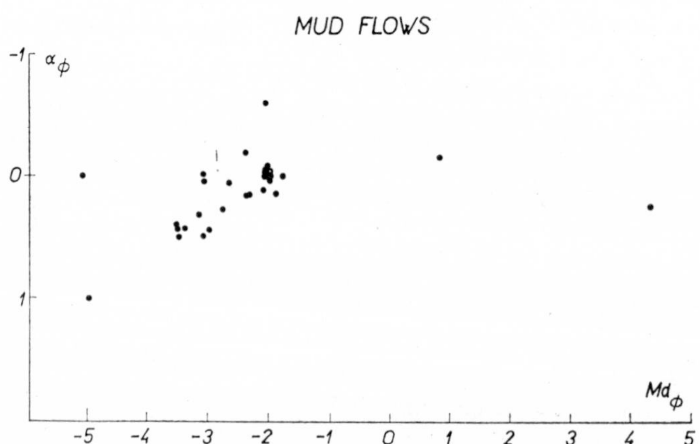

Fig. 35b. Grain-size parameters of mud-flow sediments. Relation between Phi Median Diameter and Phi Skewness Measure.

various parts. In most cases, alluvial fans attain the greatest thickness in their central parts, but fans are also known which are thickest in the upper segment. Where the subsidence is greatest below the lowest part of the fan, the greatest thickness of deposits is accumulated there.

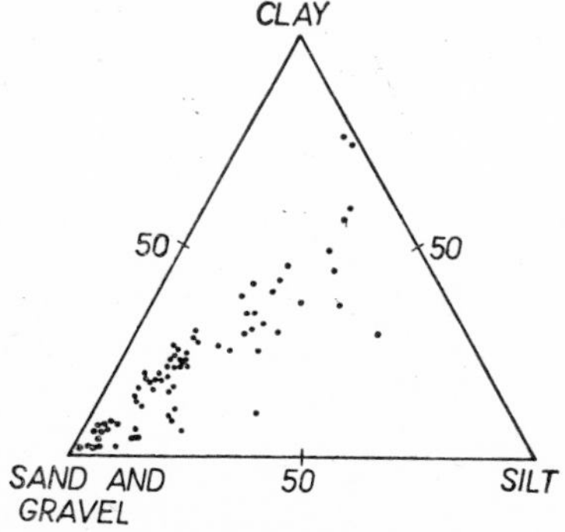

Fig. 36. Grain-size distribution in alluvial fan sediments having less than 20% gravel. Western Fresno County, California. After W. B. Bull (1964).

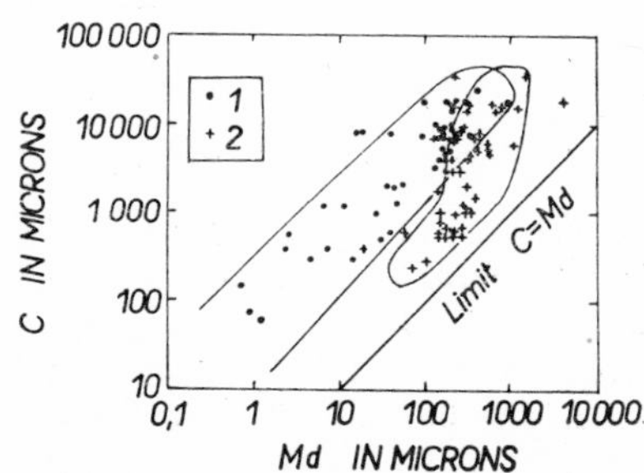

Fig. 37. Textural CM patterns of the coarsest one percentile (C) and median grain-size (Md) of alluvial fan deposits in Western Fresno County. 1. Mudflow deposits, 2. Stream channel deposits intermediate between mudflow and water-laid sediments. After W. B. Bull (1964).

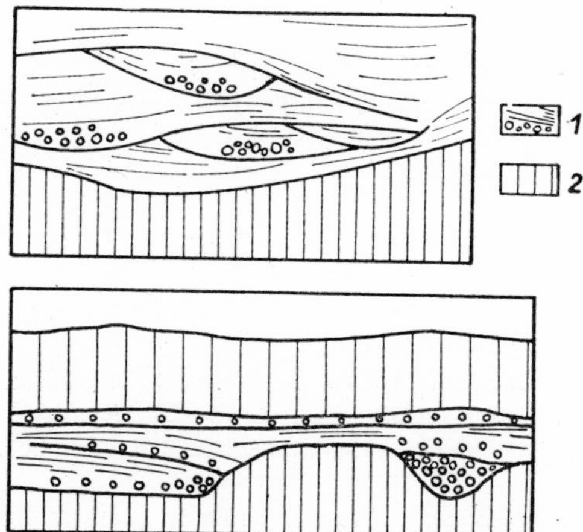

Fig. 38. Schematically illustrated structures and vertical development of alluvial fan deposits. 1. Stream channel deposits, 2. Mud-flow deposits. After E. Blissenbach (1954).

Considerable petrographic changes are observable both in the vertical and longitudinal sections of the fan. An example of a vertical section is given in Table 54.

In a longitudinal section of the alluvial fan, the grain-size decreases from the apex to the toe. Thus, for instance, the alluvial fans of Nevada reveal that particles show an exponential decrease in grain-size away from the source area. The mudflow is characterized by a steeper curve of decrease of particle size and also has a lower correlation coefficient than that of stream deposits.

Table 56

Lithological differences between deposits of alluvial fan and talus

Alluvial fan	Talus
Coarser parts are at the apex, grain-size decreases towards the lower parts	Grain-size increases towards the lower parts
Upper, middle, or lower parts of longitudinal section may be the thickest (see above)	The greatest thickness in lower parts
Characteristic features of fluviatile sedimentation are present (see above)	Features of fluviatile sedimentations are absent
Sediments may contain great amounts of clay	Clay usually absent or scarce
Some coarse particles may be well rounded	Clastic particles never well rounded

The sediments of alluvial fans can sometimes look like talus sediments which originate by gravity forces. Lithological differences are listed in Table 56.

References

Blissenbach E. (1952): Relation of surface angle distribution to particle size distribution on alluvial fans. Jour. Sedimentary Petrology, vol. 22: 25–28, Menasha.

– (1954): Geology of alluvial fans in semiarid regions. Bull. Geol. Soc. Am., vol. 65: 175–190, Baltimore.

Bluck B. J. (1964): Sedimentation of an alluvial fan in Southern Nevada. Jour. Sedimentary Petrology, vol. 34: 395–400, Menasha.

Bull W. B. (1960): Types of deposition on alluvial fans in Western Fresno County, California. Bull. Geol. Soc. Am., vol. 71: 2052, abstr., Baltimore.

– (1964*a*): Geomorphology of segmented alluvial fans in Western Fresno County, California. Geol. Surv. Prof. Pap. 352-E: 89–128, Washington.

– (1964*b*): Alluvial fans and near-surface subsidence in Western Fresno County, California. Geol. Surv. Prof. Pap. 437-A: 1–70, Washington.

– (1969): Recognition of alluvial-fan environments in stratigraphic record. Bull. Am. Assoc. Petrol. Geol., vol. 53: 710, Tulsa.

CAINE N. (1967): The texture of tallus in Tasmania. Jour. Sedimentary Petrology, vol. 37: 796—803, Menasha.

DENNY Ch. S. (1965): Alluvial fans in the Death Valley Region, California and Nevada. Geol. Surv. Prof. Pap. 466: 1—61, Washington.

ECKIS R. (1928): Alluvial fans of the Cucamonga district, Southern California. Jour. Sedimentary Petrology, vol. 21: 147—150, Menasha.

HAMILTON W. B. (1951): Playa sediments of Rosamonda Dry Lake, California. Jour. Sedimentary Petrology, vol. 21: 147—150, Menasha.

HOOKE R. L. B. (1967): Processes in arid-region alluvial fans. Jour. Geology, vol. 75: 438—460, Chicago.

PANNEKOEK A. J. (1957): Sedimentation around mountain ranges, with examples from northern Spain. In "The earth, its crust and atmosphere". P. 142—158. Leiden.

POWER W. R. (1961): Backset beds in the Coso formation, Inyo County, California. Jour. Sedimentary Petrology, vol. 31: 603—607, Menasha.

SHARP R. P. (1948): Early tertiary fanglomerate, Big Horn Mountains, Wyoming. Jour. Geology, vol. 56: 1—16, Chicago.

10. Eolian sediments

Eolian transportation and sedimentation have many common features with fluviatile sedimentation. They differ in the density and in the speed of movement of the environment. The speed of wind ranges from 1 to 30 m/sec, but the speed which should be mainly taken into account are those of 15—20 m/sec. The far smaller density of the transporting medium when compared with that of water in the fluviatile environment is the limiting factor controlling the grain-size of eolian sediments. The material is transportated by suspension, saltation and traction as it is in aqueous environments.

In eolian suspension, grains are held in the air by ascending components which are five times slower than the average speed of wind. It is recorded that wind can support in suspension grains of a maximum diameter of 3 mm, larger grains are moved by traction. In transport by saltation, the particles attain a remarkable distance of movement, which amounts up to 8 km when the grain is lifted to a height of 1 m. Eolian saltation is the main process that gives rise to ripples, as proved by R. BAGNOLD (1941). This author was the first to report that eolian ripples originate in an other ways than water-formed ripples. Under identical conditions grains of the same size are transported the same distances and fall approximately at the same place to form low ridges which join into ripple crests.

Strong wind can transport substantially large grains by traction. Dried lumps of clay of up to 25 cm in size have been found to have been moved by traction. Therefore, there are in dunes local accumulations of clay balls which have been transported from dried clays of ephemeral lakes or flood plains. Grains migrating by traction can settle down either at a change of the speed of wind or in the shadow behind or in front of an obstacle.

A characteristic feature of eolian environment is the perfect differentiation of two grain-size fractions due to the separation of suspended material fom that moved by traction. Table 57 gives a clear picture of the velocity of fall of individual grain-size fractions.

In essentials, the size of 0·05 mm forms a natural boundary; below it, grains are transported predominantly in suspension, and above it, they move by traction and saltation. From this it follows that silty and clayey fractions of the originally heterogeneous material are borne in suspension, while the sandy fraction migrates by traction at a far lesser speed. These theoretical assumptions were fully confirmed by measurement of the grain-size distribution of eolian deposits, i. e. of dust particles suspended in air and of wind-blown sands moved by traction. At a speed exceeding 6 m/sec, sand grains can also form the suspended load; they are then deposited along with silt at a sudden change of the conditions or are separated from silt by differential deposition, when the conditions remain unchanged. A mixture of silt and sand in eolian sediments is indicative of abrupt changes in the conditions, when silt and sand got into suspension by blast of wind and settled abruptly when the wind suddenly ceased.

The quantitative relationship between the various modes of eolian transport is open to discussion. It is stated that saltation is the main mean of eolian transport which carries 55—72 % of the whole of material; about 30 % is borne in suspension and only 25 % migrates by traction (creeping, slipping).

Table 57

Velocity of fall of particles of various grain-size

Diameter of particles mm	Fall velocity cm/sec
0·01	2·8
0·02	5·5
0·05	16
0·06	50
0·1	167
0·2	250
2·0	500

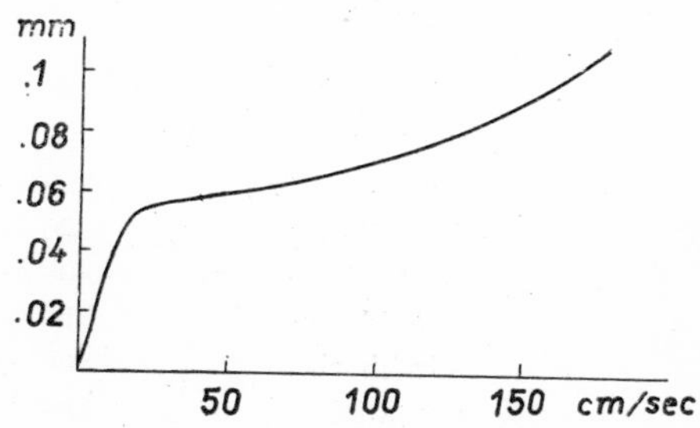

Fig. 39. Diagram illustrating the relation between particle size to its free fall velocity in the air. After L. Moldvay (1957).

The possibility that a grain will be winnowed from the rock depends on the degree of turbulence and speed of wind, on the humidity of rock, and on the size and specific weight of the grain. As is well known, initiation of movement of material is most frequently by saltation, less often by traction (creeping) or a direct lifting into suspension. Experiments have shown that grains larger than 0·11 mm are first moved along the surface, then caught by the ascending component of the air current and lifted vertically, falling ultimately at an angle of 6—8° to the ground surface.

Material of eolian deposits derives from sediments of various environments: soils, weathered rocks, fluviatile sediments, flood-plain deposits, etc., provided that the parent material is rich in silty fractions, and is incoherent and dry. K. E. LaPrade (1957) inferred from his study of materials winnowed from various environments in Texas that most material is supplied by soils, particularly during the months from November to June. Then follow (in descending order) pastures, sand dunes, bottoms of ephemeral lakes, and alluvial deposits. In the geological past, however, it is believed that most material was probably derived from alluvium, steppe, desert and glacial sediments.

The grain-size distribution and the amount of material in eolian suspension, i.e. of dust, has been studied on many places of the Earth's surface. The results have borne out the above-mentioned presumptions. F. Becke (1901) published the results of observations on the grain-size composition of eolian dust which had fallen into various parts of Central Europe (Table 58). P. M. Game (1964) supplemented these results with additional recent data (Table 59).

Table 58

Grain-size composition of eolian dust falls on various places of Central Europe (after F. Becke 1901)

Location	Prevailing diameter mm	Maximal diameter mm
Kuffstein	0·001—0·03	0·08
Zell — am See	0·001—0·02	0·08
Judenburg	0·001—0·03	
Greifenburg	0·001—0·03	0·13
Arnoldstein	0·001—0·025	0·05
Kirchbach	0·001—0·032	0·1
Pontaffel	0·001—0·03	0·11
Tarvizio	0·001—0·03	0·07
Gorizia	0·001—0·025	0·075
Hvar 1879	0·001—0·03	0·07
Hvar 1901	0·001—0·02	0·07
Rjeka	0·001—0·051	0·113

Table 59

Grain-size distribution in continental and marine dusts (after P. M. Game 1964)

Place and date of fall	Percentage of grains having mean diameters		
	< 5 μ	5—20 μ	< 20 μ
Davos, March 1936	90	10	9
Arosa, October 1942	55	42	3
Thusis, April 1941	44	51·5	4·5
Arosa, April 1944	30	60	10
S. S. Dunstan, 400 miles south-west of the Canary Islands, Feb. 1962	4	58	38

The predominant fraction consists of material with grain-sizes close to the silt/clay boundary. Grains of maximum size belong almost universally to the fine sand or somewhat coarser fractions. According to several authors, dust from the Sahara storms transported as far as to the European Continent consists of particles of 0·0013 to

Fig. 40. Distribution of particle size of the material of the dust falls in the East Atlantic, 1962. After P. M. Game (1964).

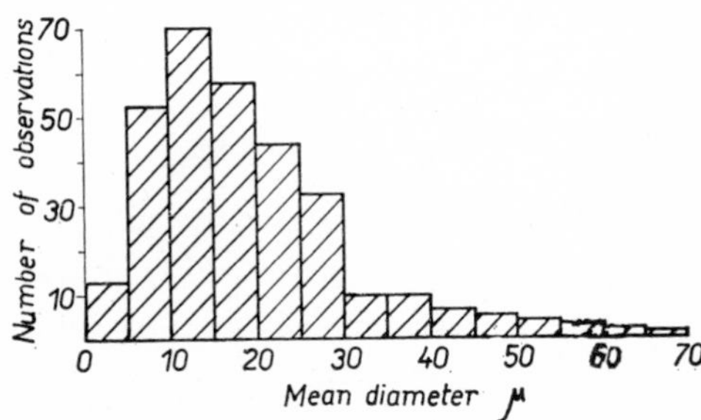

0·04 mm in size with a maximum size of 0·11 mm. Material of this provenance carried across the Mediterranean Sea and deposited near Szeged (Hungary) contained (after F. Becke 1901):

1,300 kg of sand (0·2—0·5 mm fraction)
6,200 kg of sand (0·1—0·2 mm fraction)
14,000 kg of sand (0·05—0·1 mm fraction)
248,000 kg of silt (below 0·05 mm)

As seen above, the dust contained eleven times more silt than sand.

Dust, which has not been transported a great distance consists of coarser material with a higher percentage of sand grade material. Thus, for instance, dust from the

Table 60

Grain-size of eolian suspension at various altitudes (after K. E. Laprade 1957)

Fraction mm	%	
	a	b
$\frac{1}{2}-\frac{1}{4}$	0·1	0·4
$\frac{1}{4}-\frac{1}{8}$	0·8	2·6
$\frac{1}{8}-\frac{1}{16}$	30·7	25·1
$\frac{1}{16}-\frac{1}{32}$	45·8	33·6
carbonate	14·0	29·7
organic matter	3·2	3·0

a) eolian suspension at altitude 3·5 m
b) eolian suspension at altitude 20 m

surroundings of Buffalo (A. E. ALEXANDER 1934) had a medium grain-size of 0·02 mm, maximum grain-size 0·5 mm and contained 12% sand.

Noteworthy results have been obtained by studying eolian suspension at various altitudes (K. E. LAPRADE's data from Texas, 1957, Table 60). The eolian silt, derived during dust storms from Sonora Desert and deposited in the Pacific off Northern Mexico, has two major grain-size modes, first one between 5 and 10 μ, second one between 20 and 40 μ (E. BONATTI - G. ARRHENIUS 1965). Wind-borne dust, collected at Barbados (A. C. DELANY *et al.* 1967) is also completely of silt size (95% of material falls between 2—40 μ range). The average concentration of this dust in the air allows plausible estimates for the fallout rate. Its contribution to the deep sea sedimentation of the western tropical Atlantic is about 0·6 mm/1,000 years, suggesting that a fair fraction of the deep-sea muds is transported by wind.

The analyses given above show clearly the sorting of eolian suspension on the basis of the size and shape of grains. The roundness of grains at lower altitudes is better than those at the higher ones. Some of the above-described suspensions are composed of grains at the silt/fine grained sand boundary, which indicates that they were blown out by fairly strong wind storms.

The mineral composition of eolian suspended load is, on the whole, monotonous. It contains about 70% of quartz, feldspar (the amount of which decreases with the decreasing grain-size), mica, sporadic organic remains (predominantly diatoms), aggregates of clay particles and heavy minerals. Dust from the surface

Table 61

Mineral composition of dust-fall on S. S. Dunstan
(see Table 59, after P. K. Game 1964)

Minerals	Percentage			
"Aggregate" particles	32		Dust from	S. S. Dunstan
Calcite	20		25°04′N	24°28′W
Quartz	18			
Feldspar (mainly albite)	4	SiO_2	64·1%	75·3%
Iron ore (mainly hematite)	3·5	Al_2O_3	9·8	11·5
		Fe_2O_3	5·3	6·2
		TiO_2	1·1	1·3
Amphibole	2	MnO	tr	tr
Biotite	2	P_2O_5	n. d.	n. d.
Tourmaline	1	CaO	0·8	0·9
Indeterminate particles		MgO	tr	tr
(including organic matter and		Na_2O	1·8	2·1
opaque material)	17·5	K_2O	2·3	2·7
		Ignit. loss	13·2	

of the Atlantic Ocean analysed by P. M. GAME (1964)) showed a more varied composition (Table 61).

As mentioned above, the limiting size of the fraction readily transported in suspension by eolian processes is 0·05 mm. It has also been proved by analyses of eolian deposits and dust and by laboratory experiments that the boundary between fractions which are a stable component of suspension and those which settle from suspension at the change of conditions is the size of 0·02 mm. Thus, larger grains of silt and fine sand are borne by air for a far shorter time than smaller grains which are swept upwards by the weakest ascending components of air currents. After wind storms in the Sahara Desert, material of larger grain-size than 0·02 mm fell on the Mediterranean Sea and over the entire southern Europe, whereas finer material floated several times round the Earth. Grain-size differentiation in eolian sediments can be expressed by the following scheme:

1. Original weathered rock or sediment representing the source of eolian material. After fine-fractions were blown out, material coarser than 0·05 mm remains.
2. Eolian suspension winowed from the parent material. It contains 90% particles smaller than 0·05 mm.
3. Material settling readily from eolian suspension. It contains about 75% fraction 0·05—0·02 mm in size.
4. Material remaining long in suspension. Particles are smaller than 0·02 mm.

From this point of view, eolian sediments can be divided into three large groups:

1. Eolian sands — accumulations of material transported by saltation and traction or relict deposits from which fractions were swept upwards into suspension.

2. Eolian silts corresponding mostly to loess which settles down from eolian suspension at a change of conditions.

3. Finest, omnipresent component of all sediments falling at a constant velocity on the Earth's surface.

Although there are transitions between these three types, they are clear-cut groups, as evidenced by field and laboratory studies.

Eolian sands

Eolian sands originate from various source materials, the fraction below 0.05 mm having been removed from it and carried into suspension. Grains of sandy fraction are than sorted by selective transport and accumulate in various places. The material of eolian sands derives most frequently from fluviatile, beach or glacial sediments. The study of changes accompanying the transition of fluviatile into eolian sands reveal several regularities, which are summarized in Table 62.

Most marked changes have been observed in the sorting and the roundness of grains (the recorded data refer to the transition from fluviatile sands of the Amu Darya into eolian desert sands).

Table 62

Changes accompanying the transition of fluviatile into eolian sands

Original fluviatile sands	Dune sands — product of reworking of fluviatile sands
Md 0·08—0·85 mm	Md 0·182—0·288 mm
Amount of silt and clay up to 50%	Amount of silt and clay up to 1·5%
So about 2·1	So about 1·23
Roundness of quartz grains (0·1—0·2 mm) about 0·38 (Krumbein scale)	Roundness of quartz grains of the same fraction over 0·5
Average content of quartz 58%	Average content of quartz 61%
Amount of stable heavy minerals about 3%	Amount of stable heavy minerals usually more than 5%
Average percentage of unstable heavy minerals: 5·8	Average percentage of unstable heavy minerals: 4·8
Amount of mica and chlorites over 1%	Amount of mica and chlorites about 1%

Table 63

Changes accompanying the transition of beach sands into eolian ones (after S. A. Harris 1957, 1958)

Original beach sands	Dune sands — product of reworking of beach sands
Md about 0·211 mm	Md about 0·223 mm
So about 1·33	So about 1·22
Phi Skewness (of negative value or zero)	Phi Skewness (of positive value or zero)
About 1·5% of heavy minerals	About 2·2% of heavy minerals
Roundness value of quartz grains 0·42 (Krumbein scale)	Roundness value of quartz grains 0·45 (Krumbein scale)

Beach sands passing into eolian sands exhibit less pronounced changes (Table 63, most data after S. A. HARRIS 1957, 1958).

At the transition, the main change is shown by the coefficient of skewness which passes from negative values to predominantly positive ones. The change is due to the decrease of coarser fractions which are always present in beach sands to a certain extent.

Apart from the above-mentioned sources, eolian sands can also derive from other primary materials. According to J. SEKYRA (1961), weathered gneisses of the Železné hory Mts. in the Bohemian Massif, and weathered Tertiary and Pleistocene sediments of arenaceous character in Slovakia are the main suppliers of sand.

Eolian sands are transported varying distances from the source and are usually deposited in characteristic forms. Their detailed classification, based on their genesis and form, is given by R. BAGNOLD (1941). I intend to present here only the fundamental division:

1. Accumulations related to some obstruction, originating in front of or behind it. They are subdivided into many types according to their form.

2. Accumulations not joined to an obstruction, comprising barchans, migrating dunes and shapeless drifts.

The outer form of these bodies is strongly variable, depending on the speed of wind, the stability of its direction, the substratum, the grain-size of sediment and the amount of the material carried. If the direction of wind is constant and the amount of sand appreciable, connected barchans, barchan belts and fields originate. Except for shapeless forms, all eolian accumulations are characterized by a gentle windward slope (about 12—15°) and a steeper leeward slope, the dip of which depends mainly on the grain-size of material. The slope varies most often between 25 and 30° and quite exceptionally exceeds 36°.

Barchans and other sand accumulations are usually only a few dozens of metres thick. In some areas, however, their thickness is much greater. In the deserts of Central Asia there are barchans 80—300 m high, barchan fields in the Sahara are 250—500 m high.

Petrographical characteristics of eolian sands

The development trend in the petrographical composition of eolian sands has been mentioned above. In most cases, they differ from all other deposits by a better sorting, a smaller content of clay and silt and perfect roundness of quartz grains. Parameters of dune sands are plotted on diagrams in Fig. 41 (a, b).

The median of about 90% of eolian sands varies between 0·15 and 0·25 mm. The coefficient of sorting (Trask's So) is usually lower than 1·25. Phi Deviation (σ_Φ) ranges from 0·21 to 0·26, which classes eolian sands as best sorted sediments. Phi Skewness is almost always positive. α_Φ values vary predominantly between 0·13 and 0·30.

Eolian sands are more frequently unimodal than fluviatile sands. Although they contain larger floating grains in places, unimodality (i. e. one-peak histograms) is characteristic. R. L. FOLK (1968), however, discovered very remarkable bimodality in the sands of the deflationary desert areas. These sands are mixtures of coarse sand (0·5—1·0 mm) with fine sand (0·09—0·18 mm). The diameter ratio between those two fraction ranges ordinarily from 4 : 1 to 8 : 1. This bimodality is due to the removal of the intermediate fractions which are easiest to set in saltation.

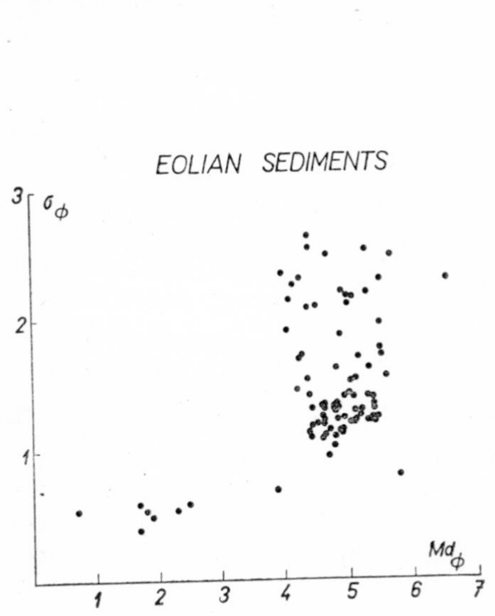

Fig. 41a. Grain-size parameters of eolian sediments. Relation between Phi Median Diameter and Phi Deviation Measure. Data compiled from various authors.

Fig. 41b. Grain-size parameters of eolian sediments. Relation between Phi Median Diameter and Phi Skewness Measure.

Although the principal features of eolian sediments are almost constant (very good sorting, absence of clay and silt, perfect roundness of grains), absolute size of grains, i. e. the median, is somewhat variable. Sands were found whose Md was 0·4 or even 1 mm on the one hand, and 0·05 and 0·1 mm on the other. These differences indicate that these sands were transported only a short distance, so that the resulting grain-size is still controlled by the grain-size of the primary material. Desert sands are generally coarsest, being derived from coarse weathered material, and sands derived from alluvial plains or ephemeral lakes are the finest.

The roundness of grains of eolian sands generally attains a high value, ranging from 0·5 to 0·7 (Krumbein's scale) for the fraction larger than 0·1 mm. It has been proved experimentally that all grains larger than 0·03 mm can be rounded by wind action. This limit is lower than with grains transported by water, owing to the different densities of the two environments. Rounding of quartz grains during transportation is very intensive. The degree of rounding has been proven by experimental evidence (PH. H. KUENEN 1959) to be directly proportional to the size of grains, the degree of primary angularity and the speed of wind. The experiments have shown that eolian abrasion is 100 to 1,000 times greater than stream abrasion. In addition to the above-mentioned factors, the original grain surface also affects the intensity of rounding. Grains with a smooth surface remain almost untouched by abrasion,

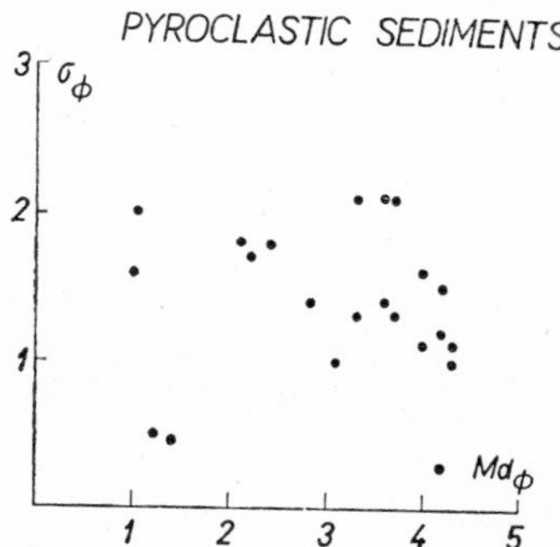

Fig. 42a. Grain-size parameters of pyroclastic sediments. Relation between Phi Median Diameter and Phi Deviation Measure. Data compiled from various authors.

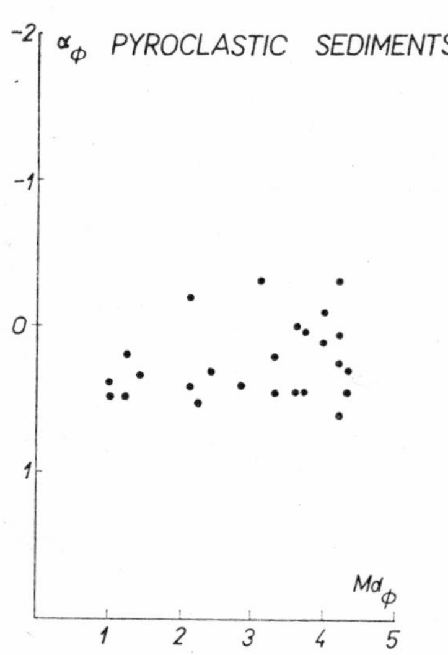

Fig. 42b. Grain-size parameters of pyroclastic sediments. Relation between Phi Median Diameter and Phi Skewness Measure.

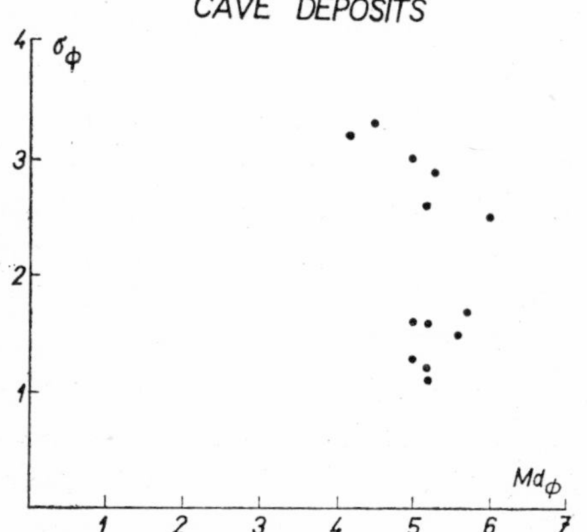

Fig. 43a. Grain-size distribution of cave deposits. Relation between Phi Median Diameter and Phi Deviation Measure. Data compiled from various authors.

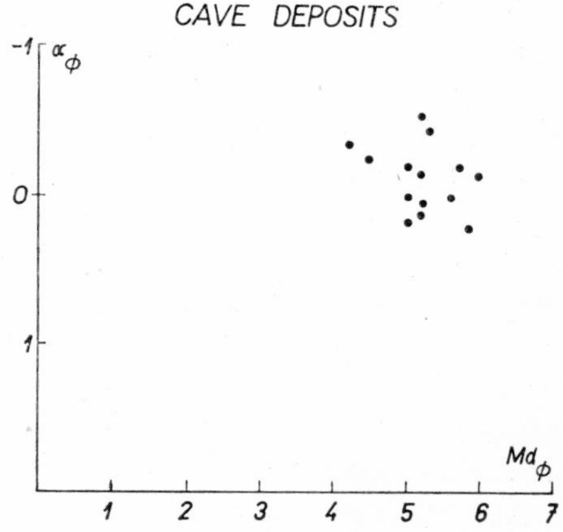

Fig. 43b. Grain-size parameters of cave deposits. Relation between Phi Median Diameter and Phi Skewness Measure.

whereas grains with a rough surface showed a greater loss in weight during the experiment. It has been likewise proved that the abrasion of feldspar and limestone grains occurs more rapidly than that of quartz grains. Feldspar grains above 0·5 mm grade are approximately three times and in finer fractions six times less resistant than quartz particles. The rounding of grains by eolian abrasion is performed by splitting-off of minute splinters from their surface.

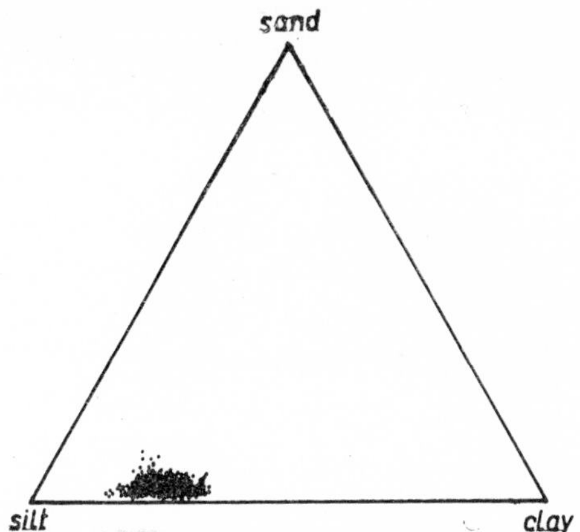

Fig. 44. Grain-size distribution of European loesses. After D. J. Doeglas (1949).

The eolian corrosion of pebbles occurs in another way. The sand blown over them grinds their exposed flat surfaces. These meet at sharp angles. These faceted pebbles or ventifacts develop within a shorter period than formerly presumed; if the sand is coarse enough ventifacts can form even in a few weeks. Under normal conditions, however, the development of ventifacts takes several tens of years.

The importance of abrasion processes in eolian transportation is probably equalled by that of selective transport. Sorting of material according to the shape of grains takes place already in eolian suspension, when grains of the same shape and size group together. Separation of grains according to the degree of roundness also occurs in the transport by saltation. Rounded grains settle earlier than the non-rounded ones. Therefore, it can be assumed that a strong rounding of grains, which were not transported far from the source, is due to selective transport and deposition of well-rounded grains.

The question of the "dull" surface of quartz grains, which is produced by microscopic unevennesses, has been decided only recently. At first the dull surface was presumed to originate only from eolian action (by impact of objects on the grain surface) and was therefore regarded as an evidence of eolian origin. Later on, doubts have been cast on the validity of this conception, because no dull surface has been achieved by imitating experimentally the eolian processes, and grains with dull surface are absent in some eolian deposits, but present in sediments of other environments. In the laboratory, dull surfaces were only produced by etching with chemical agents.

On account of these facts, the origin of dull grain surfaces has been interpreted in terms of the action of chemically aggressive waters. This disagreement was satisfactorily solved by E. W. Biedermann (1962) who arrived at the following conclusions:

1. When the unevennesses of the dull surface consist of crystallographically bounded figures, the grain can occur in any environment, as the surface was produced by chemical etching.

2. When the unevennesses are minute pits with rounded walls and bottom, the dull surface is due to eolian activity, to impact of other grains.

Only dull grains with irregular pits a few microns across can be regarded as an evidence of eolian origin.

A typical surface structures of quartz grains in eolian deposits are also the so-called percussion marks (D. H. Campbell 1963). These marks are crescentic cracks on the surface of a particle; the conoid of percussion is commonly produced by sand grain impact in the eolian environment.

The Fe-hydroxide coating of eolian grains is also one of their characteristic features. It has been found especially in desert and steppe sands. This coating is unusually stable, withstanding even a fairly long eolian, sometimes also water transportation.

Mineral composition of eolian sands

Mineral composition of sands is never indicative of their eolian origin, as it is never so characteristic as to be quite different from that of sands which have originated in other environments. Eolian sands most frequently contain 65—70% of quartz

Table 64

Mineral composition of wind-blown sands from various regions of Czechoslovakia (after J. Sekyra 1961)

Mineral	Labe River valley		Hodonín %	South Slovakia %	East Slovakia %
	1. %	2. %			
Quartz	99·1	94·6	93·5	96·9	95·9
Feldspar	0·3	2·2	0·8	0·6	0·2
Biotite	0·4	0·4	0·5	0·4	1·7
Muscovite		0·3	0·3	0·8	1·0
Opaque minerals			0·2		
Tourmaline		0·1			
Amphibole		0·4	0·1		
Rock fragments	0·2	1·8	3·6	0·1	
$CaCO_3$				12·3	3·02

and 20% of feldspars. A minimum content of micas may be typical. It can be said generally that continental eolian sands are usually richer in quartz than dune sands of coastal plains. Among heavy minerals, which are present in variable amounts (almost generally above 1%) stable minerals prevail over the unstable ones. The unstable minerals were observed to be rather sensitive to abrasion and to disintegrate already after a short eolian transport. J. SEKYRA (1961) records the mineral composition of eolian sands in Bohemian and Slovakian regions as follows (Table 64).

Coastal dunes have a greater percentage of carbonates than continental air-borne sediments, because they may contain an appreciable amount of tests of organisms redeposited from beach deposits. This redeposition of marine fauna in continental sediments can lead to a false interpretation of depositional environments of fossil sediments.

Structures of eolian sands

The internal structures of eolian accumulations is more important for the recognition of lithological properties than their external forms. The arrangement of particles produced by air currents is the same as that produced by those in water, but some features, distinctive of the eolian sedimentation may, nevertheless. appear under

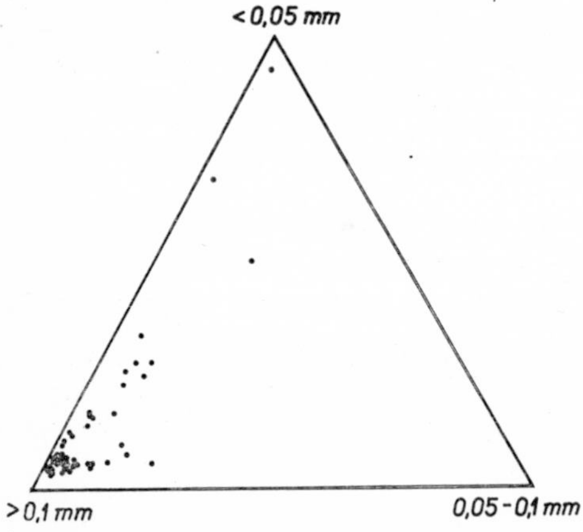

Fig. 45. Grain-size distribution of some eolian sediments of Czechoslovakia. After J. Sekyra (1961).

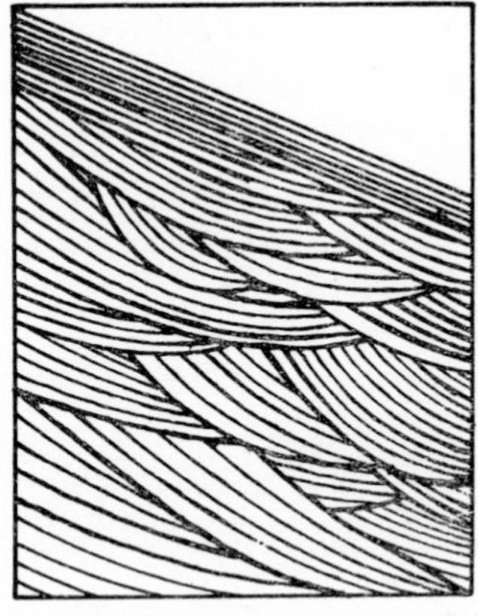

Fig. 46. Typical bedding of eolian dunes. From L. N. Botvinkina (1962).

favourable conditions. If the direction and speed of the wind were stable, sand grains would be deposited at a low angle on the windward face of the dune. These parallel, gently dipping beds, however, would be disturbed by the development of a steeper parallel stratification on the leeward side of the dune.

This ideal picture of the development of eolian bedding is complicated by a number of factors:

1. Erosive action of the wind on the windward side of the dune, which can give rise to erosional troughs with trough cross bedding.

2. Varying speed of the wind, resulting in the deposition of material of various grain-size and the change in the dip of beds and laminae.

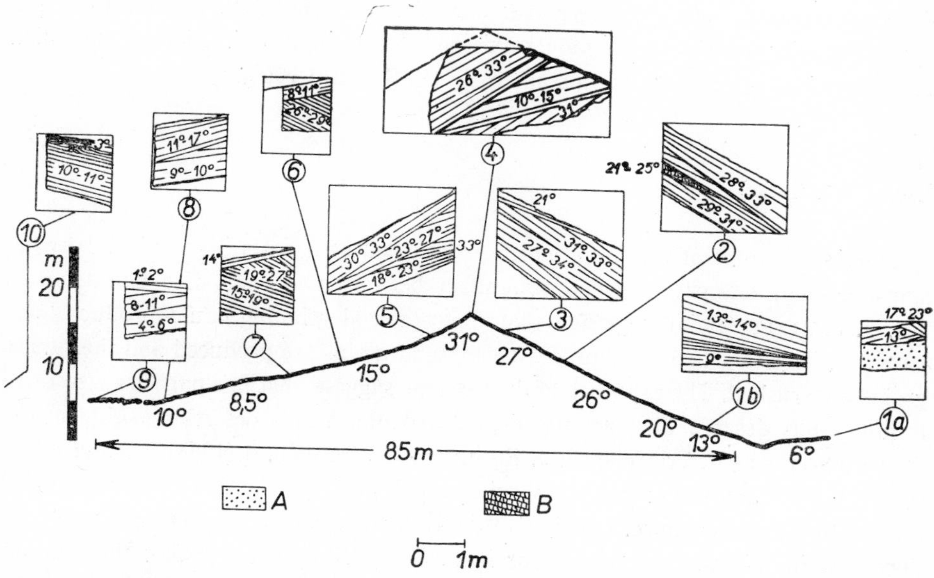

Fig. 47. Stratification in seif dune determined from test pits. Libya. A. Structureless sand, B. Near-vertical laminae between slip planes. After E. D. McKee - G. C. Tibbitts (1964).

3. Varying direction of the wind producing a shift of current bedded units. The individual sets of strata are sharply limited against over- and underlying beds and have the form of flat wedges. Owing to erosive processes the wedges are convex downwards.

Eolian sands are characterized by high-angle, wedge-planar cross stratification and through-stratification. In high-angle cross-stratification, the dip of laminae increases with the increase in grain-size. The dip of 36° is usually recorded as the maximum value: larger dips are mentioned only exceptionally (42° by L. S. Land 1964). From a large number of data, the average dip of 20—30° can be inferred. The higher values, particularly those about 30°, are a suitable guide for identification of eolian sands, as well as the dispersion of cross-bedding direction, which is smaller than that in marine of fluviatile sediments (Fig. 48, 131). The above types of cross-stratification are frequent in Recent eolian sediments and their origin has been reconstructed experimentally. Surprisingly, flat, horizontal stratification is also developed in eolian sands, as for instance, in wind-formed inter-dune deposits within the sands sea of Libya in Africa (E. D. McKee 1964).

Rounded clay galls are another characteristic structural element of wind-blown sediments. In some dunes they constitute entire laminae and beds. Compared with clay galls originated in other environments, the eolian galls are more rounded and regular.

The preferred orientation of grains is in eolian deposits less developed than in fluviatile sediments. Some authors note that only 55% of grains show a preferred orientation, with the longer axis parallel to the wind direction. On the other hand, the orientation with the longer axis parallel to the dip direction of the eolian cross-beds is very pronounced in some cases (L. S. LAND 1964).

Eolian silts (loess)

The grain-size limit of 0·05 mm, which is regarded as a natural boundary between silts and sands, is a perfect genetic boundary between two kinds of eolian deposits.

I. J. SMALLEY (1966) suggested that direct glacial grinding could produce loess particles. This is a two-stage process; first sand grains are produced and then these grains are crushed. The existence of dust storms suggests that fine particles also form in hot, sandy deserts. They are not deposited within sandy desert areas, but at the desert margins as described by I. J. SMALLEY - G. VITA-FINZI (1968). During windstorms a great amount of prevalently silt particles can be lifted into suspension and carried to a great distance. Qualitative data on the amount of silt transported and deposited in this way are given, for instance by K. LEUCHS (1932): the amount of 1,130,000 tons of dust, mainly of silty and clay fractions fell on Polish territory, on an area 970,000 km^2 large. It has been found that the material derived from the steppes of Central Asia, so that it had travelled more than 3,000 km. The area of New Zealand was covered by 100,000 tons of dust from the Australian deserts. In 1901, storms over the Sahara brought 1,000,000 tons of dust as far as Europe, to a distance of 4,000 km.

As mentioned above, the dust is composed mainly of silt fraction and small admixtures of fine-grained sand and clay.

The dust that has fallen on various places in the Atlantic Ocean has shown the following composition (after O. E. RADCZEWSKI 1939, Table 65).

The immense amount of silt in eolian dust must be conditioned by its large content in the parent rocks and sediments. Silt grains cannot be formed by the eolian abrasion of sand grains, as formerly presumed as abrasion produces particles which are mainly smaller than 2 μ.

It is common belief that during the glacial periods a large amount of silt material was blown out from glacial deposits. An analogous contemporaneous process has been observed in Greenland (W. H. HOBBS 1942). Wind storms remove large amounts of silt along with some sand from the moraines and carry it as far as several hundred kilometres from the source before depositing it in tundras, in the wind shadows of vegetation.

Table 65

Mineral composition of dust falls on various places of the Atlantic Ocean (after O. E. Radczewski 1939)

Mineral	1 %	2 %	3 %
Quartz	> 10	10	15
Feldspar	> 5	5	> 5
Mica	< 10	> 20	< 15
Aggregate grains	< 10	> 20	< 15
Sulphides	< 10	< 5	> 10
Org. SiO_2	< 2	1	3
Calcite	< 40	< 5	< 20

1. Equatorial Atlantic Ocean, 2. Atlantic Ocean 17°N, 18°W, 3. Atlantic Ocean 27°40′N, 14°57′W.

The origin of all wind-borne silts (loess) need not be restricted to the removal from a certain kind of rocks (e. g. glacial sediments), as presumed previously; it can derive from all sediments or weathered rocks which comply with the conditions stated on p. 117.

Eolian sediments whose particles belong prevalently to the silt fraction are collectively called loess, provided they show, in addition, several typical properties (high porosity, homogeneous structure, light yellowish or grey colour, high cohesion, and an admixture of calcium-carbonate). Although the term "loess" is neither descriptive nor genetic, the rocks of this character and similar genesis are so widely distributed that this term will evidently remain in usage. It is sometimes used also for sediments of another origin (e. g. fluviatile or for rocks developed by soil processes) but the prevailing part of deposits included is of eolian origin.

Normal loesses usually contain an admixture or a larger amount of local material (i. e. material transported not farther than a few dozen metres), but the bulk of particles was carried to a distance of a few hundred kilometres.

Coarser loesses (frequently at the silt-sand size boundary) deposited in hilly countries at higher elevations than 300 m above sea level contain occasionally a large amount of definitely local material brought from a distance of several to a few tens of kilometres.

Grain-size composition of loesses

The grain-size composition of loess is essentially uniform the world over. The diagram (Fig. 41 a,b) shows the Md of grain-size to be usually 0·06—0·02 mm, and the content of silt fraction 60—80%, the coefficient of sorting as Phi Deviation

(σ_Φ) 1—3, the lower value being more frequent. In most cases the Phi Skewness Measure (α_Φ) is slightly positive. Maximum grain-size is about 0·25. A perfect uniformity of the loess grain-size in certain areas was noticed, e. g. by D. J. Doeglas (1949, Fig. 44). Reference was made to loess sections where the grain-size slightly decreased upwards. Some authors considered this phenomenon as an evidence of the origin of loess by soil-forming processes. The interpretation in terms of a gradual decline in the environmental energy, however, is just as acceptable.

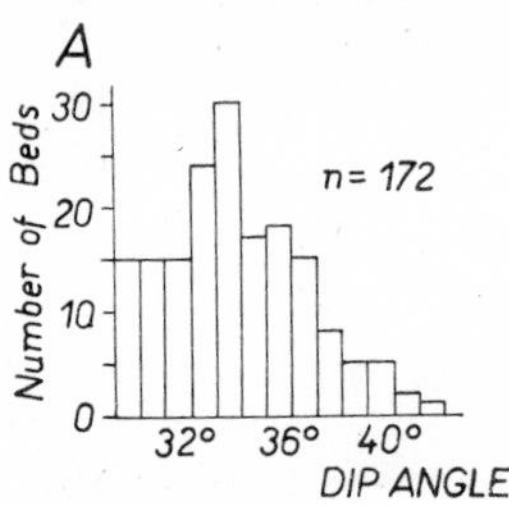

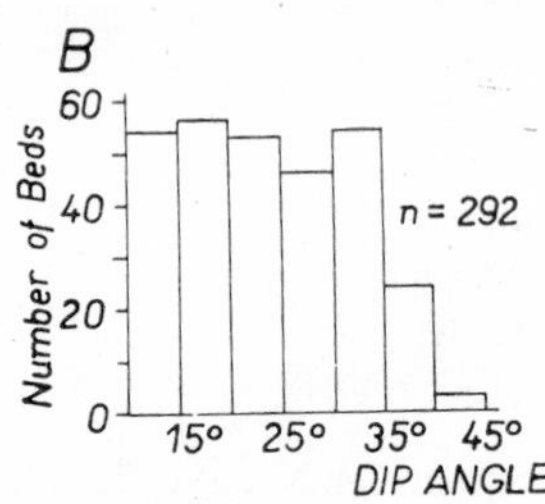

Fig. 48. Histograms of the dip high angle beds (A) and all beds excluding the 88 dune slipfaces (B) in coastal dunes of Sapelo Island, Georgia. N-sample size. After L. S. Land (1964).

The transition between the areas of sand deposition and those where silt is laid down is not so gradual as might be expected. At the boundary of these two sedimentary occurrences, the sediments are frequently bimodal. The bimodality can be recognized only by a detailed mechanical analysis, because the range of the minimum grain-size is very narrow, between 0·8 and 0·15 mm. The bimodality is caused by the mixing of two fraction of different origin: silt settled from eolian suspension and sand brought by traction and saltation. The study of loess-eolian sand transitions has shown that the changes depend on the changes of topography; the tendency to the coarsening of material is traceable already at moderate elevations.

Experiments and observations in the field have confirmed that loess is deposited most frequently at a wind velocity of 1·5—2·5 m/sec.

Mineralogical and chemical composition of loess

The mineralogical composition of loess is fairly uniform. Only the content of volcanic glass occurring in the form of sharp-edged laths of silt grade, shows some variability. The quartz grains (above 0·01 mm) attain 40—80%, feldspars 20—40%, mica 1—15% (but loesses with a far larger amount of mica have also been described). Flakes of mica, as can be expected, are larger that the quartz grains. A greater proportion of volcanic material in the loess is usually reflected in the prevalence of plagioclase over potassium feldspars; at the absence of volcanic material the relation is reversed. The amount of heavy minerals is somewhat variable, but does not exceed a few per cent.

Observations suggest that the mineralogical composition can be indicative of the origin of loess material, i. e. whether it had derived from glacial sediments or, for instance, desert regions. The former type is said to be more varied mineralogically, the latter prevalently quartzose. Loess of glacial origin is also characterized by the presence of minute rock fragments of small size. Unfortunately, the quantitative data available are not yet sufficiently detailed for a precise correlation to be made.

Thorough studies of the clay fraction of loesses have shown that it is composed of quartz, feldspars, and clay minerals whose composition is usually very variable. Thus, for instance, loesses of North America and Europe differ widely in the content of individual clay minerals. In the North American loess, montmorillonite is dominant and illite with kaolinite subordinate. In the loess of South America and that of China illite is the prevailing clay mineral.

The presence of calcium carbonate is characteristic of loesses. As referred to above, a great amount of $CaCO_3$ is transported in eolian suspension as finest silt and clay fraction, but mechanical analyses of loesses have revealed that they contain carbonate mainly in coarser fractions. This can be explained by the segregation of originally finely dispersed carbonate into larger aggregates by the activity of ground

Table 66

Chemical analyses of loesses from various parts of the world (%)

	SiO_2	Al_2O_3	Fe_2O_3	TiO_2	CaO	MgO	P_2O_5	SO_3	K_2O	Na_2O	S	$CaCO_3$
1.	47·43	9·80	4·91	1·42	16·10	3·11	0·13	tr	1·43	0·89	tr	28·61
2.	64·63	16·03	6·56	0·82	1·72	2·61	0·19	tr	2·19	1·10	0	2·62
3.	53·55	8·51	3·24	0·79	16·64	0·51	0·11	tr	1·17	0·57	tr	29·46
4.	66·92	12·71	4·34	0·90	4·36	0·77	0·13	tr	1·76	1·02	tr	7·48
5.	58·84	11·74	4·67	0·91	7·99	2·55	0·15	0	1·70	0·84	0·02	13·90
6.	43·57	5·78	2·68	1·28	22·87	2·44	0·13	tr	1·08	0·90	tr	40·69
7.	63·44	8·75	3·64	1·36	8·87	1·71	0·11	0	1·54	1·54	tr	15·61
8.	80·62	9·03	2·88	1·18	0·53	0·64	0·07	tr	1·43	1·08	tr	0·77
9.	54·05	8·17	2·83	1·07	14·60	2·09	0·11	tr	1·41	1·04	0	25·81
10.	70·62	10·41	4·19	0·45	3·98	1·16	0·05	0	1·84	0·89	0	6·98
11.	58·78	8·37	3·13	0·73	11·88	2·19	0·11	tr	1·44	0·84	0·22	20·95
12.	59·36	6·13	2·66	1·89	12·79	1·90	0·10	tr	1·45	1·03	tr	22·76
13.	59·04	7·44	2·67	1·13	12·96	1·55	0·09	tr	1·45	1·00	0	22·92
14.	51·86	6·82	2·40	0·89	17·59	1·69	0·08	tr	1·21	0·92	0	31·20
15.	72·77	7·09	2·98	1·79	5·33	1·02	0·10	tr	1·64	1·04	tr	9·94
16.	73·33	12·00	3·02	—	2·67	1·06	0·08	tr	2·25	1·48	—	
17.	72·35	13·56	2·46	—	2·04	1·77	0·14	tr	2·78	0·24	—	
18.	66·81	15·04	3·11	—	1·65	1·03	—	—	2·31	1·79	—	

1—2 Italy, 3—6 France, 7—14 Germany, 15 Belgium, 16—17 China, 18 Argentina.

waters, which also accounts for the formation of abundant calcite concretions. The $CaCO_3$ content usually varies between 6 and 15%.

The chemical composition of loesses corresponds to their mineralogical compositi-on. The average analytical values (in per cent) of loesses from various parts of the world are given in Table 66.

From this Table it follows that American loess is richer in clay than the European. The content of organic matter is generally low, predominantly 0·2—3·0% org. C. Larger amounts of organic matter are accumulated in darker layers amidst the loess series which are interpeted as buried soil horizons.

Structure of the loess

Loess is distinguished by its homogeneous structure and a high porosity (about 48—60%). The secondary segregation of calcium carbonate or organic matter is responsible for the local origin of mottled structures. Biogenic structures, due to the effects of plant roots, are also frequent. Loess deposits of eolian origin often contain interlayers of fluviatile or lacustrine sediments which show ripple bedding or lamination.

Eolian material admixtures in other sediments

The above data indicate that the finest eolian suspension of the grain-size below 0·02 mm is a stable component of the atmosphere and settles down permanently, at a constant rate, over the whole Earth's surface, so that it is virtually present in all sediments. It is obvious that it will be most abundant in those sediments that are deposited at a lowest rate. The highest possible rate of eolian sedimentation above the pelagic regions of the ocean equals approximately one-fourth of the sedimentation rate of the abyssal red clay, i. e. about 0·1 mm per 1,000 years. In minor basins or in near-shore areas coarser eolian material may also be deposited; in some continental basins, lagoons, or tidal flats it actually represents the essential part of sediments. In shallow-water environments, the maximum amount of eolian admixture can be expected in basins adjacent to deserts or areas with eolian dunes. The Persian Gulf, where a large amount of eolian material is constantly brought from the African deserts, is a good example of such an environment. The admixture contains 1·6% of fine-grained sand, 78·8% of silt, 19% of clay, and a surprisingly high percentage of $CaCO_3$ (more than 80%). The insoluble residue shows a following composition:

Quartz	62%	Olivine	2%
Feldspar	20%	Zircon	2%
Actinolite	7%	Rutile	2%
Biotite	3%	Garnet	2%

The amount of eolian material falling on the surface of the Caspian Sea makes 39·5 g/m^2/year and its grain-size does not usually exceed 0·01 mm. It was composed of 58·8% of carbonate, 35% of muscovite and 6·4% of salts. The insoluble residue contained still an admixture of quartz, minute concretions of Fe-hydroxides and pyrite. Quartz and feldspar grains larger than 0·02 mm were only sporadic. Compared with the sedimentation rate of other deposits, eolian material represents one tenth of all sediments there. A large amount of eolian material is deposited in shallow lagoons (mainly at the coast of the Gulf of Mexico), in places with broad coastal plains, and migrating dunes and prevalent offshore winds. In such cases the dunes can migrate as far as to the bay and fill it completely so that eolian sediments are the only material deposited in the basins.

It has not yet been decided to what extent the eolian material participates in the composition of abyssal sediments. Some authors consider that it is of negligible importance, whereas others attribute to it an essential role. O. E. Radczewski (1939) studied the deposits in the equatorial zone of the Atlantic Ocean and took all quartz grains with Fe-hydroxide coating for eolian matrial. He found that their amount as well as their grain-size decreased away from the shore towards the pelagic areas. The average grain-size composition of these "desert quartzes", as recorded by O. E. Radczewski, is given in Table 67.

The above data prove that the effects of eolian processes on the formation of sediments cannot be underrated in any environment. In addition to pure eolian sediments (eolian sands and loess) eolian admixture occurs also in other kinds of sedimentary rocks.

Table 67

Grain-size composition of "desert quartzes" (after O. E. Radozewski 1939)

μ	Percentage
50—100	11·1
10—50	16·7
5·5—10	18·6
1—5·5	14·4

References

Aleksina I. A. (1959): Characteristics of the eolian material of the eastern coast of Central Caspian Sea (in Russian). Doklady AN SSSR, vol. 127: 227—230, Moscow.

Alexander A. E. (1934): The dustfall of November 13, 1933, at Buffalo, New York. Jour. Sedimentary Petrology, vol. 4: 81—82, Menasha.

Bagnold R. (1941): The physics of blown sands and desert dunes. P. 1—265, London.

Beal M. A. - Shepard F. P. (1956): A use of roundness to determine depositional environment. Jour. Sedimentary Petrology, vol. 26: 49—60, Menasha.

Becke F. (1901): Mikroskopische Untersuchung der Proben von Staubschnee vom 11. März 1911. Meteor. Zeitschrift, Bd. 18: 318—321, Wien.

BIEDERMAN E. W. (1962): Distinction of shoreline environment in New Jersey. Jour. Sedimentary Petrology, vol. 32: 181—200, Menasha.

BIGARELLA J. J. (1969): Dune sediment characteristics, recognition and importance. Bull. Am. Assoc. Petrol. Geol., vol. 53: 707, Tulsa.

BLACKTIN S. C. (1934): Dust. P. 1—314, London.

BONATTI E. - ARRHENIUS G. (1965): Eolian sedimentation in the Pacific off Northern Mexico. Marine Geology, vol. 3: 337—348, Amsterdam.

BRADLEY J. (1957): Differentiation of marine and subaerial environments by volume percentage of heavy minerals, Mustang Island, Texas. Jour. Sedimentary Petrology, vol. 27: 116—125, Menasha.

CAILLEUX A. (1953): Les loess et limons éoliens de France. Bull. serv. carte géol. France. T. 51, No. 240: 437—460, Paris.

CAMPBELL D. H. (1963): Percussion marks on quartz grains. Jour. Sedimentary Petrology, vol. 33: 855—859, Menasha.

DELANY A. C. - PARRIN D. W. - GRIFFIN J. J. - GOLDBERG E. D. - REIMANN B. E. F. (1967): Airborne dust collected at Barbados. Geochim. Cosmochim. Acta, vol. 31: 888—909, London.

DOEGLAS D. J. (1949): Loess, an eolian product. Jour. Sedimentary Petrology, vol. 19: 112—117, Menasha.

FOLK R. L. (1968): Bimodal supermature sandstones: Product of the desert floor. Int. Geol. Congr., 23. Sess., vol. 8: 9—32, Prague.

FRIEDMAN G. M. (1961): Distinction between dune, beach, and river sands from their textural characteristics. Jour. Sedimentary Petrology, vol. 31: 514—529, Menasha.

GAME P. M. (1964): Observations on a dustfall in the Eastern Atlantic, February, 1962. Jour. Sedimentary Petrology, vol. 34: 355—359, Menasha.

GUDELIS V. - MINKEVICHIUS V. (1963): Lithodynamic spectra of sand drift in the coastal dunes of Lithuania (in Russian). Baltica, vol. 1: 228—230, Vilnius.

HARRIS S. A. (1957): Mechanical constitution of certain present-day Egyptian dune sands. Jour. Sedimentary Petrology, vol. 27: 421—434, Menasha.

— (1958): Differentiation of various Egyptian aeolian microenvironments by mechanical composition. Jour. Sedimentary Petrology, vol. 28: 164—174, Menasha.

HOBBS W. H. (1942): Wind — the dominant transportation agent within extramarginal zones to continental glaciers. Jour. Geology, vol. 50: 556—559, Chicago.

HOLMES C. D. (1944): Origin of loess — a criticism. Am. Jour. Sci., vol. 242: 442—446, New Haven.

INMAN D. L. (1952): Measures for describing the size distribution of sediments. Jour. Sedimentary Petrology, vol. 22: 125—145, Menasha.

JAHN A. (1950): Loess, its origin and connection with the climate of glacial epoch (in Polish, Engl. res.). Acta geol. Polonica, vol. 1: 303—310, Warszawa.

KRINSLEY D. - TAKAHASHI T. (1964): A technique for the study of surface textures of sand grains with electron microscopy. Jour. Sedimentary Petrology, vol. 34: 423—426, Menasha.

KUENEN PH. H. (1959): Sand — its origin, transportation, abrasion and accumulation. Geol. Soc. South Africa, Annex to vol. 62, No 6: 1—33.

KUHLMAN H. (1960): Microenvironment in a Danish dune area, Rabjerg Mile. Med. Dansk. Geol. Fören., Bd. 14: 253—263, København.

KUKAL Z. - SAADALLAH A. (1970): Composition and rate of deposition of dust storm sediments in Central Iraq. Čas. Mineral. Geol., Prague, in print.

LAND L. S. (1964): Eolian cross-bedding in the beach dune environment, Sapelo Island, Georgia. Jour. Sedimentary Petrology, vol. 34: 389—394, Menasha.

LAPRADE K. E. (1957): Dust-storm sediments of Lubbock area, Texas. Bull. Am. Ass. Petrol. Geol., vol. 41: 709—726, Tulsa.

LEUCHS K. (1932): Die Bedeutung von Staubstürmen für die Sedimentation. Centralblatt f. Min., Abt. B, p. 145—156, Stuttgart.

LONG J. T. - SHARP R. P. (1964): Barchan-dune movement in Imperial Valley, California. Bull. Geol. Soc. Am., vol. 75: 149—156, Baltimore.

LUKASHEV K. I. (1961): Problem of loess in the light of recent investigation (in Russian). P. 1—219, Minsk.

MATTOX R. B. (1955): Eolian shape-sorting. Jour. Sedimentary Petrology, vol. 25: 111—114, Menasha.

MCKEE E. D. (1957): Primary structures in some recent sediments. Bull. Am. Ass. Petrol. Geol., vol. 41: 1704—1747, Tulsa.

— (1964): Inorganic sedimentary structures. In "Approaches to Palaeoecology". P. 275—295, New York.

MCKEE E. D. - TIBBITTS G. C. (1964): Primary structures of a seif dune and associated deposits in Libya. Jour. Sedimentary Petrology, vol. 34: 5—17, Menasha.

MOLDWAY L. (1957): Eolian sedimentation (in Russian). Acta Geologica Acad. Sci. Hungaricae vol. 4: 271—320, Budapest.

PÉWÉ T. L. (1951): An observation on wind-blown silt. Jour. Geology, vol. 59: 399—401, Chicago.

POOLE F. G. (1964): Paleowinds in the Western United States. In "Problems in Palaeoclimatology". P. 394—405, London.

PORTER J. (1962): Electron microscopy of sand surface textures. Jour. Sedimentary Petrology, vol. 32: 124—135, Menasha.

RADCZEWSKI O. E. (1939): Eolian deposits in marine sediments. In "Recent marine sediments". P. 496—502, Tulsa.

REX R. W. - GOLDBERG E. D. (1958): Quartz contents of pelagic sediments of Pacific Ocean. Tellus, vol. 1: 153—159, Stockholm.

SCHLEE J. - UCHUPI A. - TRUMBULL J. V. A. (1964): Statistical parameters of Cape Cod Beach and eolian sands. Geol. Surv. Res. 1964 D: 118—122 (Prof. Pap. 501-D), Waschington.

SEKYRA J. (1961): Wind-blown sands. INQUA, p. 29—38. Warszawa.

SHARP R. P. (1963): Wind ripples. Jour. Geology, vol. 71: 617—636, Chicago.

SHEPARD F. P. - YOUNG R. (1961): Distinguishing between beach and dune sands. Jour. Sedimentary Petrology, vol. 31: 196—214, Menasha.

SMALLEY I. J. (1966): The properties of glacial loess and the formation of loess deposits. Jour. Sedimentary Petrology, vol. 36: 669—676, Menasha.

SMALLEY I. J. - VITA-FINZI C. (1968): The formation of fine particles in sandy desert and the nature of desert loess. Jour. Sedimentary Petrology, vol. 38: 766—776, Menasha.

STOKES W. L. (1964): Eolian varving in the Colorado Plateau. Jour. Sedimentary Petrology, vol. 34: 429—433, Menasha.

SWINEFORD A. - FRYE J. C. (1951): Petrography of the Peoria loess in Kansas. Jour. Geology, vol. 59: 306—322, Chicago.

— (1955): Petrographic comparison of some loess samples from western Europe with Kansas loess. Jour. Sedimentary Petrology, vol. 28: 164—174, Menasha.

TERUGGI M. E. (1957): The nature and origin of Argentine loess. Jour. Sedimentary Petrology, vol. 27: 322—332, Menasha.

WILLIAMS G. (1964): Some aspects of the eolian saltation load. Sedimentology, vol. 3: 257—287, Amsterdam.

11. Glacial sediments

Recent glacial sediments, just as are Recent loess, are not very widespread. However, with regard to the main purpose of this work, i. e. to provide a clue for the restoration of sedimentary conditions in the past on the basis of the present sedimentary conditions, it is necessary to deal with relatively young glacial deposits, as long as their genesis is safely proved.

The comprehensive description of the origin, morphology and movement of glaciers given in R. L. FLINT's book (1947) is more than adequate for our purposes. Only the fundamental features of ice movement and transport are common with river and eolian transports. The differences in the density of medium and internal friction are so great that only a laminar movement exists in the ice mass. It has been theoretically deduced that in the glacier mass turbulent movement could occur only when the velocity of its movement approached the speed of light.

There are many inconsistencies in glacial terminology. The term "moraine" is generally used both for the material carried by the glacier and for the resulting sediment, which does not seem logical in relation to the deposits of other environments. Therefore I use here this term only for glacial deposits, in a morphological sense. Recent glacial sediments are also called by the genetic term "till", but its use for ancient sediments is rather subjective.

The advancing glacier is sometimes compared to conglomerate cemented by ice. Rock fragments get into the glacier by falling on its surface during lateral abrasion, or by abrasion on the glacier bottom, by the plucking action of the ice. In addition, the glacier tears along a large amount of material loosened by physical weathering or periglacial processes. Because of its high density and internal friction, the glacier can move boulders of great size and weight; boulders weighing up to several thousand tons are not rare in glacial deposits. The rock particles are first transported at the base of the ice. Owing to differential movement they rise within the glacier and are borne "in suspension". The particles falling on the glacier remain on its surface only for a short time; they sink inwards at a comparative high speed, so that the clastic materials are disseminated throughout the ice mass at the end.

At the retreat or temporal halt of the glacier, the transported material forms a so-called end (terminal) moraine near the maximum extension of a glacier front. Material migrating at the base of a glacier constitutes a basal moraine. During the melting away of a glacier, an ablation moraine originates. It is composed of material originally scattered in the ice mass, and in its final stage it lies above the basal moraine.

The effects of ice are combined with the activity of waters flowing from beneath the glacier and through the tunnels in its front. They carry off and sort a great deal of sediments originally transported only by the ice. Caverns filled with melt-water can be found also inside the glacier, so that the environment for aqueous sedimentation can exist even within the ice mass. It is evident that many deposits, regarded as purely glacial are actually of fluvioglacial origin. This refers particularly to end moraines; their material is carried by ice, but their final appearance is influenced by the meltwaters that drain the glacier.

Glacier abrasion is sometimes very intensive. The mean velocity of erosion by Alpine glaciers is estimated at 1—10 mm/year. Some authors, however, do not admit such a high value, maintaining that the principal part of this abrasion is performed by glacial streams. Thus, the intensity of glacier erosion requires further study.

Petrographical composition of glacial sediments

There are few deposits whose composition shows such a variety in size and mineralogy as do glacial sediments. Their material ranges from coarse blocks, more than a few metres in size, to clay particles; moreover, the sediments of different size are mixed in various proportions, so that the pattern of size distribution is extremely varied. Glacial sediments are sometimes divided on the basis of grain size as, for instance, by S. V. YAKOVLEVA (1956, Table 68).

Table 68

Classification of glacial sediments according to grain-size (after S. V. Yakovleva 1956)

<table>
<tr><th></th><th>Gravel %</th><th>Coarse-grained medium-grained sand %</th><th>Fine-grained sand %</th><th>Silt %</th><th>Clay %</th></tr>
<tr><td>Gravel sediments</td><td>50</td><td>30</td><td>15</td><td colspan="2">5</td></tr>
<tr><td>Sandy sediments</td><td>25</td><td>35</td><td>30</td><td colspan="2">10</td></tr>
<tr><td>Fine sandy sediments</td><td>15</td><td>25</td><td>40</td><td>15</td><td>5</td></tr>
<tr><td>Sandy-silty sediments</td><td>15</td><td>20</td><td>35</td><td>25</td><td>5</td></tr>
<tr><td>Clayey sediments</td><td>10</td><td>15</td><td>30</td><td>25</td><td>20</td></tr>
</table>

From these data, the mechanical composition of moraines — from coarse gravel fraction to finest clay fraction – is apparent. No glacial deposits have been found so far which are absolutely devoid of clay material. Glacial sediments without gravel or, at least, coarse-sandy fractions, are likewise rare.

The mechanical composition of glacial sediments depends on the following factors:

1. Character of the rock substratum, on which the glacier moves.
2. Morphology and velocity of the movement of the glacier.
3. Position of the material (moraine) in relation to the glacier.
4. Subsequent reworking, fluviatile or by solifluction.

The character of the rock substratum affects not only the mineralogy and petrography, but also the grain-size of glacial sediments. The dependence is most striking in basal moraines which contain most local detritus. It has been found that glacial sediments on sedimentary rocks are richer in silt and clay that those on metamorphic rocks. The changes in the grain-size composition of glacial sediments with the increasing distance from the outcrops of metamorphic rocks are shown in Table 69.

Table 69

Changes in grain-size composition of glacial sediments with the increasing distance from the outcrop of metamorphics
(Russian Plateau, after E. V. Rukhina 1960)

Growing distance (km)	>2 mm %	2·0—0·05 mm %	0·05—0·002 mm %	<0·002 mm %
10	25·8	59·9	12·1	4·2
30	22·8	52·4	16·2	8·4
60	14·8	59·1	21·2	4·9
80	43·0	46·1	8·9	2·0
110	28·3	46·8	18·0	7·3
130	28·0	61·3	8·2	2·5
150	18·1	66·1	11·5	4·3
180	39·6	45·0	11·4	4·0
200	7·1	58·7	29·3	4·9

The tendency to enrichment of finer fractions is well seen but, nevertheless, the interference of the changes in local conditions is considerable. According to other data, glacial sediments on a sedimentary substratum usually have 3—8% of gravel fraction and twice or three times more on metamorphic rocks. It has also been ascertained that the boundary between crystalline and sedimentary complexes formed also a boundary between the predominantly erosive and the prevalently aggradational activity of the moving glacier.

The type and morphology of the glacier also considerably affects the grain-size composition of glacial sediments. Thus, for instance, it is well known that the deposits of mountain glaciers are, on the average, coarser than those of continental ice-sheets. The moving ice adapts its form to the relief of the substratum, which results in a change of morphology of ice flow and the grain-size of transported and deposited material.

Table 70

Differences in grain-size composition of basal and ablation moraines (after E. V. Rukhina 1960)

	Fraction (mm)					
	>0·70 %	0·70—0·25 %	0·25—0·10 %	0·10—0·05 %	0·05—0·01 %	<0·01 %
1. Basal moraine (Fjodor tundra, Kola)	25·75	16·20	11·90	10·40	12·35	27·80
Ablation moraine (Fjodor tundra, Kola)	25·56	24·67	16·30	12·38	8·65	12·90
2. Basal moraine (Smorodinka River, Kola)	21·90	16·50	17·50	3·40	16·10	24·00
Ablation moraine (Smorodinka River, Kola)	21·78	25·34	27·78	6·60	10·40	8·10

In many places several horizons of moraines are developed, which either represent individual developmental stages of one moraine or deposits of several successive advances and retreats of the glacier. In almost all many-layered glacial sediments, a decrease of grain-size upwards is observable, because the lowermost horizons are influenced by the underlying rocks while the upper parts contain reworked morainic material. Frequently, two main horizons are developed: the basal moraine at the base, separated sharply from the ablation moraine above (Fig. 54).

Although the difference between the basal and ablation moraine is dominantly mineralogical, the variance in grain-size distribution is also percentible. In Table 70 grain-size analyses of these two morainic types from one respective locality are given.

The above data clearly show the main difference between the basal and ablation moraines, i. e. a considerably smaller content of clay and silt fractions in ablation moraines and, consequently, their better sorting.

Reworking of glacial material takes place more or less in every glacier. Running water produces important changes in the grain-size distribution of sediments, as shown below (Table 71).

Fluvial reworking leads to a decrease in the amount of finer particles and the increase of gravel and coarse sand may result in a typical bimodal river channel sediment, rich in gravel and medium-grained sand fraction.

Table 71

Differences in grain-size composition of basal moraines and reworked moraines by running water (after E. V. Rukhina 1960)

	Fraction (mm)					
	>0·7 %	0·7—0·25 %	0·25—0·10 %	0·10—0·05 %	0·05—0·01 %	<0·01 %
1. Basal moraine (Kola)	0·08	6·02	40·20	41·90	6·80	5·00
Reworked moraine	5·56	35·27	36·30	15·87	2·00	5·00
2. Basal moraine (Kola)	21·90	16·50	17·60	3·40	16·10	24·00
Reworked moraine	26·71	24·78	44·75	2·20	0·25	0·70
3. Basal moraine (Russian Plateau)	8·84	13·26	27·72	27·19	11·40	11·59
Reworked moraine	10·43	6·31	37·40	25·31	12·55	8·00

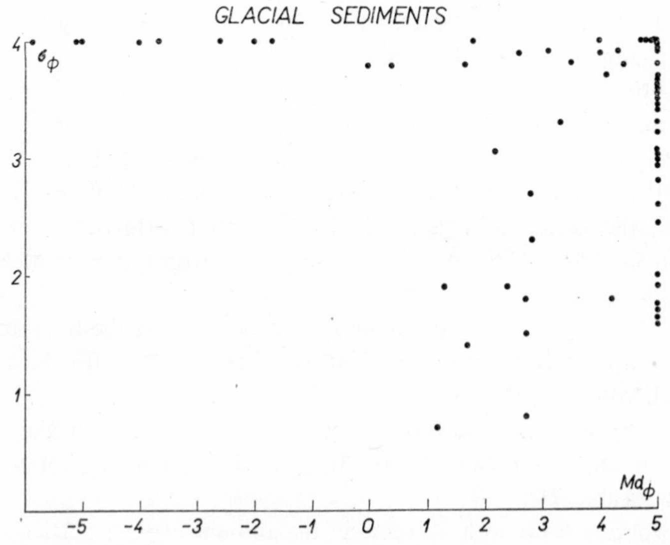

Fig. 49a. Grain-size parameters of glacial sediments. Relation between Phi Median Diameter and Phi Deviation Measure. Data compiled from various authors.

Provided that glacial sediments always contain a constant amount of coarsest fractions (coarse sand and gravel), their composition can be expressed by a triangle with sand-silt-clay as end members. The graph on Fig. 50 shows the composition of several moraines from the Russian Plateau and North America. It is evident

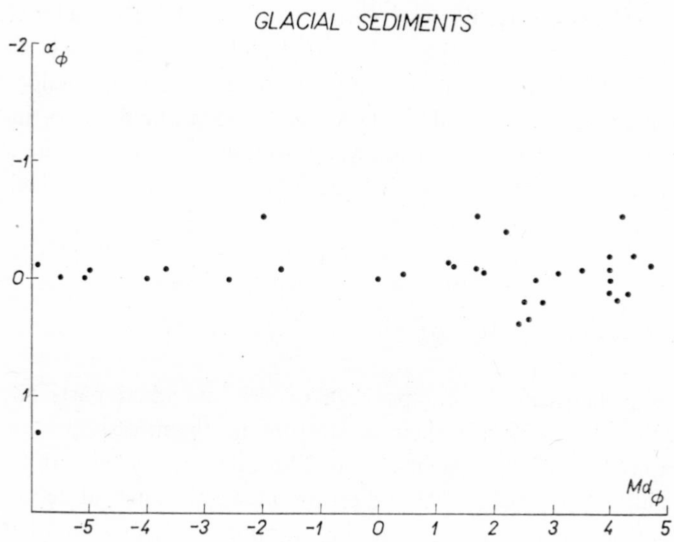

Fig. 49b. Grain-size parameters of glacial sediments. Relation between Phi Median Diameter and Phi Skewness Measure.

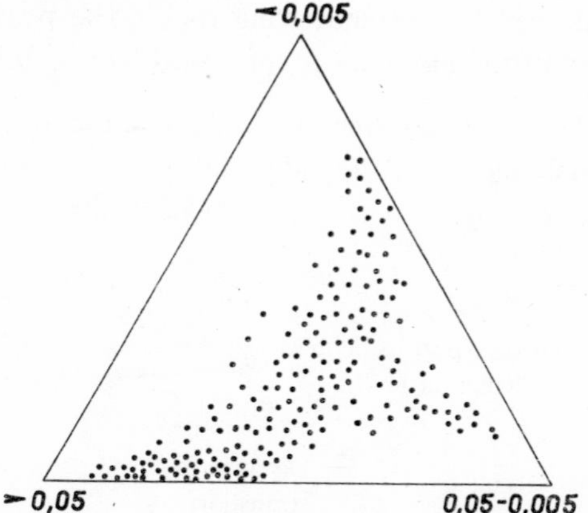

Fig. 50. Grain-size distribution of fine fractions of glacial deposits. After E. V. Rukhina (1960).

that in glacial sediments, as long as they are not reworked by running water, there is an equilibrium between the sandy, silty and clayey fractions. This most characteristic grain-size feature of glacial sediments can be formulated as follows: glacial sediments are distinguished by the constant presence of a certain amount of gravel fraction and by the equilibrium between the sandy, silty and clayey fractions.

Grain-size parameters of a number of glacial deposits are plotted on the diagrams above (Fig. 49 a,b). The Median of grain-size varies within a wide range. Points on the right margin of the diagram, belonging to extremely fine sediments, correspond mostly to glaciolacustrine deposits. A considerable number amount of points are concentrated in the area corresponding to the median between 0·5 and 0·125 mm. The sorting, expressed as σ_{Φ}, likewise varies considerably. The sediments not re-deposited by running water are very poorly sorted and their σ_{Φ} exceeds 4. Sediments with a Median 0·5—0·25 mm display a medium or good sorting analogous to that of river sediments and their σ_{Φ} is about 1. As can be seen, the poor sorting of glacial deposits cannot be generalized. Phi Skewness usually equals zero; the slightly positive values are as frequent as the slightly negative ones. The finest glaciolacustrine sediments have a low negative skewness.

Mineralogical composition of glacial deposits

The interrelations between the composition of the individual parts and kinds of moraines and the composition of their substratum are the most important questions of the petrography of glacial sediments. Their solution is connected with the establishment of the relative amount of local and erratic material in glacial deposits.

The investigation results suggest that the moraines contain somewhat less local material than material transported from a long distance. The amount of local material, not more than a few dozen kilometres distant, is estimated at 20—30% (V. C. Shepps 1958). In spite of it, the composition of basal moraines, in particular, is influenced by the underlying rocks. The petrographical and mineralogical composition of the moraines is controlled by the following factors.

1. The original source of pebbles and the direction and mode of transport.
2. Stability of the individual kinds of pebbles.
3. Composition of the rocks beneath the moraines.

Table 72

Distance of transport of pebbles
(after E. V. Rukhina 1960)

Rock	Distance of transport	Note
Quartzite	85 km	
Claystone indurated	7—8 km	Absent in fine-grained fractions
Limestone	50—100 km	Amount of pebbles decreased from 50—90% to 8—10%

Processes of transport by ice affecting the final composition of sediments are similar to those of transportation by water. The unstable pebbles and boulders disintegrate and the original material is mixed with detritus carried from subsidiary sources. The length of migration of the pebbles depends on the mode of transport within the ice mass, i. e. virtually on the size and movement velocity of the glacier, their position in the ice mass and on the substratum, but mainly on the mechanical resistance of the rock. Crumbly and little resistant boulders, such as sandstones, limestones or clayey shales, usually occur in a larger amount in the close proximity of the rock outcrops. Some data on the length of transport are given in Table 72.

It can be said generally that rocks which are scarce in the bedrock of glaciated areas are also scarce in the morainic material (see Table 73).

Table 73

Comparison between the areal extent of various bedrocks and their pebble content in three different moraines
(after E. V. Rukhina 1960)

Rocks	Areal extent in Finland km^2	Amount of pebbles in moraines		
		1. %	2. %	3. %
Granitoides	78·3	88·2	83·7	81·3
Basic intrusions	8·2	3·5	8·6	9·1
Metamorphic rocks	9·1	12·9	5·1	4·1
Quartzites and quartzeous sandstones	4·1	1·3	2·5	2·1
Limestones	0·1	0·1	0·1	0·4

Within a few minor exceptions, the content of pebbles in morainic sediments roughly corresponds to the areal extent of rock in the source area of glaciers of approximately the same stability. Apart from the different mechanical resistance of individual kinds of rocks, their tendency to disintegration into pebbles must also be taken into consideration. As is well known, some rocks tend to form pebbles of 1—5 cm in size, which subsequently agglomerate in glacial sediments, whereas others crumble into sand and their pebbles readily disintegrate. The influence of this important and sometimes underestimated factor is well demonstrated by the content of various pebbles in basal moraines right about the outcrop of the respective rock. The content of these pebbles in local morainic material is given in Table 74.

Table 74

Content of various pebbles in basal moraines around the outcrop of respective rocks (after V. Okko 1944)

Rocks and pebbles	Amount of pebbles %
Granite	73
Gneiss	62
Quartzite	47
Dolomite	46
Basic intrusions	42
Mica-schist	32

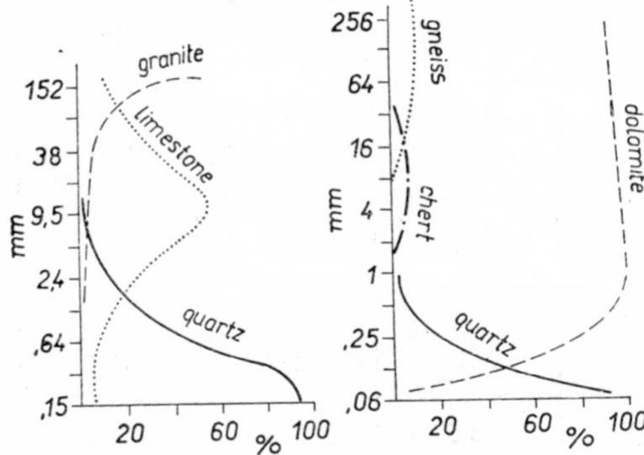

Fig. 51. Relation between composition and grain-size distribution of fluviatile (left diagram) and glacial (right diagram) deposits. After S. N. Davis (1958).

Plutonic and katazonal metamorphics show the greatest tendency to form pebbles. Mica-schists form least pebbles, as they are inclined to disintegrate into coarse sand.

S. N. Davis (1958) found a noteworthy relationship between the grain-size and mineralogical composition. In the majority of glacial sediments a considerable change in petrographical composition takes place about the limit of 2 mm. Grains under 2 mm consist predominantly of quartz and other stable material, whereas the larger fractions reflect the composition of the bedrock. Above the 2 mm boundary the changes in petrography and mineralogy are already small (generally up to 7% of the content of the individual pebbles). As will be shown below, this fact can be important distinctive feature between fluviatile and glacial gravels (Fig. 51).

Fine quartz and chert of glacial sediments is undoubtedly detritus brought from a great distance, whereas coarse material can be either local or transported. The petrographical composition of morainic sediments changes with the position in the glacier and the type of moraines. Basal moraines are considerably richer in local material than the overlying ablation moraines. The petrographic composition of glacial sediments in relation to the chemical composition of the bedrock is given in Table 75 (after E. V. Rukhina 1960).

Table 75

Comparison of chemical composition of bedrocks and basal moraines (after E. V. Rukhina 1960)

	SiO_2	TiO_2	Al_2O_3	Fe_2O_3	FeO	MnO	MgO	CaO	Na_2O	K_2O	P_2O_5
Moraine	69·49	0·50	13·29	2·34	1·89	0·05	1·38	2·38	3·26	1·82	0·30
Leptite	62·10	0·79	14·75	1·19	5·04	0·09	2·34	7·33	2·74	1·84	0·09
Difference	7·19	−0·29	−1·46	1·15	−3·15	−0·04	−1·05	4·95	−0·52	−0·02	0·21
Moraine	72·21	0·42	9·05	1·23	1·06	0·04	0·47	1·87	2·30	3·50	0·10
Rapakivi	72·57	0·25	12·62	1·45	2·41	0·02	0·46	1·34	2·30	6·10	0·02
Difference	2·64	0·17	−3·57	−0·22	−1·35	0·02	0·01	0·53	0·00	−2·60	0·08
Moraine	61·88	1·25	12·36	6·64	1·98	0·13	3·61	2·11	2·61	1·45	0·12
Basic intrusives	51·48	0·70	15·93	3·66	6·97	0·14	7·01	9·49	2·39	1·08	0·12
Difference	10·40	0·55	−3·57	2·98	−4·99	−0·01	−3·40	−7·38	−0·79	0·37	0·00

A pronounced tendency of SiO_2 increase towards the inside of the moraine is caused by the mixing of local material with quartz detritus transported in suspension from greater distances. Glacial sediments have a surprisingly lower content of Al_2O_3 than has the bedrock. This can be interpreted in terms of predominantly mechanical weathering of rocks and outwashing of clay fraction by melt waters. As can be deduced theoretically, the amount of FeO declines considerably. The remaining elements do not show any marked tendency.

The composition of the finest clay fractions of glacial sediments has been a subject of controversy. Earlier authors thought that they were formed not of clay minerals but chiefly of the finest rock debris called "rock flour", and some plausible evidence has been found to substantiate this opinion. It has, however, been ascertained recently that in almost all glacial sediments the clayey fractions are in fact composed of clay minerals. In all the deposits studied, illite was dominant, chlorite and some kaolinite with montmorillonite being a subordinate admixture.

The content of heavy minerals in glacial sediments has been rarely studied in detail (e. g. R. F. SITLER 1963). In morainic sediments formed of local material, the content of heavy minerals equals roughly that of the underlying parent rocks. According to some data, their content ranges from a few to 35 per cent. The overwhelming part of them consists of unstable minerals of igneous rocks, such as biotite, decreases with the length of transport, because the increase in the content of stable minerals does not level out the decrease in the percentage of unstable minerals which far prevail in the parent rocks.

The content of carbonates has been studied through the entire profiles of some morainic sediments and preliminarily in many other glacial deposits. In the opinion of some authors, the study of $CaCO_3$ content is very important for the solution of the stratigraphy of glacial sediments. Carbonates of glacial sediments originate in two ways: either by the crushing of limestone pebbles and their disintegration into the finest fractions or by infiltration processes and secondary precipitation of carbonate in pores.

The content of carbonate (disregarding limestone pebbles near the outcrop) generally increases with the increase of fine silt and clay fractions. The $CaCO_3$ content in morainic sediments usually varies between 1 and 3% of carbonate, although far larger amounts have been found incidentally. On the basis of the $CaCO_3$ content, A. DREIMANIS (1960) differentiated moraines of mountain glaciers from those of continental ice sheets in North American glacial deposits. The former had on the average 21% of $CaCO_3$, whereas the latter only about 10%.

The shape of clastic particles of glacial sediments

The moving glacier wears the clastic particles in a similar way to the aqueous currents. Owing to a high internal friction, the glacial abrasion of pebbles is fairly intensive. The lower limit of its activity, however, is much higher than that of the water environment. The observations suggest that the glacier cannot round particles smaller than a few millimetres in size. Larger pebbles can be perfectly rounded by glacial activity, so that the change of the shape need not be ascribed to fluvioglacial processes. The greater part of pebbles, particularly of local origin, remains unrounded. The pebbles are frequently crushed and broken in result of the high internal friction inside the ice mass and thus a considerable pressure acting on them. These processes can be even stronger than the rounding of edges. The tendency for glacier-borne pebbles to develop a disk shape is characteristic. It seems that every pebble would assume this shape in the course of transport unless it changed its orientation within the ice mass or got broken. The particles flowing round the pebbles blunt the edges particularly at the opposite sides. All investigations made have shown that glacial sediments are on the average richer in pebbles of disk shape than river sediments. CH. K. WENTWORTH (1936) determined the following amounts of various pebbles in the moraines of North America (Table 76).

Table 76

Shape of pebbles in the moraines of North America

(after Ch. K. Wenthworth 1936)

Shape	%
Discoidal	71
Rod-shaped	22
Bladed	2
Irregular	1
Pyramidal and bipyramidal	3
Others	1

Although the prevalence of disk-shaped pebbles is very distinctive, it is not a safe evidence for the glacial origin of the deposits studied. Under favourable conditions, a great amount of disk-shaped pebbles may occur also in marine or fluviatile sediments.

The pebbles in glacial sediments are on the whole less rounded than pebbles in river and marine sediments, but in some parts of glacial, particularly glaciofluvial, sediments, they may get perfectly rounded. Computations have shown that about 10% of pebbles of morainic sediments attain a higher roundness than 0·4 (according to the Krumbein scale).

The striation of pebbles has long been considered as a diagnostic feature of glacial origin. Recent investigation, however, has revealed that it is not such an unmistakable sign; well-developed parallel or sub-parallel striation can be found only on little resistant pebbles (e. g. on limestone pebbles, where it proved to be ten times more frequent than on those of other rocks). Striated pebbles can be found also in mudflow deposits, as described by E. L. WINTERER - C. VON BORCH (1968). The striae were probably cut during a relatively dry phase of the flow, when most grains were in contact. The cutting tools were silt and fine sand grains, which can constitute the bulk of such deposits. Some authors record (C. D. HOLMES 1960) that in many morainic sediments they did not find a single pebble with typical glacial striation. According to other data, the investigated glacial deposits of North America show 28% of pebbles with surface structures, but only a small percentage of those which in fossil stage can be safely interpreted as evidence of glacial origin. The striation is perfectly parallel only in a small number of cases, being usually subparallel or forming irregular intersecting patterns. The pebbles bear abundant markings. CH. K. WENTWORTH (1936) examined 2035 pebbles of different morainic sediments of North America and found the following surface structures (Table 77).

Table 77

Surface textures of 2035 pebbles of different morainic sediments of North America
(after Ch. K. Wenthworth 1936)

Surface texture	Number of pebbles
Parallel striation	3
Subparallel striation	335
Irregular patterns	791
Short striae, irregularly distributed on the surface	123
Small pits	753

The particle orientation in glacial deposits has been studied by a number of authors. The survey of the literature shows that the particle orientation in moraines is affected by a great many factors. The dynamics are more difficult to explain than, for example, in running water. In spite of the many investigations, the fabric conditions cannot yet be regarded as having been definitely clarified. A general rule, however, seems to be that, when particles are immersed in the transporting medium, the internal shearing stress of the moving medium causes parallel alignment of particles, analogous to particle movements above the bed in running water or within a gravitating mass. As soon as they make a certain contact with a hard high-friction floor, the resulting orientation tends to be transverse to flow. Heavy collisions between particles seem to cause different divergencies from the general trajectories, usually producing secondary orientation maxima. The orientation of finer particles in the ice groundmass is very characteristic: some larger grains are often flowed round by fine grains and mica flakes. The orientation and imbrication of disk- and blade-shaped pebbles is usually irregular, although some authors mention their preferred orientation against the direction of glacier movement, similarly as in the beds of rivers. The imbrication, when it exists, is relatively best visible in basal moraines. The pebble orientation is secondarily affected by solifluctuation and fluviatile processes. In the opinion of some authors, the preferred orientation and imbrication were produced secondarily by fluviatile processes.

Structures of glacial deposits

The glacial deposits themselves do not show any marked structure. Because of the alternation of coarse and finer sediments, the structure is generally chaotic. Occasionally, the above-mentioned concretional structure is observable, which originated

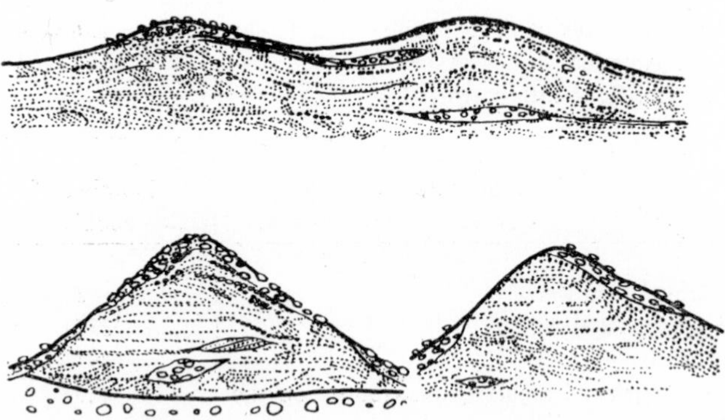

Fig. 52. Sedimentary structures of some glacial deposits. After M. Epstein (1957) from L. N. Botvinkina (1962).

by encircling of larger clasts by finer particles. As, however, the majority of glacial sediments was affected by running water, the bulk of these deposits has characteristic structures of coarse-grained fluviatile sediments. The structures indicate that eskers are also composed of glaciofluvial sediments (Fig. 52). The sediments of kames are frequently current-bedded, diagonal- and ripple-bedded or streaked. A characteristic feature of kames is a cap of coarse-grained unstratified material. Drumlins formed mainly of better sorted coarse gravels usually show diagonal bedding.

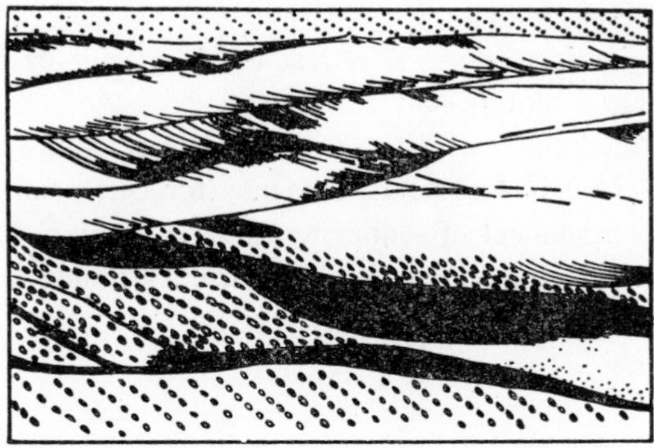

Fig. 53. Type of stratification of fluvioglacial deposits. After L. N. Botvinkina (1962).

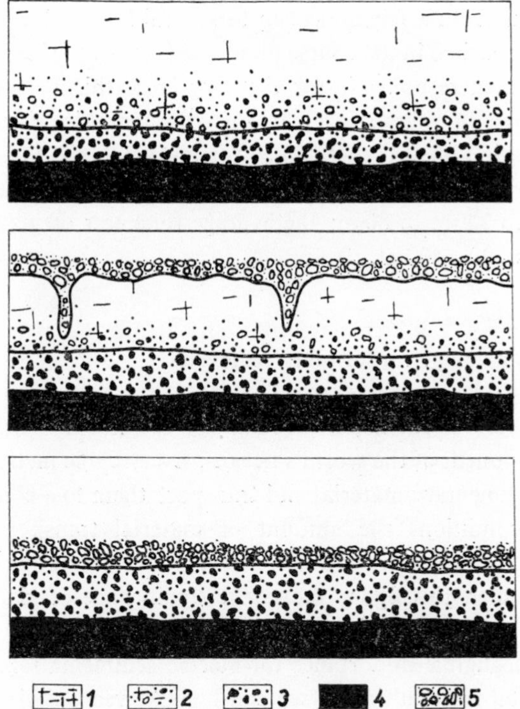

Fig. 54. Schematical illustration showing the origin of basal and ablation moraines. 1. Moving glacier, 2. Basal moraine in moving glacier, 3. Basal moraine, 4. Basement rocks, 5. Ablation moraine. After R. L. Flint (1947).

Glaciofluvial sediments

Glaciofluvial sediments can be divided into:

1. Sediments originating during the movement of glacier, inside and under the ice mass or at its sides.

2. Sediments originating after the retreat of the glacier by fluviatile reworking of the frontal and basal moraines.

The influence of fluviatile processes on the change of grain size composition has been given above. There is a constant tendency for better sorting by removal of finest fractions and enrichment of sediments by sand fraction. Moreover, rounding and orientation of coarse particles as well as current bedding are produced by the action of running water.

Thickness of glacial sediments

The thickness of glacial sediments are most varied, depending on the relief of the terrain, the thickness and character of the glacier, and the amount of material carried. A thickness ranging from a few dozen up to several hundred metres is recorded for glacial sediments deposited in the course of one advance of the ice sheet; the recurrence of ice advance is then responsible for the very large thickness of glacial deposits, indeed. The thickness of sediments left by the continental ice sheet is largely influenced by the modellation of the relief. Thus, for instance, in Lithuania their thickness is about 250 m on the elevations and 350 m in the depressions; in Karelia these differences make up 20 up to 80 m; the average thickness of glacial deposits on the Russian Plateau is 200—400 m.

Glacial material as a component of other sediments

At present, ice covers about 20,000,000 km^2 of the Earth's surface, which equals approximately one fourth of the world's oceans. Every cubic metre of ice can contain 100—300 kg of sedimentary material and transport them to a considerable distance. Under optimum conditions the amount of material transported and deposited annually by ice can thus be estimated at 3×10^{12} tons. Although the actual amount is substantially smaller, this figure shows the significance of glacial transport even nowadays.

River ice is of negligible importance for marine sedimentation. Sedimentation of ice-rafted continental material in the sea can be observed only in the case of north-bound rivers (e. g. the major streams of Siberia).

Sea ice plays a more important role in glacial sedimentation, even though it can entrap material only from near-shore, shallow areas. The depth involved is usually

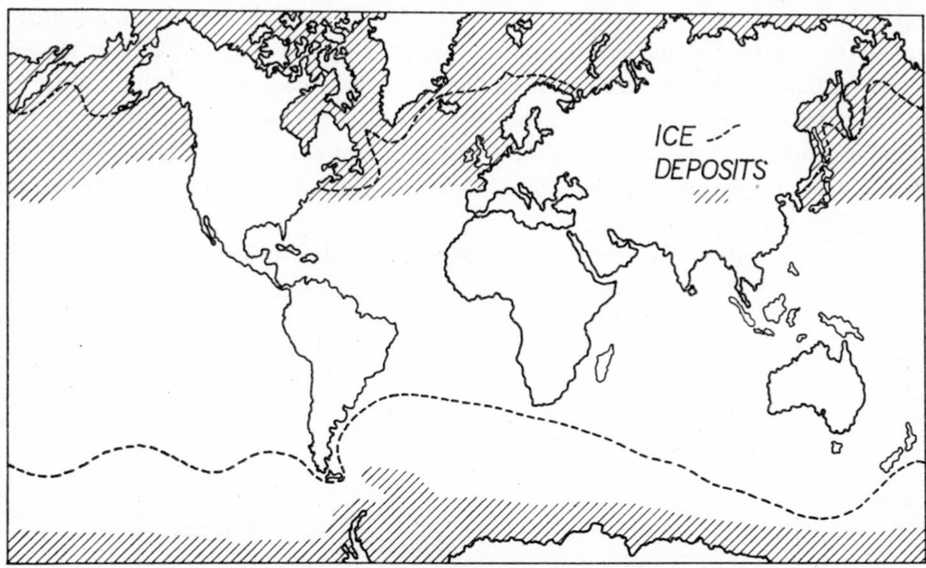

Fig. 55. Area of greatest importance for transporting coarse sediments by ice. Dashed line indicates equatorial limit of present-day floating ice; cross-hatching shows areas where bottom sediment may be classed as glacial marine origin. After K. O. Emery (in M. N. Hill 1963).

Table 78

Grain-size distribution of glaciomarine sediments (after A. P. Lisicyn 1961)

Fraction mm	Bering Sea %	Pacific Ocean %
50	4·72	1·19
50—25	9·02	8·67
25—10	28·10	35·71
10—5	22·12	18·65
5—2·5	16·09	6·34
2·5—1	15·00	3·26

up to 10 m, although material from a depth of 50 m is also known to occur in ice floes. Therefore, they include mainly coarse-grained beach and near-shore sediments. A significant component is the biogenic material, mainly tests of diatoms. In one cubic metre of sea ice up to 1,000 g diatoms have been found.

The most active carriers of clastic material are the icebergs, which are detached parts of glaciers capable of floating to immense distances. They have been observed not only in moderate but also in arid climatic zone. They enclose terrigenous detritus just as do the continental ice sheets; the material is dispersed throughout the ice mass, particularly in its basal part, and may attain a considerable concentration. In the environments where the ice-borne material is deposited, an increase in the gravel fraction is ascribed to it. In the Bering Sea and the north-western part of the Pacific Ocean, glacial material has shown the following grain-size distribution (Table 78).

The glacial material is generally well sorted (Trask coefficient So ranges from 1 to 2). Table 79 gives data on the sorting of glacial sediments in various seas.

Table 79

Sorting coefficients (Trask's So) of glacial sediments in various seas (after A. P. Lisicyn 1961)

So	Okhotsk Sea %	Caucasian coast of Black Sea %	Bering Sea %
1·0—1·5	75	64	21
1·5—2·0	25	18	44
2·0—2·5	0	9	28
2·5—3·0	0	0	3
3·0—4·0	5	9	4
>4	0	0	1

According to A. P. Lisicyn (1961) sediments of floating icebergs cover about 1,410,000 km^2 of the present day sea areas. They are divided into purely glacial sediments of icebergs (nearly 100% of glacial material) and sediments of other origin but containing disseminated glacial material. The latter type shows the following size distribution (after A. P. Lisicyn 1961):

Median	Fraction mm	%
	200—100	0
	100— 70	0
	70— 50	22
	50— 25	65
	25— 10	9
	10— 5	3
	5— 2,5	1
	2·5— 1	0
	<1	0

Coefficient of sorting (Trask's So)	Value	% of cases
	1·0—1·5	0
	1·5—2·0	10
	2·0—2·5	31
	2·5—3·0	18
	3·0—4·0	21
	>4	20

The striation of pebbles of decidedly glacial origin has been found roughly in one in a thousand pebbles.

References

BOTVINKINA L. N. (1962): Op. cit. on page 106.

CROWELL J. C. (1964): Climatic significance of sedimentary deposits containing dispersed megaclasts. In "Problems in Palaeoclimatology". P. 81—85, London.

DAVIS S. N. (1958): Size distribution of rock types in stream gravel and glacial till. Jour. Sedimentary Petrology, vol. 28: 87—94, Menasha.

DREIMANIS A. (1960): Morphology of the glacial pebbles. Jour. Sedimentary Petrology, vol. 30: 332—338, Menasha.

DREIMANIS A. - REAVELY G. H. (1953): Differentiation of the lower and the upper till along the north shore of Lake Erie. Jour. Sedimentary Petrology, vol. 23: 238—259, Menasha.

DROSTE J. B. - WHITE G. W. - VATTER A. E. (1958): Electron micrography of till matrix. Jour. Sedimentary Petrology, vol. 28: 254—350, Menasha.

FLINT R. L. (1947): Glacial geology and Pleistocene epoch. P. 1—571, New York.

GILLBERG M. (1965): A statistical study of till from Sweden. Geol. Fören, Förhdl., vol. 87: 84—108, Stockholm.

GLEN J. W. - DONNER J. J. - WEST R. G. (1957): On the mechanism by which stones in till become oriented. Am. Jour. Sci., vol. 225: 194—205, New Haven.

GRIPP E. (1953): Tracing of glacial boulders as an aid to ore prospecting. Econ. Geology, vol. 48: 714—725, Lancaster.

HARRISON P. W. (1957): A clay-till fabric: its character and origin. Jour. Geology, vol. 65: 275 - 307, Chicago.

HOLMES C. C. (1960): Evolution of till-stone shapes, Central New York. Bull. Geol. Soc. Am., vol. 71: 1645—1660, Baltimore.

JOHANSSON C. E. (1965): Structural studies of sedimentary deposits. Geol. Fören Förhdl., vol. 87: 3—61, Stockholm.

KARCZEWSKI A. (1963): Morphology, structure and texture of ground moraine area of West Poland (in Polish with English res.). Poznanskie Tow. Przyjaciój Nauk, T. 4: 1—111.

KELLER W. D. - REESMAN A. L. (1963): Glacial milks and their laboratory-simulated counterparts. Bull. Geol. Soc. Am., vol. 74: 61—76, Baltimore.

KRINSLEY D. H. (1969): Recognition of Pre-Pleistocene glacial environments. Bull. Am. Assoc. Petrol. Geol., vol. 53: 727, Tulsa.

LANDIM P. B. - FRAKES L. A. (1968): Distinction between tills and other diamictons. Jour. Sedimentary Petrology, vol. 38: 1213—1223, Menasha.

LISICYN A. P. (1961): Glacial transport and distribution of coarse grained material (in Russian). In "Recent sediments of seas and oceans". P. 232—284, Moscow.

— (1963): Bottom sediments of the shelf of Antarctica. (in Russian with English res.). In "Deltaic and shallow marine sediments". P. 82—88, Moscow.

OKKO V. (1944): Moränenuntersuchungen im westlichen Nordfinland. Bul. Comm. Geol. Finland, No 131: 1—112, Helsingfors.

RUKHINA E. V. (1960): Lithology of moraine deposits (in Russian). P. 1—141, Leningrad.

— (1962): Classification of moraine deposits on the basis of lithological signs (in Russian). Voprosy litologii i paleogeografii. Učenyje zapiski, No 310, vyp. 12: 139—146, Moscow.

SCHWARZBACH M. (1964): Criteria for the recognition of ancient glaciation. In "Problems in Palaeoclimatology". P. 81—85, London.

SHEPPS V. C. (1953): Correlation of the tills of Northeastern Ohio by size analysis. Jour. Sedimentary Petrology, vol. 23: 34—48, Menasha.

— (1958): Size factor, a means of analysis of data from textural studies of till. Jour. Sedimentary Petrology, vol. 28: 482—485, Menasha.

SITLER R. F. (1963): Petrography of till from Northeastern Ohio and Northwestern Pennsylvania Jour. Sedimentary Petrology, vol. 33: 365—379, Menasha.

— (1968): Glacial till in oriented thin sections. Int. Geol. Congr., 23. Sess., vol. 8: 283—295, Prague.

STANLEY D. J. (1968): Reworking of glacial sediments in the Northwest arm, a fjord-like inlet on the southeast coast of Nova Scotia. Jour. Sedimentary Petrology, vol. 38; 1224—1241, Menasha.

WENTWORTH CH. K. (1936): An analysis of the shape of glacial cobbles. Jour. Sedimentary Petrology, vol. 6: 85—96, Menasha.

WINTERER E. L. - BORCH C. von (1968): Striated pebbles in a mudflow deposit, South Australia. Palaeog., Palaeoclim., Palaeoecol., vol. 5: 205—211, Amsterdam.

YAKOVLEVA S. V. (1956): On the glacial pebbles on Russian Plateau (in Russian). Materialy po četv. geologii i geomorfologii, p. 18—43, Moscow.

ZINGG T. (1935): Beitrag zur Schotteranalyse und ihre Anwendung auf die Glazialschotter. Schweiz. Min. Petr. Mitt., Bd. 15: 39—140, Basel.

12. *Lacustrine sediments*

It is somewhat difficult to include lacustrine sediments into one group, because the sediments of large lakes are petrographically almost indiscernible from marine ones and quite different from the deposits of small lakes originated under different hydrological and biological conditions. Therefore, the composition of Recent lacustrine sediments shows an extraordinary variety.

General information on the origin and kinds of lakes is given in all current textbooks of physical geography and geology especially in the special book by C. C. REEVES (1968) dealing also with lacustrine sediments. The classification of lakes can be based on many criteria, such as:

1. Climatic zones where the lakes are situated.
2. Origin.
3. Hydrological and hydrochemical conditions, or the thermal regime of their waters.
4. Sediments.
5. Drainage (with through drainage, drainless, etc.).

As the climate affects to a large extent other factors also, the first type of classification is mostly used. Lakes occuring in the same climatic zone frequently show also the same hydrodynamical, hydrochemical and hydrobiological conditions and thus have also sediments of similar character. The climatic and geographical classification of lakes should evidently take into account also the elevation above the sealevel. On this climatic basis lakes are grouped into several zones:

1. Zone of subtropical and tropical fresh-water lakes.
2. Zone of salt and brackish lakes to the north and south of the former zone.
3. Zone of mountainous lakes.

In addition, they are the so-called extrazonal lakes, whose character is influenced by particular geological and hydrological conditions.

In discussing the hydrochemical properties of lake waters, the following scale is used for the classification of lakes of different salinity:

	Per mille salinity
Fresh-water lakes	0·3—1·0
Brackish lakes	1·0—24·7
Salt lakes	above 24·7

According to the dissolved solids, lakes can be divided into carbonate lakes with dominant HCO_3' and CO_3'' anions, sulfate lakes with predominant SO_4'' anions, and chloride lakes with Cl′ anions. The majority of lakes, just as river water, belong to the carbonate type. With the increasing concentration of salts, as, for instance, by water evaporation, they alter successively to the sulfate and chloride types. There is a general rule that lakes with weakly mineralized water belong to the carbonate type, while those with strongly mineralized water are sulfate and chloride lakes. The salinity of lake water changes highly even with a moderate oscillation of the water level, as in large lakes a small drop of water level can result in a large decrease of the water volume a thus an increase of salt concentration. Table 80 shows the differences of salinity of the Great Salt Lake in the course of forty-four years.

Table 80

Differences of salinity of the Great Salt Lake in the course of 44 years (after A. J. Eardley 1938)

Ions	Sea water	Amount of salts in water of Great Salt Lake in the years %							
		1869	1877	1879	1889	1892	1904	1907	1913
Cl′	55·29	55·09	56·21	55·57	56·54	55·69	55·63	55·11	55·48
SO_4''	7·69	6·57	6·89	6·86	5·97	6·52	6·73	6·66	6·68
Na·	30·59	33·15	33·45	33·17	33·39	32·92	34·65	32·97	33·17
K·	1·11	1·69		1·59	1·08	1·70	2·64	3·13	1·66
Ca··	2·00	0·17	0·20	0·21	0·42	1·05	0·16	0·17	0·16
Mg··	3·73	2·52	3·18	2·60	2·60	2·10	0·57	1·96	2·76
Salinity	3·50	15·00	13·80	15·70	19·60	23·00	27·70	23·00	23·30

As a considerable change in salinity and, consequently, in sedimentation occurs within a relatively short interval, the lacustrine sediments alter both vertically and laterally far more quickly than marine sediments.

Salinity also changes within one lake, depending on the supply of fresh river water. Thus, for instance, salinity of the western part of Lake Balkhash, at the mouth of river Ili, is 1,260 mg/l, whereas in the eastern part the content of dissolved solids is as much as 5,200 mg/l.

The thermal condition of lake water is a very important factor in the development of organisms and sedimentation. In most lakes, three vertical thermal zones exist almost continually. The upper warm zone is called epilimnion, the middle transitional zone — the metalimnion, and the lower cool zone – the hypolimnion. The photic epilimnion is sufficiently supplied with oxygen, so that it is rich in organisms. The metalimnion represents a boundary zone for the vertical movement of plankton. In the hypolimnion, the stagnation sometimes causes a great lack of free oxygen. The thermal zonation of the majority of lakes changes in annual cycles. In autumn, a cooling of the upper water layer gives rise to vertical zoning (homothermy). In winter, lake waters show a reverse thermal zoning, the upper layers being cooler that the lower ones. The thermal zonation of lake waters differ according to the climatic zones.

1. Tropical lakes with mean temperature above 4 °C. They show a thermal zoning, small differences between the upper- and lower-water layers and lack the metalimnion (i. e. transitional zone).

2. Lakes of moderate zone with normal thermal zoning in summer and reverse zoning in winter; in spring and autumn homothermal conditions exist.

3. Polar lakes with a mean temperature below 4 °C throughout the year; except for a short homothermy period in spring, a reverse stratification is developed.

The character of lacustrine sediments, especially in small lakes, is greatly affected by the influx of fresh river water. It has been found, for instance, that three-quarters of the sediment in the Salton Lake in California is formed of suspended material from the Colorado River. On the other hand, lakes without river mouths are silted up chiefly by organic or chemical sediments.

The change of the water level in lakes and the resulting changes of salinity always affect sedimentation, both indirectly (through the changes of physico-chemical properties of water) and directly (e. g. vertical oscillation of boundary between fine-grained biochemical calcareous sediments and coarse organogenic deposits). The direct changes are observable, for instance in the Caspian Sea, the indirect effects in the Great Salt Lake, where clay deposition changes into the carbonate one as a result of the increased concentration of salts.

On the basis of the integrate character, i. e. the hydro-chemical, hydrobiological, sedimentological and genetic conditions, lakes are divided into three large groups:

1. Oligotrophic, 2. Euthropic, 3. Dystrophic

These three types of lakes differ essentially in the content of oxygen and nutrients in water and in the amount of organic production (Table 81).

Table 81

Lakes	Quantity of	
	oxygen	nutrients
Oligotrophic	great	small
Eutrophic	small	great
Dystrophic	small	small

Eutrophic lakes have a high production of plankton and their waters are rich in organic matter, which stains them brown and makes them less translucent than oligotrophic lakes. The production of plankton in the latter is lower. In addition to these purely hydrological factors, the regime of the two types of lakes is influenced by the bedrock in the bottom and the vicinity of the lake. Eutrophic lakes develop, namely, particularly on porous sediments rich in organic matter or on soils, which are easily leached and enrich the lake water with organic matter. The typical oligothropic lakes, however, originate on solid igneous or metamorphic rocks, on moraines etc.

Other properties of the three main types of lakes are sumed up below:

O — Oligothropic. E — Euthropic, D — Dystrophic lakes.

Distribution:

O — Mountains (Alps), North Germany, North American lowland.
E — Lowlands in North Germany, Baltic coastal pains, in surroundings of Alps.
D — Skandinavian lowlands, in some Central European highlands, in northern parts of Russian Plateau in association with marshes.

Morphology:

O — Average depth over 18 m. Thickness of hypolimnion is greater than of epilimnion layer.
E — Mean depth below 18 m. Flat relief. Thickness of epilimnion is greater than of hypolimnion.
D — Usually small lakes, mostly very shallow, passing into marshes. Minimal extent of drainage area.

Water colour:

O — Blue to green.
E — Yellow and yellow-green, eventually yellow-brown.
D — Brown.

Transparency of waters:

O — Over 10 m.
E — Only several metres.
D — Up to 2—3 m.

Chemical composition of water:

O — Poor in nitrates and phosphates, poor or rich in carbonates.
E — Rich in phosphates and nitrates, mostly rich in carbonates, rich in electrolytes but poor in humus.
D — Poor in carbonates and electrolytes, but rich in humus.

Suspension in water:

O — Mostly clayey suspension, but only small amounts.
E — Abundant seston and tripton in suspension.
D — Suspension rich in humus colloids.

Sediments:

O — Mainly mineral deposits, without sapropel. Occasionally gyttja, in form of littoral occurence of algal gyttja, partly clayey gyttja.
E — Mainly gyttja, in greatest depths also sapropel, lake marls and other calcareous gyttjas.
D — Dy passing into littoral sediments, transitional members between dy and gyttja, diatomaceous deposits and lake iron ores.

Oxygen:

O — Waters in hypolimnion saturated with O_2 up to 70%. Strong oxidation.
E — Waters in hypolimnion saturated with O_2 up to 40%. Weak reduction.
D — Oxygen most frequently absent. Strong reduction.

Littoral vegetation:

O — Not abundant, only local in some coastal parts.
E — Very abundant, mainly in epilimnion.
D — Poor.

Benthos:

O — Very abundant, but poor in species. In small lakes great amount of pelecypods. In Alpine lakes, however, pelecypods are scarce.
E — Poor in species, pelecypods very abundant.
D — Poor in benthos, pelecypods absent.

Trends of development:

O — Through eutrophic stage into dystrophic lake, eventually direct to dystrophic one.
E — Change into dystrophic lakes.
D — Change into peat bogs.

Lacustrine sediments

As mentioned above, lacustrine sediments are of great variety, their genesis and composition depending on the conditions referred to above. It can be roughly said that all known Recent sediments develop in lakes; they can be divided into:

1. Mechanical or clastic sediments.
2. Chemical sediments (carbonates, salts).
3. Biochemical sediments comprising deposits formed by the physiological activity of organisms.
4. Organic sediments, including sediments originated from minerogenic parts of organisms and those built up of unstable parts of organic bodies.

Clastic lacustrine sediments

Sedimentary conditions in large lakes are roughly the same as in seas; therefore we range some large continental basins, such as the Caspian or Aral Sea, to the seas and their sediments are dealt with in the chapter on continental seas.

The grain-size distribution of clastic lacustrine sediments are controlled by the size and hydrodynamical conditions of the lake, topographical differences, i. e. the relief of the area and the depth of the lake, the gradient of the lake bottom, the supply of material by rivers. Wind action is a subordinate factor. The character and petrographical composition of the surrounding country is of great importance. Thus, for instance, in lakes whose bottom and banks are formed of glacial moraines coarse-grained sediments are laid down. In the littoral and deeper zones no Recent sediments occur, because all fine particles are washed to a greater depth. The bulk of coarse-grained sediments consists of redeposited morainic material from the bottom and closest vicinity. This is a specific feature of Recent lakes, because through the greater part of geological history such a quantity of coarse-grained glacial sediments could not be available. Therefore, it seems likely that fossil lacustrine sediments will generally be finer grained than the Recent ones.

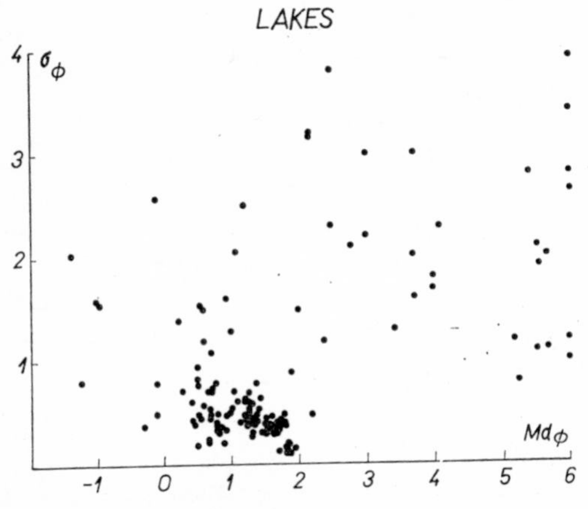

Fig. 56a. Grain-size parameters of lacustrine sediments. Relation of Phi Median Diameter and Phi Deviation Measure. Data compiled from various authors.

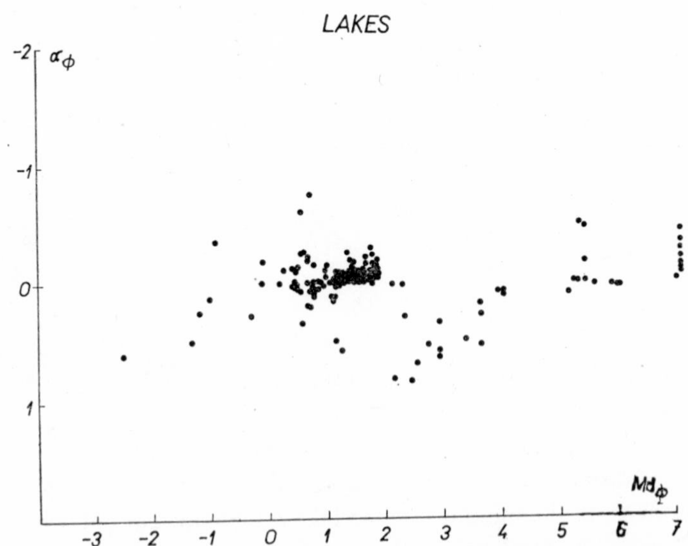

Fig. 56b. Grain-size parameters of lacustrine sediments. Relation of Phi Median Diameter and Phi Skewness Measure.

Large lakes furnish a good example of the interdependence between the grain-size of sediments and the relief of the lake bottom. A considerable coarsening of detritus is brought about even by small elevations. In Lake Michigan, for instance, small topographical differences induce a change of the Md from 0·05 mm in depressions up to 5 mm on elevations. In small lakes, the grain-size depends rather on the distance from the shore, than by the unevennesses of the bottom in the central part of the lake.

Gravels usually occur on beaches extending to a small depth (with the exception of glacial lakes). Coarse gravels can extent to greater depth when they are brought by mountain rivers (e. g. in Lake Issyk-kul down to a depth of 34 m). If gravels are produced from the abrasion of coast, they are confined to shallow zones only. The differences between gravel deposits of stream and abrasion origin are well seen from Table 82.

Table 82

Differences between gravel deposits of stream and abrasion origin

	Gravel deposits with abrasion material	Gravel deposits with river-borne material
Thickness of gravel deposits	very small, up to several metres	may be great, in some cases more than 100 m
Shape of gravel deposits	blanket	fan
Grain size	Md usually between 3 mm and 1 cm, according to hydrodynamic conditions	values of Md more variable
Sorting	usually well sorted, So between 1·2 and 2·0	moderately to badly sorted material, So most frequently between 1·5 and 3·0
Petrographic composition	composition of pebbles corresponds to the composition of adjacent coast; most frequently uniform composition	composition more varied, pebbles do not correspond to the composition of adjacent coast
Pebble orientation	imbrication of flat pebbles towards the lake	preferred orientation occurs rarely

Sands are the coarsest sediments of many lakes. Their presence is so characteristic that the difference between lakes with and without sandy sediments is quite marked. The depth of sand deposition varies with the topography and slope of the bottom and the presence of inflowing streams. In the deeper lakes of mountainous and piedmont areas sands reach to a depth of several tens of metres (in Lake Issyk-kul to a depth of 100 m). On the other hand, shallow lakes in lowlands with a gentle slope of the bottom have sand deposits to a maximal depth of a few metres. It has been observed that the bathymetric distribution of coarse sediments in some lakes is affected by the predominant direction of winds. On the windward side of lakes sands extend deeper than cn the leeward shore. This difference has been found to be a few metres in North German lakes.

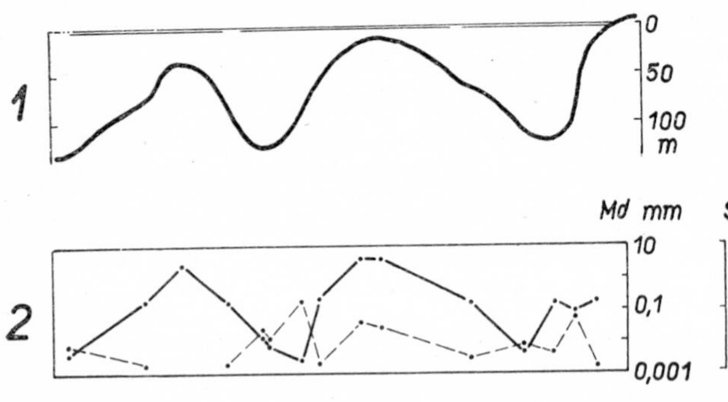

Fig. 57. Relation of bathymetry and grain-size distribution of sediments of Lake Michigan. Median diameter is represented by heavy line. After J. E. Moore (1961).

Lacustrine beach sands can be well sorted and their clastic grains perfectly rounded. The Md of grain size is on the average somewhat lower than those of marine beach sands. Towards the central part of the basin, sands gradually become finer and grade into fine grained sandy sediments with the Md between 0·05 and 0·1 mm. The coefficient of sorting, being dependent largely on the grain size of sands is not uniform. Best sorted sands have the Md about 0·2 mm; the degree of sorting decreases towards the coarser and finer fractions. The coefficient of skewness of lacustrine sands is mostly slightly negative on account of the small percentage of coarser fractions always present in the sediments. Diagrams on Fig. 57 show grain-size analyses of sediments of the Lake Michigan. From the correlation of the above analyses with mechanical analyses of marine sediments interference can be made that there is no substantial difference between the grain size composition of sediments in large lakes and in shallow zones of the seas.

Clay sediments are the most abundant lacustrine deposits. They can grade into clayey-silty sediments or into a large and varied group of chemical and organic deposits. Clays occur also in the ephemeral lakes — playas. Black mud with silt-clay ratio of 3/7 was described from Wilcox Playa, Arizona (B. W. Pipkin 1967).

The grain-size of sediments in lakes with major flows or through-drainage corresponds to the grain-size of the suspended load. Lake Salton in California, through which flows the Colorado River, can serve as a suitable illustration. The major part of the bottom is covered with clay-silty sand of Md = about 0·09 mm, which corresponds to the grain size of the suspended load of the river.

Chemical sediments

As lacustrine chemical sediments are closely connected with organic sediments, it is necessary to deal with them jointly. We do not know so far a more reliable criterion which would enable us to distinguish, for instance, some carbonate sediments of purely chemical origin from carbonates of biochemical origin. This is also true for salts, iron ores, phosphates and other less common components of lacustrine deposits.

A typical chemical carbonate deposition occurs chiefly in tropical salt lakes, where the evaporation of water produces an increase in the concentration of salts and their progressive precipitation. Two best known examples of such carbonate sedimentation are afforded by the Great Salt Lake and the lakes of south-eastern and eastern Australia. In the Great Salt Lake carbonate sediments are represented mainly by aragonite and calcite, Mg calcite, huntite, dolomite and probably also magnesite and vaterite. In Australian lakes, the precipitation of carbonate has been observed directly, because their water is continuously turbid by the uninterrupted precipitation of minute $CaCO_3$ crystals. Magnesian calcites and calcian dolomites have been observed in active simultaneous precipitation from shallow saline

Table 83

Composition of clayey and carbonate sediments of Great Salt Lake (after A. J. Eardley 1938)

Sediment	Percentage
River-borne clays and silts	25
Wind-borne clays and silts	32
Organic matter	1·5
Aragonite	15
Dolomite	5
Parasepiolite	7
Montmorillonite and colloid particles	8

waters with the same ionic proportions as sea water. During precipitation a pH range of 8·9–9·2 and a salinity range of 1·57–14·14% were recorded in two shallow isolated lakes. The composition of the fine-grained calcite ranged from $Ca_{0.77}Mg_{0.23}(CO_3)$ to $Ca_{0.98}Mg_{0.02}(CO_3)$. The dolomite composition ranged from apparently stechiometric dolomite to protodolomite containing 6 mole percent $CaCO_3$ in excess of the stechiometric formula. Dolomite was also described from pluvial basins in West Texas, in Salt Flat Graben, Texas and in Deep Spring Lake, California (C. C. Reeves 1968). Protodolomite has weak or absent ordering *x*-ray diffraction peaks, excess of Ca ions, and diffuse *x*-ray diffractions from planes perpendicular to the *c*-axis. It is apparently formed before well-ordered stechiometric dolomite. Such ordered stechiometric dolomite was found in pluvial lakes in West Texas, being associated with volcanic ash falls (C. C. REEVES 1968). In some other lakes dolomite horizons can represent past periods of lakes desiccation, as in the Lake Mead (C. C. REEVES 1968). Small amounts of celestite precipitate concomitantly with the carbonates. Active chemical sedimentation takes place in water whose salinity is half that of normal marine water. The periodicity and maximum intensity of chemical sedimentation in periods of most vigorous growth of vegetation corroborates the fact that this kind of sedimentation is supported by biochemical precipitation.

The composition of clay-carbonate sediments of the Great Salt Lake is given in Table 83. Clay of the deeper parts of this lake contains 26% of aragonite and 4% of dolomite, whereas the nearshore clay in the western part of the lake has 13% of aragonite and 6% of dolomite.

Calcareous ooids in lacustrine sediments are probably of purely chemical origin, although biochemical processes can contribute to their formation. A classic example of lacustrine oolite occurrence is provided by oolitic sediments of the Great Salt Lake, where they are deposited on the flat windward coast to a depth of four metres. The best developed ooids of ideal ovoid shape occur in places of strongest surf. Under quiet water conditions, the ovoids have a more irregular from and an imperfect concentric structure. The ooids of the Great Salt Lake have this chemical and mineral composition:

CaO	46·6%	CO_2	38·9%
MgO	1·8%	Clay	5·6%
	Nucleus	7·0%	

Carbonates are represented by 85% aragonite, 10% calcite and 5% dolomite. The core of an ooid generally consists of a quartz grain of silt size, a fragment of test or faecal pellet. Small crystals of carbonate precipitated round the core, entrapping and enclosing clay particles, so that ooids are never free of insoluble residue. Most authors agree that ooids were formed by physico-chemical processes, mainly by evaporation of water in near-shore shoals. Other opinions, however, also appeared in literature, claiming, for instance, that ooids are produced by calcareous algae. Oolitic calcareous sediments have also been found in the Australian lakes and large continental basins, such as the Caspian and Aral Lakes.

All lakes of the arid zone, both Recent and ancient, underwent a characteristic development consisting of several stages.

1. Calcite-dolomite stage; the concentration of solutions is still comparatively low and evaporites do not precipitate; calcite and dolomite precipitate.

2. Dolomite stage; concentration increases. Gaylussite — $Na_2Ca(CO_3)_2 . 5 H_2O$ precipitates during summer and its crystals dissolve again in winter. Calcite and dolomite precipitate.

3. Stage; gaylussite in association with trona is stable in sediments.

Recent gypsum deposition is abundant in some lakes, but it is not comparable to ancient gypsum lacustrine deposition. Lake Eyre, Australia, is estimated to contain at least 4,000 milion tons of gypsum (C. C. Reeves 1968). Whether gypsum will be precipitated in lakes depends on the concentration of sulfate, the concentration of calcium and also on the presence of hydrogen sulfide.

Biochemical sediments

Biochemical sediments occur in abundance in lakes. Recently, biochemical origin is also ascribed to some sediments which were formerly regarded as a purely chemogenic product. Biochemical deposits are divided into:

1. Primarily cohesive sediments.
2. Primarily non-cohesive sediments.

The former group includes chiefly abundant crusts produced by algal activity. In many lakes, especially on rocky shores, algal crusts reach up to a depth of six metres. They occur in the photic, well-aerated region with fairly warm water. The occurrences of crusts on a soft clayey bottom are, on the whole, exceptional. Algae are important contributors to the formation of lithoid tufa in the Mono Lake (California). Calcium carbonate as calcite, aragonite and high-magnesium calcite has been deposited here in the form of pinnacled masses of tufa. Crusts frequently occur on pieces of wood or other foreign material. They usually occur to a depth of a few metres, but sporadically reach even some dozen metres deep (e. g. 60 m in the Constance Lake). The thickness of algal crusts is generally of the order of millimetres, less frequently attaining a few centimetres. The algae *Schizotrix fasciculata* and *S. lateritia* are the main crust-builders. The composition is generally monotonous: 50—80% of carbonate and up to 25% of organic matter. Clay entrapped amidst the algal filaments makes up a few per cent. The algal crusts in the Constance Lake show a depth zonation: they are built up of *Schizotrix* to a depth of 10 m, of the genus *Aegropropila* at a depth of 25—35 m. The amount of clay material increases with depth. In the upper zones they contain 1—3% insoluble residue which below the 30 m limit rises to 22 per cent. The algal crusts are found both in fresh-water and salt, even hypersaline lake waters (the Kara-Bougas-Gol Bay or the lakes of the Australian steppes).

Algal balls, freely deposited or attached to the bottom, have been recorded from many lakes. They are of cm- to dm-size, strongly porous. Some of them have a distinct core usually formed of a pelecypod test. In the cross-section, the nodules show a radiate arrangement of algal filaments. Recent and Subrecent algal balls of North American lakes have the following chemical composition (Table 84).

As the algal balls are confined to moving water, they occur mainly in vegetation-free well-aerated lakes. They very often disintegrate into algal sands by the activity of boring algae and mechanical effects of currents.

Flat varieties of these nodules (1—2 cm × 10—20 cm in size) occur in North

Table 84

Composition of Recent and Subrecent algal balls from North American lakes

	Recent algal balls %	Subrecent algal balls %
Organic matter	10–15	1–12
H_2O	1	1
SiO_2	12	12
$CaCO_3$	60–70	70–80
Fe	1	2
Al	traces	traces
Mg	traces – 1	traces – 1

American lakes and are known as "algal biscuits". It is of interest that they are found in periodically evanescent fresh-water lakes. Their perfect concentric structure is explained as due to the recurrent withering and growth of algal generations owing to dessication and flooding by water. Some writers ascribe their growth partially to the activity of bacteria. The genera *Gleocapsa*, *Gleotheca*, *Microcystis*, *Aphanocapsa*, *Oscillatoria*, *Rivularia*, *Nostoc*, *Chroococcus* are the most frequent components. Diatomaceae also occur occasionally in abundance. It should be mentioned that algal nodules are also formed in rivers and streamlets in the same way.

Biochemical loose (non-cohesive) sediments are far more widespread, particularly those of a calcareous composition. Unfortunately, there is so far no uniform petrographical and genetic nomenclature for their designation. In German literature the terms Seekreide and Alm have been introduced. "Seekreide" (lacustrine chalk—lake marl) is usually used for all pure calcareous loose lacustrine sediments; when they are covered by humus and vegetation, they are in German called "Wiesenkreide". The term "Alm" was meant originally for calcium carbonate deposits located in the proximity of springs opening into lakes or swamps. Recently it has been extended to cover fine-grained carbonate sediments rich in humic matter. The transition between lacustrine chalk and alm is gradual and the boundary subjective.

The composition of chalks is variable. In places they are extremely porous, in other they pass into strongly argillaceous varieties or, sporadically, into gypsum lake marls. Several analyses of lake marls are given below (Table 85).

Shells of pelecypods are a constant admixture of lacustrine chalks, attaining more than 1% of the total volume. The variable content of organic matter colours the sediment brownish or greyish.

The bathymetric range of lacustrine calcareous sediments is not the same everywhere. They are developed from minimum depth up to 5—6 m under the water level, where they pass into clayey sediments. In some cases, they are the only deposits present in the lake. In large lakes they generally form a broad strip between coarse

Table 85

Chemical composition of lake marl

	Lake Michigan (average composition)	Lake Schönau (North Germany)		
		depth 3 m	depth 7 m	depth 10 m
SiO_2	2·80	8·98	1·40	5·73
Al_2O_3	2·58	traces	traces	0·57
Fe_2O_3		0·65	1·06	2·85
CaO	49·43	42·34	46·53	40·26
MgO	0·15	0·20	0·10	0·38
K_2O		0·48	0·42	0·53
Na_2O		1·00	1·11	1·14
SO_3	0·78	0·69	0·86	1·37
P_2O_5		0·25	0·31	0·20
CO_2	38·98	32·53	35·86	31·31
N		0·69	0·66	0·78

littoral sediments and clayey sediments in the centre, whereas in certain small lakes they are spread over the whole bottom.

Lacustrine chalk is a typical polygenetic sediment: it can develop in different ways under different conditions. However, it is not yet known precisely which form of genesis is most important and most widerspread. The factors controlling the genesis of calcareous lacustrine sediments can be ranged most likely in this order (from the most to the least important):

1. The activity of algae and water mosses.
2. Inorganic factors, i. e. escape of carbon dioxide, decrease in solubility of calcium carbonate, increase in temperature of water, influence of springs issuing in the bottom of lakes, etc.
3. Mechanical disturbance of the shells of pelecypods and other organic remains.

With regard to the safely proved great influence of plant photosynthesis, the first mode of origin seems to be the most frequent. The formation of fine calcium carbonate by the activity of plants has been imitated also experimentally. The total influence of the inorganic factor is questionable thus far; it can obviously change from place to place. The disintegration of organic remains into minute carbonate particles has frequently been observed both in nature and laboratory. One fact, however, does not allow to place this process among the main factors controlling the formation of the lacustrine chalk. In every deposit of lacustrine chalk a number of absolutely intact tests of pelecypods are found and it is not easy to explain why some tests are completely disarticulated and others remain undisturbed. Nevertheless, this mode of lake marl origin must not be underestimated.

Lake ores (iron-rich sediments)

The petrography and mineralogy of lake iron ores have so far been scarcely studied. Petrographically, they are divided into:

1. Iron sands and iron crusts.
2. Iron clays comprising ochres, brown clay and black sulphide clays.
3. Iron ochres without clay admixture, i. e. decay forms of Fe-hydroxydes.
4. Concretions of Fe-hydroxydes of various forms and size.

Sometimes only the third and fourth types of sediments are meant by the name lake iron ores.

Lakes containing iron ores occur at different places of the Earth's surface. They can be grouped into a few geographical and climatic entities:

1. Lakes in the area of crystalline complexes in the northern part of the moderate zone (Scandinavia).
2. Lakes in the Central European plain.
3. Lakes of the calcareous Alps and their piedmont.
4. Lakes and lagoons of the humid tropical zone, especially Brazil.

Table 86

Properties of waters of a typical iron-ore lake (Lake Pinnus-Järvi in Karelia)

Depth m	Temperature °C	pH	HCO_3' mg/1	O_2 mg/1	Fe tot. mg/1
0	19·96	5·9	12·4	6·60	6·00
1	19·87	6·0	12·9	6·52	4·32
2	18·43	6·0	11·8	6·46	4·60
3	12·30	5·5	18·3	2·10	6·94
4	8·46	6·24	57·1	0	40·80
5	7·05	6·5	129·3	0	136·60

Some iron-ore lakes in Scandinavia and in the north-western parts of the Russian Plateau have been studied most thoroughly. Iron ores develop particularly in the strongly humified lakes without any major organic production. Table 86 lists the characteristics properties of waters of a typical iron-ore lake (Lake Pinnus-Järvi in Karelia).

The absence of free oxygen already at a depth of 4 m and the strong increase in Fe concentration with depth are distinctive. In addition, a weakly acid reaction of waters caused by the presence of CO_2 and humic acids can be observed.

The characteristic occurrences of lake ore intervene between the littoral coarse sediments and deeper zones with clay sedimentation. They develop most frequently on the flat bottoms at a depth of 1—5 m. At first, an interrelationship between the composition of lake banks and the presence of lake ores was presumed, but later investigations have proved that ores can exist in lakes of various types of banks. An intensive deposition of oxidation iron compounds, particularly Fe-hydroxydes is also observable at the issue of springs on the bottom of lakes or near the mouths of streamlets.

Lake iron sediments are usually enriched in manganese. Unfortunately, the forms of its occurrence and the relation to other lacustrine sediments have not yet been studied. In some lacustrine deposits the enrichment in nickel is supposed to be of organic origin.

Organic lacustrine sediments formed of minerogenical particles of organisms

This group of sediments comprises very abundant sediments consisting of the tests of pelecypods. Their distribution in lakes is governed by the properties of waters, the character of the lake bottom, depth movement of water, light and amount of nutrients. The influence of the substratum is considerable but cannot be treated here

Table 87

Bathymetric distribution of various species of pelecypods in Lake Mendota (after J. Pia 1933)

Species	Depth (m)						
	0—1	1—2	2—3	3—5	5—7	>7	optimum depth
Amnicola limosa	3	5	5	5	5	4	1— 4
Physa heterostropha	2	3	1				1— 2
Physa ancillaris	1	2	3				2— 3
Valvata tricarinata	1	2	4	2	2	1	2— 3
Anodonta	2	2	3	3	1	1	2— 3
Sphaerium occidentale	2	4	4	4	4	2	2— 3
Ancylus	1	2	2	2	2		2— 5
Physa gyrina	2	2	4	4	4	4	2— 6
Pleurocera elevatum	1	2	2	2	2	2	3— 5
Planorbis bicarinatus	2	1	2	4	5	4	4— 9
Limnaea stagnalis	1	1	1	2	3	3	4—12
Pisidium					4	6	6—25

Numbers 1—5 indicate proportional abundance of species
1 — most scarce
5 — most abundant

in detail. It can only be mentioned that some pelecypods, such as *Limnaea*, *Physa* or *Planorbis* live on rocky bottom, and consequently, are found right on the rocky substratum. Most pelecypods favour agitated water but surf conditions are deleterious, as the tests are broken and carried to greater depths. There are a number of detailed data available on the depths in which various mollusc genera occur. Those from Lake Mendota in Wisconsin are given here as an example (Table 87).

The optimum depth for pelecypods living in lakes ranges from 2—6 m. Empty tests are found in greater depths than living organisms. Thus, for instance, tests of the species *Pisidium abyssorum* and *Pisidium solitudum* were found even deeper than 100 m. On the other hand, tests of terrestrial molluscs (e. g. *Succinea*) are sometimes washed down into lakes.

The dependence of the occurrence of some pelecypod species and genera on the rock substratum is shown in Table 88.

Table 88

Dependence of the occurrence of some pelecypod species and genera on the rock substratum (after J. Pia 1933)

	Coarse gravel	Fine gravel	Sand	Mud	Vegetation
Amnicola limosa	3	4	Op	4	Op
Physa heterostropha			Op		
Physa ancillaris			Op	Op	
Valvata tricarinata			Op	1	
Anodonta			Op	1	
Sphaerium occidentale	2	3	Op	2	
Ancylus	1	1	1	Op	
Physa gyrina			Op	Op	
Pleurocera elevatum			Op		
Campeloma			2	Op	
Planorbis parvus	1		3	3	Op
Planorbis companulatus		3	Op	Op	Op
Limnaea stagnalis			Op	Op	Op
Pisidium				Op	

Numbers 1—5 indicate proportional abundance.
Op — optimum occurrence.

Table 89 presents a total distribution pattern of living pelecypods and empty tests or their fragments at different depth zones of the lake.

Empty tests and their fragments form a larger percentage of Recent lacustrine sediments than tests of living pelecypods. Table 89 also clearly shows that the tests are accumulated deeper than the organisms. A large amount of tests up to a depth

Table 89

Abundance of pelecypod shells in various depths of Great Plöner Lake (Germany) (after J. Lundbeck 1929)

Depth m	Shells of living pelecypods g/m^2	Empty shells g/m^2	Total weight of shells including fragments g/m^2
6·5	36·00	146·67	500
10	11·34	299·33	600
11	10·30	469·11	800
12	1·52	1,591·11	2,000
13·5	8·72	447·56	700
14	1·34	147·11	200
17	1·16	19·56	40
20	2·12	28·29	40
23·5	1·16	6·22	20
26	1·42	18·22	25
29·5	0·72	28·00	30

of 12 m indicates that in this interval they are the main component of lacustrine deposits. Generally, it has been observed that at these depths the tests are accumulated in several clear-cut bathymetric zones. In an ideal case the zones are as follows:

1. The beach girland originated at a level to which the surf ejects empty tests. In this zone the genera *Limnaea* and *Sphaeria* are commonly predominant; depending on local conditions also other genera may prevail.

2. The transitional zone, which usually shows a less pronounced development. It originates in places where the advancing waves meet the waters returning down the slope. In those places smaller tests which are neither ejected on the shore nor transported deeper into lake accumulate.

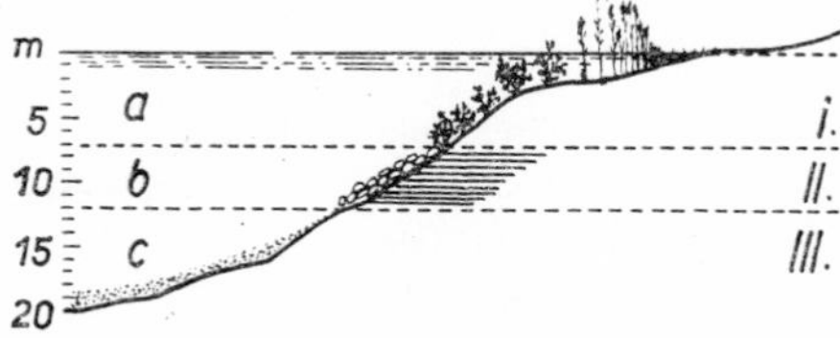

Fig. 58. Conditions of occurrence of shell zone (dashed layer) in littoral zone of lakes. a. epilimnion, b. metalimnion, c. hypolimnion.

3. At the depths of 5—7 m and 10—12 m the shell zones proper occur, forming the boundary between the littoral and profundal. In the North German lakes, which in this respect have been studied most thoroughly, these zones are formed chiefly of *Dreissensia polymorpha* and ubiquitous *Valvata* and *Bythinia*. In other lakes, tests of *Unio*, *Neritina* and *Paludina* may also be dominant.

The amount of tests of individual species in these zones depends on:

1. The amount of living organisms.
2. Resistence of tests to mechanical disturbance.
3. Living conditions of individual species.
4. Weight and shape of tests which control the distance of transportation. The capacity of transport is increased by the presence of air or gases (developing on the decay of organic matter) in the tests. This accounts for a movement to a distance of several kilometres into the deepest parts of lakes.

Table 90

Weight % of shell fragments from total weight of shell carbonate
(Great Plöner Lake, Germany, after J. Lundbeck 1929)

Depth m	% shell fragments
5	60
6·5	13
8·9	12
12	2

The ratio of tests and their fragments varies with depth. The amount of fragments increases towards the shore, because there the surf and wave action contribute most to their formation (see Table 90).

In the main zones the proportion of tests in the total volume of sediments usually amounts to 40—90%.

It has been found that, like the coarse-grained terrigenous sediments, the shell zones, too, occur deeper on the windward side of lakes than on the leeward side. Otherwise, the absolute depth of the zone depends on the intensity of surf and wave activity which in turn are directly proportionate to the size of the lake. In lakes of various types the depth of the main shell zone is as follows (see Table 91).

Table 91

Depth of occurrence of the main shell zone

Dimension and type of lake	Depth range of main shell zone m
Great inland basins (Black and Caspian Seas)	10—18
Great lakes	6—12
Lakes of medium size	4—6
Shallow, flat lakes	about 2
Small lakes	1—2, often poorly developed
Humus lakes	absent

In the lakes where few pelecypods exist, there is only as indication of the shell zone. The Alpine lakes are a typical example.

The origin of the shell zone has already been discussed by a number of authors. All theories agree in the explanation of the genesis of the two upper zones. As mentioned above, the uppermost zone is formed in places to which the surf can eject the tests. The second zone originates where the advancing wave meets the returning water. The opinions on the origin of the main zone vary; it has only been objectively established that universally it lies deeper than the zones inhabited by individual species. Two principal theories suggest either a biological or a mechanical factor as responsible for the zone distribution.

1. According to the first theory, the difference between the depth distribution of living pelecypods and empty tests is caused by the present-day migration of pelecypods to smaller depths as a result of the change of oligotrophic lakes into eutrophic. It has also been suggested that the depth of zones corresponds to the depth of the issues of springs which are supposed to bring plenty of nutrients and wash out clay particles from the sediments.

2. The theory of mechanical origin attributes the bathymetric distribution of the shell zones to the activity of waves and currents. The back-current is tentatively indicated as the main factor, because the main zone is formed at a depth where the back-current weakens.

So far it is not safely known which of these two hypotheses is closer to reality, but the mechanical factors seem to play the main role in the final formation of the shell zone.

The incessant redeposition of empty tests leads to their mechanical disintegration. Experiments have shown that every test is completely broken when exposed to a continuous water movement for two months. In addition to mechanical factors, biogenic factors contribute to disintegration, as for instance, the activity of boring algae, bacteria and birds. As a result, in Recent sediments shells disintegrate even in tranquil water. The shell zones are frequently preserved in Subrecent and ancient sediments. In Subrecent lacustrine sediments they are sharply limited against the overlying calcareous or organic sediments. They occasionally form beds inside fine-grained calcareous sediments.

Diatomaceous sediments

Lacustrine diatomaceous sediments are, on the whole, confined to a minor part of lakes, but occur in a great abundance in them. The deposition of diatomaceous sediments in lakes requires the following conditions to be fulfilled:

1. Suitable conditions for the growth of diatoms.
2. Absence of disturbing and masking conponents.

A low temperature of water complies with the first requirement, because it supports the production of diatoms and restrains the growth of bacteria. Therefore, lakes with diatomaceous sediments occur mainly in higher geographical latitudes and higher altitudes above sea level. Mountain lakes with cool clear water furnish an ideal environment for the growth and deposition of diatoms. Another important condition is the presence of SO_2. Diatoms can originate already at the presence of 1—5 p.p.m. of dissolved SiO_2 and their maximum development occurs at 5—20 p.p.m. of SiO_2 in water. The springs on the lake bottom, quartz sand, volcanic ash, etc. are the sources of silica. Finally, the development of diatoms needs a certain amount of nutrients, i. e. phosphates and nitrates, which are supplied by ground waters and tributaries or originate by the regeneration of organic matter. The composition of some Recent diatomaceous sediments is given in Table 92.

Table 92

Composition of some Recent diatomaceous sediments

Lake	SiO_2 %	Org. matter %
Grassy	20	80
Irving	35	65
Key	73	27

Well-known occurrences of diatomaceous sediments are in the Alpine lakes, especially in Lake Luzern. In the U.S.S.R., numerous diatomaceous lakes occur in the lowlands near the White Sea. Provided that all the above-mentioned conditions are fulfilled, a continuous carpet of diatom tests is formed on the surface of some lakes during summer. After attaining thickness of a few millimetres the carpet sinks. In this way layer of diatomaceous sediments up to 30 m thick can originate. Their usual thickness varies between a few and 15 metres. It is interesting that the depths of diatomaceous lakes are at present smaller than the thicknesses of these deposits.

Lacustrine diatomaceous sediments show the following composition: diatom tests, plant remains, a variable amount of coarse (sandy and silty) terrigenous detritus and clay. Some diatomaceous sediments also contain gypsum crystals, which originated by the oxidation of pyrite and do not indicate an increased salinity, as could be erroneously interpreted.

Organic sediments composed of soft unstable organic components

These organic lacustrine sediments are of particular importance, mainly because they do not occur in any other environment in such a well-developed form. They are designated by various names (gyttja, sapropel, dy, afja, förna, etc.) but their classification has not yet been firmly established. The present state is so confused that many authors disregard all classifications and use the above terms haphazardly. We use the classification based on the division of E. Wasmund (1930) and K. Hansen (1959).

The basic organogenic lacustrine sediments are sapropel and gyttja. These two terms, however, have two different meanings: on the one hand they express a different primary composition, and on the other a different mode of alteration. Sapropel is defined as the final stage of the alteration of the primary material called förna. The alteration is induced by bacterial processes under reducing conditions. The primary material consists of macrophytes, it is more frequently autochthonous than allochthonous and contains macroscopically indiscernible plant and faunal remains. Gyttja is the final product of the alteration of material called afja under oxidation conditions. The primary material is formed chiefly of planktonic organisms rich in fats and proteins.

Table 93

Summary of differences between sapropel and gyttja

	Original material	Kind of decomposition
Sapropel	plant material, mainly macrophyta, autochthonous and allochtonous (förna)	Decomposition during reduction, mainly bacterial
Gyttja	fine phytoplankton and zooplankton, always allochthonous (afja)	Decomposition during oxidation, coprogenous

The following scheme is valid for ideal conditions (Table 93). From the above survey it follows that the primary material can be affected by different types of decomposition. Thus, for instance, if the primary material corresponding to gyttja undergoes an reduction process, the transitional stage between sapropel and gyttja originates. It cannot be decisively stated whether the primary composition or the mode of alteration are of more importance. The resulting sediment is sometimes controlled rather by the primary composition, at another time by the type of secondary alterations. At present, the authors favour the view that the difference between gyttja and sapropel is due mainly to the different forms of alteration. The term "dy" is used rather loosely for organogenic sediments of allochthonous origin, whose particles are of colloidal dimensions. It is usually stated that the dy is gyttja mixed with unsaturated humic colloids brought into lakes from swamps. The term tyrphopel is sometimes used as a synonym for the dy. The name, frequently met with is saprocol, which designates a diagenetically modified and consolidated sapropel.

The determination of organic C/organic N ratio is very important for the establishment of the character of organic matter. It has been statistically determined that when the above ratio is lower than 10 humus is of neutral nature and the sediment is gyttja. When the ratio exceeds 10, gyttja has an admixture of acid humus, and the sediment is dy (tyrphopel). The increased org. C/org. N ratio indicates the increase of acid humus resulting in the formation of peat and sphagnum marsh.

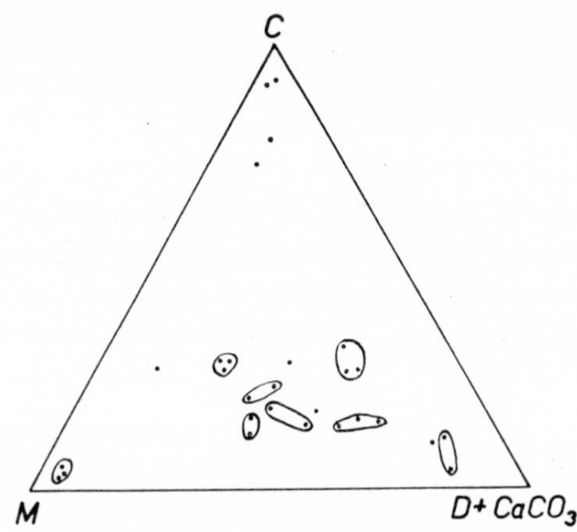

Fig. 59. Composition of Danish lacustrine sediments. C. organic carbon, M. minerogenic compounds, D ± $CaCO_3$. chemogenic compounds and diatom frustules. After K. Hansen (1959).

Table 94

Plant association of peat bogs

Depth m	Species
< 1	*Carex gracilis*
	Allsa plantago
	Sagittaria
2–3	*Scircus lacustris*
	Phragmites communis
	Equisetum heleochantis
4–5	*Nymphaea alba*
	Nuphar luteum
	Potamogeton natans
	Spargentum natans
	Ceratophyllum demersum
	Miriophyllum
	Calliergon giganteum
	Chara algae
	Transition between peat and sapropel

From the above it follows that the composition of organic lacustrine sediments is complicated and far from solved in the petrographical respect. At present, the approach to the study of the composition of these sediments is not uniform; it is investigated from the chemical, pedological and petrographical points of view, the last of which, although most relevant for geologists, is unfortunately, least advanced.

The freshwater lakes and their organogenic deposits are closely connected with swamps and marshes.

The gradual overgrowing of lakes with vegetation results in the development of peat bogs, which consist of typical associations of hydrophilous plants. The association alters with depth, following mostly the scheme in Table 94.

The lower zones consist solely of green algae, *Vaucheria*, *Clarophora* and at the base there are blue-green algae and diatoms.

Peat bogs are classified on the basis of their mode of occurrence and morphology, with due regard to pedological, geographical and technological criteria. A discussion of these classifications, however, is beyond the scope of this book.

Of special importance for the study of ancient sediments is the investigation of littoral marshes, the deposits of which are most frequent also in ancient deposits. They are typically developed at several places in the world, as for instance, on the coast of Florida, Louisiana and Sumatra. Algae and many other plants accumulate in the shallow and stagnant water of these swamps. In the Netherlands, the cutting of swamps on the sinking sea shores by stream channels is to be seen. Fig. 34 illustrates the development of delta swamps on the alluvial plain of the Mississippi. The thickness of individual peat layers varies between 50—100 cm, but can be even greater in places of a fairly rapid subsidence. Swamps originate mainly in bays, between the river branches or at the margins of alluvial plains. The following plant associations are characteristic of various environments:

Flood-plain:

Quercus virginiana, Populus deltoides, Celtis laevigata, Salix nigra, Zizaniopsis miliacea, Typha latifolia, Scirpus americanus, Phragmites communis, Spartina alterniflora,

Fresh-water marsh:

Scirpus americanus, Typha latifolia, Spartina alterniflora, Phragmites communis, Panicum repens, Alternathera philoxeroides.

Brackish marsh:

Spartina patens, Juncus roemerianus, Scirpus olneyi, Ruppia maritima, Potamogeton foliosus.

Salt marsh:

Spartina patens, Spartina alterniflora, Juncus roemerianus.

Regularities of the distribution and development of lacustrine sediments

It can be said that lacustrine sediments have probably the most varied composition of all deposits. They comprise all clastic sediments, carbonates and evaporites as the representatives of chemogenic sediments, and various organogenic deposits. In no lake, however, are all of them present concurrently. The most logical division of lakes according to their sedimentary filling is the simple differentiation of lakes with coarse clastic sediments (gravels and coarse-grained sands) and lakes devoid of this material. In lakes of the first type, coarse shallow-water sediments at certain depths pass into clayey, or calcareous deposits. In lakes of the second type, predominantly clayey sediments with a variable content of carbonate and with occasional admixture of organic matter are deposited. Lakes of the third characteric type have deposits of organic origin; in these the shallow-water zone of carbonate sands composed of fragments of algae or pelecypod tests can, but need not, be developed. The major part of the lake bottom is covered by various kinds of gyttja.

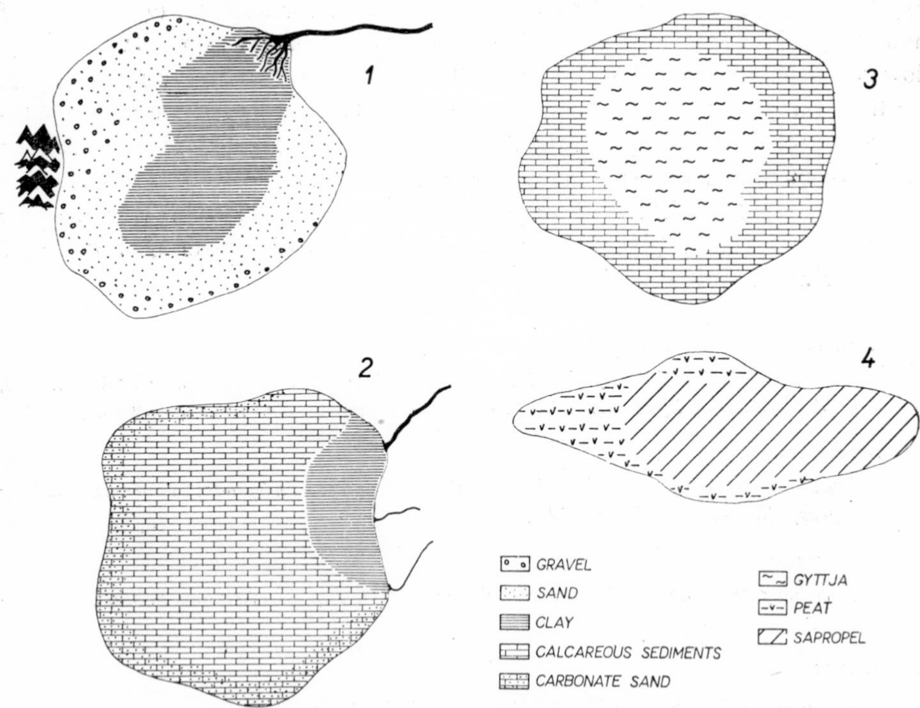

Fig. 60. Schematical representation of four lake types according to their deposits. 1. Lakes with terrigenous deposits, 2. Lakes with carbonate sediments, eventually with narrow nearshore belt of terrigeneous clastics, 3. Lakes with carbonate sediments and gyttja, 4. Lakes with sapropel passing into peat.

The sapropel lakes, representing the fourth type, contain only autochthonous macrophyta resulting from shallow-water growths. In shallow-water conditions peat develops passing into sapropel towards the deeper parts.

Almost every lake has its history. When the development depends only on internal factors, without the influence of external agents, it generally shows an analogous tendency. There is, however, some difference between the behaviour of the shallow-water and the deeper part of the lake. The littoral zone shows the following development:

1. Subsequently to its origin, the lake undergoes the oligotrophic stage with a typical deposition of coarse-grained clastic sediments.
2. With the proceeding euthrophy, calcareous clays or pure calcareous sediments with tests and algal fragments begin to be deposited.
3. The sedimentation of peat represents the final stage.

In the deeper parts of lakes the development occurs in these stages:

1. The initial oligotrophic stage when silt, sand or pure clay are deposited.
2. The eutrophization of the lake is accompanied by the deposition of organogenic clay, silt or sand with diatoms.

The next stage can occur as:

3a. Gyttja development, ending with the deposition of diatomaceous clays or other gyttjas, or as

3b. Sapropel development, ending with the sedimentation of sapropel.

Thus there is a general tendency in lakes from non-calcareous clastic sediments to biochemical and organogenic carbonate deposits which alter into the gyttja or sapropel sediments with the proceeding organic production. Iron and manganese ores fall into the transitional stage between calcareous and organic sedimentations.

References

AHROCK R. R. - HUNZICKER A. A. (1935): A study of some great basin lake sediments of California, Nevada and Oregon. Jour. Sedimentary Petrology, vol. 5: 9—30, Menasha.

ALDERMAN A. R. (1959): Aspects of carbonate sedimentation. Jour. Geol. Soc. Australia, vol. 6: 1—10, Adelaide.

ARNAL R. E. (1961): Limnology, sedimentation, and microorganisms of the Salton sea, California. Bull. Geol. Soc. Am., vol. 72: 427—478, Baltimore.

BOGOSLOVSKIJ B. B. - MURAVENSKIJ C. D. (1960): Fundaments of limnology (in Russian). P. 1—175, Moscow.

COAKLEY J. P. - RUST B. R. (1968): Sedimentation in an arctic lake. Jour. Sedimentary Petrology, vol. 38: 1290—1300, Menasha.

EARDLEY A. J. (1938): Sediments of Great Salt Lake, Utah. Bull. Am. Assoc. Petrol. Geol., vol. 22: 1305—1411, Tulsa.

EARDLEY A. J. - STRINGHAM B. (1952): Selenite crystals in the clays of Great Salt Lake. Jour. Sedimentary Petrology, vol. 22: 234—297, Menasha.

FISK H. N. (1960): Recent Mississippi River sedimentation and peat accumulation. IV. Congr. strat. geol. Carbonif., p. 187—199, Heerlen.

GORHAM E. (1960): The relation between sulphur and carbon in sediments from the English lakes. Jour. Sedimentary Petrology, vol. 30: 466—470, Menasha.

HANSEN K. (1959): Sediments of Danish lakes. Jour. Sedimentary Petrology, vol. 29: 38—46, Menasha.

HOUGH J. L. (1935): The bottom deposits of southern lake Michigan. Jour. Sedimentary Petrology, vol. 5: 57—80, Menasha.

HUTCHINSON G. E. (1957): A treatise on limnology. P. 1—1015, New York.

LUNDBECK J. (1929): Die Schalenzone der norddeutschen Seen. Jhrb. Preuss. Geol. Landesanstalt, p. 229—333, Berlin.

MANN J. F. (1951): The sediments of lake Elsinore, Riverside County, California. Jour. Sedimentary Petrology, vol. 21: 151—161, Menasha.

MINDER L. (1939): Der Zürichsee als Eutrophierungsphenomen. Geol. d. Meere u. Binnengew., Bd. 2: 284—299, Berlin.

MOORE J. E. (1961): Petrography of northeastern Michigan bottom sediments. Jour. Sedimentary Petrology, vol. 31: 402—436, Menasha.

MORTIMER C. H. (1963): Frontiers in physical limnology, with particular reference to long waves in rotating basins. Great Lakes Research Divis., Publ. No 10: 9—42, Ann Arbor.

NAUMANN E. (1922): Die Bodenablagerungen des Süsswassers. Arch. f. Hydrobiologie, Bd. 13, p. 1—97, Stuttgart.

PIA J. (1933): Die rezenten Kalksteine. P. 1—418, Leipzig.

PICARD M. D. - HIGH L. R. (1969): Lacustrine criteria. Bull. Am. Assoc. Petrol. Geol., vol. 53: 736, Tulsa.

PIPKIN B. W. (1967): Mineralogy of 140-feet core from Wilcox Playa, Cochise, Arizona. Bull. Am. Assoc. Petrol. Geol., vol. 51: 478—479, Tulsa.

REEVES C. C. (1968): Introduction to paleolimnology. Development in Sedimentology, vol. 11: 1—228, Amsterdam.

REID J. R. (1963): Geology of bottom sediments from Burt Lake, Cheboygan County, Michigan. Jour. Sedimentary Petrology, vol. 33: 304—313, Menasha.

SAPOZNIKOV D. G. - VISELKUNA M. A. (1960): Recent sediments of Issyk-kul Lake (in Russian). Trudy inst. geol. rudnych mestoroždenij, vyp. 36: 1—160, Moscow.

SAUNDERS G. W. (1963): The biological characteristics of fresh water. Great Lakes Research Divis., Publ. No 10: 245—257, Ann Arbor.

SEMENOVICH N. N. (1958): Limnologic condition of the sedimentation of Fe-rich sediments in lakes (in Russian). Trudy labor. ozerovedenija, vol. 6: 1—186, Moscow.

SCHOLL D. W. - TAFT W. H. (1964): Algae, contributors to the formation of calcareous tufa, Mono Lake, California. Jour. Sedimentary Petrology, vol. 34: 309—319, Menasha.

SKINNER H. C. W. (1963): Precipitation of calcium dolomites and magnesian calcites in the southeast of South Australia. Am. Jour. Sci., vol. 261: 449—472. New Haven.

SLÁNSKÁ J. (1965): Environments of Recent lake basins and their sediments (in Czech, unpublished Thesis). P. 1—126, Prague.

SWAIN F. M. - MEADER R. W. (1958): Bottom sediments of southern part of Pyramid lake, Nevada. Jour. Sedimentary Petrology, vol. 28: 286—297, Menasha.

TEICHMÜLLER R. (1955): Über Küstenmoore der Gegenwart und die Moore des Ruhrkarbons. Geol. Jahrbuch, Bd. 71: 197—220, Hannover.

TWENHOFEL W. H.-MCKELVEY V. E. (1941): Sediments of fresh-water lakes. Bull. Am. Assoc. Petrol. Geol., vol. 25: 826—849, Tulsa.

USDOWSKI H. A. (1967): Die Genese von Dolomit in Sedimenten. P 1—105, Stuttgart.

WASMUND E. (1930): Bitumen, Sapropel und Gyttja. Geol. Fören. Förhdl., vol. 52: 315—350 Stockholm.

— (1931): Lakustrische Unterwasserböden. Handbuch d. Bodenlehre. Bd. 5: 97—189, Berlin.

13. Deltaic sediments

The deltaic environment is regarded as a transition between continental and marine environments. The deltaic environment and sedimentation are treated in detail, amongst others, in the papers by L. M. J. U. van STRAATEN (1960) and the present author (Z. KUKAL 1962); therefore, we shall give here only a concise survey of this topic. The term "delta" designates a body of sediments transported by stream and deposited at the boundary of the continental and of the marine (or lacustrine) environments. Part of this body bears signs of continental sediments, part typical features of marine deposits. The largest part of the deltaic body is deposited below the surface of the sea (or lake) and part can rise above water. Similarly as with other geological designations, the term "delta" also developed from the two-dimensional to the three-dimensional geological conception.

From the geological and particularly sedimentological points of view, deltas can be divided into deltas of small and those of major rivers. Surprisingly, the boundary between the two types is quite sharp, and the differences in the character of sediments and in the mode of deposition are considerable.

1. Deltas of small rivers are characterized by a discontinuous sedimentation. The streams often shift their channels and deposit the load only on the shelf. They only exceptionally transport the material over the narrow shelf depositing it on the continental slope only sometimes, especially under flood conditions. The streams flowing from a near mountain range can lay down gravels and sands, the lowland rivers deposit clay possibly with a subordinate amount of silt or fine-grained sand. In deltaic deposits of small streams no other environment is usually distinguishable.

2. Deltas of major rivers, which are the main subject of this discussion, are generally a site of long-term sedimentation. The deltaic sediments of major streams spread over both on the shelf and the continental slope. They consist predominantly of silt and clay with a small amount of fine-grained sand. They are distinguished by a conspicuous lateral and vertical differentiation, so that a series of sub-environments is discernible in them.

The distribution of deltaic sediments of major streams is controlled by the following factors:

1. The manner of the influx of fresh-water into the sea.
2. The presence of currents which originate in the mouths of rivers owing to the differences in densities of the waters mixed, or to the effects of hydraulic factors at the outlet of running water into still water.
3. Tidal currents.
4. Wave movement caused by wind action.
5. Longshore currents caused by wind action.

6. Currents of regional character caused by wind activity.
7. Turbidity currents and slumps.

Near the influx of fresh-water into the sea, a water layer of lower density is formed at the surface, which can often be traced several tens to hundreds of kilometres seaward. At the mouth of the Mississippi the layer is about 5 m thick. This process plays an important role in transporting the river suspension into distant parts of the basin. The influence of regional and ocean currents on the distribution of sediments in deltas is also considerable. For instance, the waters of the Amazon River are driven by oceanic currents northwards, those of the Mississippi stretch in the direction corresponding to the water circulation in the Mexico Gulf, i. e. chiefly east- and southeast-ward.

The morphology of deltas

The size of delta depends primarily on the size of the suspended load and the intensity of currents, especially in front of the river mouth. When the most favourable factors coexist, the larger deltas are formed. Table 95 summarizes both the favourable and the unfavourable factors for the formation of deltas.

Table 95

Summary of favourable and unfavourable factors for the formation of deltas

Favourable factors	Unfavourable factors
Great amount of river-borne suspension	Small amount of suspension
Stability or slight subsidence of coast (up to 3 cm/1,000 years)	Rapid subsidence of the coast (over 3 cm/1,000 years), exceptions occur
Flat coastal relief	Strong tides, strong tidal currents
Wide flat shelf before river mouth	Narrow shelf, with adjacent mountain ranges

A direct relation between the size of the above-water parts of deltas (and thus also of the whole deltas) and the amount of material imported by river is documented by Table 96.

The above-water portions of deltas are bodies of fine-grained sediments cut through by distributaries and crevasses, which carry away flood waters. Otherwise, their morphology is similar to that of alluvial plains.

Table 96

Relation between the area of the subaerial parts of deltas and the material imported by rivers

River	Area of subaerial parts of delta km^2	Amount of suspension load in mil. of tons/year
Amazon	100,000	1,000
Amu Darya	1,000	105
Dnieper	350	0·8—2·0
Danube	3,500	83
Euphrates	1,000,000	?
Colorado	8,500	160
Ganges and Brahmaputra	80,000	1,800
Hwang-Ho	2,000	620
Indus	8,000	400
Kolyma	3,000	4,7
Lena	28,500	12
Mississippi	80,000	500—730
Niger	40,000	67
Nile	20,000	?
Seine	0	1·3
Volga	12,000	25·5
Zambezi	8,000	100

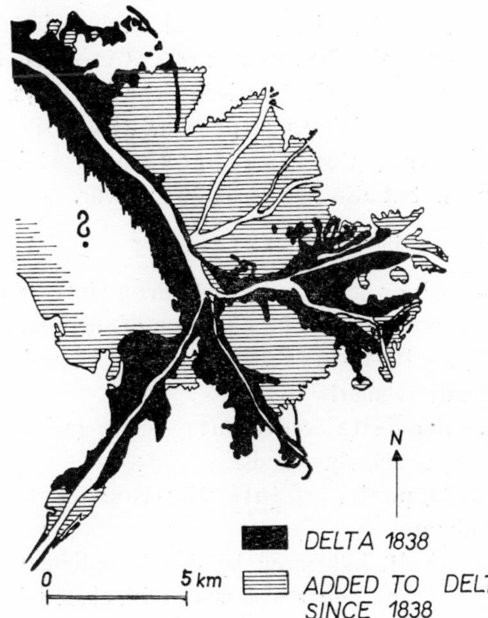

Fig. 61. Illustration of the large growth of the Mississippi Delta during the last century. After F. P. Shepard (1961).

The submarine portions of deltas are either simple with a flat fan-shaped form, or are strongly differentiated, both in morphology and petrography. They often extended into the delta-front platform, which is a more gently inclined outer delta slope. The decrease in slope takes place at a depth of about 5—20 m. The width of the delta-front platform is fairly great (that of the Orinoco River is 35 km).

The generally used classification of deltaic environments of major streams is based mainly on the conditions of the Mississippi and Niger River delta:

Subaerial portions of the delta	Submarine portions of the delta
a) Channel	a) River mouth bars
b) Flood plain	b) Delta-front environment
c) Marsh	c) Pro-deltal environment
d) Crevasse	d) Interdistributary bay environment
e) Other sub-environments between distributaries	e) Shelf influenced by deltaic sedimentation

Deltaic sediments

The following sediments are predominant in different environments:

Subaerial parts:

a) Channels — fine-grained, medium-grained, sometimes also coarse-grained sands, passing into silty or clayey sands. Usually poor in $CaCO_3$ (amounts up to 15%), content of organic matter negligible.

b) Flood-plain — fine-grained deposits, silts with intercalations of fine-grained sands prevail. The amounts of $CaCO_3$ range from 0 to 25%, org. C from 1 to 5%.

c) Marsh — fine-grained deposits, clays, eventually silty clays and clayey silts prevail, containing sometimes mottles of silts and fine-grained sands.

d) Other deposits between distributaries — fine-grained sediments prevail, but also fine gravel may occur.

Submarine parts:

a) River mouth bar — fine-grained sands, usually with clay and silt admixtures, poor sorting. Intercalation of silty clays.

b) Delta-front sediments — fine-grained clayey and silty sediments prevail. Usually regularly bedded.

c) Pro-deltal sediments — fine grained clayey and silty sediments. Some beds of clayey and silty sands.

d) Deposits of interdistributary bays — clays with silt and fine sandy admixtures.

e) Shelf, influenced by deltaic sedimentation — mostly coarser deposits, calcareous sands, quartzose sands with glauconite, with intercalations of clayey silts.

Gravels are rather rare in the deltas of major streams, occurring usually at the base in the form of coarse-grained fluviatile relict conglomerate with pebbles up to 10 cm in size. In some deltas, intraformational conglomerates formed of fragments of consolidated and eroded clay sediments occur approximately in mid-sequence. Gravels are abundant in the deltas of minor mountain rivers, as for instance the sand and gravel delta of the river Var in the Alps. Gravel banks (bars) are deposited also at their mouths and may reach in the lake or sea to a depth of several tens of metres.

Sandy deposits may be either primary, i. e. those which settle directly from suspension and are not reworked and sorted by wave and current action, or secondary, those which originated by sorting and secondary accumulation into ridges, etc.

Primary sandy deposits are found only in the deltas of streams which carry large amounts of sand fraction, e. g. the lowland rivers (the Volga, the Ural) or mountain rivers (in the Alps). The sands are poorly sorted and contain a high percentage of clay and silt. Lacustine deltas are on the average sandier than the marine ones, because clay is not flocculated at the point of inflow into the lacustrine fresh waters as it is when the waters flow into salt water, so that it can be more easily separated from the coarser fractions. Secondary sandy deposits are generally better sorted and accumulate into sand bars, ridges or spits. Their common grain-size ranges from 0·1 to 0·2 mm. They can settle to a depth of a few metres only; for instance, the lower boundary of their deposition in the Rhône delta is the —8 isobath.

Silt as well as sand is the important constituent of Recent deltas, the former being occasionally even more abundant. This is the case with some mountain rivers, such as the Amu-Darya and the Colorado River, etc. In the largest deltas, the sediments deposited right at the mouths of individual distributaries are generally richest in silt. This decreases seaward and is accompanied by an increase in clay content. The amount of silt also decreases rapidly towards the bays between distributaries. In deltas of a simpler structure (e. g. the Orinoco) silt forms broad belts round the points of inflow of river water; the greater their width, the lower is the gradient of the outer shelf slope. Sediments of the sub-aerial environments, such as swamps and flood-plains, are also frequently characterized by a large content of silt.

The clay fraction predominates especially in the deltas of lowland rivers (the Dniepr, the Orinoco, the Amazon, the Hwang-Ho, the Nile, the Guadelupe). The study of clay minerals has shown than the composition changes more or less from the mouth seawards. In the Gulf of Mexico the amount of montmorillonite decreases and the proportion of illite and chlorite increases seaward. Montmorillonite, forming about 10% in the proximity of the mouths of distributaries, drops to zero after a few kilometres of travel. These mineralogical changes are thought by the majority of authors to be the manifestation of early diagenetic changes which can take place at the boundary of fresh and salt waters. In the Paria Bay the relations are quite the reverse: Montmorillonite increases and illite decreases from the river mouth into the bay. This pattern has been interpreted in terms of mechanical differentation, because illite, which usually forms large particles, could have settled nearer to the mouth than montmorillonite, the smaller particles of which were transported further into the bay.

The content of carbonate is very variable in deltaic sediments, so that carbonate-poor and carbonate-rich deltaic sediments can be differentiated. Table 97 sums up the factors which affect the deposition of carbonates in river deltas.

Table 97

Summary of favourable and unfavourable factors for the deposition of carbonate in river deltas

Favourable factors	Unfavourable factors
Great amount of carbonate in suspension load (cold and moderate climate and abundance of carbonate rocks in drainage basin)	Small amount of carbonate in suspension load
River mouth with normal marine or higher salinity and water with alcalic reaction	River mouth in bay or semi-isolated basin with brackish water with acid reaction
Great amount of carbonate organisms	
Deltaic sediments formed predominantly by finer fractions	Coarser fractions prevail in deltaic sediments

The regime of $CaCO_3$ deposition in deltaic sediments is generally consistent with that in river sediments, particularly because in both environments this is associated chiefly with the fine clay fractions.

In the subaerial parts of deltas the highest percentage of $CaCO_3$ has been found in the flood-plain and swamp deposits. In the submarine parts, the shelf deposits

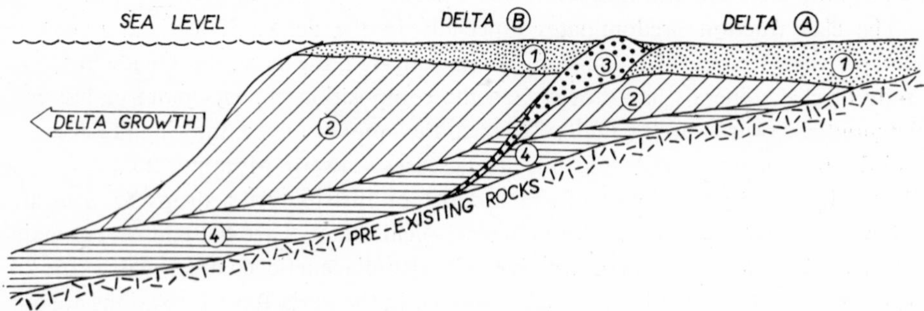

Fig. 62. Typical delta profile with greatly exaggerated slope in foreset beds. 1. Topset beds, 2. Foreset beds, 3. Destructional deposits, 4. Bottomset beds. After P. C. Scruton (1960).

influenced by deltaic sedimentation are richest in $CaCO_3$. The origin of calcite cement of some barrier sands is prevalently explained by solution of calcareous tests and migration into interstitial pores.

Sedimentary structures of deltaic deposits

The individual environments of the Mississippi and the Niger deltas, which are probably two of the most thoroughly studied deltas of the world, are characterized by the following sedimentary structures:

Channels: similar to common river channels. Many types of diagonal and cross-stratification. In fine-grained sediments irregular streaky bedding.

Flood-plain: Regular bedding and lamination, graded bedding and sometimes current ripple-bedding.

Marsh: Mainly uniform fine organic rich clayey silts and silty clays with roots and root mottles.

Delta-front platform: Mainly even bedding and lamination (sandy silts — silty clays). In shallower parts sometimes mottles of fine sand.

Pro-delta slope: Regular bedding and lamination, coarser layers ripple bedded. Sometimes abundant mottles. In deeper areas uniform clayey silts and silty clays.

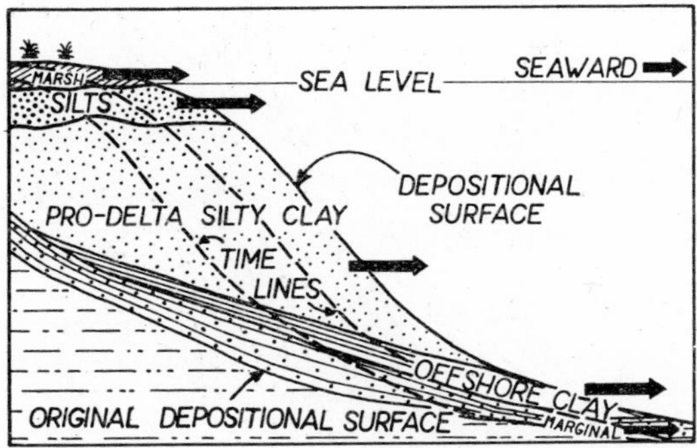

Fig. 63. Seaward migration of depositional environments. With delta growth, the different depositional environments migrate seaward and extend the relatively homogeneous sediment unit. After P. C. Scruton (1960).

Regularly stratified fine-grained sands — silty sands (or other fairly fine-grained sediment) are distinctive for pro-deltal parts of major streams. As far as it could be established, the beds of coarse sediments decrease in number from the mouth seaward. In some cases (e. g. in the Rhône delta) the beds can be traced from a minimum depth to a depth of over 100 m. It is thought that this stratification is caused by the differences in the grain-size of the suspended load. The coarser laminae or

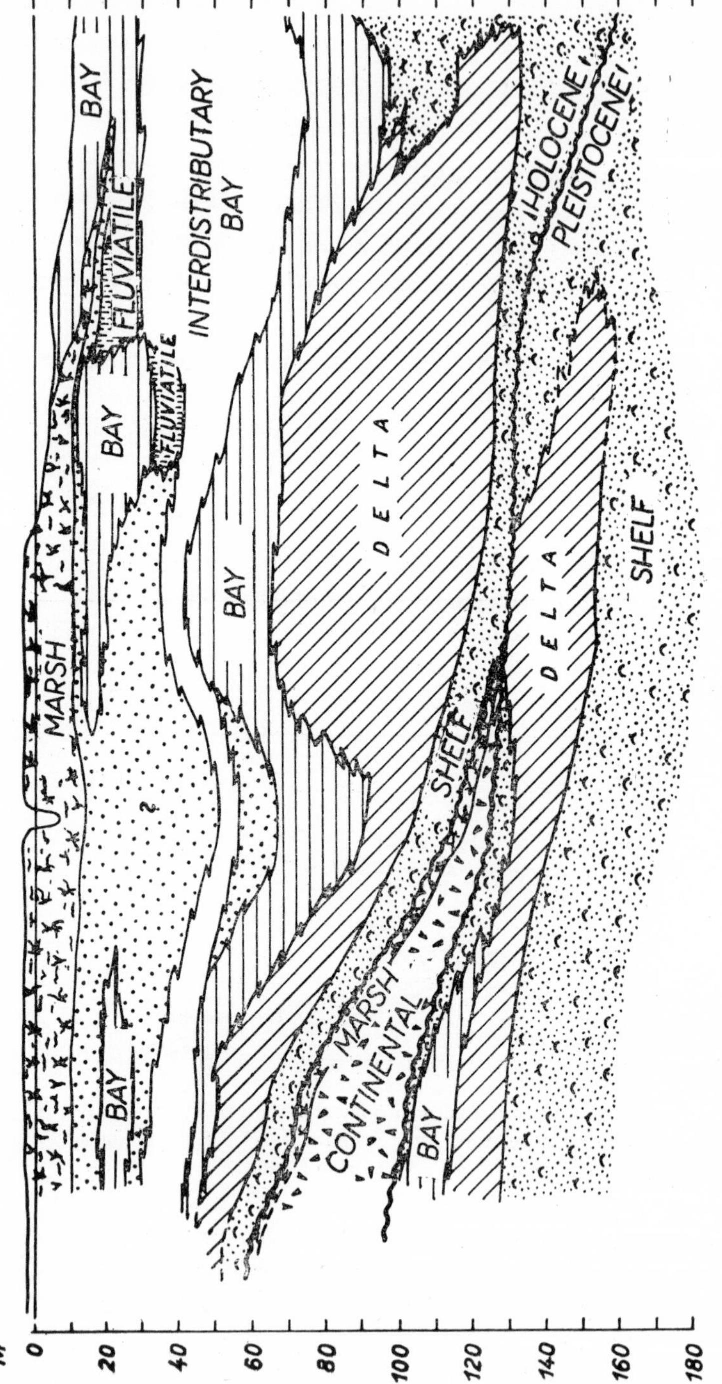

Fig. 64. Reconstruction of environments in N — S section of Holocene Mississippi Delta sediments. (×100.) After R. L. Lankford - F. P. Shepard (1961).

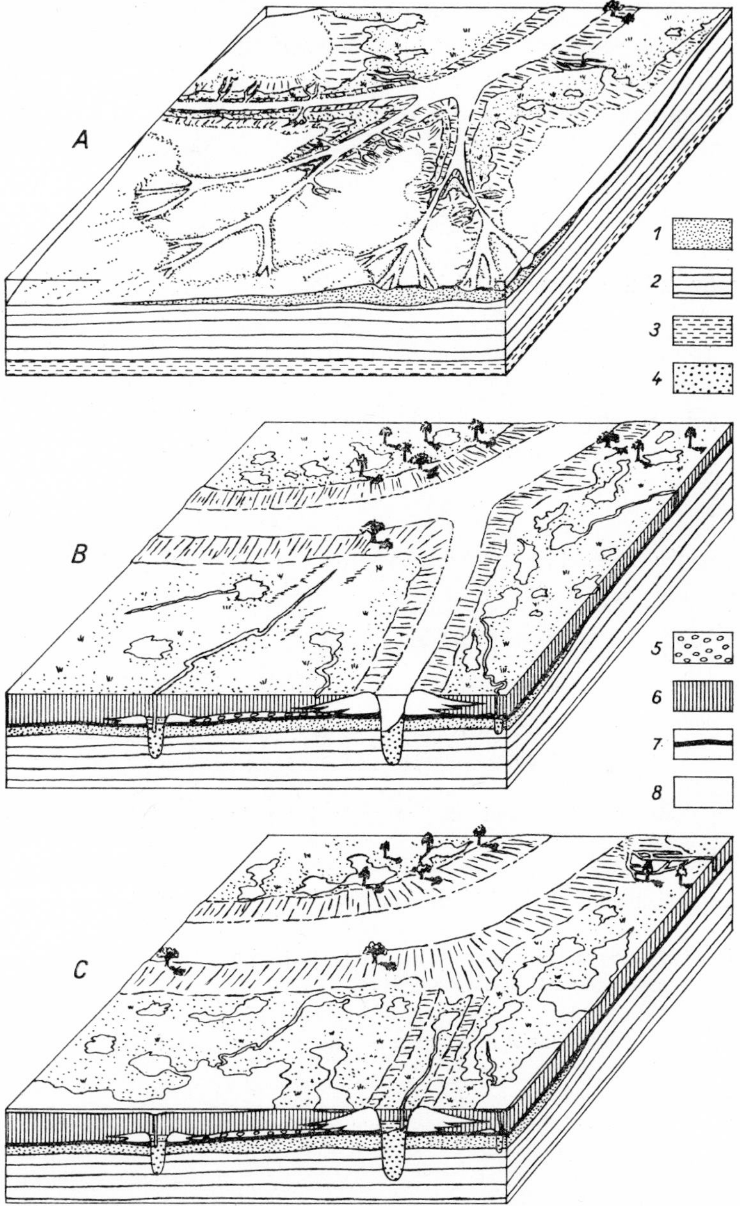

Fig. 65. Three stages of development of a part of the Mississippi Delta. 1. Delta-front sands, 2. Delta-front silty clays, 3. Prodeltal clays, 4, 5. Coarse channel sediments. 6. Marsh sediments, 7. Marine bay sediments, 8. Natural levee. After H. N. Fisk 1961.

beds correspond to floods which supply more coarser material. This phenomenon, however, has also been explained as due to the periodical outwashing of fine fractions from originally homogeneous sediments. The stratification described is usually frequent in ancient sediments and it is probable that the deltaic sediments represent their recent analogy. Therefore, it can be presumed that ancient sediments with analogous structures may often be of deltaic origin.

Stratigraphy and development of deltaic deposits

There are some differences between the mechanism of Recent and ancient sedimentation. This statement is valid also for deltaic deposition. In earlier geological times, rivers sometimes deposited material right on the continental slope, because the broad shelf did not exist. The rapid rise of the sea level after the last glacial period

Table 98

Bore log from the subaerial part of the Rhône delta

Thickness m	Deposit	Environment
11	clayey silts with intercalations of silty clay, peat laminae, plant remains	flood-plain and marsh
5	coarse-grained and medium-grained poorly sorted sands, current bedded	channel
1	clays, silty clays with sand laminae	flood-plain
0·5	coarse-grained sand with sharp contact with underlying and overlying beds	crevasses
9	silty clays, uniform, sometimes with laminae of silts	flood-plain
2·5	clays and silts with carbonate admixture, brackish fauna	alluvial lakes
13	medium-grained quartzose sands, well sorted	sand bars
28	clayey silts and silty clays, regularly and streaky bedded, laminae of silty sands	delta-front and prodeltal sediments
4	coarse-grained gravel quartzose sand with marine fauna and glauconite grains	channel shelf

brought about special conditions for deltaic sedimentation. The inflows of rivers were pushed inland as the sea rose and only after the rise of the sea slowed down did deltaic deposits re-advance into the basins. This sequenced events account for the presence of fluviatile gravel out on the shelves. The gravels are generally overlain by a varied sequence of Sub-recent and Recent deltaic sediments of various environments, with subaerial deposits at the base and subaqueous sediments in the upper parts. Such a sequence is, naturally, an ideal pattern, which in reality is modified by a number of circumstances, as for instance, by the rate of growth of the delta or by the speed of the regional subsidence. The log of a borehole from the subaerial part of the Rhône delta is a good illustration of a sequence of deposits encountered in the deltaic environment (Table 98).

The fauna of deltaic sediments

R. H. PARKER (1958) has described the invertebrate fauna of the Mississippi delta and its dependence on the sedimentary environments. He has found that the character of sediments is a decisive factor in the distribution of some species. With a few exceptions, the sandy bottoms are richest in organisms. In the delta-front and prodeltal parts, where clays and sands are deposited at a high rate, fauna is usually rare. In places of slow deposition or non-sedimentation, as along sand bars, mounds and ridges, assorted shells accumulate.

The fauna of the subaqueous parts of deltas differs widely from that of the subaerial parts. Whereas the latter have continental fauna, the former yield mostly typical marine fauna, even though the salinity of this environment is considerably lowered.

References

ALLEN J. R. L. (1965): Late Quarternary Niger delta, and adjacent areas: sedimentary environments and lithofacies. Bull. Am. Assoc. Petrol. Geol., vol. 49: 547—600, Tulsa.

ANDEL TJ. VAN (1955): Sediments of Rhône delta. II. Sources and deposition of heavy minerals. Verhandel. v. koninkl. Nederl. Geol. Mijnb. Gen., Geol. Ser., vol. 15: 515—556, Amsterdam.

— (1967): The Orinoco delta. Jour. Sedimentary Petrology, vol. 37: 297—310, Menasha.

BANU A. C. (1963): Some consequences of the secular rising of the Black Sea level on the morpho-hydrographic development of the Danube delta. Hidrobiologia, vol. 4: 109—128, Bucarest.

BERNARD H. A. (1965): Résumé of river delta types. Bull. Am. Assoc. Petrol. Geol., Abstr., vol. 49: 334, Tulsa.

COLEMAN J. M. (1967): Deltaic evolution. In: Fairbridge R. (Editor): Encyclopedia of Earth Sciences, 255—261, New York.

COLEMAN J. M. - GAGLIANO S. M. - FERM J. C. (1969): Deltaic environment. Bull. Am. Assoc. Petrol. Geol., vol. 53: 712, Tulsa.

COLEMAN J. M. - GAGLIANO S. M. - WEBB J. E. (1963): Minor sedimentary structures in a prograding distributary. Marine Geology, vol. 1: 340—358, Amsterdam.

DUBOUL-RAZAVET CH. (1956): Contribution a l'étude géologique et sédimentologique du delta du Rhône. Mém. Géol. Soc. France. N. S., T. 35: 1—234, Paris.

FISK H. N. (1956): Nearsurface sediments of the continental shelf of Lousiana. Proc. of the 8th Conf. on soil mechanics and foundation engineering, p. 1—36, Austin.

— (1959): Padre Island and the Laguna Madre flat coast. 1. South Texas. 2nd Coastal Geography Conference, p. 103—152.

— (1961): Bar finger sands of Mississippi delta. In "Geometry of sandstone bodies". Am. Assoc. Petrol. Geol., p. 29—52, Tulsa.

JOHNS W. D. - GRIM R. E. (1958): Clay mineral composition of Recent sediments from the Mississippi river delta. Jour. Sedimentary Petrology, vol. 28: 186—199, Menasha.

KUKAL Z. (1962): Recent deltaic sediments (in Czech). Anthropozoikum, vol. 10: 249—271 Prague.

LAGAAIJ R. (1965): Sediments and fauna of the Rhône delta. Bull. Am. Assoc. Petrol. Geol., vol. 49: 347 (Abstr.), Tulsa.

LANKFORD R. R. - SHEPARD F. P. (1960): Facies interpretation in Mississippi delta borings. Jour. Geology, vol. 68: 408—426, Chicago.

MABECONE J. M. (1963): Depositional environment and provenance of the sediments in the Guadeleto estuary (Spain). Proc. 6th Internat. Sediment. Congres., p. 82—88, Amsterdam.

MATTHEWS W. H. - SHEPARD R. P. (1960): Sedimentation of Fraser River delta, British Columbia. Bull. Am. Assoc. Petrol. Geol., vol. 46: 1416—1438, Tulsa.

NOTA D. J. G. (1958): Sediments of the Western Guiana shelf. Diss. P. 1—98, Wageningen.

PARKER R. H. (1956): Macro-invertebrate assemblages as indicators of sedimentary environments in east Mississippi delta region. Bull. Am. Assoc. Petrol., vol. 40: 295—276, Tulsa.

SCRUTON P. C. (1955): Sediments of the eastern Mississippi delta. Soc. Econ. Pal. and Min., Spec. Publ., 3: 21—50, Tulsa.

SHEPARD F. P. (1959): The earth beneath the sea. 3rd ed. P. 1—275, Baltimore.

SIOLI H. (1957): Sedimentation in Amazonasgebiet. Geol. Rdsch., Bd. 45: 608—633, Stuttgart.

SOLIMAN S. M. (1964): Primary structures in a part of the Nile delta sand beach. Deltaic and shallow marine sediments. P. 379—387, Amsterdam.

STRAATEN L. M. J. U. VAN (1960): Some recent advances in the study of deltaic sedimentation. Liverpol Manchester Geol. Journal, vol. 2: 411—442.

TRICART J. (1961): Notice explicative de la carte géomorphologique du delta du Sénégal. Mém. Bul. Rech. Géol. et Min., No 8: 1—223, Strassbourg.

14. *General characteristics of the marine environment*

Marine sedimentation is intimately associated with the properties of sea-water. Whereas hydrodynamic differs with environments, the physico-chemical and biological conditions are controlled by the same laws in all regions. Therefore, a good knowledge of the properties of marine water is a necessary prerequisite for the adequate understanding of marine sedimentation.

Chemical composition of sea water

The solubility of salts and gases, their mutual reactions and equilibrium states present a series of most complicated problems. These are dealt with in a number of special papers. In relation to sea-water, chemical elements can be divided into two groups:

1. Thalassogenic, which are carried from the adjancent land as solid particles without passing into solution as a result of their character and the properties of their compounds.

2. Thalassophilous, the compounds of which dissolve in water and enrich thus the marine environment.

A list of the elements in sea water is given in Table 99 (mainly after K. KALLE 1943, complemented by data of other writers).

The ionic potential of thalassogenic elements varies between 3 and 10. The typical elements of this group are Sc, Fe^{3+}, Al, Mo and Si; their concentration in sea-water is minimal. The group of thalassophilous elements comprises the halogenes, sulphur, boron, alkalis, alkaline earth except Ba and the initial elements of some other groups. The deposition of these elements occurs principally by sorption on clayey particles, the intensity of sorption increasing with the rising charges. Recently, particular attention has been paid to the concentration of radioactive isotopes in sea-water. The respective data are listed in Table 100.

The following elements are of special importance in sedimentation:

B — this is present in sea-water as boric acid, for the major part undissociated. In some marine plants its concentrations are considerably increased.

Fe — the concentration of this element in sea-water is very variable, ranging from 1 to 60 mg/m^3. Two varieties of Fe may be generally differentiated in water. Fe in true solution (i. e. that which passes through molecular filters) and colloidal Fe (i. e. that which is entrapped on molecular filters). According to some data, in sea-water at pH 8 there is only from 4×10^{-7} to 3×10^{-8} Fe in true solution. The content of dissolved Fe on the average increases with depth. In shallow-water areas a seasonal

Table 99

Element	Amount in sea-water per mille	Percentage of amount brought to sea, which remains in sea-water	Ratio element concentration in sea-water/element concentration in sediments
H	108·0		
Li	7×10^{-5}	0·23	0·14
Na	10·75	65	111
K	0·39	2·6	1·5
Rb	2×10^{-4}	0·11	0·05
Cs	2×10^{-6}	0·033	0·02
Ba	0·002		
Mg	1·295	10·5	8·3
Ca	0·416	1·97	1·86
Sr	0·013	5·3	7·7
Ba	$5{\cdot}4 \times 10^{-5}$	0·023	0·014
Ra	10^{-13}	0·017	0·001
B	0·005	280	3·1
Al	0·00012	0·00025	0·00015
Sc	4×10^{-8}	0·0014	0·0008
Y	3×10^{-7}	0·0016	0·006
C	0·029	15	0·40
Si	0·001	0·0006	0·00037
Pb	4×10^{-7}		
N	0·001		
P	0·00006	0·013	0·006
As	0·000015	0·52	0·38
O	857·0		
S	0·90	300	45
Se	0·000004	0·97	0·67
F	0·0014	0·96	
Cl	19·345	6900	
Br	0·066	1900	
I	0·00005	0·25	
Cu	0·000005	0·0086	0·0025
Ag	3×10^{-7}	0·5	
Au	4×10^{-9}	0·67	
Zn	0·000005	0·0042	0·012
Hg	3×10^{-8}	0·01	
Ga	5×10^{-9}		
Th	0·0000004	0·003	0·004
V	3×10^{-7}	0·0005	0·004
Mo	7×10^{-7}	0·0078	
U	0·000002	0·069	0·047
Mn	0·0000005	0·0009	0·00056
Fe	0·00005	0·00017	0·00011
Ni	8×10^{-8}	0·00013	
Ce	4×10^{-7}	0·0015	

Table 100

Concentration of radioactive isotopes in sea water

Isotope	Concentration g/cm^3	Specific activity (number of desintegrations on cm^3/sec)	Total amount in oceans megatons	Total activity in oceans megacuries
K^{40}	$4 \cdot 5 \times 10^{-8}$	$1 \cdot 2 \times 10^{-2}$	63,000	460,000
Rb^{87}	$8 \cdot 4 \times 10^{-8}$	$2 \cdot 2 \times 10^{-4}$	118,000	8,400
U^{238}	$2 \cdot 0 \times 10^{-9}$	$1 \cdot 0 \times 10^{-4}$	2,800	3,800
U^{235}	$1 \cdot 5 \times 10^{-11}$	$3 \cdot 0 \times 10^{-6}$	21	110
Th^{232}	10^{-11}	$2 \cdot 0 \times 10^{-7}$	14	8
Ra^{226}	$3 \cdot 0 \times 10^{-6}$	$3 \cdot 0 \times 10^{-5}$	$4 \cdot 2 \times 10^{-4}$	1,100
C^{14}	$4 \cdot 0 \times 10^{-17}$	$7 \cdot 0 \times 10^{-6}$	$5 \cdot 6 \times 10^{-5}$	270
H^3	$8 \cdot 3 \times 10^{-20}$	$2 \cdot 5 \times 10^{-5}$	$1 \cdot 5 \times 10^{-9}$	12
Nb^{93}			1·4	
Ca^{144}			560	
Ru^{106}			0·014	
Sr^{90}			10^{-7}	
Ca^{138}			700	
Kr^{58}			420	
Sn^{157}			56	

fluctuation of the content of dissolved Fe with the oscillation of the production of phytoplankton has been ascertained.

Si — is present in sea water as silicate, predominantly in true solution. The Si content is extremely variable. In the upper water layers it is extracted chiefly by diatoms and its concentration can drop below 20 mg SiO_2/t. With depth it can be increased eventually up to 7,000 mg SiO_2/t. The mean concentration of SiO_2 in the waters of the Pacific Ocean is 500—4,500 mg/t, in the Indian Ocean 30—3,000 mg/t and in the Atlantic ocean 300—2,000 mg/t. The roughly regular changes of the concentration of dissolved SiO_2 with depth has been found in all the oceans. The concentrations increase at a fairly high rate to about 2,000 m under the surface, at which depth the amount of dissolved SiO_2 is maximal, then gently decreases to 5,000 m; from this level down to the bottom the content is already constant. The seasonal fluctuation has been observed in shallow seas. In the English Channel the waters containt 200—400 mg SiO_2/m^3 in winter and less than 10 mg in spring. Silica is one of the compounds whose concentration greatly depends on the biological production.

Organic matter — sporadic analyses have shown that marine water also contains a minor amount of dissolved organic substances. In the waters of the Pacific Ocean 2·36 mg org. C/l have been determined.

Cl content and salinity — chlorinity and salinity are the criteria of salt concentration in sea water. The term chlorinity is used to express the total amount of

Cl`, Br` and I`(in g) in kg of sea water, on the presumption that Br` and I` ions are replaced by Cl` ions. The chlorinity of normal marine water is 19·4 per mille. Salinity can be computed from chlorinity after Knudsen's formula:

$$\text{Salinity (per mille)} = 0{\cdot}030 + 1{\cdot}8050 \times \text{chlorinity (per mille)}$$

The relation of individual elements to chlorinity is generally stable; some departures have been observed in brackish and highly saline waters, particularly as concerns Ca, either due to its consumation by organisms, or to the solution of freshly deposited carbonate on the bottom. For practical computation the so-called chlorosity, i. e. chlorinity expressed in g/l water at 20 °C, is used. The composition of sea water is also characterized by its alkalinity, which expresses the content of cations bound on weak anions. As, besides small amounts of BO_3', CO_3'' forms the only other weak anions of sea water, this value represents in fact the amount of dissolved carbonates. It is expressed as the number of milli-equivalents of H· ions needed for the liberation of ions of weak acids in that amount of water which at 20 °C has a volume of 1 m³.

Gases in sea water

The content of dissolved gases is often omitted in the data on the composition of marine water, although these can appreciably affect the properties of waters. H. Sverdrup *et al.* (1942) give the following concentrations of gases in marine waters (Table 101).

Table 101

Concentration of gases in sea water (after H. Sverdrup *et al.* 1942)

Gas	Concentration ml/l
Oxygen	0–8·5, sometimes more
Nitrogen	9·4–14·5
Carbon dioxide	34–56
Argon	0·2–0·4
Helium	$1{\cdot}2\times 10^{-4}$
Neon	$1{\cdot}8\times 10^{-4}$

Oxygen — the concentration of oxygen in sea-water depends on the current system of water masses. The descending currents which were in contact with the atmosphere are usually saturated with oxygen, while the ascending currents are poor in oxygen. In the uppermost water layers the content of oxygen is near the limit of saturation, in places of intense photosynthesis even oversaturation is possible. The oversaturation of Antarctic waters can be up to 12%. In the upper water layers day and night changes in the content of oxygen have been commonly observed; they are ascribed to photosynthesis and respiration of phytoplankton. The waters are richer in oxygen during the day, when photosynthesis is active, than at night. Below the upper, oxygen-rich layer the concentration of this gas decreases everywhere. The minimal content of dissolved oxygen has been ascertained in all oceans between 200 and 1,000 metres. The origin of this, on the whole, regular layer of minimum O_2 concentration present in all oceans has provoked much discussion. Some writers explain it by the reducing processes at the sinking of organic substances from the upper layers to this level; others believe that the oceanic circulation of water masses is mostly responsible for the development of this zone. It is thought to represent a huge water mass of arctic origin which has lost oxygen on its way southwards. The temperature also affects the amount of oxygen in sea water. Dissolved oxygen generally increases in amount with the decline of temperature; therefore polar waters are always richer in oxygen than tropical waters.

Nitrogen — sea-water is usually moderately undersaturated with nitrogen. The content is influenced by temperature and its concentration depends on chlorinity.

Inert gases — argon and other inert gases form 2·7% of the total content of nitrogen. Their amount varies with the concentration changes of this gas. In some parts of water masses the contents of helium and neon are moderately increased.

Other gases — fairly great amounts of hydrogen sulphide and methan occur in stagnant waters. Traces of gaseous hydrogen have been found in shallow littoral waters.

The amount of gases in marine waters depends on:

1. Temperature and salinity.
2. Biological activity, particularly life activity of phytoplankton.
3. Marine currents and the mixing of waters.

The composition of marine water, as given above, is only found in an open system, i. e. in oceans and seas with free water circulation. The conditions are somewhat different in places with a limited circulation. In stagnant waters, devoid of circulation, the composition is controlled chiefly by the life activity of anaerobic bacteria.

The concentration of hydrogen ions. In the upper layers of oceanic waters the concentration varies within narrow limits — between 8·0 and 8·3, dropping only rarely below this value. Deeper below the surface, particularly in the zone poorest in oxygen, where organic substances are decomposed, the environment is more acid, pH range between 7·6—7·8. The value of pH increases with the increase of oxygen content and decreases with the increase of CO_2. There is also an interdependence between pH, temperature, and pressure; pH is indirectly proportionate to pressure and the production of phytoplankton and directly proportionate to temperature. With the increasing activity of phytoplankton the content of CO_2 in water decreases and with the increasing content of oxygen the value of pH also increases. The lowest limiting value of pH in basins of normal salinity is 7·5 (these low values have so far been determined in some deep places in the Pacific Ocean) as sea water cannot accomodate a higher percentage of CO_2 which would lower the boundary still more. In isolated or semiisolated basins the presence of hydrogen sulfide can lower the pH value to 7. The pH increases somewhat just below the water surface and drops downwards to a minimum at a depth of 400—800 m. Below this level it increases moderately to a constant value which is persistent up to the greatest depths. Some data indicate that the deep waters of the Pacific Ocean have a lower pH than those of the Atlantic Ocean.

The CO_2—H_2CO_3—HCO_3'—CO_3'' system. The equilibrium conditions of this system in marine waters are extremely intricate; this system can be expressed schematically by the following relation:

$$\begin{matrix} CO_2 \\ \upharpoonleft\downharpoonright \\ H_2CO_3 \end{matrix} \rightleftharpoons HCO_3' + H \rightleftharpoons CO_3'' + H$$

With the increase of acids, the equilibrium moves towards carbon dioxide; with the increase of alkalis, it moves towards the carbonate ion. None of these changes, however, produces any change of pH, because the carbonate system is a perfect buffer. The equilibrium of the system and the number of end members depend on the temperature, pressure and concentration of hydrogen ions. With the increase in temperature free CO_2 escapes and bicarbonate ions pass into carbonate ones. On the other hand, with the increase in pressure the amount of free CO_2 increases and carbonate ions pass into bicarbonate ones, in other words the carbonate dissolves. With the increasing pH value the amount of free CO_2 declines and the content of carbonate ions rises. The maximum amount of free carbon dioxide in the oceans is within the zone between 200 and 1,000 m, i. e. in the layer containing a minimum amount of oxygen. Henceforth it drops slowly to a depth of about 2,000 m and then remains constant. At a small height above the bottom, an increase of concentration has occasionally been observed, which is accounted for by the dissolution of freshly deposited calcium carbonate. The deep water can also be enriched by carbon dioxide originated by the bacterial decomposition of organic matter sinking from the upper water masses. There exists an equilibrium and interchange between atmospheric carbon dioxide and CO_2 contained in the uppermost water layer of the world oceans. In the lower geographical latitudes the atmosphere takes CO_2 from the water and transports it by circulation to higher geographical latitudes where it is absorbed by water and transferred to the equator by oceanic currents. The solubility of $CaCO_3$ and the factors affecting it can be summed up as follows:

1. The increase in concentration of carbonic acid, i. e. the decrease of pH increases the solubility of $CaCO_3$.

2. The increased content of Ca^{2+} ions from other sources causes the decreased solubility of $CaCO_3$.

3. The content of neutral salts:

a) affects the dissociation of carbonic acid, which is manifested by a slight decrease of solubility of $CaCO_3$;

b) affects to a minimum extent the product of $CaCO_3$ solution in that it increases its solubility by about one-hundreth;

c) reduces the solubility of CO_2 and thus decreases also the product of $CaCO_3$ solution.

4. Temperature affects the solubility of $CaCO_3$ in three different ways:

a) With the increase of temperature the degree of dissociation of carbonic acid increases, which results in the decrease of $CaCO_3$ solubility;

b) the value of the product of $CaCO_3$ solubility decreases with the increasing temperature (in contrast to other salts);

c) the solubility of CO_2 decreases with the increasing temperature which in turn produces a decrease in the solubility of $CaCO_3$.

From the above it is obvious that the question of the solubility of calcium carbonate is very complicated and depends on a whole series of factors. Practically, the main controlling factors are pH, temperature and partial pressure. According to H. WATTENBERG (1933), the upper water layers, especially in lower geographical latitudes, are strongly oversaturated with $CaCO_3$. In the Atlantic Ocean, in the proximity of the equator, the oversaturation amounts to 300%, and is of 100% value at 60 degrees of the northern and southern geographical latitudes. Most authors, however, do not regard this oversaturation as a proof of the chemical deposition of carbonates, because calcium carbonate shows a high tendency to the oversaturation in solution. Unless the nuclei around which precipitation takes place are present, it does not necessarily precipitate. Recently, some facts (not as yet safely substantiated) suggest the presence of a fair amount of calcium carbonate in the colloidal state in sea waters. There is a general consensus of opinion that deep oceanic waters are undersaturated with calcium carbonate. Therefore, chemical sedimentation of calcium carbonate in deep oceanic realms can hardly be assumed.

The relation of chemism to the biology of sea-water

Organisms appreciably modify the chemical composition of sea-water; on the other hand, the organic production is influenced by the presence of certain components acting as the main nutrients, which are necessary for the life activity of organisms—the compounds of nitrogen and phosphorus, silicon, and iron. The concentration of nitrogen and phosphorus is independent of salinity but depends on the organic production, the current system and other subsidiary factors. In the world oceans four layers with different concentrations of phosphates and nitrates can be distinguished:

1. The surface layer of a low concentration of nutrients.
2. The subsurface layer with a rapidly increasing concentration of these substances.
3. The layer of maximum phosphate and nitrate concentration, reaching from 500 to 1,500 m.
4. Oceanic abysses with a high and almost constant concentration of nutrients.

The surface layer of a low concentration of nutrients is thickest in low geographical latitudes, the thickness dropping to zero in the areas of ascending currents. Away from the equator, the thickness of the surface layer decreases and to the N and S of the 50° of latitude this layer is absent.

The direct proportion of the phosphorus and nitrogen contents can be expressed by the equation $N : P = 15 : 1$.

The total amount of nutrients is controlled by the equilibrium between the introduction of nutrients and their consumption by organisms. Owing to the introduction from the landmass a high concentration of phosphates and nitrates is usually present near the shore. This leads to considerable organic production in the near-shore marine zones. In the upper water layers of the open seas the concentration of nutrients are not high because these are soon consumed by organisms unless their reserves are renewed. The nutrients accumulate in the sheltered parts of the basin with a restricted circulation. Thus, for instance, in the Norwegian fjords the concentration of phosphates and nitrates is on the average twenty times higher than in open seas. Seasonal changes in the concentration of nutrients have frequently been observed in shallow littoral waters. The concentration decreases to a minimum at the time of the intensive production of phytoplankton and increases rapidly in the winter months. An analogous cycle and an even shorter one (of 24 hours) exist in the upper layers of the oceans but it is not so clear-cut.

Phosphates: In all oceans the zone with the maximum content of phosphates extends between 750 and 1,000 m below the surface. At this depth the organic matter sinking from the upper plankton-rich layers decomposes. The maximum concentration of phosphorus is there up to 80 mg/m^3. The regeneration of organic phosphorus compounds to inorganic compounds can occur both directly and indirectly. The direct regeneration occurs by the decomposition of faecal pellets, the indirect one by the decomposition of the bodies of organisms.

Nitrogen: The cycle of nitrogen compounds in marine waters is far more complicated than the phosphorous cycle, and consequently, it is less known. Unlike phosphorus, which in an inorganic form occurs only as phosphate, nitrogen forms a series of widely distributed inorganic compounds: nitrates, nitrites and ammonia. The bulk of nitrogen in marine water is in the form of nitrates, whose distribution is similar to that of phosphates. In the Arctic seas, at a depth of 300—1,000 m, sea water contains 270 mg N in nitrate form in 1 m^3 of water, in the Northern Atlantic

245—250 mg/m^3 at the surface, and in the Sargasso sea 250 mg/m^3 at a depth of 2,000 metres. Similarly as with phosphates, the highest concentrations of nitrates have been measured in the Arctic seas. Waters of the Pacific and the Indian Oceans are richer in nitrates than the Atlantic Ocean. A minor amount of nitrogen occurs in the form of nitrites, which in the surface water layers attain only 3—10 mg/m^3. Their content moderately increases with depth. The concentrations of ammonia are not precisely known as yet. According to sporadic data, the Arctic waters bear 30—50 mg N-ammonia in 1 m^3 near the water level, whereas in the Pacific Ocean up to 90 mg/m^3 have been found.

Neither the cycle of nitrogen nor the equilibrium of its three main components are as yet completely understood. The opinion prevails that decaying organic matter is converted first into ammonia by mineralization, and then it is oxidized to nitrite and nitrate. As is universally known, bacteria which obtain the necessary energy by oxidation of nitrogen compounds play an important role in this process.

The biology of sea water

There is an intimate relation between the organic production of the sea and the presence of nutrients. The interrelationship of the abundant occurrence of phytoplankton and the ascending marine currents has long been known. Because of the life activity of phytoplankton the upper water layers are persistently impoverished in nutrients which ultimately sink and accumulate in the lower layers. These, however, are deficient in light, and this prevents the existence of phytoplankton. The maximum development of phytoplankton is possible in those places where the prerequisites for its life (light and nutrients) are present, i. e. where the deep waters ascend to the surface. This may happen in the following ways:

1. By convection currents, the mechanism of which consists in the instability of the water column and in the changes of density produced by surface cooling. Therefore, they should be taken into consideration mainly in polar regions.
2. By the compensating deep currents which are a component of oceanic circulation (upwelling). The western coasts of Africa and South America afford a classic example of this.
3. By the ascent of horizontal currents across ridges.
4. By the effects of horizontal surface whirls which make possible the ascent of deep waters.
5. By turbulence which is of importance particularly in shallow waters.

The organisms affect the chemical composition of sea water and the circulation of some elements. K. Kalle (1943) computed the factors by which the percentage of elements should be multiplied to get their respective content in marine organisms (Table 102).

Table 102

Table indicating the factor by which the percentage of element in sea water should be multiplied to obtain their respective content in marine organisms
(after K. Kalle 1943)

Element	Enrichment factor	Element	Enrichment factor
Na	0·28	K	2·5
Mg	0·73	Ca	1·2
C	1,100	Si	20,000 (diatoms)
N	17,000	P	20,000·2
S	1·1	Cl	0·2
Cu	500	Zn	4,000
B	0·4	V	1,000
As	6·5	Mn	400
F	0·7	B	0·04
Al	8·5	Ra	40
Fe	2,000 (phytoplankton)	Fe	800 (zooplankton)

Some organisms and groups preferentially concentrate certain elements from sea water, as shown by the following examples.

Si — Diatoms and radiolaria
Sr — Radiolaria
Ba — *Xenophora* (protozoa)
Fe-Mn — Bacteria
S — Sulphur bacteria
Cu — Molluscs
I — Algae and fungi

Phytoplankton and the conditions of its evolution

Of practical importance for the Recent and ancient sediments are diatoms and some flagellates. As mentioned above, light and the presence of nutrients are decisive for the development of phytoplankton. On the basis of light energy several layers can be distinguished in a vertical profile through the ocean.

1. Euphotic layer which extends roughly to a depth of 80 or more metres. Provided the nutrients are present in a sufficient amount, it provides the most suitable conditions for the growth of phytoplankton.
2. Dysphotic layer reaches from 80 to 200 m. The production of phytoplankton is absent from it and only the decomposition of planktonic detritus, sinking from the upper layers, takes place here.
3. Aphotic layer extends from a depth of 200 m to the bottom.

Photosynthesis alters the chemical composition of sea water, which becomes enriched in oxygen and improverished in carbon dioxide. This results in the decreased concentration of H˙ ions and the shift of the carbonate system equilibrium towards carbonate ions (the decreased solubility of calcium carbonate affords possibilities for its chemical precipitation). On the other hand by the respiration occuring during lack of light, water is enriched in CO_2 and this exerts a reverse influence on the equilibrium of carbonate system. At a certain depth below the surface, called the compensation zone, there is a suitable amount of light for the photosynthesis and respiration to become balanced. It has been determined experimentally that in some parts of oceans and seas phytoplankton has enough light available for photosynthesis at a depth of as much as 100 m (e. g. in the Mediterranean or Sargasso Seas). Owing to the increasing amount of suspended material this depth decreases landwards; it is only a few metres in estuaries.

In addition to the above factors, other agencies of subsidiary significance also effect the production of phytoplankton. For instance, it has been ascertained that, all optimum conditions given, the production of phytoplankton in an open sea can be decreased by the lack of iron. Some writers think the lack of Mn and probably also of Mo and Ga has a rather unfavourable influence. Recent experiments have shown that the production of phytoplankton is not fully stopped even by the total absence of phosphates and nitrates; yet it is proved to be strongly suppressed and organisms built their cells without these components. Turbulence affects very unfavourably the production of phytoplankton by dragging the organisms into lower water layers where they perish. In this way some shallow-water areas, in spite of otherwise propitious conditions, are impoverished in phytoplankton.

The ecology of diatoms, the recognition of which facilitates the understanding of Recent deposition of diatomaceous oozes, will be dealt with in greater detail. In addition to light and nutrient, it is the decreased salinity and lower temperature which contribute to the evolution of diatoms, in contrast to the zooplankton, which demands undiminished salinity and warm water. The abundance of diatom tests in polar and some other oceanic and marine regions is immense. Thus, for instance, in the North Pacific there are 220,000 diatoms in 1 litre of water; some nearshore waters are 50 times more productive. The average content of diatom tests in the Atlantic Ocean is 1,000 tests in 1 lit. water. Their amount increases away from the equator, rising slowly to the lat. 20°N and then rapidly to 60°N, where it attains the maximum value. It is estimated that phytoplankton produces 5 tons of hydrocarbons on 1 km^2 of ocean surface.

Marine zooplankton

Marine zooplankton, which contributes largely to the Recent deep-water sedimentation, consists mainly of foraminifera, radiolaria, and in tropical latitudes also of pelecypoda. As mentioned above, their living conditions differ from those of the phytoplankton. Both foraminifera and radiolaria live to a depth of 100 m but can occur even deeper (living planktonic foraminifera have been found in depths below 2,000 metres). As is well known, planktonic foraminifers are far more plentiful but poorer in species than benthic foraminifers. In tropical regions about twentyfive species of planktonic foraminifers have been determined. Their population density is considerably lower than that of diatoms. Thus, for instance, in the Gulf of Mexico more than 73 tests have been recovered from 1 m^3 of water, but in the open oceans the density is lower. In the Kuril-Kamchatka trench the amount of zooplankton changes with depth as shown in Table 103.

Table 103

Changes of zooplankton amount with depth in the Kuril-Kamchatka trench (after L. A. Zenkevich - J. A. Birnstein 1956)

Depth m	Amount of zooplankton mg/m^3
0— 50	297·6
50— 100	320·4
100— 200	246·6
200— 500	228·0
500—1,000	59·3
1·000—2,000	21·8
2,000—4,000	9·3
4,000—6,000	2·64
6,000—8,000	0·84

Table 104

Changes of amount of benthos with depth in the Kuril-Kamchatka trench (after L. A. Zenkevich - J. A. Birnstein 1956)

Depth m	Amount of benthos mg/m^3
950—4,070	6·94
5,070—7,230	1·22
8,330—9,950	0·32

Flagellates are important in the production of calcareous oozes. In the Mediterranean Sea up to 24,000 individuals of *Coccolithus fragilis* have been obtained from 1 litre of water. Tests of these organisms are, however, very minute and are poorly resistant so that they soon disintegrate both chemically and mechanically and only about 1/20 of the original production gets into abyssal oozes.

Recent sedimentation is also affected by the presence of benthos. The amount of benthos decreases with the depth, but by means of submarine photographs abundant trails of crawling benthic organisms have been established even deeper than 4,500 m. The decrease of marine benthos with depth in the Kuril-Kamchatka trench is well seen from Table 104.

Bacteria

The importance of bacteria in the development of the chemistry of marine water has not yet been fully appreciated. There are two kinds of marine bacteria:

1. Autotrophic, forming carbohydrates and carbon dioxide from inorganic salts, similarly as does phytoplankton;

2. Heterotrophic, obtaining their energy from the oxidation of organic compounds. Most bacteria are heterotrophic.

Although marine bacteriology is a very young science, it was already recognized in 1939 that 1,335 species of bacteria occur in sea-water and sediments. Bacteria affect both the Eh and pH of sea water. The pH value is decreased by the following processes.

1. Production of carbon dioxide as a result of respiration.
2. Production of organic acids by the decomposition of carbohydrates.
3. Oxidation of hydrogen sulfide or free sulfur to sulfuric acid or acid sulfate.
4. Reduction of free sulfur to hydrogen sulfide.
5. Formation of nitrite or nitrate by the oxidation of ammonia.
6. Assimilation of ammonia as the nitrogen source or its oxidation as the source of energy.
7. Liberation of phosphate from organic compounds.

The increase of pH is caused by:

1. Consumption of CO_2 by chemosynthesis or photosynthesis of autotrophic organisms.
2. Oxidation or decarboxylation of salts of organic acids.
3. Reduction of sulfates to sulfur.
4. Reduction of nitrates and nitrites to ammonia.
5. Formation of ammonia from nitrogen compounds, such as aminoacids, proteins, urea, etc.

The distribution of bacteria in sea water is not uniform. They usually occur in swarms or are accumulated round particles of organic matter or tests of planktonic organisms. Most bacteria occur in shallow littoral waters. Sea-port waters may contain more than 1,000.000 bacteria in 1 ml of water. In inland seas their content is about 500/1 ml and in open seas roughly one-half of this amount. The controlling factor of the number of bacteria in sea-water is the amount of dissolved organic matter.

References

Buch K. (1939): Beobachtungen über das Kohlensäuregleichgewicht und über den Kohlensäureaustausch zwischen Atmosphäre und Meer im Nordatlantischen Ozean. Acta Akad. Aboensis, Math. et Phys., vol. 11: 1—32, Abo.

Harvey H. W. (1945): Recent advances in the chemistry and biology of sea water. P. 1—164, Cambridge.

Hentschel E. (1932): Die biologischen Methoden und das biologische Beobachtungsmaterial. Ergebnisse Dtschl. Atlant. Exped. Meteor, Bd. 10: 1—137, Berlin.

Hill M. N. (Editor, 1963): Op. cit. on page 10.

Kalle K. (1943): Der Stoffhaushalt des Meeres. P. 1—263, Leipzig.

King C. A. M. (1951): Depth of disturbance of sand on sea beaches by waves. Jour. Sedimentary Petrology, vol. 31: 131—140, Menasha.

Rankama K. - Sahama P. (1950): Geochemistry. P. 1—906, Chicago.

Revelle R. (1934): Physico-chemical factors affecting the solubility of $CaCO_3$ in sea water. Jour. Sedimentary Petrology, vol. 4: 103—110, Menasha.

Schäfer W. (1956): Gesteinsbildung im Flachseebecken, am Beispiel der Jade. Geol. Rundschau, Bd. 45: 71—84, Stuttgart.

Shepard F. P. (1963): Op. cit. on page 10.

Smith P. V. (1953): Studies on origin of petroleum: Occurrence of hydrocarbons in Recent sediments. Bull. Am. Assoc. Petrol. Geol., vol. 38: 1285—1299. Tulsa.

Sverdrup H. U. - Johnson M. W. - Fleming R. F. (1942): The oceans. P. 1—1060, New York.

Turekian K. K. (1969): Oceans. P. 1—252, New Jersey.

Wattenberg H. (1933): Das chemische Beobachtungsmaterial und seine Gewinnung. Kalciumkarbonat und Kohlensäuregehalt des Meerwassers. Wiss. Ergebnisse Dtschl. Atlant. Exped. Meteor, Bd. 8: 1—233, Berlin.

Wüst G. (1958): Über Stromgeschwindigkeit und Strömungen in der Atlantischen Tiefsee. Geol. Rundschau. Bd. 47: 187—195, Stuttgart.

Zenkevich L. A. - Birnstein J. A. (1956): Studies in deep water fauna and related problems. Deep Sea Research, vol. 4: 54—64, London.

15. Beach sediments

The morphology, development and terminology of the littoral zones are elaborated in detail especially in C. A. M. King's book (1959) and comprehensive chapters dealing with this problem can be found in F. P. Shepard's (1963) book and the Symposium "The Sea" (editor M. N. Hill).

The beach is the zone of unconsolidated material extending landward from the mean low-water line to the place where there is a change in material or physiographic form as, for example, the zone of permanent vegetation, or a zone of dunes or a sea cliff. The upper limit of the beach usually marks the effective limit of storm waves (F. P. Shepard 1963).

The composition and the mode of deposition of beach sediments are affected by these factors: wave action, littoral currents, tides, petrographical composition of the shore rocks and the general morphological and geological character of the shore. The origin, composition and preservation of beach sediments are usually the result of a precise equilibrium between these factors.

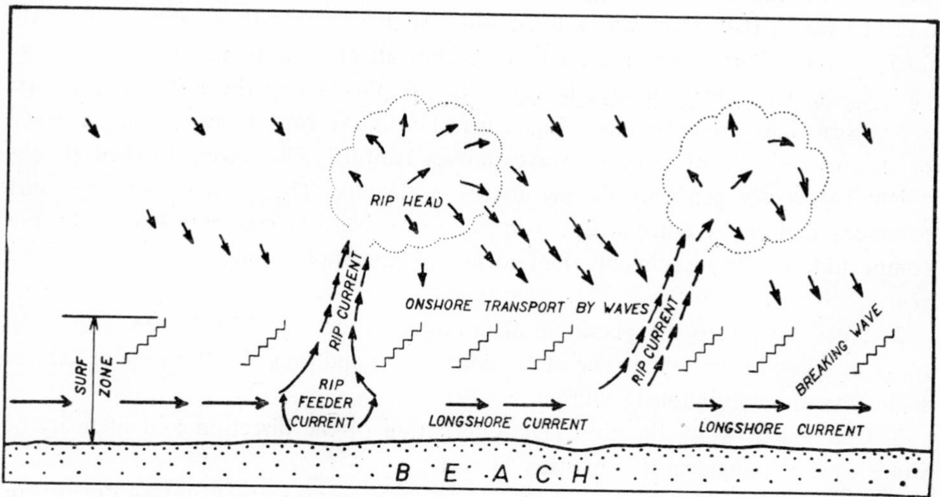

Fig. 66. Nearshore circulation systems and related terms. After F. P. Shepard 1963.

The intensity of the activity of waves and currents depends on the exposure of the shore which in turn depends on the size of the water body an the direction of the prevailing winds. The wave action in the littoral zones is affected by the submarine topography of the sea bottom, for several kilometres from the shore. Thus, for instance, marine elevations induce a concentration of wave energy, so that the littoral erosion is then many times stronger then under normal conditions. The longshore currents, generated by waves breaking at an angle to the shore line, are of great intensity. They are important for the littoral processes. Their normal speed varies between 15—17 cm/sec, but can attain a value of more than 125 cm/sec. They often transport an immense amount of detritus so that they can be compared to a stream rich in suspended and bed load. On the Atlantic coast of the United States longshore currents transport as much as 112,500 up to 369,750 m^3 of material annually. In the Gulf of Mexico somewhat lower values have been measured and in the area of the Great Lakes amounts smaller by about two-thirds. Rip currents, which originate probably by the accumulation of excess water at various places along the shore and its concentrated return, carry material offshore to deposit it where they slacken. They are most likely responsible for the very frequent forms called beach cusps, which are rather uniformly spaced ridges projecting into the sea. The troughs between ridges are places of the backflow of rip currents into the sea.

A beach is formed where the supply of material into the littoral zone is in excess of its removal. In a simplified manner the origin of a beach can be conceived as follows: If the development of the shore profile and of beach sediments was affected by the surf and other factors of the unchangeable intensity, three main zones would soon develop: zone of shoaling waves, zone of turbulence and breakers and zone of swash and backswash. Where the waves break, the coarsest material is accumulated and because at that point waves loose most of their energy they carry higher only finer material. Part of the material, sometimes all of it, is returned to the sea by backswash. This ideal picture, hewever, is complicated by the following factors:

1. Storm and extraordinarily high surf. During storms, there are larger wave orbits and the zone of breaking waves moves seaward. Also owing to their greater height the waves penetrate further to the backshore. The anomalously high surf produced mainly by autumn and winter storms exerts a greater influence on the composition of beach sediments than normal processes prevailing during most of the year.

2. Tides. Owing to the persistent fluctuation of the water level even within the range of several metres, the zone of breaking waves and thus also the zone of coarser sediments are continuously shifted.

3. Changes in wind direction. The alternation of the direction and intensity of winds causes changes in the intensity of surf.

4. The action of longshore currents. This causes an accumulation of sediments in sheltered places and their removal from exposed places.

The interaction of all those factors is so complicated that the investigation of their equilibrium is the subject of special study.

The shape of beaches and the composition of beach sediments depend largely on the geological composition and topography of the shore. The best developed and largest beaches occur in bays, whereas by the cliffs, and other exposed parts of the shore only poorly developed beaches can be found.

The source of beach material is mostly the detritus imported by rivers. It is true that in some places a considerable amount of material is loosened by intensive erosion of the shore, but wherever the intensity of marine abrasion has been compared with that of stream transport, the amount supplied by streams proved to be many times larger. It has been computed that littoral abrasion supplied about 5% of clastic material. Precise values obtained from part of the beach of the Black Sea gave 1,963,000 km^3 gravel imported by streams and only 430,000 km^3 gravel brought by small streams or littoral abrasion. Clastic material is deposited either directly on the beach or, more frequently, it is laid down farther seaward and is even brought back by waves and longshore currents.

The slopes of beaches are controlled mainly by backswash action. The finer the sediment, the less permeable it is and the more readily it succumbs to erosion. Therefore, fine-grained beaches are levelled by backswash and develop gentle slopes. Coarse sediments maintain steeper slopes because they are strongly permeable and the water of backswash percolates largely through their pores. It has been found by experiments and field observations that definite slopes of beaches correspond to following grain-sizes (Table 105).

Table 105

The relation between the slope of beach surface and grain-size of beach sediment

Grain-size mm	Angle of slope of beach surface degrees
$\frac{1}{16}$— $\frac{1}{8}$	1
$\frac{1}{8}$— $\frac{1}{4}$	3
$\frac{1}{4}$— $\frac{1}{2}$	5
$\frac{1}{2}$— 1	7
1— 2	9
2— 4	11
4— 64	17
64—256	24

These values give maximum angle of repose, but in reality the slopes are mostly somewhat gentler.

The intensity of wave action in the littoral zone is high enough to redeposit and sort gravels of all grain-sizes. Thus, it is evident that the composition of beach sediments is not affected only by wave action, but depends largely on the kind of clastic material supplied to the basin, as shown by the difference between gravel and sand beaches. Gravel beaches are less abundant than sand beaches because detritus of this grain-size is not everywhere available.

Gravel beaches

Gravel beaches can develop in those places where:

1. Gravel is supplied by mountainous streams (the Mediterranean Sea in the piedmont of the Alps, the Black Sea in the piedmont of the Caucasus).

2. The shore is built up of rocks which on being abraded readily form pebbles such as cherts and other silicites, granites, gneiss and other solid rocks, glacial sediments, etc. — for instance cliff shores of southern England, cliff shores of the Baltic Sea).

Gravel beaches are never very wide and their thickness is usually rather small. The mean grain-size is difficult to determine, but it is roughly estimated at 5—10 cm. Gravel beaches provide an adequate evidence of the importance of exceptionally high surf during storms for the origin and morphology of this kind of shore deposit. They are, namely, coarsest in backshore and get finer-grained towards the foreshore, changing sometimes into sands. The steep slope of gravel beaches (about 10°) is responsible for their small width, which most frequently does not exceed a few tens of metres.

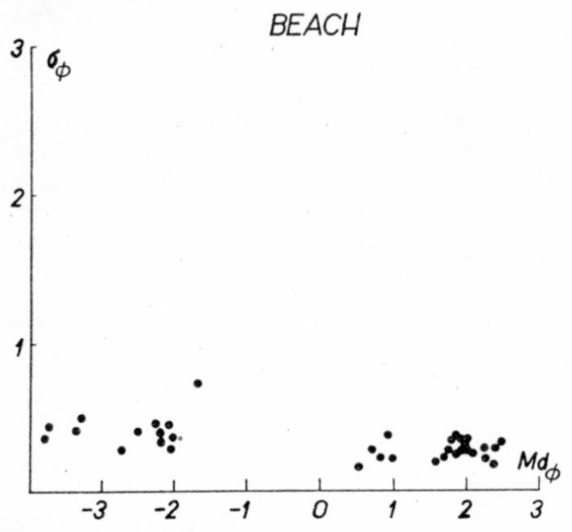

Fig. 67a. Grain-size parameters of beach sediments. Relation of Phi Median Diameter and Phi Deviation Measure. Data compiled from various authors.

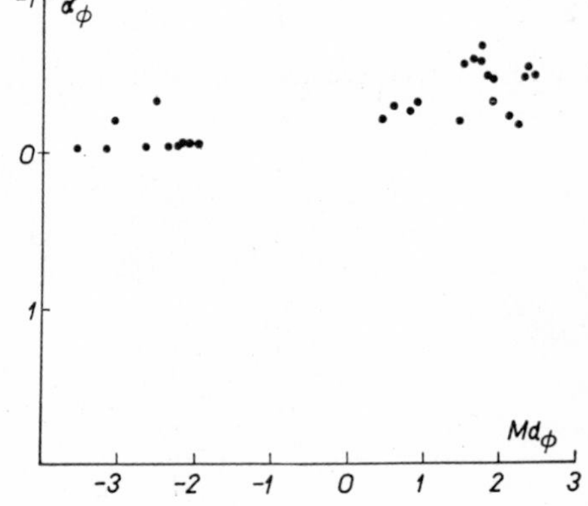

Fig. 67b. Grain-size parameters of beach sediments. Relation of Phi Median Diameter and Phi Skewness Measure.

The grain-size parameters of gravel beaches are given in Fig. 67 a,b. The values of sorting coefficient are very low and, as a result, beach gravels are the best sorted sediments of all gravel-grade deposits. After burial, however, the overlying finer sediments can fill in the voids and thus modify the primary character of the deposit. The Phi Skewness (α_Φ) of beach gravel usually fluctuates within moderately negative values, which indicates a minor prevalence coarser material. The rounding of beach pebbles is generally perfect and is nearly always greater than Krumbein's values 0·6—0·7. It has been statistically proved that the pebbles are mostly of disk shape and show a conspicuous preferred orientation with the long axis perpendicular to the shore line. Besides, they inclined at a small angle (about 10°) seawards; but cases of the inverse imbrication have also been observed. The petrographical composition of beach gravels can be most varied. Gravels supplied by streams are more variable in composition and do not correspond to the composition of the adjacent coast. Gravels generated by abrasion are of the same composition as the shore cliffs and are usually monomictic. The differences between abrasion and denudation gravels are analogous to those ascertained for lake gravels (see p. 163).

Where the beach sediments are in equilibrium with the present-day hydrodynamic conditions, the sorted beach gravels do not reach deeper than 10 m. Where they represent the extension of large alluvial cones washed down the nearby mountains directly to the sea, the gravel deposits (which are more poorly sorted) can stretch to depths of a few hundred metres. The thickness of gravel beaches is generally of several metres; a greater thickness has been ascertained only when the beach is a part of an alluvial cone.

Gravel beaches can be divided into movable and immovable ones according to their relation to the present-day hydrodynamic conditions. Movable gravel beaches correspond to the recent hydrodynamic conditions and as these change continuously, the grain-size and morphology of beaches also change. They are persistently shifted by the longshore current along the shore and move alternately land- and seaward by other shore processes. Immovable beaches are no longer in equilibrium with present hydrodynamical conditions and the grain-size composition of their sediments does not become greatly modified.

Gravel material can also accumulate on the bars and ridges as these are the places of the primary wave breaking and thus the most exposed part of the shore zone. When fractions coarser than the sand grade are present in bars, they are always concentrated on the upper seaward side.

Sand beaches

Sand beaches are far more common than gravel beaches. It is estimated that sand makes up more than 95% of all beach sediments. Sand beaches sometimes cover areas of immense extent and pass into widespread low sandy plains; their width can reach up to tens of kilometres. The grain-size distribution of sand beaches is usually nearly uniform all over the area or coarser sediments are found in two sections: in the zone of wave breaking, as mentioned above, and in the zone of maximum water level at high tide, where coarse material brought by the surf is partly left behind. The grain-size parameters of sand beaches are plotted in diagrams in Fig. 67 a,b.

Md frequently ranges from 0·3 to 1·0 mm. A boundary can be drawn between beaches with sediments of finer and coarser grade than 0·20 mm, because these types

Table 106

Differences between the beaches formed of finer-grained and coarse-grained deposits

Beaches with coarse-grained deposits (Md >0·20 mm)	Beaches with finer-grained deposits (Md <0·20 mm)
Frequently several tens of metres wide. Steeper slopes	width up to several km, slight slope
Beaches of oceanic coasts	beaches of smaller seas, bays, lagoons
Beaches on the exposed parts of coasts	beaches in sheltered bays
Beaches on seaward side of bars, barriers and spits	landward or lagoonward sides of bars and barriers

Table 107

Changes in grain-size of beach in California
(after D. L. Inman - G. A. Rusnak 1956)

Date	Md μ	Md_{Φ}	σ_{Φ}	α_{Φ}
17. 7. 1953	153	2·71	0·38	−0·03
23. 7. 1953	146	2·78	0·38	−0·05
12. 8. 1953	146	2·78	0·42	−0·05
14. 8. 1953	145	2·79	0·45	0
25. 8. 1953	152	2·72	0·45	0
31. 8. 1953	151	2·73	0·42	−0·07
11. 9. 1953	147	2·77	0·42	−0·07
15. 10. 1953	137	2·87	0·45	−0·09
21. 10. 1953	137	2·87	0·45	−0·09

somewhat differ in the mode of occurrence and of deposition. The differences are apparent from Table 106.

The sorting of beach sediments is nearly always good and the values of Standard Deviation (σ_{Φ}) rarely exceeds 0·5 (or Trask's value So 1·2). The perfect sorting is produced by persistent movement of the sediment and the removal of fine fractions. Beach sediments are characterized by a moderately negative Phi Skewness (α_{Φ}), because of the small percentage of coarser material.

The grain-size distribution of movable sand beaches changes continuously, depending on the continuously changing hydrodynamical conditions. The changes of all grain-size parameters can attain considerable values even at one place, as evidenced by the studies of beaches in California (D. L. INMAN-G. A. RUSNAK 1956) and of the changes in grain-size with time (Table 107).

The existing analyses of beach sands give a picture of the grain-size distribution of the living sediment which, being at equilibrium with its environment, adapts itself to it with a certain small retardation. These sediments are continuously reworked and their composition alters every moment. On the other hand, there are much fewer analyses of "dead" beach sediments that are already in a state in which they would be preserved in geological history.

Beach sediments very often pass gradually into eolian dune deposits, which are distinguishable from the former only to a certain extent by grain-size parameters. Some generalizations can be drawn from the differences ascertained between beach and dune sands in the Gulf of Mexico (Table 108).

Table 108

Differences between beach and dune sands in the Gulf of Mexico

	Md mm	σ_{Φ}	α_{Φ}
Beach sands	0·142	0·309	0·03
Dune sands	0·138	0·273	0·14

Table 109

Changes in composition of sands on various parts of beaches (Azov Sea, after N. V. Logvinenko - I. N. Remizov 1964)

Fraction mm	Backshore %	Foreshore %	Offshore %
>10			
10—7			
7—5			
5—3		0·2	
3—2		0·3	6·2
2—1		0·5	0·1
1—0·5		3·8	tr— 0·5
0·5—0·25	tr— 5·3	1·0—41·2	4·0— 8·1
<0·25	35·0—94·7	6·7—54·0	11·0—91·1
Md	0·166	0·234	0·182
So	1·63	2·34	1·82

From the above and other data it can be inferred that beach sediments are somewhat finer-grained than sediments of the adjacent dunes. Beach sediments have a slightly poorer sorting and their Phi Skewness tends to show negative values. Table 109 shows that in the cross-section of sand beaches some changes in the grain-size composition are observable (after N. V. LOGVINENKO - I. N. REMIZOV 1964).

The rounding of beach sand grains is almost always perfect. Most sand grains of 0·1—0·25 mm fraction have a higher value than 0·7 (Krumbein scale). According to S. V. MARGOLIS (1968) sand grains from beaches with low wave activity exhibit oriented etch pits, attributed only to the solution by sea water. By contrast, quartz sand grains from beaches with higher wave energy show a combination of chemical etch figures and phenomena caused by grain-to-grain impact. It is noteworthy that in spite of the action of longshore currents most grains show a preferred orientation with the long axis perpendicular to the shore line (J. R. CURRAY 1956).

Under present conditions, sorted beach sands occur in oceans and seas generally to a depth of 10—12 m, in inland seas to 5—8 m and in minor basins and bays to 3—5 m. Where beach sediments are found at present at greater depths, they undoubtedly represent the relics of glacial sedimentary conditions, when the sea level was much lower. The transition of beach sediments into shallow-water deposits is usually very slow. With the increase of silt and clay, the sands grade into silty and clayey sands and finally with the increasing depth into finer sediments. The zones of transition depend on the slope of the bottom; where the slope is gentle, they can be even several kilometres broad.

Coastal sediments are also made up of deposits of spits, bars, barriers, sand ridges, transverse, longitudinal and crescent ridges, which corresponds in grain-sizeand other properties to beach sands. Coarser sediments occur on the windward side of the barriers and when these are still a few metres under the water-level, coarse material covers their crests. When, however, they reach the sea-level or close to it, the coarsest sediments migrate precisely on this windward side of the barriers.

The mineral composition of beach sands is on the whole uniform. Sands contain, as a rule, about 80% quartz. There are exceptions to this rule, as for instance, beach sands with 75% of basic plagioclases or sands consisting of augite-olivine mixture, but these are of rather minor and local occurrences in places where basic rocks undergo intensive abrasion and the beaches are supplied with grains of their minerals. Normally, the content of plagioclases in beach sands corresponds roughly to their content in the source material, i. e. in the material imported by streams. The decrease in the content of feldspars away from their source is explained rather by the mixing with quartz material than by disintegration of feldspar grains. There is a direct relationship between the mineral and grain-size composition of sands. The percentage of feldspars decreases towards the finer fractions, so that they are very scarce under the limit of 0·1 mm. Fragments of rocks sometimes appear already in fractions above 0·25 mm. Beach sands are occasionally characterized by a high mica content, which is associated with the silt fraction and enclosed in the pores between larger grains. The content of mica is never large but helps to distinguish beach sands from dune sands which bear mica only exceptionally.

The content of carbonate in beach sands is variable; it is restricted usually to the coarser fractions because almost all carbonate derives from calcareous tests of organisms. The amount of tests and their fragments attains as much as 50%. The maximum concentration of tests is found either on the backshore or on the crests or landward sides of foreshore ridges. Owing to the sorting activity of waves and currents, the tests themselves can build up whole ridges; however this is more characteristic of tidal flats and will be dealt with below.

Some Recent and Subrecent beach sands are also cemented by inorganic cement. The precipitation of carbonate is induced by the alternation of physico-chemical conditions during day and night. The chemical composition of beach sands is usually as follows:

SiO_2	87·8700—99·72%	(after J. H. C. MARTENS 1939 and others)
Al_2O_3	0·0027— 6·60%	
Fe_2O_3	0·0700— 1·20%	
MgO	0·0200— 0·54%	
CaO	0·0500— 1·30%	
Na_2O	0·0080— 1·30%	
K_2O	0·0070— 1·70%	

The study of the representation of elements in various fractions has brought interesting results: The Al_2O_3 and K_2O components increase surprisingly towards the coarser fractions; their maximum is in 0·177—0·25 mm fraction, which means that they are joined mainly to feldspars just as Na_2O. The increase of Fe_2O_3 both towards the coarser and finer fractions suggests that its presence is connected with heavy minerals and, on the other hand, with the finest silt or clay fractions. In the backshore parts of beaches foreign coarse material is occassionally found, as for instance, abundant fragments of volcanic glass, pumice, or amber. In extreme cases these accessory components can be sorted, separated out and transported into bars.

Natural concentrates of heavy minerals are characteristic of beach sediments. Contrary to stream concentrates, the content of heavy minerals can attain as much as 100%. Their average amount varies between 1 and 2%. Heavy minerals are concentrated whenever periods of accretion alternate with periods of erosion; when these processes alternate recurrently, the deposition of light and heavy fractions and the removal of the light fraction result in a large concentration of the heavy fraction. Concentrations of economical importance originate in those places where the erosion strongly cuts into older sand accumulations. At the retreat of the shore line, sediments enriched in heavy fractions are left behind. A weak concentration of heavy minerals is observable already on the ripple crests; the shifting of the crests produces their concentration into laminae and beds. In a similar way accumulations develop on subaqueous ridges. They originate on the seaward sides of the ridges, which are most exposed to erosion, and by the lateral shifting of the ridges quite extensive layers enriched in heavy minerals are generated. The concentration also originates in the backshore and by undercutting of eolian dunes.

In general, it can be summed up that the greater the concentrations of heavy minerals are, the more the older sand deposits were exposed to erosion.

Structures of beach sediments

Beach sediments show different kinds of stratification. The hydrodynamical conditions alternate very rapidly in this environment and, as a result, the grain-size alternates abruptly in vertical direction and the dip of laminae and sets of laminae are changed.

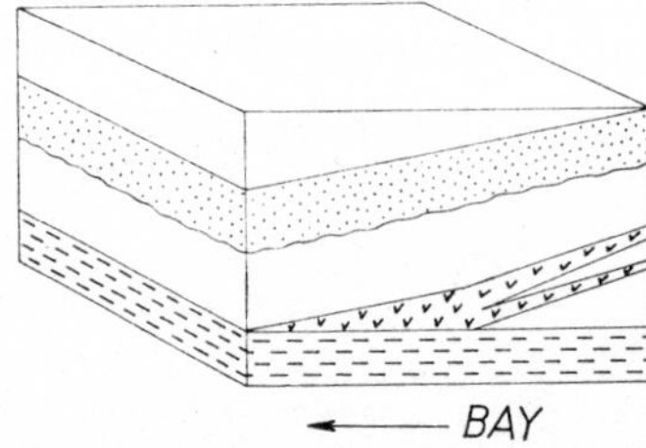

Fig. 68. Stratification of beach sands (white — fine grained-sands, dots — medium-grained sands, wedges — shell bed). After E. D. McKee (1957).

In the foreshore and backshore two principal types of cross bedding are distinguishable:

Foreshore: low-angle simple or planar cross-strata. The angle of dip is controlled in part by the slope of the shelf on which it forms and in part by the type of sediment of which the beach is composed. According to E. D. McKee (1957) quartz sand beaches of the Gulf Coast are typically very low angle, whereas those of the Pacific Coast of North America are somewhat steeper; shell beaches are commonly considerably steeper.

Backshore: A typical is trough cross-bedding. Backshore beaches commonly contain buried channels which are roughly parallel to the beach crest or berm. Such channels are typically irregular, and the sand that fills them is deposited with irregular structure. Also common are such unsorted materials as pieces of charcoal, shells, or debris, and, in places, intraformational conglomerate formed of weakly coherent, laminated lumps of beach sand, all of which are distributed here and there in the channelfill.

Sand bars (offshore): emergent offshore bars have seaward slopes which actually are beaches superimposed on bars. The crossstrata of these are low angle like those of ordinary beaches, but these strata are associated with steeper dipping beds of the bar that face shoreward. The dip of these strata is mostly between 18 and 28°.

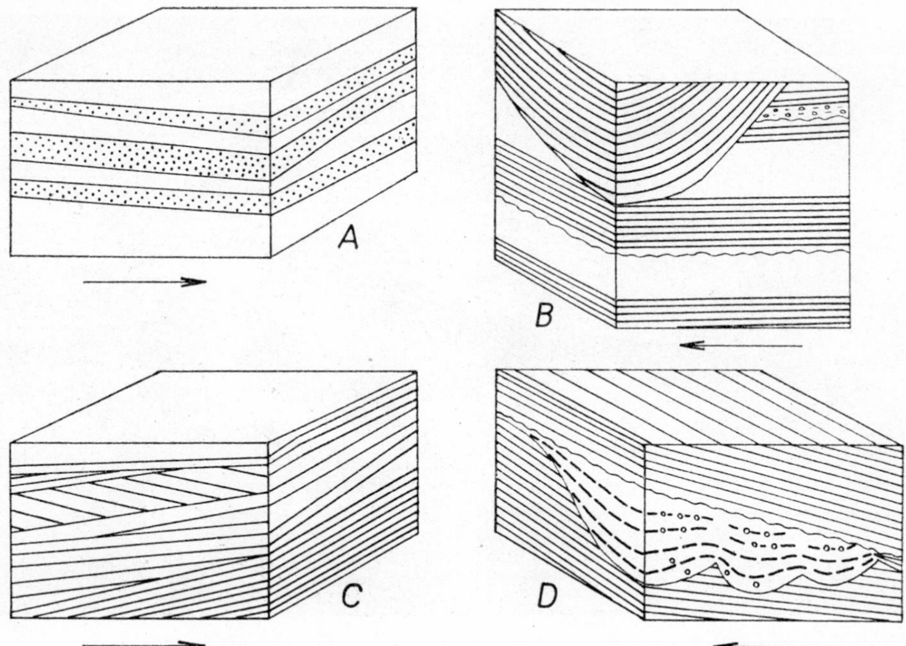

Fig. 69. Various types of stratification of beach deposits. The arrows indicate offshore direction. After E. D. McKee (1957).

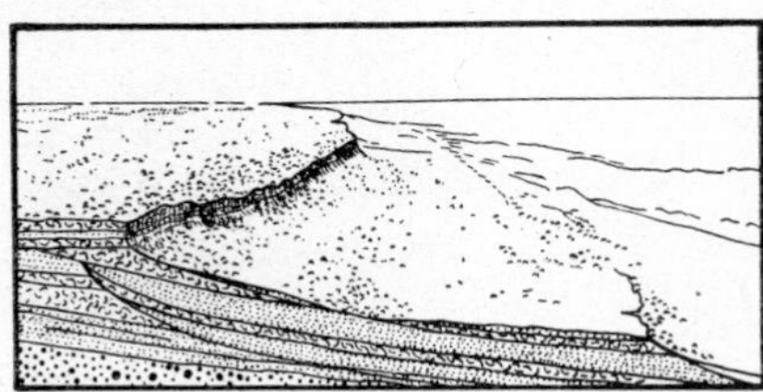

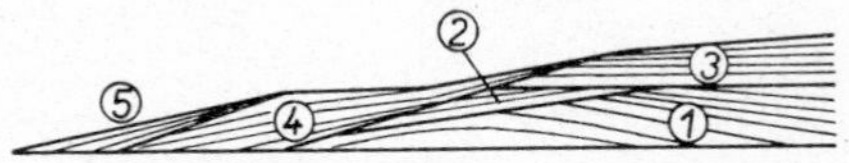

Fig. 70. Cross bedding of beach sediments. Azov Sea, spit Obitochnaya. 1, 2, 3. Sets formed by storms and strong waves, 4, 5. Sets formed by weak waves. After N. V. Logvinenko - I. N. Remizov (1964).

In finer-grained beach deposits an irregular distortion of laminae is often observed, particularly on beaches of lagoons, on leeward beaches of spits and in similar environments. The distortion is explained by the escape of air from a rapidly deposited waterlogged sediments. These structures resemble convolute structures.

Ripple marks are the most frequent surface structure of beaches. They are formed in all sediments coarser than silt, whenever the speed of the orbital movement of water is 10—100 cm/sec. The ripple marks are developed in most varied types and,

Photo 1. Upper photo: Barrier islands, lunate sandkey, and lunate bar at Stump Pass along the west coast of Florida near Fort Myers. Islands in background are old barriers beyond which new barriers have been built and the sandkey has formed since 1939. Various transverse bars are shown faintly through water. Lower photo: Cuspate spits built from barrier island into lagoon along the northwest coast of Florida. Pattern of reticulated bars is shown between two spits. After F. P. Shepard (1952).

therefore, the beach environment provides the basis for their classification and the description of the basic types. They occur in the foreshore in the backshore as well as in the offshore. In the foreshore, ripple marks cover large areas of several square kilometres in size.

About 80% of beach ripple marks are symmetrical; their wave length ranges from 3·5—20 cm and their amplitude roughly equals one fourth to one tenth of the length. The size of ripple marks depends on the grain-size, both the length and the amplitude increase with the increase of grain-size. On open, exposed beaches the crests of ripples are fairly regular. On beaches of deep gulfs the ripple marks show rather a tendency to develop an irregular arrangement of crests; they can even pass into linguoid ripples. The steepness of ripples also changes with the environment. Steeper ripples with a larger amplitude originate rather in sheltered environments (on the beaches of lagoons and gulfs), where the waves are lower and their orbital speed smaller. Most ripples are oriented with their crests parallel to the shoreline. It is estimated that 63% of ripples are orientated parallel to the shore line, 10% at a right angle, and the remainder diagonally to it. Rhomboid ripples are a frequent phenomenon of beaches. They are roughly diamond- or rhombohedral-shaped and are usually elongated in the direction of current movement. They are found in about equal numbers on the upper and lower foreshore. These ripples are also present on the backshore but are less often seen due to their irregular and infrequent formation and to modification and burial by wind action. They are formed by the wave backwash flowing seaward in area of gentle beach slope (0·5—2°). There is a direct relation between the length-width ratio of rhomboid ripples and the slope of the beach surface (Fig. 134). So these marks, which are abundant on modern beaches, are also useful environmental indicators.

Rill marks, another particular structure of beaches, occur both in the upper foreshore and the backshore. In the latter environment they are usually preserved; they originate, similarly as rhomboid ripples from backswash, the seaward flow being concentrated into rills. Where the top of the wave comes to a halt the water sinks into the sand and leaves behind lines of beach debris or mica, called swash marks. These are destroyed by each wave that extends to a higher level. Another type of markings develops where the backswash encounters obstacles. These obstacles divide the flow and leave streaks of dark sand in a diagonal pattern called backswash marks.

Bioglyphs (tracks and trails of animals) occur abundantly on the surface of beaches. The great variety of organic life on the beaches results in a great variety of forms and occurrence of bioglyphs, on the foreshore as well as the backshore. Some of them, which are significant for the restoration of the sedimentary environments of fossil deposits, are described on p. 452, 453.

The thickness and evolution of beach sediments

Beach sediments never attain a great thickness; rather they form a veneer on the underlying solid sediments. They are generally several metres up to a few tens of metres thick and do not increase appreciably with the positive movement of the water level either. Barrier sands, on the other hand, can attain a thickness of even several tens of metres.

The greatest thickness of beach sediments has been ascertained in the offshore zone. Landwards, they thin out or pass into continental deposits, most frequently of eolian origin. Laterally, they can grade into the sediments of lagoons, bays or they may wedge out. Seawards, they gradually become finer-grained and pass into shelf sediments of fine-grained deposits of lagoons and bays. In geological history, beach sediments are either the basal member of some formations or occur in the middle of the stratigraphical sequence. They are often disturbed both by positive (increased action of waves and currents) and negative (continental erosion) oscillation of the sea level, so that they are preserved only when buried by delta deposits or alluvial cones. At present, there are known instances of beaches or parts of them being suddently covered by fine-grained deltaic sediments as a result of the shift of a river mouth, or by alluvial cones of short ephemeral streams. Beach sediments can also be preserved when the positive movement of the sea level is so rapid that there is not enought time left for the destruction of the beach. This was the case at the Pleistocene-Holocene boundary, so that the relics of Pleistocene beaches have been preserved and now exist at greater depths on the sea floor. Also, Subrecent beaches buried by silts and clays of tidal flats are apt to persist.

The fauna of beach sediments

Beach sediments are remarkable for their large content of organisms. The populations of those species which live in the sediment attain an immense amount. Thus, for instance, about 100,000 individuals of the worm *Thoracophelia mucronata* have been computed in 1 m^3 of the sediment. The qualitative and quantitative compositions of fauna living on beaches show a marked zoning.

Owing to the activity of organisms, a dark-coloured reduction zone, rich in organic matter, develops in fine-grained beach sands. The zone has a negative Eh value and pH higher than 8·0, which indicates that organisms on the beaches do not only affect the mechanical alternation of sediments, but also induce a change of physico-chemical conditions during the early diagenesis.

References

BERRYHILL H. L. - DICKINSON K. A. - HOLMES Ch. W. (1969): Criteria for recognizing ancient barrier coastlines. Bull. Am. Assoc. Petrol. Geol., vol. 53: 706—707, Tulsa.

BLUCK B. J. (1969): Particle rounding in beach gravels. Geol. Magazine, vol. 106: 1—14, London.

BOTVINKINA L. N. (1962): Op. cit. on the page 106.

CURRAY J. R. (1956): Dimensional grain orientation studies of Recent coastal sands. Bull. Am. Assoc. Petrol. Geol., vol. 40: 2440—2456. Tulsa.

EMERY K. O. - NEEV D. (1960): Mediterranean beaches of Israel. Geol. Surv. State of Israel, Bull. No 36: 1—24.

FLEMMING N. C. (1964): Tank experiments of the sorting of beach material during cusp formation. Jour. Sedimentary Petrology, vol. 34; 112—122, Menasha.

GARDNER D. E. (1955): Beach-sand heavy mineral deposits of eastern Australia. Dpt. of Nat. Development, Bur. of Min. Res., Geol. Geophys., No 28: 1—103, Canberra.

GESSNER F. (1957): Meer und Strand. P. 1—426, Berlin.

GORSLINE D. S. (1964): Beach studies in West Florida, USA. In "Deltaic and shallow marine deposits", p. 144—147, Amsterdam.

GOTTHARD R. - PICARD K. (1965): Anreicherungen von Schwermineralien an den Küsten Schleswig-Holsteins. Geol. Mitteilungen, Bd. 4: 249—272.

HILL M. N. (Editor) (1963): The sea, ideas and observations on progress in the study of the seas (Symposium). Vol. 3: 1—963, New York, London.

HOYT J. H. - HENRY V. J. (1963): Rhomboid ripple mark, indicator of current direction and environment. Jour. Sedimentary Petrology, vol. 33: 604—608, Menasha.

— — (1964): Development and geologic significance of soft beach sand. Sedimentology, vol. 3: 44—51, Amsterdam.

INGLE J. C. (1966): Movement of beach sand. Developments in Sedimentology, vol. 5: 1—221, Amsterdam.

INMAN D. L. (1960): Shore processes. Encyclopedia of Science and Technology. P. 300—306, New York.

JOHNSON J. W. (1956): Dynamics of nearshore sediments movement. Bull. Am. Assoc. Petrol. Geol., vol. 40: 2211—2232, Tulsa.

KAMEL A. M. (1963): Littoral studies near San Francisco using tracer technique. Beach Erosion Board, Techn. Memo 131 : 2—88.

KELLER W. D. (1941): Size distribution of sand in some dunes, beaches and sandstones. Bull. Am. Assoc. Petrol. Geol., vol. 29: 215—221, Tulsa.

KING C. A. M. (1951): Depth of disturbance of sand on sea beaches by waves. Jour. Sedimentary Petrology, vol. 31: 131—140, Menasha.

— (1959): Beaches and coasts. P. 1—409, London.

LEONTIEV O. K. (1963): Fundaments of marine geology (in Russian). P. 1—463, Moscow.

LOGVINENKO N. V. - REMIZOV I. N. (1964): Sedimentology of beaches on the North coast of Sea of Azov (in Russian with English res.). Deltaic and shallow marine deposits, 244—252, Moscow.

MACCARTHY G. R. (1933): Calcium carbonate in beach sands. Jour. Sedimentary Petrology, vol. 3: 64—67, Menasha.

MARGOLIS S. V. (1968): Electron microscopy of chemical solution and mechanical abrasion features on quartz sand grains. Sedim. Geology, vol. 2: 243—256, Amsterdam.

MARTENS J. H. C. (1939): Beaches. Recent marine sediments, p. 207—218, Tulsa.

MCKEE E. D. (1957): Primary structures in some recent sediments. Bull. Am. Assoc. Petrol. Geol., vol. 41, Tulsa.

MILLER R. I. (1959): A study of the relation between dynamics and sediment pattern in the zone of shoaling wave, breaker and foreshore. Eclog. Geol. Helv., vol. 51: 542—551, Basel.

OTVOS E. G. (1964): Observations on rhomboid beach marks. Jour. Sedimentary Petrology, vol. 34: 683—687, Menasha.

PANIN N. (1967): Structure des dépôts de plage sur la côte de la mer Noire. Marine Geology, vol. 5: 207—219, Amsterdam.

PRICE W. A. (1958): Environment and history in identification of shoreline types. Quarternaria, vol. 3: 151—166, Rome.

RAO C. B. (1957): Beach erosion and concentration of heavy mineral sands. Jour. Sedimentary Petrology, vol. 27: 143—147, Menasha.

RUSNAK G. A. (1956): Changes of sand level on the beach and shelf at La Jolla, California. Beach Erosion Board. Techn. Mem., No 82: 520—585.

SHEPARD F. P. (1952): Revised nomenclature for depositional coastal features. Bull. Am. Assoc. Petrol. Geol., vol. 36: 1902—1912, Tulsa.

— (1959): The earth beneath the sea. P. 1—275, Baltimore.

— (1963): Submarine geology. 2nd edition. P. 1—557, New York.

STEWART J. H. (1958): Sedimentary reflection of depositional environment in San Miguel lagoon, Baja California, Mexico. Bull. Am. Assoc. Petrol. Geol., vol. 42: 737—788, Tulsa.

TANNER W. F. - EVANS R. G. - HOLMES Ch. W. (1963): Low energy coast near Cape Romano, Florida. Jour. Sedimentary Petrology, vol. 33: 713—722, Menasha.

THOMPSON W. O. (1937): Original structures of beaches, bars, and dunes. Bull. Geol. Soc. Am., vol. 48: 723—752, Baltimore.

TREFETHEN J. M. - DOW R. L. (1960): Some features of modern beach sediments. Jour. Sedimentary Petrology, vol. 30: 589—602, Menasha.

ZENKOVICH V. P. (1962): Fundaments of sea coasts development (in Russian). P. 1—710, Moscow.

— (1964): Formation and burial of acumulative forms in littoral and near-shore marine environment. Marine Geology, vol. 1: 175—180, Amsterdam.

16. *Shallow-marine sediments*

Under the term "shallow-marine sediments" we include here all marine deposits extending from beach sediments approximately to the —200 m isobath, i. e. to the edge of the continental shelf. Being comparatively easily accessible to study, they have been thoroughly studied. They have also attracted the attention of geologists because their ancient analogues are probably very frequently encountered in the geological column. The term of shallow-water sediments (eventually neritic sediments) includes, however, sediments differing widely both in genesis and composition, so that their only common feature is the deposition at relatively small depths. The natural division of Recent neritic sediments is based on the differentiation of the open and sheltered shelves. These two environments can be further subdivided as follows:

1. The environment of open shelf:
   a) Inner shelf to a depth of 20 m;
   b) Central shelf between 20 and 100 m;
   c) Outer shelf between 100 and 200 m;
   d) Shallow-water flats and shelves of islands.
2. The environment of sheltered shelf:
   a) Bays, lagoons, tidal flats, fjords, etc.;
   b) Depressions in the shelf.

The continental shelf is most likely a special feature of the present geological period, having originated from the interplay of accumulation and erosion processes in Pleistocene and at the Pleistocene/Holocene boundary. Therefore, it should be carefully considered for which of the present-day shallow-water environments an analogy could be found, and those for which analogues are unlikely to have occurred in geological history.

Hydrodynamical conditions

The grain-size distribution in the clastic sediments of shallow-water environments is controlled by the hydrodynamical conditions, i. e. the action of waves and currents. The wave action is effective only to a small depth; the question of its range has been often discussed and studied by experiments and direct observation, but no decisive answer has been obtained. Two exact observations, for instance, have brought quite contradictory results. The oscillation of water particles produced by wave action has been measured to a depth of 200 m, but on the other hand, precise measurements carried out in many shallow-water environments have shown that under normal marine conditions the waves stir the sediments to a depth of a few metres only. From all the data obtainable the following generalization can be drawn, allowing, of course, for a considerable schematization (Table 110).

Table 110

Environment	Waves stir unconsolidated fresh clayey sediment
Oceans	up to 90—100 m, sometimes deeper
Inland seas	up to 30—50 m (lack of data)
Bays, lagoons tidal flats	up to several metres (with exceptions)
	Waves able to rework unconsolidated sand
Oceans	up to 15 m
Inland seas	up to 8 m, sometimes deeper
Bays, lagoons, tidal flats	up to 1 m, or several decimetres only

The greater the orbits, length and amplitude of the waves, the deeper the action. All these circumstances depend on the strength of the wind and the size of the water body. Numerous direct measurements of the wave energy have proved that, on the average, it is much greater in the oceans than in the inland seas (values are given on p. 12).

The currents can be divided into the following groups:

1. Local currents, influenced by local differences in temperature and salinity.
2. Larger-scale, semi-permanent currents induced by the predominating direction of the wind. They are characteristic of the circulation in some inland basins, some parts of the oceans or large gulfs.
3. Currents of the world-ocean circulation.
4. Tidal currents.

All these kinds of currents can affect the deposition in shallow-water environments and are operative indisputably to a depth of 200 m. Their mean velocity ranges from a few cm/sec to several tens of cm/sec. Some local currents attain a considerable velocity particularly in the straits, where they are capable of transporting coarse sand and fine gravel to a depth of even several hundred metres. Tidal currents can also attain a velocity of several tens of metres near the bottom at the edges of shelves. The edges of shelves are the sites of more rapid bottom currents than are the central and inner parts of shelves. All the above-mentioned currents are surface currents which can but need not affect deeper water layers. Besides, in shallow-water environments, there also exist local deep currents and the deep oceanic circulation runs out to the margins of shelves. Mention should be made particularly of some zones of convergence, where the ascending currents reach the surface and wash the shores, as for instance on the western coast of Africa and the western coast of South America. These currents attain a velocity of about 1 cm/sec and in places can prevent the deposition of clay. Every, even slight, elevation is the source of local deep currents, which show the highest velocity (of up to several tens of cm/sec) above their crests.

Sediments of the open shelf

Open shelves of regular topography are very frequent on the Earth's surface; they pass landwards into the littoral zones and seaward into the continental slope. Their slope is uniform, usually broken by several more gently inclined stages. The width and the slope of shelves vary appreciably. Off mountainous coastlines the width is of a few kilometres only, but elsewhere, as for example, in the Antarctic it can be up to 1,300 km. The total picture of shelf sedimentation is well known, but the regularities of deposition have not yet been ascertained in detail everywhere.

The uneven surface of the shelf is due to various causes. Some shelves are furrowed by a series of troughs and ridges. Terrace levels, which are a world-wide phenomenon, are the product of recurrent movements, and stabilization of the sea level during a progressive transgressions and regressions. In the Pacific Ocean the 20 m terrace

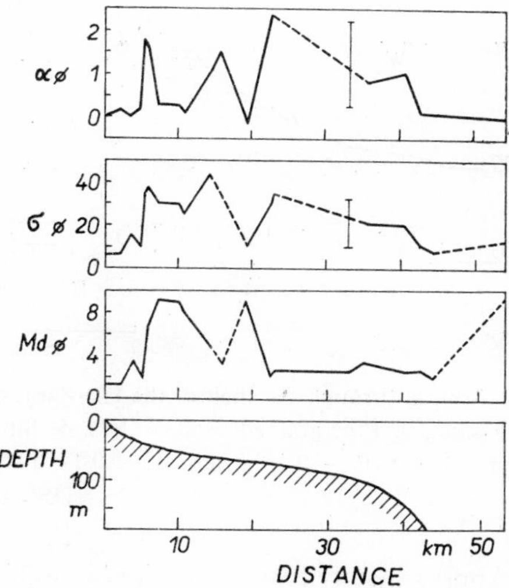

Fig. 71. Variation of size distribution in sediments of the Visakhapatnam area (continental shelf of India). After M. S. Rao (1964).

is widely distributed whereas in the Atlantic Ocean the 40 m and 60 m terraces are most widespread. Local unevennesses are represented by salt domes, buried bioherms and other features. The outer shelf can also be disturbed by submarine canyons. Sedimentation is affected even by small topographical differences, because they produce local currents.

The general pattern of sedimentation on open shelves can be summed up as follows:

1. Shelf sediments of glaciated areas differ considerably from those of other areas. They contain a large amount of glacial material transported by sea ice and icebergs. Their petrographical composition is given above.

2. At lower latitudes the shelf sediments usually do not show a decrease in grain-size with depth; on the contrary, increase has very often been observed. On the inner and central shelves, the distribution of sediments is irregular, depending on local factors; on the outer shelf a sudden coarsening of deposits is very frequent. This phenomenon is explained either by the transportation of coarser material by storms or in terms of relict Pleistocene sediments.

3. In many places, large parts of shelves are covered by the relics of Pleistocene sediments, whose size distribution does not correspond to the present hydrodynamic conditions. They occur on the surface because the persistent currents prevent clay material being laid down.

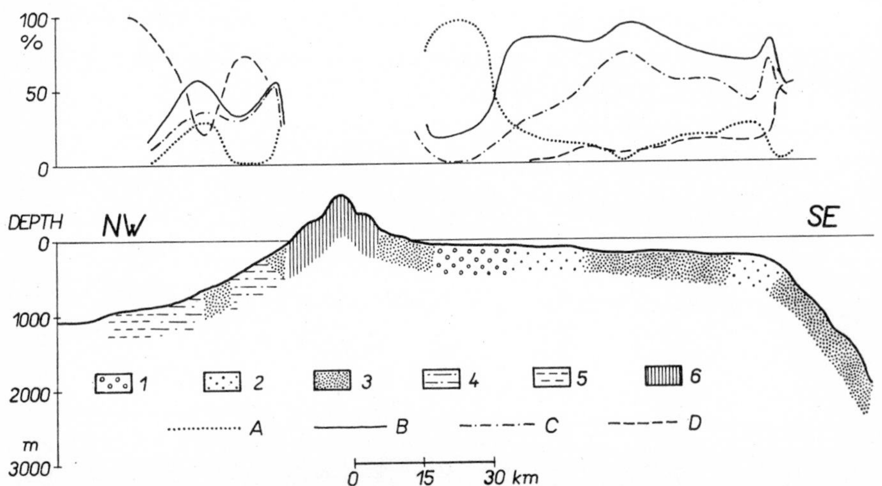

Fig. 72. Section through the shelf of the Isle Paramshir (Northern Kuriles). 1. Gravel, 2. Coarse-grained sand, 3. Fine-grained sand, 4. Silt, 5. Silty clay, 6. Rock outcrops and pebbles. A — fraction >1 mm, B — 0·1—1 mm, C — 0·1—0·25 mm, D — <0·1 mm. After I. O. Murdmaa (1963).

4. Approximately from the latitude of 40°—50°N and S, towards the equator the shelves bear a larger content of carbonates which have accumulated particularly on the outer shelf, being frequently represented even by calcarenites. The content of carbonate in the sediments of central and inner shelves is variable.

5. Close to the mouths of major streams, finer-grained sediments are more widespread, extending to the central and outer parts of shelves, or even to the continental slope.

6. There is a general rule that in places of active sedimentation muddy sediments are found, and at places of non-sedimentation sandy and coarser sediments — relics of Pleistocene deposits — occur.

Gravels can occur at the shelf-littoral zone boundary and the outer shelf-continental shelf boundary, as well as on the crests of isolated elevations, but compared with the finer-grained sediments, they are generally scarce.

The Md of shelf sediments generally varies between 0·08 and 0·01 mm, which points to the fine sand or silt grade. Trask's coefficient So ranges from about 1·5 to above 3·0. Phi Skewness is zero or near to it. The above-values are valid for Recent fine-grained sediments corresponding to the contemporary hydrodynamical conditions. The Md of relict sediments on the outer margin of the shelf frequently exceeds 0·5 mm and their So value is reported to be roughly 1·5—1·8.

The size of those shelf parts which are the site of present day sedimentation coincides roughly with the size of shelves covered by Pleistocene sediments, not including the glacial shelves. The Atlantic shelf of North and South America and the eastern shelf of India are typical shelves covered with abundant relict Pleistocene sediments. Shelves with active sedimentation are universally developed round the deltas of major streams. The grain-size zonation on shelves with relict sediments is quite irregular and often reverse; the coarsening of sediments proceeds with the distance from the landmass. On shelves with a rapid rate of sedimentation of prevalently fine-grained detritus, contrarywise, the grain-size diminishes with the distance from the mainland and with the depth. On the basis of the data established for the California shelf this relationship has been expressed by the equation:

$$Md_{\Phi} = 0{\cdot}0208D + 3{\cdot}30$$

where D = depth given in fathoms.

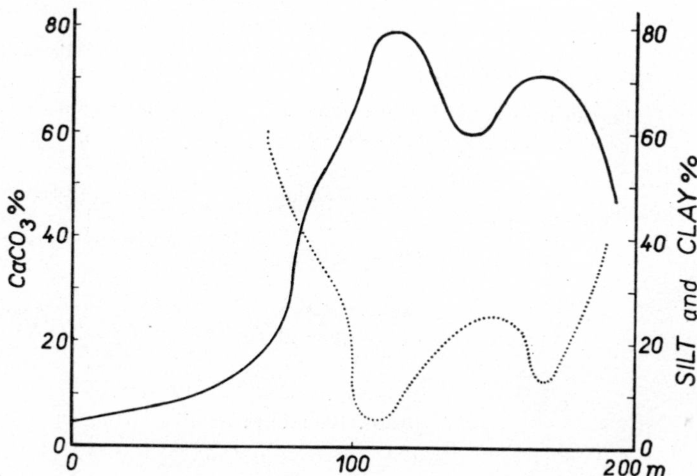

Fig. 73. Content of silt, clay and carbonate in shelf sediments of India and their changes with depth. Carbonate = dotted line. After M. S. Rao (1958).

Table 111

Classification of marine shelf sediments off the Bay of Bengal
(after M. S. Rao 1964)

Zonal classification	Composition	Texture	$CaCO_3$ %	Age of formation
Sands (0–15 fathoms)		beach sands		
		fine silty sands (0–5 fath.)	negligible	Recent, probably subject to seasonal variation
		nearshore sands (5–15 fath.)		
Clays (15–30 fathoms)	clastic sediments	clayey sands (15–22 fath.)	< 20	
		sand-silt-clay (30–40 fath.)	20–40	
		clayey sand (40–45 fath.)		
Shells (30–70 fathoms)	mixed clastic-calcareous sediments	clayey sand (about 45 fath.)	50	
		clayey oolitic sand (45–55 fath.)	50–60	relic (Pleistocene)
	calcareous sediments	oolitic sand (55–80 fath.)	> 60	
		clayey foraminiferal sand (80–100 fath.)		Recent
		sand-silt-clay (about 100 fath.)	30–50	
		silty clays (below 100 fath.)	10–15	

The influence of the local unevennesses of the surface is not invariable and cannot be generalized. On the shelf of the Barents Sea, the coarsening of sediments from 0·06—0·12 mm is produced already by an elevation only few metres high, whereas on the Californian shelf depth difference of 300 m results in the increase of the Md by 49μ only. Results published from this region show that the grain-size in places depends rather on the distance from the shore than on the local elevations of the sea bottom, even if the latter are of a substantial size. The relationship under consideration is affected by several factors, such as primary deposition from suspension or secondary outwash of fine sediments from the summit parts of elevations. In places where there is a rapid supply of material and thus of deposition of suspended load, a slighter dependence on the topography of the bottom can be expected than in those where the amount of material is small and the fine fractions are completely washed out.

The effect of climate on the regional distribution of various fractions was underlined especially by M. O. Hayes (1967). Mud fractions are very sensitive and accumulate mostly on the shelves of humid tropics.

Mineralogical composition of sediments

Shelf sediments contain on the average a high percentage of calcium carbonate, which depends largely on the rate of deposition. In places of rapid recent deposition, where prevalently fine grained material is laid down; the shelf deposits are poor in carbonate. On the other hand, in places of slow sedimentation and of predominantly coarse deposits, carbonate-rich or pure carbonate sediments are usually developed. Examples of shelf sediments with the respective contents of carbonates are listed in Table 112.

Table 112

Carbonate contents of shelf sediments

Region	Amount of carbonate
Guyana shelf	central shelf 30—40% $CaCO_3$ outer shelf 30—90% (calcarenites)
East Asia shelf	28% on the average
Barentz Sea	about 3—5%, evenly distributed
California shelf	10—80%, sediments passing into calcarenites
Shelf of East India	up to 80—90% on the shelf edge
Shelf of Antarctica	1—10%
Shelf of Alascan Bay	most frequently below 5%, sometimes up to 20%
Shelf of NW Pacific Ocean	below 10%

The boundary between the continental shelf and continental slope is frequently manifested by a rapid seaward decrease in the content of carbonate.

P. D. TRASK (1936) records the mean content of organic matter in all shallow-water sediments to be 2·5% of org. C. This percentage is, of course, increased by a higher content of organic matter in sediments of sheltered shelves (in fine-grained deposits of the inner shelf it attains a value only about 1% of org. C).

Fig. 74. Orientation of sand grains on the coast of Texas Bay. The arrows indicate orientation of long axes of sand grains. After J. R. Curray (1956).

The coarsest sediments of the outer shelf — calcarenites — contain commonly only 0·1—0·2% of org. C. The seaward drop in org. C/org. N ratio which is clearly observable, is due to the decrease of organic matter of terrestrial origin and the increase of organic matter of planktonic origin.

The content of organic matter increases not only with the decrease in grain-size but also towards higher geographical latitudes, where the cooler water retards their decay.

The presence of glauconite is characteristic of open shelves. Its amount ranges from 1 to 70%; the statistical evaluation of all samples examined indicates 3—10% to be the most frequent value. Glauconite occurs mainly in sandy sediments of central and outer shelves and decreases into the fine-grained sediments to be replaced finally by pyrite. There is a direct relation betweeen the content of sand fraction and the percentage of glauconite. This mineral occurs in most varied forms: as clastic rounded grains, as metacoloidal nodules with reniform surface, as filling of tests of foraminifers and other organisms and as the alteration product of biotite flakes and other mineral particles. In addition, it forms the material of faecal pellets and replaces metasomatically the carbonate tests of organisms. Generally, the amount of glauconite is indirectly proportionate to the rate of sedimentation.

Phosphates, an important component of shelf sediments, frequently occur in association with glauconite. They form small grains or larger (up to $^1/_2$ m) nodules, coatings on large shells and pebbles or replace metasomatically the carbonate of tests. The finds of broken and recemented nodules prove their authigenic origin. The phosphates bear traces of the boring activity of organisms. Large nodules are found on the elevations of the sea bottom where finer grained particles are not deposited. They show a concentric structure, the individual zones being formed of different phosphatic minerals. The presence of phosphates is indicative of slow sedimentation. Their origin is caused by special hydrological factors. Generally, they develop in places of ascending deep currents which having reached the shelf, meet the littoral waters rich in iron and manganese.

In the sand fraction, the percentage of feldspars varies in relation to the quartz grains, and depends on the provenance of clastic material. Generally, feldspars do not make up more than 25% of the sandy fraction.

The chemical composition of sediments varies considerably and corresponds to the mineralogical composition. Table 113 represents the chemical composition of various sands of the Atlantic shelf in the United States.

Table 113

Chemical composition of various sands of the Atlantic shelf in the United States (after D. S. Gorsline 1963)

	All samples %	Carbonatic sands %	Glauconitic sands %
SiO_2	52·04	1·07	22·76
Al_2O_3	0·27	0·47	0·90
Fe_2O_3 tot.	0·86	1·09	5·28
MnO	0·11	0·23	0·13
MgO	0·92	1·68	1·45
CaO	22·75	49·13	35·09
K_2O	0·18	0·33	0·98
P_2O_5	3·80	5·12	4·04
Sr	0·27	0·50	0·35
CO_3	18·80	40·40	29·02
Ca/Mg	31·10	36·60	30·20
Sr/1,000 Ca	7·60	6·60	6·40

Structures of sediments of the open shelf

The surface structures are probably not so abundant in shelf as in beach sediments. Yet ripple marks occur even to maximum depths and, as far as it is known, asymmetrical ripples somewhat prevail over the symmetrical ones, just as ripples with regular ridges prevail over the irregular ones. At a depth greater than 20 m, however, they never cover a large surface.

For the study of the internal structure of shelf sediments comparatively numerous cores of Recent deposits have been available. It has been ascertained that the structure of these sediments is prevalently homogeneous, presumably owing to the very intensive reworking by benthic organisms and the disturbance of primary structures, which are generally not preserved. In places of active sedimentation, in fine-grained desposits, mottled structures passing into irregular bedding are usually developed. The following relationship between the rate of sedimentation and the presence of structures is valid: in places of rapid sedimentation the organisms have not enough time for disturbing the primary bedding. The slower is the sedimentation, the more readily the regular bedding passes into an irregular and through a mottled to a homogeneous structure. In the shelf sediments bioglyphs are also frequent but their amount and variety are lower than in the beach sediments. They are most abundant on the fine-grained sandy or silty bottom.

Horizons, corresponding to the former periods of subaerial or submarine weathering, appear in shelf sediments as oxidation zones formed of Fe-hydroxides and oxidation compounds of manganese. The upper boundary of the subaerial weathering zone is commonly sharp, whereas that of the submarine weathering zone is rather gradual. These zones are preserved only when they were soon buried by other deposits and thus an accelerated lithification took place. Under a slow sedimentation, a recurrent reduction of ferric and other compounds could take place.

Thickness and mode of occurrence of sediments of open shelf

As compared with littoral sediments, the shelf sediments attain a great thickness. As the morphology of shelves originated by the combined sedimentary and erosive processes in the Pleistocene and at the Pleistocene-Holocene boundary, the sediments particularly on those shelves that are within the reach of the supply of river detritus are of immense thickness. Thus, for instance, about 6,000 m of sediments were laid down on the shelves of the Gulf of Mexico during the Late Tertiary and Quarternary. 1000 m of unconsolidated sediments have been measured also on the Atlantic shelf of the United States, centred in "pockets" on the shelf edge. The large thickness of shelf deposits is particularly remarkable considering that at present the shelves are often sites of nonsedimentation or even erosion. The shelves of erosive origin bear a thin veneer of sediments on a solid substratum.

Sediments of sheltered shelf

The term "sheltered shelf" is very wide covering a varied complex of depositional environments: widespread sheltered shelf seas, bays, estuaries, lagoons, tidal flats, fjords, shallow-water flats. This complex of depositional environments indicates

Table 114

Principal differences between the sediments of open and sheltered shelves

Sheltered shelf	Open shelf
Mostly rapid contemporaneous accumulation of Recent sediments	Slow accumulation or non-sedimentation
Clayey and silty sediments prevail	Sandy sediments prevail
Great variability of sea-water salinity, brackish and hypersaline environments occur	Generally normal marine salinity
Fresh-water, brackish and marine fauna occur	Fauna only marine
Average depth only some tens of metres	Depth from 10 up to 200 m
Sedimentary environments greatly differentiated	Sedimentary environments more uniform

Table 115

Differences in the sedimentary conditions of the open and sheltered shelves of East Asian shelf and adjacent gulfs
(after H. Niino - K. O. Emery 1961)

	Continental shelf open	Yellow Sea	Tonkin Bay
Average depth	60 m	50 m	50 m
Annual variability of temperature	17—28 °C	5 – 26 °C	23—28 °C
Annual variability of salinity	3·3—3·4%	3·1—3·2%	3·2—3·3%
Md	0·12 mm	0·06 mm	0·04 mm
So (Trask's value)	1·5	2·5	3·0
Percentage of foraminiferal tests	14%	6%	7%
Percentage of other tests	14%	2%	5%
Org. C	0·3%	0·8%	0·6%

that the sediments were deposited in a shallow sea and have some features in common which distinguish them from open shelf deposits. The principal differences are summed up in Table 114.

The East Asian shelf and the adjacent gulfs furnish the best illustration of the differences in the sedimentary conditions of the open and sheltered shelves (Table 115).

It should be emphasized that the sheltered parts of a shelf have on the average finer-grained sediments, a higher content of organic matter and a lower content of carbonates as a result of a smaller amount of calcareous shells. No less interesting is the comparison of the sedimentary conditions in two continuous different environments. In Table 116 sediments of the open shelf of the Gulf of Mexico are compared with those of minor bays and lagoons of Texas bay, and sediments of the open Guayana shelf with the sheltered shelves of Paria Bay.

Table 116

Composition of sediments of open shelf environment and adjacent bays and lagoons

Bays and lagoons	Open continental shelf
Pyrite prevails in authigenic compounds	In authigenic fraction prevails glauconite and phosphorite; pyrite occurs only in larger grains and globular or nodular aggregates
Md of sediments is most frequently between 0·03 and 0·1 mm; generally great portions of clay fraction occurs, silt is more scarce, mixed sediments clay-fine-grained sand are abundant	Md is mainly between 0·05 and 0·15 mm; lack of clay, mixtures sand-silt abundant
Carbonate occurs in organic fraction and in smallest fractions, its amount rises towards the central parts of bays and towards straits and inlets	Carbonate occurs mainly in coarser fractions
Amount of organic carbon above 1%: its content increases towards the central parts of bays	Content of organic carbon is generally below 1%
High roundness of quartz grains especially in dunes and sand bars	Generally slighter roundness of quartz grains
Admixture of eolian material, originating on coastal dunes and sandy islands	Eolian material scarce
Great variability in composition of clay minerals; changes in the direction from river mouths towards bay centres are distinguishable	Composition of clay minerals is more uniform

Bays and lagoons

In this paragraph, bays and lagoons are not distinguished from one another, because from the sedimentological point of view they are very similar. Neither are the geological or geomorphological boundaries between them sharp and clear-cut. As compared with bays, lagoons are separated from the sea by synsedimentary forms, such as spits, sand bars and berms, so that they are more closed than bays. A narrow strait is frequently the only connection with the open sea.

The composition and distribution of sediments of bays and lagoons depend on the following factors:

1. The size and depth of bays.
2. The width of the connecting straits.
3. The inflow of river water to the bay and the resulting hydrological and hydrobiological conditions.
4. The topography of the coast.

The size of bays is a very important factor. In large and deep gulfs the activity of waves is more intensive and the average velocity of currents greater, so that a wide beach, i. e. a belt of coarse-grained sediments can develop there. Yet there are many gulfs (the Baltic Sea, the Bay of Bengal, the Gulf of Paria, etc.), where coarse-grained sediments reach to smaller depths than in minor bays. Therefore, the size of bays is not necessarily the controlling factor of the mean composition of sediments.

The depth of bays is another important agent. Deeper bays are a site of deposition of finer sediments. There is a generally valid rule that below the — 5 m isobath the diminishing of grain-size with depth can be observed.

Bays, even a small size, have well developed beaches (particularly when the prevailing direction of wind is perpendicular to the shoreline), when they are connected with the open sea by a wide water area. Bays connected only by a narrow straits with the sea behave as separate basins. They show a characteristic coarsening of sediments towards the straits, in which the fine-grained sediment is washed out even to great depths and also, occasionally coarse gravel deposits reach to a depth of several tens of metres. In addition, coarse organogenic carbonate, in the form of broken shells, bryozoans and corals increases in amount towards to narrows, as will be mentioned below.

The inflow of fresh water is decisive for the hydrological conditions of bays and the supply of material. Bays, into which major streams flow themselves, are rapidly filled by fine-grained sediments; their water is brackish, the salinity increasing away from the mouth of stream. A coast with numerous bays and lagoons is typical of areas with a positive movement of the sea level. The rising base of erosion results in the aggradation of rivers, so that they transport only fine detritus to the basin.

The topography of the coast is decisive for the vertical distribution of coarse-grained sediments. The effects of topography are manifested chiefly in large bays and gulfs, because small bays are usually adjacent to a flat coast. The climate affects the sedimentation in bays in that it controls the weathering of rocks in the source, especially the coastal area. If dune sheets are developed in the vicinity of bays, these are rapidly filled in by eolian sandy sediments; if the coast is overgrown by woods, mangrove forests and swamps, fine-grained sediments even of the shallowest parts of bays will be rich in organic matter. The acid waters of these bays will also be little suitable to the organic life and the precipitation of carbonate. Under arid conditions, especially in the bays limited against the sea, the concentration of salts in the water can increase up to a point where they precipitate.

From this review it follows that in discussing the sedimentation in bays all the above factors must be taken into consideration.

Their influence on the deposition is shown on several examples below. The sedimentary environments and deposits of a few thoroughly studied bays are described.

Texas Bays

The shore consists of a number of minor bays and lagoons of small size and depth, which are separated from the sea by sand bars and ridges. In the coastal areas there are large areas covered by eolian dunes. Minor and large streams flow into some bays. From these characteristics the composition of sediments can be inferred. Large parts of the bays are filled by fine-grained river silt-clay sediments with a considerable admixture of eolian sandy material.

Table 117

The variability of grain size parameters in small bays based on the data from Southern California

Subenvironment	Md range	Md average	So range	So average
Beach	0·12—0·14 mm	0·15 mm	1·2—2·7	1·35
Coastal dunes	0·10—0·18 mm	0·13 mm	1·2—1·6	1·28
Nearshore parts of bays	0·09—0·75 mm	0·16 mm	1·2—2·6	1·30
Central parts of bays	0·03—0·23 mm	0·08 mm	1·2—2·8	1·60
Straits	0·09—4·00 mm	1·24 mm	1·2—4·2	2·10

The Md varies between 0·008 and 0·12 mm, the content of carbonate ranges from a few per cent in the central parts of bays to 80% in the straits. The content of organic matter varies from 3% of org. C in fine-grained sediments in the centre to several tenths of per cent in coarser deposits at the margins of the bays.

The Gulf of Paria

A gulf is bordered by a lowland on one side and by mountain ridges on the other. In the central part it is more than 100 m deep. The supply of river water is high and the connection with the sea quite narrow. Hydrogeologically, it is a separate basin with various sedimentary environments. The sediments are most varied, ranging from gravels in the mountain piedmont to the clay deposits in the deltas and the central parts of the gulf. On the open shore, there are narrow sandy beaches, and silty clays are deposited on the sheltered shore. The content of carbonate ranges from several tenths of per cent in the delta to 30% in the straits connecting the gulf with the sea. The content of organic matter varies between 1·5% of org. C in the delta and 1% in the centre of the gulf; in the coarse-grained shallow-water deposits it makes up a few per cent.

The Persian Gulf

A large, not very deep gulf (the mean depth —25 m; maximum depth —100 m) connected by a narrow strait with the sea, with an immense inflow of fresh water. The coast is of a lowland relief. The sedimentation is influenced partly by terrigenous fine-grained material supplied by streams and abundant eolian sand and silt brought from the adjacent deserts. Over wide areas where there is no incursion of terrigenous river material, an intensive chemical and biological deposition of carbonate takes place partly as a result of strong evaporation and increased salinity of waters. Coastal sabkhas of the Persian Gulf are the places of sedimentation of penecontemporary dolomite together with gypsum.

The Yellow Sea

A large and deep gulf connected by a broad strait with the sea; a large inflow of fresh water and an abundant supply of suspended material mainly by the river Hwang-Ho. The coastal relief is low. The clays with a silty admixture are predominant. The content of carbonate is low, of a few per cent only, the maximum content of organic matter is 0·8% of org. C. The average Md is 0·006 mm, the mean content of carbonate is 10%, of which only a small part is formed by biological carbonate.

The Bay of Naples

A relatively small but fairly deep bay, open to the sea, with a negligible inflow of fresh water. The relief of the coast is moderate to high. The material deposited is coarse-grained, with an appreciable admixture of calcium carbonate, derived chiefly from organic remains.

The grain-size of sediments: coarse-grained beach sand to a depth of 3—5 m; sand and silt to a depth of 15 m; clay and silt-sediments of deeper parts (below 30 m).

In places, the sand reaches as deep as 50 m. As a result of slow sedimentation, the grain-size gradually decreases with the increasing depth. A small inflow of fresh water and a small supply of stream suspension produce a moderate increase in salinity and concentration of biological carbonate.

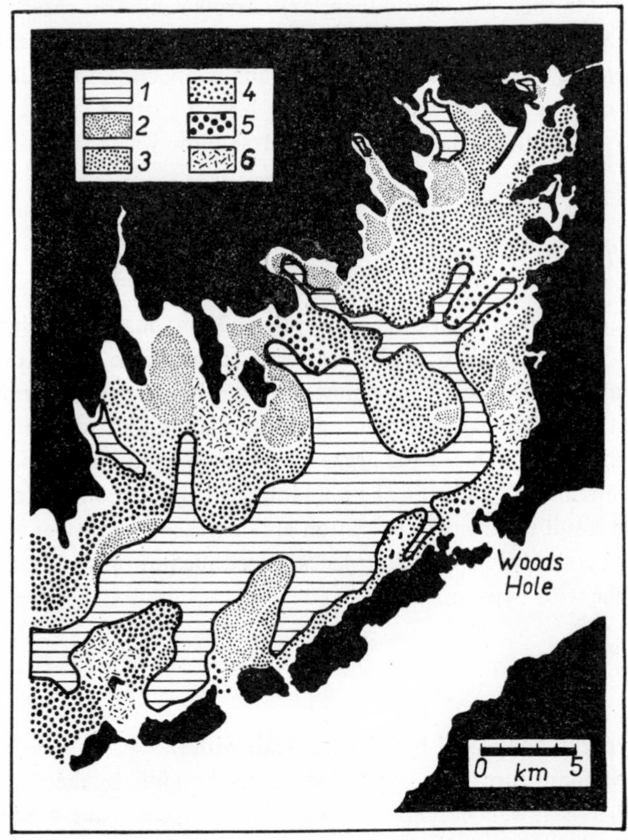

Fig. 75. Distribution of sediments in Buzzard Bay, Massachusetts. 1. Silt, 2. Fine-grained sand, 3. Medium-grained sand, 4. Coarse-grained sand, 5. Very coarse-grained sand, 6. Gravel. After J. R. Moore (1963).

The Bay of Kiel (*The Northern Sea*)

A bay of medium size; the connection with the sea fairly wide. The average depth between 10 and 25 m; a small inflow of river water; the relief of coastal area subdued. The grain-size of sediments varies considerably: from gravels in the narrows between the islands up to silty clays in the central parts. The content of carbonate increases towards the coarser sediments, which contain many calcareous shells and fragments, up to 50% of $CaCO_3$. Silty clays bear a negligible amount of carbonate. The content of organic matter is small.

The Lagoon of Terminos, Campeche, Mexico

A lagoon of medium size (about 60 × 30 km), connected by a narrow strait with the sea; the coast has a flat relief. The lagoon is very shallow and the influx of the river water considerable. Fine-grained sediments of prevalently silt-clay grade prevail. A large part of the bottom is covered by silty-sandy deltaic sediments. The content of carbonate is highest in the straits (more than 60%) and is fine sediments of the central parts (above 50%). It is predominantly of organogenic origin.

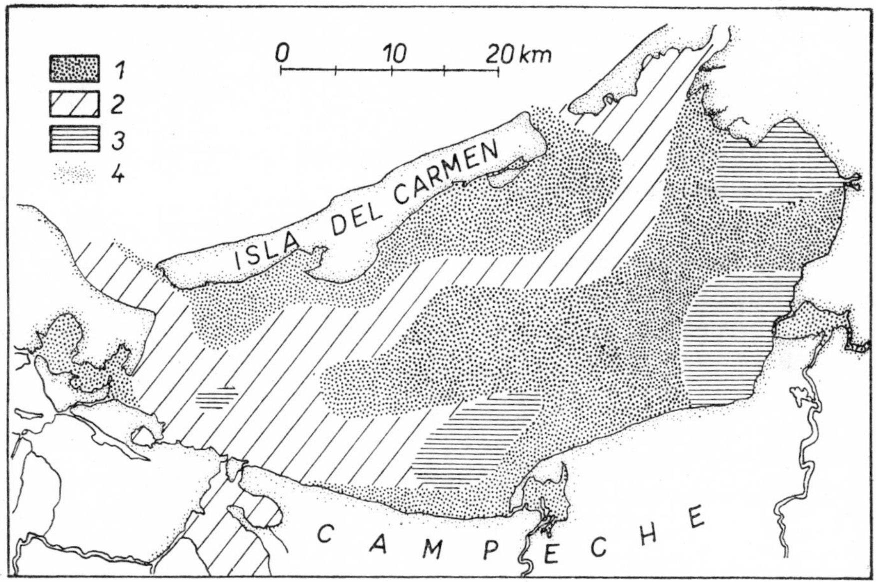

Fig. 76. Distribution of Recent sediments in Laguna de Terminos (Campeche, Mexico). 1. Unsorted sands, 2. Clayey silts, 3. Silty clays, 4. Beach sands. After A. Yañez (1963).

Buzzard Bay, Massachusetts

A small bay (approx. 50 × 20 km) without a significant influx of river water and a small depth (less than 10 m). The topographical differences of the adjacent area are small. The composition of sediments is affected by the presence of moraines on the coast and intensive tides. The grain-size is mostly variable: silts are laid down in the centre of the basin and coarse-grained sands and gravels on the margins. The content of calcium-carbonate is very low, usually not exceeding 5%. It attains a higher value only at several places where the tests of organisms occur. The content of organic carbon is also low.

The Gulf of California

A large and deep gulf with a rich supply of fresh water and clastic material. The coast has a crude relief. As a result of a large import of detritus by the stream, silty and sandy clays are deposited in the major part of the gulf. At a greater distance from the shore, the immense production of phytoplankton is the cause of the sedimentation of diatomaceous clays. Carbonate deposits originate in places which are beyond the river import in waters of normal salinity. Carbonate occurs chiefly in the form of shells of organisms.

Tampa Bay (Florida)

A small bay with maximum depth of about 15 m, without an influx of fresh water, the coast is of low relief. It is bordered by mangrove forests and eolian dunes. Owing to a large eolian admixture, the sediments are sandy in the greater part of the bay. Only in the vicinity of mangrove growths are they silty-sandy, with a clay admixture. The content of carbonate is largest in straits and shallow-water coarse sediments. Carbonate occurs almost entirely in the form of tests of organisms.

The Azov Sea

A bay of medium size with a large inflow of fresh water and a rich supply of suspended load. The relief of the coastal area is low. The water is brackish.

The sedimentation bears signs of a high supply of suspended material and an immense organic production. The deposits consist prevalently of silty clay with a high content of organic matter and a small admixture of calcium carbonate. The amount of clay varies between 10 and 90%, increasing gradually with depth. It attains a maximum value in the centre of the sea. The content of $CaCO_3$ ranges from 0·5 to 29·8% (on the average 2·1%) and drops with the decrease of grain size. The content of org. C averages 1·4% (0·44—2·94%).

The above examples illustrate the influence of all recorded factors on the sedimentation in bays and lagoons. From the size and depth of bays, the supply of sediments and the inflow of fresh water, from the morphology of the coastal area and the climatic zone we can infer with a satisfactory probability the mode of deposition, distribution and the plausible composition of sediments. On the other hand, the composition and distribution of ancient sediments make it possible to determine the unknow factors mentioned above.

Grain-size composition of bay sediments

Gravel deposits are very scarce in bays. Gravel beaches or relict gravel and gravel fractions consisting of tests of organisms can occur only in large bays or in straits.

The following conditions are favourable to the deposition of sandy sediments in bays and lagoons:

1. A not very large and deep lagoon.
2. The presence of eolian dunes in the coastal area or a wide and thick sand barrier separating the bay from the sea. Sand transported from dunes or barriers forms the bulk of bay or lagoon sediments.

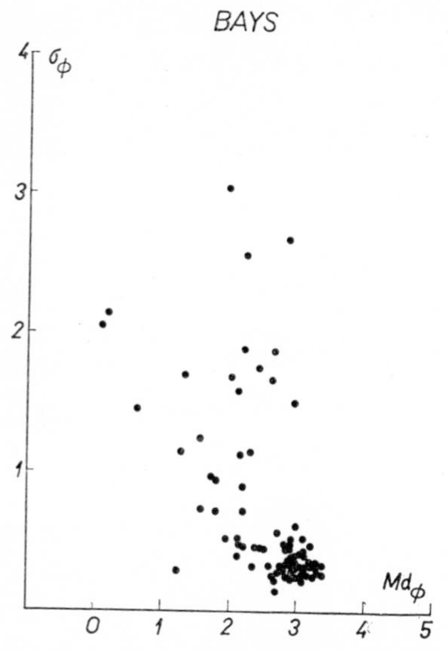

Fig. 77a. Grain-size parameters of bay sediments. Relation of Phi Median Diameter and Phi Deviation Measure. Data compiled from various authors.

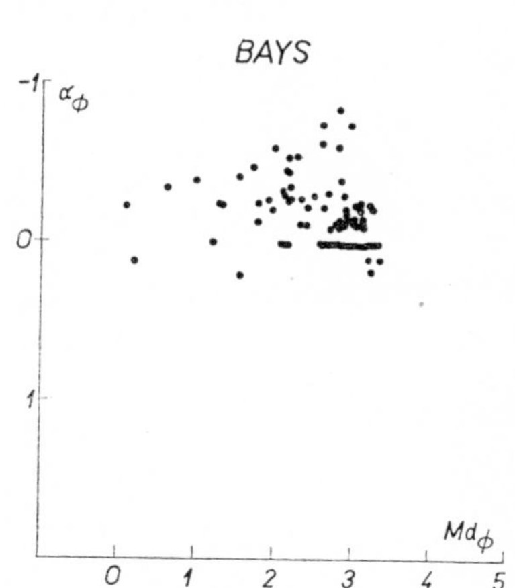

Fig. 77b. Grain-size parameters of bay sediments. Relation of Phi Median Diameter and Phi Skewness Measure.

3. The presence of old relict Pleistocene sediments and Recent non-sedimentation conditions. This factor, which is so abundant on an open continental shelf, is exceptional in bays and lagoonal environments.

Beach sands of lagoons and bays are finer-grained than sands of marine and oceanic beaches, whose Md varies between 0·08 mm and 0·15 mm. Unlike these, usually well-sorted sands, those of bays are occassionally clayey, unsorted and deposited directly from the stream suspension.

The sedimentation of silts and clays which represent generally the bulk of bay sediments is intimately connected. The conditions of their deposition are reverse to those of the sedimentation of sands.

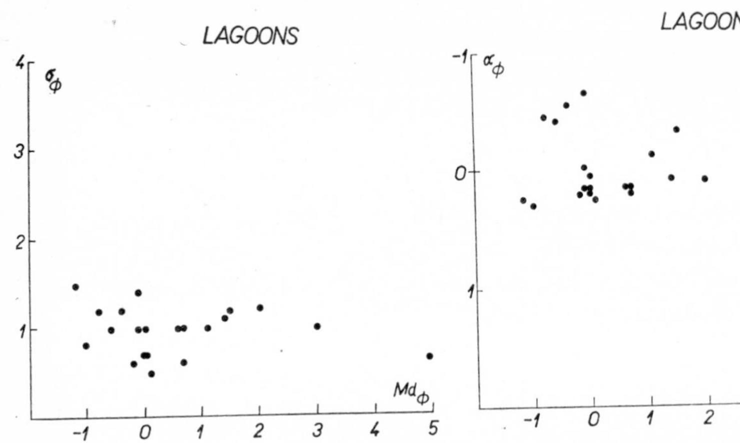

Fig. 78a. Grain-size parameters of sediments of lagoons. Relation of Phi Median Diameter and Phi Deviation Measure. Data compiled from various authors.

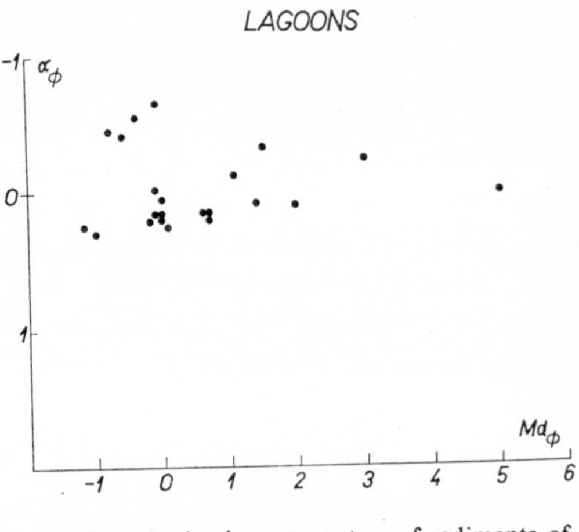

Fig. 78b. Grain-size parameters of sediments of lagoons. Relation of Phi Median Diameter and Phi Skewness Measure.

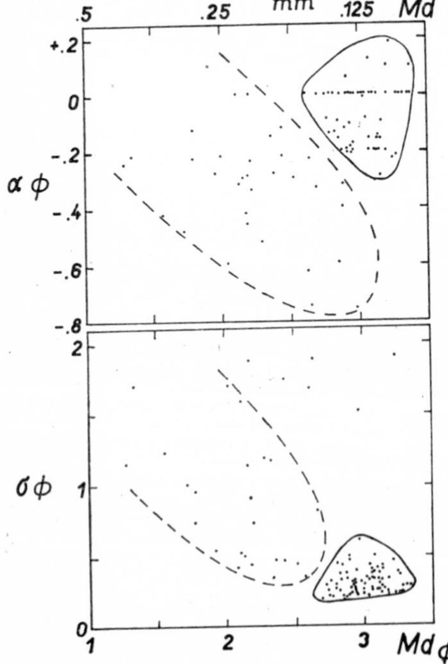

Fig. 79. Grain-size parameters of sediments of Californian coastal bays and their mutual relation. The field of fine-grained sediments (encircled by full line) responses to sediments deposited from suspension. The coarser sediments of the field limited by dashed line were deposited by undirectional currents or by wave processes. After F. B. Phleger *et al.* (1962).

Bay sediments provide a most favourable occasion for the study of the grain-size, i. e. its relation to the mode of deposition and distribution of individual grades and for genetic considerations. The division of bay sediments on the basis of the Md/So relation has proved most convenient for the genetic investigations.

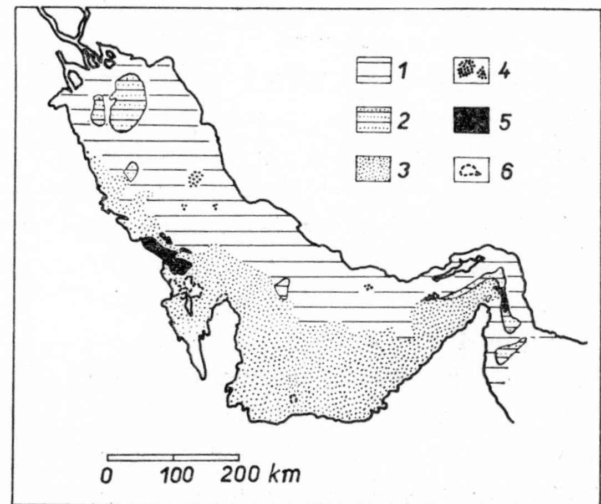

Fig. 80. Recent sediments of the Persian Gulf. 1. Clays and silty clays, 2. Sandy clays, 3. Sands, 4. Gravels, 5. Rock outcrops, 6. Coral reefs. After K. O. Emery (1956).

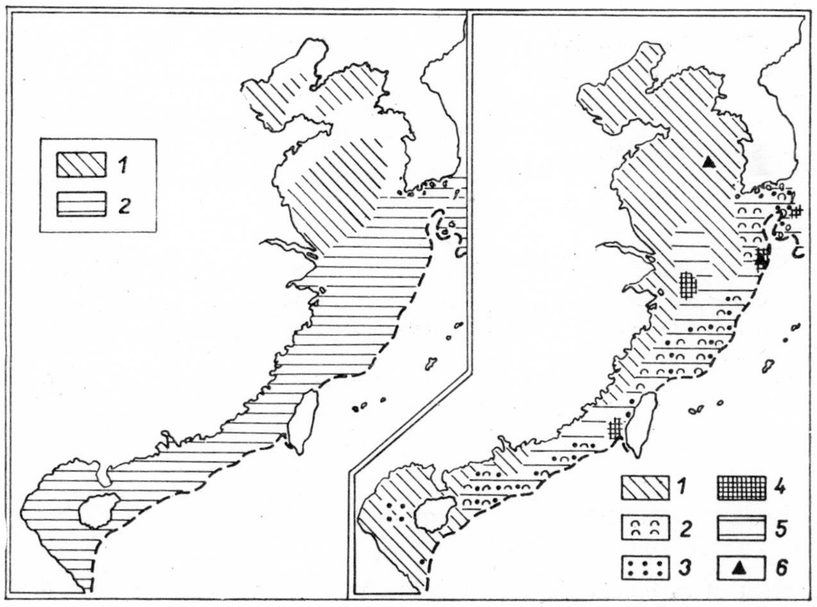

Fig. 81. Chart representing tectofacies (left picture) and sediments (right picture) of East China Sea, South China Sea and Tonkin Bay. Left picture: 1. Zeugogeosyncline, 2. Shelf. Right picture: 1. Recent detrital sediments, 2. Recent organogenous sediments, 3. Recent authigenic sediments, 4. Residual sediments, 5. Relic sediments, 6. Volcanic sediments. After H. Niino *et al.* (1961).

This division makes it possible to differentiate several genetically diverse groups of sediments:

1. Sediments originated by similar processes as river sediments.
2. Sediments whose deposition is affected by wave action.
3. Sediments originating by a slow deposition in quiet-water environment.

These three groups differ in the Md, sorting and skewness. Surprisingly, sediments of the first group are widespread in bays, which implies that during floods the river suspension is brought far into bays and laid down there without further sorting. Coarser and best sorted deposits affected by wave action originate in the shallowest parts of bays, whereas sediments deposited in quiet water occur mainly in the deepest central parts of bays, unaffected either by wave action or river supply. These three types of sediments defined and stated by the study of Recent deposits can be applied to deposits of other environments and those found in ancient deposits.

In a small bays on the Californian coast the following parameters for sediments of various environments have been established (Table 117).

Coarser fractions of coarser sediments are made up prevalently of the tests of organisms. In the Bay of Naples the individual environments show the following parameters of grain-size (Table 118).

Table 118

Grain size of sediments of various environments in the Bay of Naples (after G. MÜLLER 1958)

Subenvironments	Md	So
Coarse beach sands	0·12 —1·5 mm	1·5—2·5
Coarse sands on swells	0·12 —1·5 mm	1·5—2·5
Nearshore finer sediments	0·035—0·08 mm	2·5—3·0
Finer sediments on swells	0·035—0·065 mm	2·5—3·5
Sediments of the deepest parts of the bay	0·035—0·065 mm	2·7—3·0

The content of carbonate in bay sediments

The content of carbonate depends chiefly on the hydrochemical and hydrological conditions of bays, particularly on pH and salinity. Both these factors are influenced by the presence or absence of influx of fresh water and the width of the connection with the sea. The sedimentation and the distribution of carbonate in bay sediments are governed by the following laws:

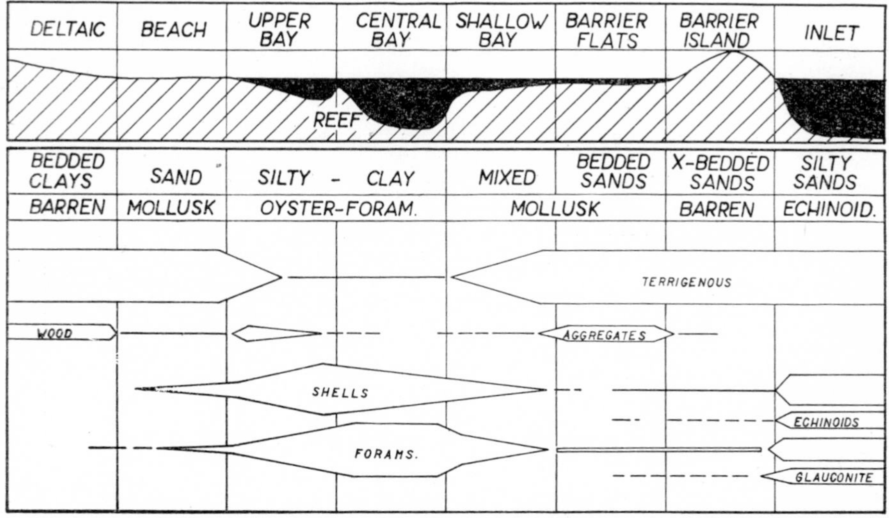

Fig. 82. Diagrammatic representation of some typical structural, textural and compositional features which characterize bay facies. After G. A. Rusnak (1960).

Table 119

Content of carbonate in Sebastian Viscaino Bay and Gulf of Naples (after K. O. Emery *et al.* 1957 and G. Müller 1958)

Environments	$CaCO_3$ %
1. Sebastian Viscaino Bay	
Beaches	12·0 (local accumulation of shells)
Coastal shallow parts of bay	10·0
Central parts of bay	21·0
Straits	65·0
2. Gulf of Naples	
Coarse-grained beach sands	1—20
Coarse-grained sands from submarine elevations	20—40
Coastal medium-grained sediments	5— 6
Medium-grained sediments from submarine elevations	10—15
Sediments from the deepest parts of bay	5—10

1. The sediments of bays with brackish water (pH lower than that of sea water) contain carbonate only in coarser fractions in the form of test fragments or organic remains. The maximum percentage of $CaCO_3$ in coarser sediments is roughly 80%. The amount of $CaCO_3$ rapidly decreases in the fine-grained sediments, but increases towards the straits connecting the bays with the sea, i. e. to those places where the bay conditions pass into the normal marine ones. There, the sediments

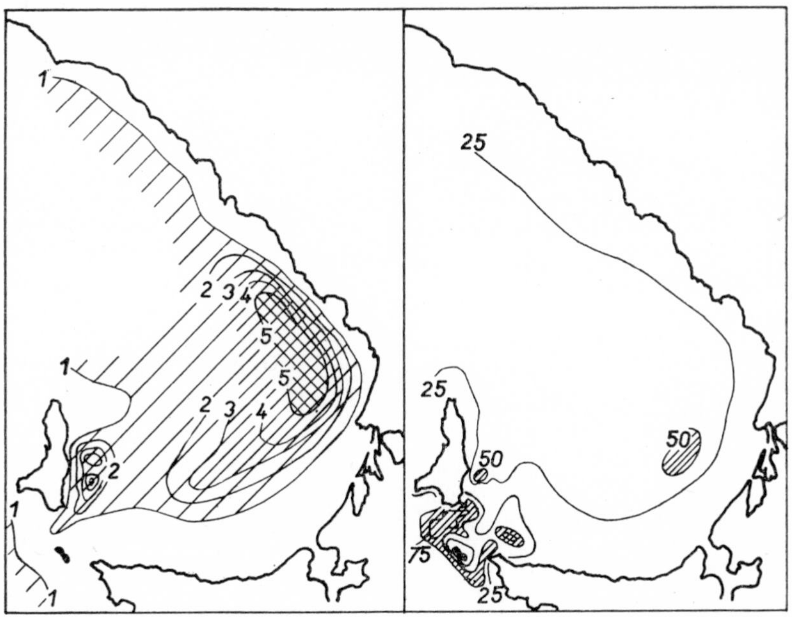

Fig. 83. Distribution of organic matter (left picture) and $CaCO_3$, (right picture) in Sebastian Viscaino Bay (California). The number indicates percentage of org. C and $CaCO_3$. After D. S. Gorsline (1957).

are generally extremely rich in $CaCO_3$ (sporadically up to 100% has been ascertained), which is almost completely of biological origin. Concomitantly, the grain-size of sediments also increases.

2. In the sediments of bays with hypersaline waters, without the influx of fresh water, or completely separated from the sea and showing a more alkaline reaction of waters, the carbonate can occur both as coarse biological material and as fine muddy material produced by chemical or biochemical precipitation of calcium carbonate. In an extreme case, the amount of carbonate can increase toward the finer-grained sediments of the central parts of bays.

Table 119 presents some detailed data on the content of carbonate in sediments of various environments.

In the tropical and subtropical lagoons the sedimentation is strongly carbonatic and a great part of the carbonate is of biological origin. Algal mats are present together with pelleted lime mud, which is deposited mainly in the intertidal zone (F. J. LUCIA 1968). Aragonite pisolites and lumps contribute also to the sedimentation. Persian Gulf and South Bonaire (Netherlands Antilles) are good examples of this type of sedimentation.

Organic matter

Organic matter is very sensitive to the grain-size composition of sediments and its amount increases in the finer fractions. Clays of the central parts frequently contain more than 5% of org. carbon. The composition of organic matter sometimes changes away from the mouths of streams, along with the decrease of the org. C/org. N ratio. Thus, for instance, in small bays on the Australian coast, the C/N ratio varies between 7·35 and 26·1, the amount of org. carbon increasing simultaneously from 2·4 to 40‰. In the bays of Florida, the C/N ratio decreases from 37 to 17 away from the mangrove growths towards the free bays, but in places where the organic matter derives rather from the marine benthos and plankton, it drops again to 10. In the different environments of the Sebastian Viscaino Bay on the Californian coast the content of organic carbon shows the following values (Table 120).

Dark sands of bays, which contain about 10—20% of clay fraction but a large amount of organic matter, are noteworthy for their origin. It has been stated that the clay originated by the disintegration of faecal pellets produced by organisms that had lived originally in pure sand.

Table 120

Content of organic carbon in the sediments of Sebastian Viscaino Bay (after K. O. Emery *et al.* 1957)

Environments	Org. C %
Beaches	< 1·0
Coastal shallow-water parts of bay	about 1·0
Central parts of bay	3·5
Straits	< 1·0

Mineralogical composition of sediments

The sands and gravels of bays are of a similar composition as those of other sedimentary environments. More interesting and important indicators of the environments are the authigenic minerals of clay sediments. Whereas glauconite and phos-

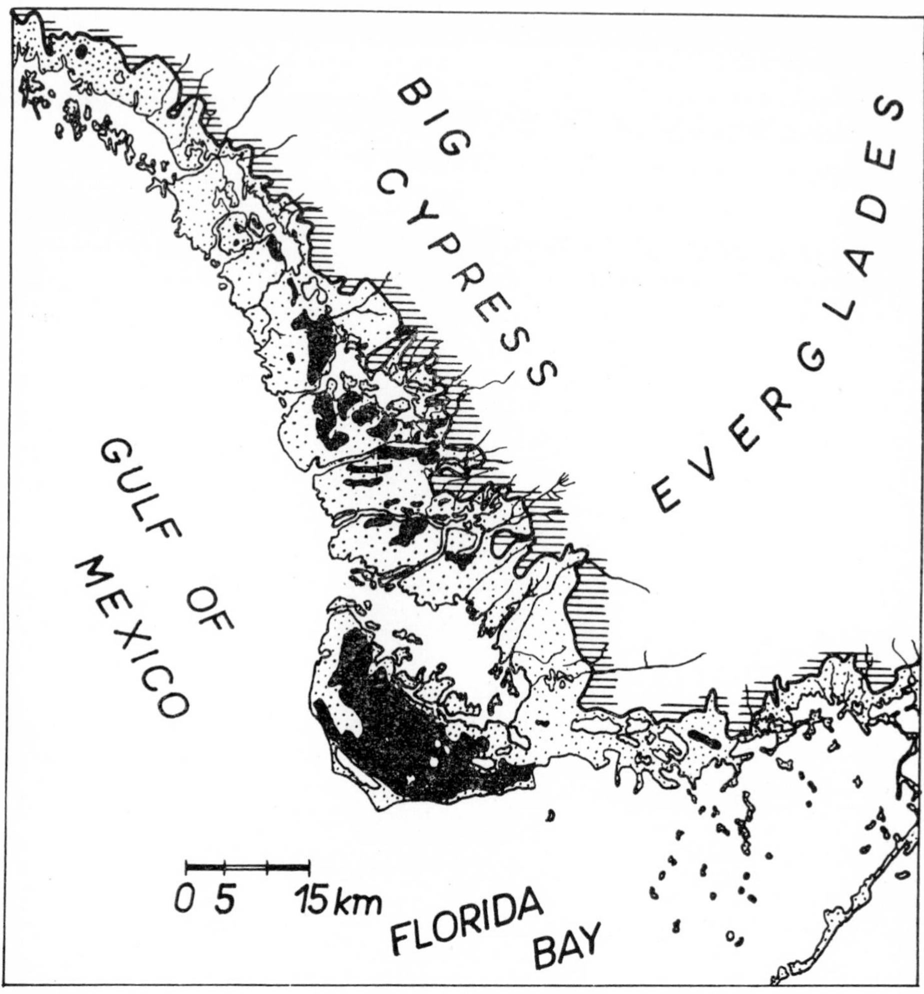

Fig. 84. General map of south west Florida coast with extreme development of mangrove forests. Area of coastal mangrove forest is shown by dotting. Dark areas in forests are salt marshes. After D. W. Scholl (1964).

phates are the prevalent authigenic minerals of open shelves; pyrite is most abundant in bay sediments. It often occurs also in the top layer of sediments where there is already a moderately reducing environment, owing to a large amount of organic matter and an imperfect exchange of waters. The amount of pyrite increases with depth but its concentration is sometimes maximum a few centimetres under the surface of sediments.

The detailed study of the composition of clay minerals in the finest sediments led to contradictory results. Some authors record that montmorillonite increases and illite decreases in amount away from the mouths of streams baywards (e. g. the

Gulf of Paria), but the results from the Gulf of Mexico are quite contrary. The former case has been interpreted in terms of mechanical differentiation, the latter is explained by diagenetic processes, or even syngenetic processes on the fresh water/salt water boundary. The possibility of these diagenetic transformations, particularly of montmorillonite into illite (as will be given in detail on p. 418), has been proved by experiments and direct observations on Recent sediments. In all bays, there is generally a moderate prevalence of illite, montmorillonite being the next in amount. Kaolinite is predominant in some tropical bays or mangrove swamps, but on the average, it is subordinate. The presence of minerals with mixed structures has not yet been clarified; they have so far been recorded from a few bays (e. g. mixed illite-chlorite structures from the Adriatic Sea). The carbonate minerals are described in a special chapter.

Table 121

Composition of sediments of various environments in the San Miguel Lagoon (after J. H. Stewart 1957)

Environment	Shell fragments %	Quartz %	Heavy minerals %	Rock fragments %	Mica %	Plants %
Open oceanic shelf	0·3	68·5	6·5	3·2	17·7	0·4
Coastal dunes		66·5	21·9	11·4	0·2	
Lagoon in proximity to ocean		79·9	10·1	5·3	3·4	0·1
Lagoon in proximity to river		76·2	6·6	11·8	4·3	0·1
Inner parts of lagoon	0·4	62·8	8·4	2·4	28·6	1·3
Marsh		28·9	5·7	2·6	17·5 (12·0 of aggregates)	33·3

The composition of sediments of various environments in the San Miguel lagoon in California is presented in Table 121.

The composition and origin of lagoonal evaporites were recently summarized by D. J. J. Kinsman (1969).

Biological sediments and the fauna of bays

As a result of immense organic production biological sediments are the principal sediments of some bays. First rank the diatoms which when they bloom form a continuous layer on the surface of the water and after decay they represent the bulk of sediments. It has been ascertained that in diatomaceous bay sediments as many as 45,000 tests can occur in 1 cm^3 of material. The tests of diatoms are fairly resistant to solution in bay waters saturated with SiO_2. SiO_2 can be supplied by river or deep marine waters, and sometimes also by volcanism. Very detailed information is available, for instance, on the diatomaceous sediments of the Gulf of California. They are mostly laminated and consist of regularly alternating light-coloured (diatom-rich) and dark-coloured (clay-rich) laminae, approximately 2 mm thick. The mean contents of opal (as diatom frustules) are: in the light-coloured laminae 52·4% of opal and in dark-coloured laminae 26·5% of opal (S. E. CALVERT 1966). They consist of about 20% of $CaCO_3$, a small percentage of terrigeneous material and spicules of fungi. Their grain-size distribution is given in Table 122.

Table 122

Grain-size distribution of diatomaceous sediments of the Gulf of California (after J. V. Byrne - K. O. Emery 1960)

Fraction μ	%
< 2	43·5
2— 4	16·6
4— 8	18·9
8—16	5·0
16—32	5·7
32—64	5·0
> 64	6·4

As a result of immense production of diatoms, the deposition of diatomaceous sediments is roughly as rapid as that of terrigenous deposits. This explains why they exist as a separate sediment also in places of normal terrigenous sedimentation.

Oyster bioherms are another very frequent bay sediment, developed particularly at the margins of bays and in the proximity of straits. They have been described, for instance, from many places of the Texas Bay. The sedimentation is influenced even by bioherms varying in extent several tens of metres.

The tests of organisms constitute particularly the coarse fraction of sediments, the individual groups being represented to various degrees. Thus, for instance, in the Japanese bays different genera make up following proportions of coarse fractions (Table 123).

The presence of individual organisms depends largely on the type of sediments, as is well seen from the amounts of foraminifers in different kinds of sediments (Table 124).

F. B. PHLEGER and G. C. EWING (1962) tried to characterize the environment by the ratio of living foraminifera to the total number of individuals (see Table 125).

Table 123

Proportion of tests of organisms in the coarse fraction of sediments of Japanese bays (after H. Niino 1950)

Genera	Maximal percentage of shells in coarse fractions %
Pelecypoda	87·7
Gastropoda	61·6
Scaphopoda	31·8
Pteropoda	10·4
Brachiopoda	28·3
Porifera	4·6
Anthozoa	35·0
Echinodermata	24·8
Bryozoa	14·3

Table 124

Amounts of foraminifers in different kinds of sediments in coastal lagoons of Baja California (after F. B. Phleger - G. C. Ewing 1962)

Sediment	Percentage of foraminifers in sandy fraction		Percentage foraminifers in total sediment	Prevailing species
	range	average		
Coarse-grained sand	0— 1	0·25	0·22	
Medium-grained and fine-grained sand	0— 2·2	0·46	0·4	benthonic
Fine-grained sand up to silt	0— 6	2	1·0	mixture
Clayey silt	0—46	11	1·4	planktonic
Silty clay	9—64	31	0·8	planktonic

Table 125

Ratio of living foraminifers to the total number of individuals in the sediments of Baja California (after F. B. Phleger - G. C. Ewing 1962)

Environment	Ratio of living foraminifers to their total number
Central parts of lagoons	45
Marsh	17
Tidal flat	83—97

In shallow parts of bays, the sedimentation is under the influence of sea grasses *Thalassia testudinum* and *Diplanteria wrighti*, as well as locally growing *Hypnea*, *Acanthopora* and *Ulva*. These growths are densely inhabited by molluscs. In other shallow, prevalently clayey parts of bays, the whole bottom was lined by a gelatinous mantle formed of algae and especially eggs of invertebrates and vertebrates.

Table 126 sums up the results of TJ. VAN ANDEL - H. POSTMA's study (1954) of the dependence of faunal association on sedimentary environments.

Characteristic of the lagoons and bays is the presence of normal marine fauna, which was transported over the barrier into the basin where it did not survive long

Table 126

Macrofauna associations in various sedimentary environments in the Gulf of Paria (after Tj. van Andel - H. Postma 1954)

Description of macrofauna association	Present environment
Ostrea banks, thick accumulations of mainly Ostrea. No debris.	Shallow, mud bottom, clear water; marine-brackish, strong variation.
Burrowing shells, isolated. No debris.	Shallow, mud bottom; marine-brackish strong variation.
Single mud dwelling species; isolated speciminas. Living.	Shallow, mud bottom; fresh brackish.
Abundance of species, mainly gastropods. Dead and abraded. Abundant debris.	Very deep to littoral; mud and sand bottoms: marine-brackish strong variation.
Mixture of mud dwellers and gastropods, latter of restricted size, abundance of debris; all dead and bleached.	Shallow; mud bottom; marine-brackish, strong variation.
Small specimens, thin shelled, few living, most dead but fresh, abundant debris.	Shallow; sand and rock; clear water; marine.
Abundance of species, shells, Bryozoa; calcareous worms; all dead; abundant debris.	Shallow; sand bottom, marine — almost fresh, strong variation at the surface, at bottom possibly less.
Mixture of debris of branching corals and reef fauna with mud dwelling shells. All dead. Abundant debris.	Shallow; mud bottom; muddy water: marine-brackish; strong variation
Branching corals, broken, dead, abraded.	Beach to very deep water; mud and sand; marinebrackish, strong variation.

owing to different environmental conditions. The foraminiferal associations of lagoonal sediments form likewise a mixture of normal marine forms with typical brackish forms.

The study of faunal associations from the bay and lagoonal sediments of the Gulf of Texas makes it possible to generalize tentatively the interrelationship between the associations and the sedimentary environments:

1. Sediments of deltas and salty swamps which are associated with bays: plentiful wood pieces, plant fragments and characteristic moluscs *Neritina reclinata* and *Littorina.*

2. Margins of deltas and river mouths of appreciably decreased salinity: characteristic pelecypods *Rangia cuneata, R. flexuosa, Macoma mitchelli* and *M. tageliformis* in association with a small mollusc *Littoridina sphinctostoma.*

3. Nearshore areas of large lagoons in the proximity of the mouths of major streams with decreased salinity and rapid sedimentation; characterized by not very numerous pelecypods, particularly *Mulinia lateralis, Nuculana eborea, Abra liocia* and gastropod *Nassarius acutus.*

4. Minor bays slightly influenced by the influx of river water separated by barriers; typical bay—inhabiting species.

5. Straits between barriers with unusually strong currents: abundant redeposited sessile organisms; pelecypods are represented by *Crassinella marticensis, Trachycardium muricatum, Chione cancellata,* gastropods *Natica pusilla, Abachis avara semiplicata, Olivella mutica, Canthares cancellarius* and by various species of *Turbonilla.*

6. Barrier sides of lagoons; distinctive pelecypods *Mercenaria compochiensis texana, Aequipecten irradians amplicostatus* and gastropods *Cerithium floridanum, Cerithium muscarum* and *Cerithium variable.*

7. In places of slow sedimentation tests of organisms from the above-mentioned environments are accumulated.

Sedimentary structures of bay deposits

Bay sediments have abundant and varied structures. The three fundamental structural types of sediments of shallow bays and lagoons are:

1. Homogeneous structures.
2. Regular stratification and lamination.
3. Stratification and lamination secondarily disturbed by:
   a) inorganic factors,
   b) biological factors.

There is a general rule that the slower the sedimentation and the more abundant the benthonic organisms, the more homogeneous are the sedimentary structures.

A perfect stratification originates only in two optimum cases: because of the absence of benthonic fauna or because sedimentation is so rapid that the original stratification cannot be disturbed by organisms.

Homogeneous structures develop in coarse sediments (sands) and clays. In sands, this structure is almost invariably of secondary origin; it may be primary in some clays from the deeper parts of bays. Mottled structures are likely to be most abundant in bay sediments. The mottles formed of coarser sediments are sharply limited against the surrounding finer-grained sediments or pass gradually into them.

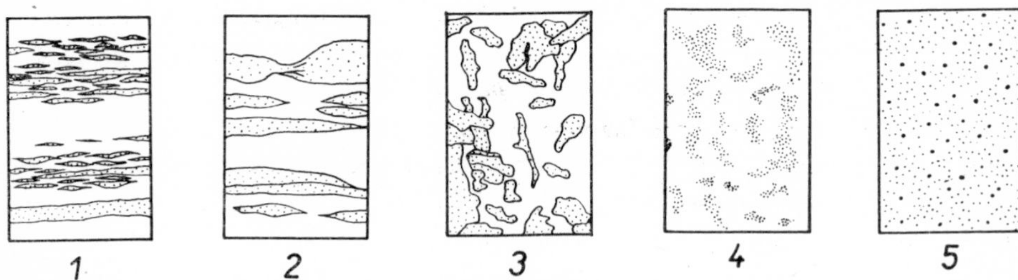

Fig. 85. Fundamental structures in shallow-water sediments. 1. Regularly bedded sediments, 2. Irregularly bedded, 3. Mottled sediments with sharply defined mottles, 4. Mottled sediments with mottles gradually passing into surrounding sediments, 5. Homogeneous sediment. After D. G. Moore *et al.* (1957).

Regular lamination of bay sediments originates either by the alternation of material of different grain-size or by the alternation of biological and chemical material. The latter type of lamination is far more abundant in bay sediments than the former, which is rather rare. E. Seibold (1958) records a typical lamination from the Adriatic Sea, which is due to the alternation of light carbonate-rich laminae with darker laminae, which are richer in organic matter. The average thickness of laminae is 0·25 mm and the mean grain-size is 13—26 μ. The lighter laminae contain 60% of $CaCO_3$, 8% of organic material, 30% of terrigenous clastic components and 2·5% of Fe-sulphides. Darker laminae bear only 30—50% of $CaCO_3$. The lamination evidently reflects the seasonal changes: the light laminae correspond to the summer periods and the darker ones to autumn and winter times. A well-developed lamination has been found in diatomaceous clays deposited in the Gulf of California, where about 0·5 mm-thick dark laminea, rich in terrigenous material, alternate with lighter laminae of the same thickness which are richer in diatom frustules. In the light-coloured laminae a graded bedding was observed in places. The light laminae were laid down after the summer bloom of diatoms when a "rain" of tests fell onto the bottom. The rate of deposition of diatom tests has been calculated to be twice as great as that of terrigenous clays and silts (92 mg/cm^2 against 53 mg/cm^2/year).

The regular lamination is preserved also owing to the lack of benthonic fauna in parts of this area.

The secondary disturbances of stratification and lamination in bay sediments are very widespread, produced either by the activity of organisms, or by the escape of air bubbles from water-logged fine-grained sediments (J. H. STEWART 1956). The mode of origin, as described above, has been reproduced in the laboratory, so that it may be regarded as safely proved. In lagoonal fine-grained sands of the Californian coast, dark and lighter, disturbed continuous or interrupted laminae alternate. The composition of the two types of laminae is given in Table 127.

The difference in colour is caused by a different content of dark heavy minerals.

In recurrently desiccating parts of bays and lagoons mud cracks often develop, to which breccia-like structures correspond in the vertical section through the sediment.

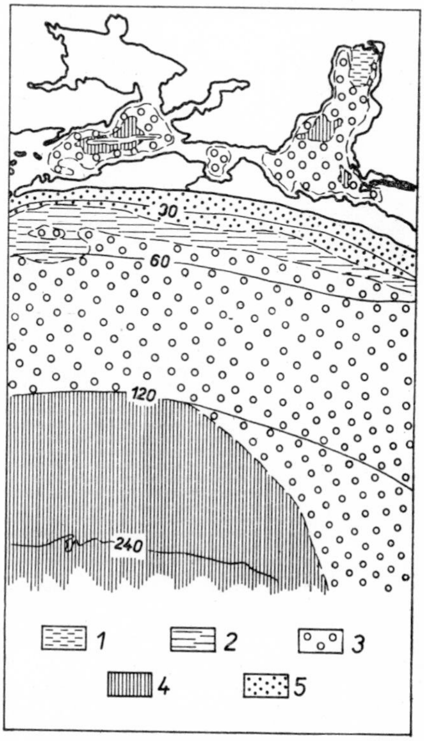

Fig. 86. Regional distribution of fundamental sedimentary structures in nearshore environ ment of Texas Bay. 1. Regular bedding, 2. Irregular bedding, 3. Mottled structure, 4. Homogeneous sediment (clays and silts), 5. Homogeneous sediments (sands). Depth in feet. After D. G. Moore *et al.* (1957).

Table 127

Composition of light and dark laminae of laminated sediments in lagoons (after J. H. Stewart 1956)

	Light lamina	Dark lamina
Md	0·196 mm	0·172 mm
So_{Φ}	0·30	0·32
Heavy minerals	15·3%	74·0%

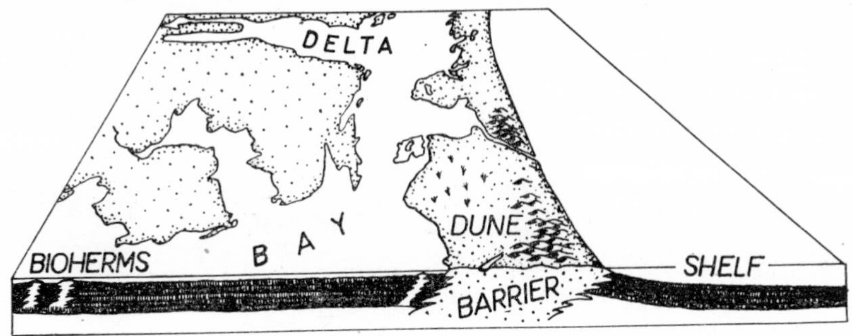

Fig. 87. Lateral and vertical distribution of shallow-water environments in Texas Bay. After F. P. Shepard (1958).

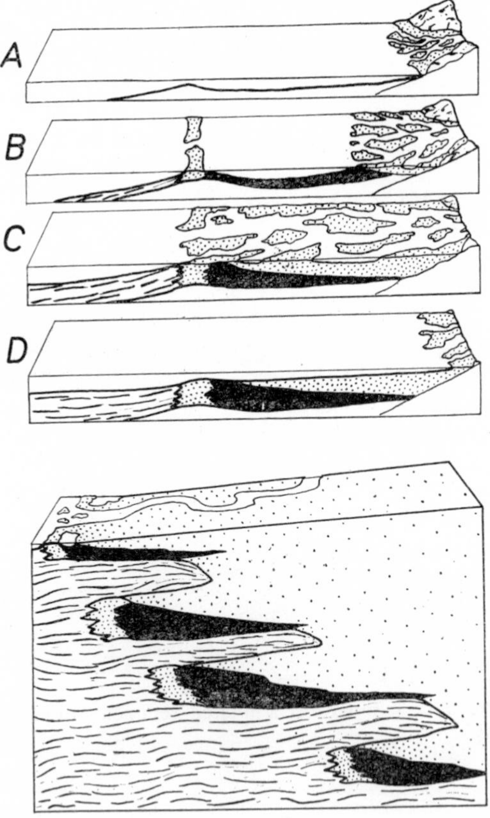

Fig. 88. The vertical development of shallow-water marine environments in the course of marine transgression. A. Immediately after transgression, slow filling of lagoon, B. Sand bar is forming, C. Advance of sand bar into the lagoon and filling of lagoon with sands. D. New transgression. Lower picture: Vertical distribution of sediments having originated by processes described. Black: lagoonal fine-grained sediments: loose dots: continental sediments; dense dots: marine sediments. After E. Tongiorgi *et al.* (1958).

Mode of deposition of bay and lagoonal sediments

The sediments of bays and lagoons occupy a precisely defined position among Recent sediments and their significance in geological history cannot be overlooked. Similarly as deltaic sediments, they represent a transition between continental and marine deposits. Where marine sediments are the only component of ancient series, the lagoonal deposits are their marginal facies. On the other hand, they can form the marginal facies of continental sediments. On the continental side, they adjoin the beach deposits or sediments of swamps or the continental dunes. On the marine side, they grade into the sediments of sand barriers or into beach deposits of the islands. Laterally, they frequently pass into deltaic deposits. All bay and lagoonal environments are very sensitive even to small oscillations of the sea level. Abundant reference is made to the vertical alternation of fine-grained lagoonal sediments and coarser barrier sands, which results from the oscillation of the sea level and the subsequent equalizing of conditions by lateral migration of the individual environments. Several alternatives postulated for the origin of this vertical alternation are shown diagramatically in Fig. 88. Provided that the positive movement of the sea level occurs at a constant velocity, a sand barrier first develops and the lagoon thus formed is being filled by fine grained, mainly sandy sediments. During the following period, the sand barrier migrates landwards and fills the lagoon by sand. The whole process repeats itself during the next positive movement of the sea level. As a result, marine sediments, barrier sands, clayey lagoonal and continental (mainly fluviatile) deposits can alternate through the vertical section.

The thickness of Recent bay sediments may vary considerably. As mentioned above, the bay environment is mostly a site of rapid sedimentation owing to an abundant supply of detritus by streams. Where no intensive subsidence occurred, the thickness of Recent bay sediments is about a few tens of metres, corresponding to vertical oscillations of the sea level from the end of Pleistocene till the present. In places of strong subsidence it can attain several hundred metres.

Sediments of mangrove swamps

Mangrove swamps represent a particular sedimentary environment which is generally connected with lagoons and bays. They occur in the tropical climatic zone, in lagoons and bays protected from surf effects. The coasts of Florida, Cuba, long sections of the Atlantic coast of Brazil and parts of the coast of the East Indian archipelago, particularly of the Sumatra, are the typical sites of their growths. The general picture of mangrove swamps differs from place to place; it can be developed as a swamp with isolated trees or an impenetrable mass of trees and shrubs. They pass seawards into growths of marine grasses. Mangroves can exist in brackish, normal marine or weakly hypersaline waters. They are not confined to a special type of sediments but they themselves affect the composition of sediments to a large extent. As long as the substrate is not coarser than medium-grained sands, they can vegetate on sand, clay and silt or even coral-sand. The mangrove flora alters with the change in the salinity of water, induced, for instance, by an increased inflow of river water.

Quasi-primordial mangrove swamps occur on coral islands, on fine grained sediments bordering the lagoons. The pioneer of the mangrove vegetation is always *Rhizopora* which works the way to other floras. *Rhizopora mucronata*, *R. conjugata*, *Bruguiera gymnorhiza*, *B. criopetala* and *Avicena oficialis* are the most widespread mangrove species.

Mangrove growths affect the sedimentation only when they are dense enough for the sediments to be entrapped. In this case, their effects are considerable, especially within the range of tidal oscillations, because the detritus brought by high-tide current remains entrapped on the vegetation. The low-tide currents cannot reach it and wash it back. The sediments of varied composition correspond to the sand-silt-clay grade the equilibrium shifting from silt-clay fraction to sand fraction (chiefly in coral islands and on limestone coast). The content of carbonate is variable. On the Florida coast and coral reefs of limestone coasts, they contain much detrital carbonate and the content of $CaCO_3$ sometimes exceeds 50%. In other places the content of $CaCO_3$ is quite small. The proportion of organic matter is extremely high. Thus, for instance, in the Florida mangrove swamps, the percentage of organic matter varies between 6 and 50% and the org. C/org. N ratio between 7 and 37. Sediments are characterized by a high content of pieces of wood and plant remains. In the mangrove swamps on the coast of Brazil, fragmentary oxidic iron ore are formed of minute hydroxide pieces, which originate by breaking of originally solid oxidation crust (B. v. FREYBERG 1930).

Sedimentary structures are represented chiefly by the mottled structure formed by rafts of fine sand or silt with faint margins, which are embedded in darker and finer-grained silty clay. The mottled structure can originate also by the uneven distribution of organic matter.

The macrofauna consists of 80—90% of molluscs. The areal distribution of their assemblages tends to parallel the coastal area.

The thickness of mangrove sediments is not large on the tectonically stable coasts, where it equals the difference between the mean water-level of the high and low tides. On the other hand, the sediments can attain a great thickness of a subsiding shore, where sediments are caught in the persistently overgrowing mangroves. They occassionally grade laterally into fine-grained lagoonal sediments or interfinger with sandy barrier sediments.

References

ANDEL TJ. VAN - POSTMA H. (1954): Recent sediments of the Gulf of Paria. Verhandel. Koninkl. Nederl. Akad. Wetenschap., afd Natuurkunde, Erste Reeks, D. 20., No 5: 1—244, Amsterdam.

ANDEL TJ. VAN - SHOR G. G. (Editors) (1960): Marine geology of the Gulf of California (Symposium). Am. Assoc. Petrol. Geol., Mem. 3: 216—310, Tulsa.

BYRNE J. V. - EMERY K. O. (1960): Sediments of the Gulf of California. Bull. Geol. Soc. Am., vol. 40: 2354—2383, Baltimore.

CALVERT S. E. (1966): Origin of diatom-rich, varved sediments from the Gulf of California. Jour. Geology, vol. 74: 546—565, Chicago.

DOEGLAS D. J. (1951): The transport of sedimentary material. CR 3. Congr. Strat. Géol. Carbonif., T. 1: 147—148, Heerlen.

EMERY K. O. (1954): Some characteristics of Southern California sediments. Jour. Sedimentary Petrology, vol. 24: 50—58, Menasha.

— (1956): Sediments and water of Persian Gulf. Bull. Am. Assoc. Petrol. Geol., vol. 40: 2354—2383, Tulsa.

EMERY K. O. (1968): Relict sediments on continental shelves of world. Bull. Am. Assoc. Petrol. Geol., vol. 52: 445—464, Tulsa.

EMERY K. O. - BUTCHER H. S. - GOULD H. R. - SHEPARD F. P. (1950): Submarine geology off San Diego, California. Jour. Geology, vol. 60: 511—548, Chicago.

EMERY K. O. - GORSLINE D. S. - UCHUPI E. - TERRY R. D. (1957): Sediments of three bays of Baja California, Sebastian Viscaino, San Cristobal and Todos Santos. Jour. Sediment. Petrology, vol. 27: 95—115, Menasha.

EMERY K. O. - HÜLSEMANN J. (1962): The relationship of sediments, life and water in marine basin. Deep Sea Research, vol. 8: 165—180, London.

FISK H. N. - MCCLELLAND B. (1959): Geology of continental shelf off Louisiana: its influence on offshore foundation design. Bull. Geol. Soc. Am., vol. 70: 1369—1394, Baltimore.

FREYBERG B. v. (1930): Zerstörung und Sedimentation an der Mangroveküste Brasiliens. Leopoldiana, Bd. 6: 1—116, Halle.

FÜCHTBAUER H. (Editor 1969): Lithification of carbonate sediments. Special issue of "Sedimentology", vol. 12: 1—244, Amsterdam.

FÜTTERER D. (1969): Die Sedimente der nördlichen Adria vor der Küste Istriens. Göttinger Arb. Geol. Paläont., vol. 3: 1—57, Göttingen.

GERSHANOVICH D. E. (1963): Shelf sediments in the Gulf of Alaska and the conditions of their formation (in Russian). Deltaic and shallow-water littoral marine sediments. P. 32—38, Moscow.

GOODELL H. D. - GORSLINE D. S. (1961): A sedimentological study of Tampa Bay, Florida. Report of the 21 sess. Internat. Geol. Congr., Norden, p. 23: 75—88.

GORSLINE D. S. (1957): The relation of bottom sediment type to water motion: Sebastian Viscaino Bay, Baja California, Mexico. Rev. Geograph. Phys. et Géol. Dyn., vol. 1: 83—92, Paris.

— (1963): Bottom sediments of the Atlantic shelf and slope off the southern United States. Jour. Geology, vol. 71: 422—440, Chicago.

GREENMAN N. N. - LEBLANC R. J. (1956): Recent marine sediments and environments of Northwestern Gulf of Mexico. Bull. Am. Assoc. Petrol. Geol., vol. 40: 813—847. Tulsa.

GRIPP K. (1958): Rezente und fossile Flachmeer-Absätze petrologisch betrachtet und gedeutet. Geol. Rundschau, Bd. 47: 83—99, Stuttgart.

GROSS M. G. - MCMANUS D. A. - HAIN-YI LING (1967): Continental shelf sediments, northwestern United States. Jour. Sedimentary Petrology, vol. 37: 790—795, Menasha.

HARRISON W. - LYNCH M. P. - ALTSCHAEFFL A. G. (1964): Sediments of lower Chesapeake Bay, with emphasis on mass properties. Jour. Sedimentary Petrology, vol. 34: 727—755, Menasha.

HAYES M. O. (1967): Relation between coastal climate and bottom sediment type near the inner continental shelf. Marine Geology, vol. 5: 111—132, Amsterdam.

HOYT J. H. - SMITH D. D. - OOSTDAM B. L. (1965): Sediment distribution on the Inner Continental Shelf, West Coast of Southern Africa. Bull. Am. Assoc. Petrol. Geol., vol. 49: 344, Abstr., Tulsa.

INMAN D. L. (1959): Environmental significance of oscillatory ripple marks. Ecolog. Geol. Helv., vol. 51: 522—523, Basel.

INMAN D. L. - FILOUX J. (1960): Beach cycle related to tide and local wind wave regime. Jour. Geology, vol. 68: 225—231. Chicago.

INMAN D. L. - CHAMBERLAIN T. K. (1956): Particle size distribution in nearshore sediments. In "Finding ancient shorelines". P. 106—127, Tulsa.

KAGAMI H. (1961): Submarine sediments off Sakata, Yamagata, Japan. Japanese Journal of Geology and Geography, vol. 32: 297—409, Tokyo.

KINSMAN D. J. J. (1969): Modes of formation, sedimentary associations, and diagenetic features of shallow-water and supratidal evaporites. Bull. Am. Assoc. Petrol. Geol., vol. 53: 830—840, Tulsa.

KLENOVA M. V. (1960): Geology of Barents Sea (in Russian). AN SSSR, 1—368, Moscow.

KOFOED J. W. - GORSLINE D. S. (1963): Sedimentary environments in Apalachiola Bay and vicinity, Florida, Jour. Sedimentary Petrology, vol. 33: 205—223, Menasha.

LISICYN A. P. (1963): Bottom setsediments on the shelf of Antarctic continent (in Russian). Deltaic and shallow water marine sediments, 82—88, Moscow.

LUCIA F. J. (1968): Recent sediments and diagenesis of South Bonaire, Netherlands Antilles. Jour. Sedimentary Petrology, vol. 38: 845—858, Menasha.

— (1969): Recognition of evaporite-carbonate shoreline sedimentation. Bull. Am. Assoc. Petrol. Geol., vol. 53: 729, Tulsa.

LUCKE J. B. (1935): Bottom condition in a tidal lagoon. Jour. Paleontology, vol. 9: 101—107.

MOORE J. R. (1963): Bottom sediment studies, Buzzard Bay, Massachusetts. Jour. Sedimentary Petrology, vol. 33: 511—558, Menasha.

MOORE D. G. - SCRUTON P. C. (1957): Minor internal structures of some recent unconsolidated sediments. Bull. Am. Assoc. Petrol. Geol., vol. 41: 2723—2751, Tulsa.

MURDMAA I. O. (1963): Sedimentation on the shelves of Kurile Islands (in Russian). Deltaic and shallow water littoral marine sediments. P. 39—42, Moscow.

MÜLLER G. (1958): Die rezenten Sedimente im Golf von Neapel. Geol. Rundschau, vol. 47: 117—149, Stuttgart.

— (1961): Das Sand-Silt-Ton Verhältnisse in rezenten marinen Sedimenten. N. Jhrb. Min., Mh, H. 7: 148—165, Stuttgart.

— (1964): Die Korngrösseverteilung in den rezenten Sedimenten des Golfes von Neapel. Deltaic and Shallow Marine Deposits, 282—292, Amsterdam.

NAYUDU J. R. - ENBYSK B. J. (1964): Bio-lithology of Northeast Pacific surface sediments. Marine Geology, vol. 2: 310—342, Amsterdam.

NIINO H. (1950): Bottom deposits at the mouth of Wakasa Bay, Japan, and adjacent continental shelf. Jour. Sedimentary Petrology, vol. 20. 37—54, Menasha.

NIINO H. - EMERY K. O. (1961): Sediments of shallow portions of East China Sea and South China Sea. Bull. Geol. Soc. Am., vol. 72: 731—762, Baltimore.

NOTA D. J. G. (1958): Sediments auf dem West-Guyana Schelf. Geol. Rundschau., vol. 47: 167—177, Stuttgart.

PHLEGER F. B. - EWING G. C. (1962): Sedimentology and oceanography of coastal lagoons of Baja California, Mexico. Bull. Geol. Soc. Am., vol. 73: 145—182, Baltimore.

Powers M. C. - Kinsman B. (1953): Shell accumulations in underwater sediments and their relation to the thickness of the traction zone. Jour. Sedimentary Petrology, vol. 23: 229—234, Menasha.

Pratt W. L. (1962): Origin and distribution of glauconite and related clay aggregates on sea floor of Southern California. Bull. Am. Assoc. Petrol. Geol., vol. 46: 275, Abstr., Tulsa.

Prattje O. (1931): Die Sedimente des Kurischen Haffes. Fortschr. d. Geol. u. Pal., 1—142, Berlin.

Rao M. S. (1960): Organic matter in marine sediments off east coast of India. Bull. Am. Assoc. Petrol. Geol., vol. 44: 1705—1714, Tulsa.

— (1964): Some aspects of continental shelf sediments off the east coast of India. Marine Geology, vol. 1: 59—87, Amsterdam.

Seibold E. (1956): Wasser, Kalk, und Korngrössenverteilung in einem Adria und Bodensee-Sedimentkernen. N. Jhrb. Geol. Mh., H. 10:451—470, Stuttgart.

— (1958): Jahreslagen in Sedimenten der mittleren Adria. Geol. Rundschau, vol. 47: 100—117, Stuttgart.

Scholl D. W. (1963): Sedimentation in modern coastal swamps, Southwestern Florida. Bull. Am. Assoc. Petrol. Geol., vol. 47: 1581—1603, Tulsa.

Shepard F. P. (1932): Sediments of continental shelves. Bull. Geol. Soc. Am., vol. 43: 1017—1040, Baltimore.

— (1939): Continental shelf sediments. Recent marine sediments. P. 220—230, Tulsa.

— (1958): Sedimentary environments of the Northwest Gulf of Mexico. Eclog. Geol. Helv., vol. 51: 598—608, Basel.

Shepard F. P. - Phleger F. B. - Andel Tj. van (Editor) (1960): Recent Sediments, Northwest Gulf of Mexico (Symposium). P. 1—394, Tulsa.

Sindowsky K. H. (1957): Bodengestalt und Bodensediment im Nordteil des Golfes von Neapel. Geol. Jahrbuch, vol. 73: 595—612, Hannover.

Stetson H. C. (1939): Summary of sedimentary conditions on the continental shelf off the coast of United States. Recent marine sediments. P. 230—244, Tulsa.

Stewart J. H. (1956): Contorted sediment in modern coastal lagoon explained by laboratory experiments. Bull. Am. Assoc. Petrol. Geol., vol. 40: 813—847, Tulsa.

— (1958): Sedimentary reflection of depositional environment in San Miguel Lagoon, Baja California, Mexico. Bull. Am. Assoc. Petrol. Geol., vol. 42: 2567—2618, Tulsa.

Stewart R. A. - Gorsline D. S. (1962): Recent sedimentary history of St. Joseph Bay, Florida, Sedimentology, vol. 1: 256—286, Amsterdam.

Straaten L. M. J. U. van (1959): Minor structures of some littoral and neritic sediments. Geol. Mijnbouw, 21 e Jg, P. 197—217, Amsterdam.

Tanner W. F. (1960): Shallow water ripple mark varieties. Jour. Sedimentary Petrology, vol. 30: 481—485, Tulsa.

Tongiorgi E. - Trevisan L. (1959): Le rôle des lagunes dans la sédimentation rythmique des bassins subsidentes. Eclog. Geol. Helv., vol. 51: 767—774, Basel.

Vigneaux M. (Editor, 1969): Océanographie dans le Golfe de Gascogne. Bull. Inst. Geol. du Bassin d'Aquitaine. No 6: 1—367, Bordeaux.

Werner F. (1964): Sedimentkerne aus den Rinnen der Kieler Bucht. Meyniana, Bd. 14: 52—65, Kiel.

Yañez C. A. (1963): Batimetria, salinidad, temperatura y distribucion de los sedimentos recientes de la laguna de Terminos, Campeche, Mexico. Boletin Universidad Nacional Autonoma de Mexico, No 67: 1—57, Mexico.

17. Tidal flats (waddens, wades)

Tidal flats are sometimes grouped with the bay environment, but their specific features justify us in describing their deposits separately. At present, tidal flats are defined as bodies of sediments aggraded by the sea into bays whose water surface is maintained by tidal currents approximately at the mid-level between the high and low tide (H. E. Reineck 1955). The former, rather geographical conception regarded tidal flats as that part of the flat coast between the high and low water levels. The following factors should be now taken as distinctive for tidal flats:

1. Periodical desiccation of large areas of sediments during the low tide.
2. Marine origin of sediments.

The great importance of the latter factor is mostly underestimated. In deltas, for instance, a large proportion of sediments is also above the water level during low tides, but their material is of continental origin in contrast to that of tidal flats. It has been ascertained that sands of the North Germany tidal flats are derived from the North Sea floor. The clayey deposits of tidal flats also have a different composition than clays transported by adjacent rivers.

The typical and most thoroughly studied tidal flats are the waddens of the coast of Northern Germany, the Netherlands and Denmark. They are also recorded from the Atlantic coast of Canada, the Pacific coast of Yucatan and Guatemala, from the Red Sea, from the coasts of San Salvador and Chile, from southern and south-eastern Africa, from minor bays near the mouth of Thames in England (especially the Wash) and from the Bay of Fundy near Newfoundland.

Typical tidal flats occur today only on flat coasts with a constant or recurrent positive movement of the sea level (the rate of this movement, is for instance on the North German coast 3·7 cm in 100 years). The subsidence of the coast is thus one of the controlling factors of the morphology and development of tidal flats. The tides are another significant factor. Of importance is the difference between the mean sea level at low and high tides (in North Germany this difference amounts to 3·75 m). At high tide, the waters distribute over coastal flats, concentrating during low tides into minor channels or creeks (so-called "Prielen") which lead into the central channels. The main influence on the deposition of material above the mean high-tide level is exerted by anomalous high tides (the so-called "Storm tides") produced by strong winds. The maximum recorded high tide on North German wadden exceeds this level by 4 m. The mean velocity of the ebb-tide current is about 50 cm/sec, whereas that of the flood-tide current is 80 cm/sec. This difference in the velocities causes the so-called "tidal-stratification" (Gezeitenschichtung) which will be dealt with in detail below. The velocity of low-tide current in central discharge channels is much greater — up to 100 cm/sec.

Photo 2. Erosion of semiconsolidated clay deposits in tidal flats. Origin of clay balls and clay galls. Jadebusen, North German wades. After H. Ehrhardt (1937).

Photo 3. Tidal creek (Priel) on tidal flat near Wilhelmshaven (North Germany). After H. Ehrhardt (1937).

In the opinion of some authors, wave action is more important than the tides. Wave action gives rise to ridges (Spülsäume) and redeposition of fine-grained sediments to greater depths and may prevent the deposition of fine-grained sediments on some parts of tidal flats.

Some European tidal flats occur behind chains of sand islets. Some of these islets are encompassed by broad sand flats lying at the height of the high-tide water level. In narrow straits separating individual islets the tidal currents attain maximum velocities, the erosion in these channels being thus effective to considerable depths (of 50 m). The branching channels towards the tidal flats is a characteristic morphological feature of tidal flats on bathymetric maps.

The above mentioned channels (tidal creeks — Prielen) resemble rivers by their meandering course and system of tributaries. Their width ranges from several centimetres or metres to a few kilometres. All these channels are joined by tributaries; they widen and open into straits between the sand banks. They are cut into unconsolidated sediments, so that, both the vertical and lateral erosion is very rapid. It is estimated that the channel moves laterally 20—30 m yearly and in extreme cases even more than 100 m. When the water suddenly abandons its channels during storm tides, these are filled with sediments within a few days or weeks. Downcutting

Photo 4. Tidal flat covered by sand waves. North German wades. After H. Ehrhardt (1937).

is also strong, as indicated by their depth of about several tens of metres. Discharge channels differ from the adjacent tidal flats by the composition of sediments and faunal associations. They are characterized by an abundance of intraformational conglomerates in the initial state of development, diagonal bedding and slump structures. The direction of tributaries, and their widening and deepening suggest that the configuration of discharge channels is modelled wholly by ebb-tide currents. The remaining area of tidal flats has usually a moderate relief. It is mostly differentiated into sandy flats of varying shape and size and surface height corresponding to the high-tide water level and into shallow depressions filled by finer grained sediments. The interrelationship of these two morphological elements frequently changes during one high-tide low-tide interval. The sand flats are often almost permanently covered by regular, rather flat waves, whose length varies between several tens and a few hundred metres.

Tidal flats show some other characteristic phenomena of minor size but of great genetic interest, as for instance, ridges of coarse material (Spülsäume). These are comparatively high ridges formed prevalently of biogenic material sorted by transport, which are aggraded to the height of or above the high-tide level. They are called "kitchen midden of the sea". Their origin is due to the activity of advancing and high-tide currents which sort the material and deposit it where their transport power dies out. In the large Jade Bay (North Germany) the ridges develop at the borders of channels, where waves and currents loose their intensity and in places of abrupt curving of the shoreline. Their organic material — prevalently fragments of organisms — is sorted according to size and shape. Ridges of echinoderm shells afford a good example of perfect sorting. Several kilometres long and approximately 1 m broad and 10 cm high ridges are fomed of rounded shells devoid of spines, which were washed away and deposited elsewhere in separate ridges. Reference is made of ridge of medusal bodies, pelecypod shells and crustacean bodies. A ridge of beetles, 1·5 km long, 20—50 cm broad and several centimetres high has also been recorded, but similar to all ridges, it succumbed to the next stormy high tide. A slighty sinuous course of some ridges recalls the course of beach cusps. The ridge formed of earth-worm bodies, found in North Germany wades, and the ridge built up of pieces of wood, described from the mangrove coast of Brazil, were of this form.

K. LÜDERS (1930) reconstructed experimentally the origin of aggraded sand ridges capped by coarse-grained material, most frequently of pelecypod shells. Aggraded ridges are undoubtedly one of the most interesting phenomena of tidal flats, attesting to the remarkable sorting capacity of a shallow sea. Their fossil analogies are probably more numerous than presumed so far.

Sediments of tidal flats

Most detailed petrographical information has been gathered on sediments of tidal flats in the Netherlands and in the Wash on the eastern coast of England. Geologically, two kinds of sediments are usually distinguished in tidal flats: sand and

Table 128

Parameters of the sediments of the Wash
(after G. Evans 1965)

Subenvironment	Md_{Φ}		Phi Standard Deviation		Phi Skewness	
	Mean	Range	Mean	Range	Mean	Range
Salt marsh	6·70	5·80—7·60	—	—	—	—
Higher mud flat	4·51	4·00—6·75	1·56	0·60—2·70	+0·62	+0·29 to +0·89
Inner sand flat	3·52	3·40—3·70	0·51	0·45—0·60	+0·35	+0·17 to +0·44
Arenicola sand flat	3·39	2·70—3·80	0·38	0·30—0·45	+0·15	+0·15 to +0·44
Lower mud flat	3·57	3·20—3·80	0·85	0·50—1·60	+0·42	+0·20 to +0·64
Lower sand flat	3·08	2·90—3·50	0·53	0·25—1·25	+0·40	+0·20 to +0·64

"Schlick" (silt with clay and organic particles in different ratios). Gravels are very rare; they occur only as gravels composed of pebbles of Subrecent and Recent consolidated clay and silt sediments. They are solely of local origin and are not found at a great distance from the place of their formation. Therefore, they develop mainly in channels, by the erosion of their walls, or within the reach of high tide current by the erosion of Subrecent consolidated silty clay. These pebbles form in places minor ridges. Where well-rounded, they form the so-called armoured mud balls, because of their coating of sand or tests. Their size does not exceed 5 cm, being commonly about 1 cm.

Extensive sand deposits of tidal flats built up either their central parts or aureoles around dune islands. Parameters of tidal flat sands of the Wash and size distribution of other sediments are shown in Fig. 90 and Table 128. Sands of Danish tidal flats are characterised in the following patameters:

	Range	Average
Md	0·08—0·6 mm	0·13 mm
So	1·20—1·9 mm	1·30 mm

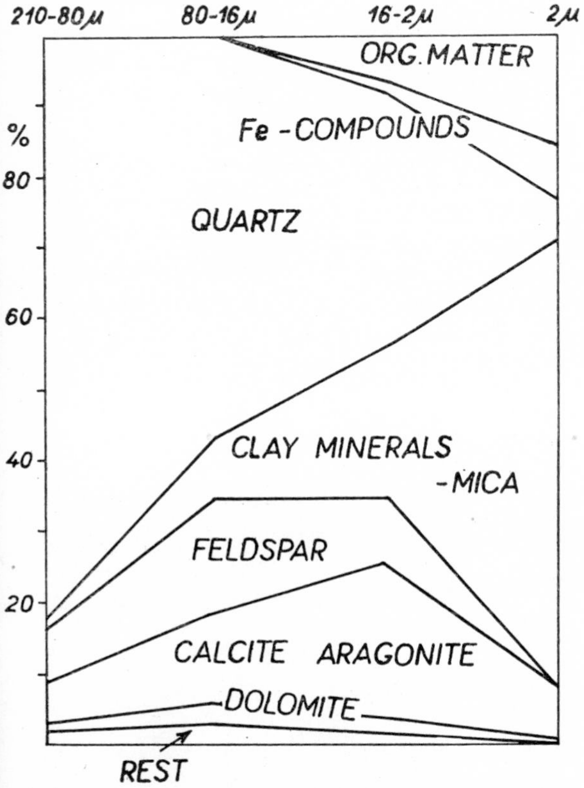

Fig. 89. Grain-size distribution of sediment-forming compounds of tidal flat sediments. After L. M. J. U. van Straaten (1959).

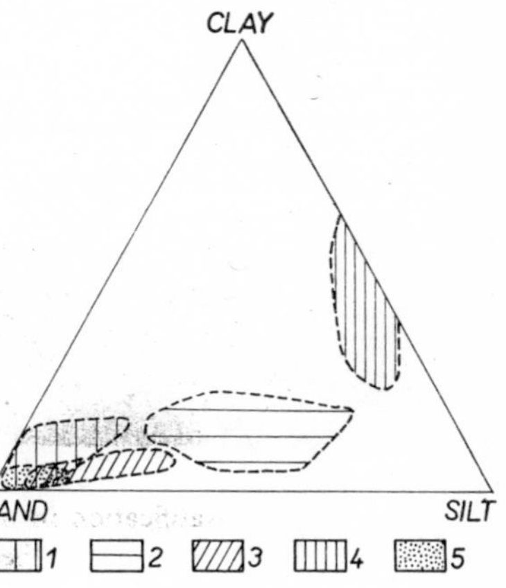

Fig. 90. Grain-size of the sediments of tidal flats of the Wash (England). Triangular diagram showing sand, silt and clay ratio. 1. Lower sand flat deposits, 2. Higher mud flat deposits, 3. Lower mud flat deposits, 4. Salt marsh deposits, 5. Arenicola sand flat deposits. After G. Evans (1965).

From the Tables the similarity of wadden sands to the fine grained sands of lagoonal beaches is well seen. Their mineralogical composition is likewise analogous. Heavy minerals form in places concentrations the development and occurrence of which are governed by the same rules as those in beach sediments. There are no gradual transitions between the sands and the finer-grained sediments of tidal flats. The Md of the latter falls almost invariably into the silt fraction which is nearly always more abundant than the clay fraction. In the North German tidal flats the silt fraction makes up more than 60% and in the Wash it averages 72% of fine-grained deposits. The parameters of fine sediments of Danish tidal flats are listed below:

	Range	Average
Md	0·01—0·05 mm	0·03 mm
So	2·90—4·40 mm	3·30 mm

The high percentage of silt grains is connected with the presence of quartz grains, but especially with a large content of organic particles. Samples of "Schlick" collected from various tidal flats yielded 90% of organic origin (especially diatom frustules, coprolites and aggregates of organic matter) and only 10% of terrigenous components.

The content of calcium carbonate in the sediments of tidal flats varies greatly: the minimum amount which derived from organic tests, has been ascertained in

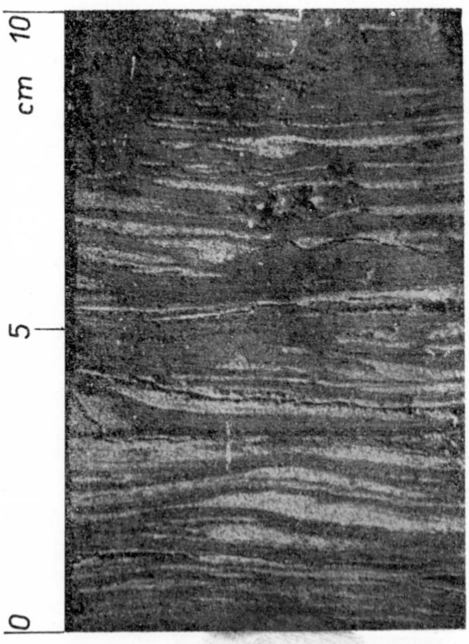

Photo 5. Types of stratification in tidal flat sediments (North German wades). Lenses of ripple bedded sand in silty clays.

Photo 6. Some cores from sediments from tidal flats in the Wash. 1. Outer salt marsh deposits, showing a mud crack and infilled plant root cavities, 2. Outer salt marsh deposits with infilled plant root cavities, 3. Inner salt marsh deposits, 4. Creek levee deposits, 5, 6. Outer higher mud flat deposits, with burrows. After G. Evans (1965).

sands; the average amount of 5·56% has been found in "Schlick" (the North German coast). The content of carbonate, the grain-size of sediments and the salinity of water show an interesting relationship. The percentage of carbonate increases with the rise in the content of finer-grained fraction, which implies the probability of its largely biochemical origin. In brackish waters, near the river mouths, sediments are very poor in carbonate, the amount of which rises towards the centre of bays. The increased precipitation of carbonate in the "Schlick" can occur already during the early diagenesis, soon after their deposition. Thus, for instance, Subrecent "Schlick" containing about 35—50% of $CaCO_3$ has been found.

Organic matter is a substantial component of sediments of tidal flats. Whereas sands are poor in organic matter (although they are richer than those of other environments, containing even more than 1% of org. C) fine-grained sediments have usually above 10% of org. C. Organic matter derives partly from soft organic bodies, partly from faeces; of the former, fats, waxes and other resistant components are preserved, cementing occasionally parts of deposits in tidal flats. In places, fine-grained sediments are rich in faeces of sea-gulls, in which bones, fish scales, fragments of tests of calcareous organisms and diatom frustules are enclosed.

"Schlick" of tidal flats is often packed with diatom frustules. CH. BROCKMANN (1935) calculated that in 1 g of dried "Shlick" from the port Wilhelmshaven there are 2,538,000 diatom frustules, in the mouth of Jade Bay 75,000 frustules, whereas at the transition from the bay into the open sea there are only 23 frustules per gram. The increasing percentage of diatom frustules landwards is explained by the increasing amount of nutrients in marine waters. In some places of tidal flats, the number of diatom frustules is so large as to form a thin solid film on clay deposits (called the local name "Wattenpapier"). This immense amount of frustules will undoubtedly contribute to the production of silica during diagenesis, which can migrate and precipitate in other places.

Tidal flats are the habitats of many species of pelecypods and other organisms with calcareous shells, which afterwards become a component of sediments or form themselves a distinct sediment. The accumulations of pelecypod shells are rather frequent and in Germany they are called "Schill", or "Bruchschill", when fragments of shells are prevalent. On the coast of the Netherlands and Germany the pelecypod shells form bands in tidal flats, up to 3 m thick, hundreds to thousands of metres long and 15—30 m broad. The bands were accumulated by tidal currents and occur therefore at about the high tide level. They are occasionally distributed also along the discharge channels and can migrate along the coast. The pelecypod shells genera *Cardium*, *Mytilus*, *Mactra*, *Donax* and *Venus* are the main constituents producing up to 91% $CaCO_3$.

Owing to a large content of organic matter, fine-grained sediments of tidal flats have almost invariably an increased percentage of sulfides. Most writers believe that by bacterial activity organic matter is altered into hydrogen sulfide which reacts with ferrous compounds of marine water and so gives rise to sulfides. A high content of metacoloidal hydrotroilite contributes to the black colouring of fine-grained sediments of tidal flats.

Distribution of sediments in tidal flats

The distribution of sediments in tidal flats follows a certain design and corresponds roughly to morphology of tidal flats. Flats that lie not very deep under the high-tide level are usually formed of sands and the contiguous depressions of finer-grained sediments. Fig. 91 shows the distribution of environments in the Wash on the English coast. Otherwise, the following general succession of environments can be assumed for all tidal flats:

1. Clayey and silty sediments above the high-tide level, especially at the margins of bays. According to PH. H. KUENEN - L. M. J. U. VAN STRAATEN (1957, 1958), the intensive clay and silt deposition is made possible by the supply of fine material by flood-tide currents. The ebb-tide currents are not capable of removing these particles into deeper parts of the bay.

2. Belts of sandy flats developed below the high-tide level. They are separated by channels and depressions filled frequently with coarse sediments.

3. The deepest parts of tidal flats under the low-tide level, where fine-grained sediments are prevalently deposited.

4. Belts of dune islands and adjacent sand flats.

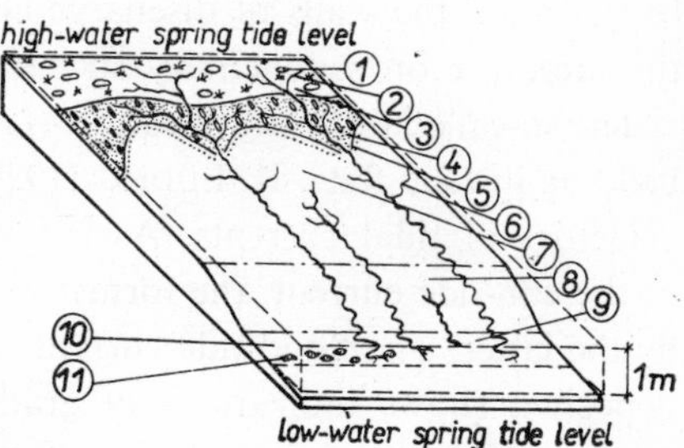

Fig. 91. The Wash. The subenvironments of deposition in tidal flat. 1. Salt pans, 2. Creek levee, 3. Salt marsh, 4. Higher mud flat, 5. Inner sand flats, 6. Creek borders, 7. Arenicola sand flats, 8. Lower mud flats, 9. Lower sand flats, 10. Giant ripple-marks, 11. Mussel banks. After G. Evans (1965).

5. Tidal Creeks ("Prielen") with a characteristic deposition of sands, "Schlick" and clay pebble gravels.

Each of these environments is distinguished by peculiar dynamics of sedimentation, which is summed up in Table 129.

Table 129

Schematic representation of the dynamics of sedimentation in the intertidal zone of the Wash (after G. Evans 1965)

Subenvironment	Dynamics of sedimentation
Salt marsh	No reworking by waves or organisms, slow sedimentation, sediment bound by plants. Filter effect of plants.
Higher mud flats	Very limited reworking by waves and organisms, fairly rapid sedimentation, sediment bound by algae.
Inner sand flats	Limited reworking by waves, extensive reworking by Corophium, slow sedimentation.
Arenicola sand flats	Extensive reworking by waves and Arenicola, slow sedimentation.
Lower mud flats	Little reworking by waves and organisms, rapid sedimentation
Lower sand flats	Little or no organic reworking, minor wave action, strong-tidal currents, slow sedimentation.

Sedimentary structures

Tidal flats afford the best possibility for the study of sedimentary structures, both of mechanical and biological origin. Twice a day, exposed surfaces of sand and "Schlick", corresponding to fossil bedding planes, reveal a large amount of most varied markings in an unconsolidated state. The origin of various types of stratification can be traced on the walls of discharge channels and artificial cuttings. No wonder that the literature on sedimentary structures of tidal flats is very comprehensive.

The so-called tidal-stratification (Gezeitenschichtung) is the characteristic type of bedding in tidal flats. K. LÜDERS (1930) proved its development from his study of the velocities of tidal currents. As the velocity of the flood-tide current exceeds that of the ebb-tide current, the former lays down coarser material than can be removed by the latter. The flood-tide current deposits material of different grain-size; the bed deposited shows indications of graded bedding, as the velocity of the flood-tide current gradually decreases. The following ebb-tide current washes away only the upper finer-grained parts of the bed, so that only a lamina of coarsest material remains. In the final phase of the low-tide, this coarse lamina is covered by fine-grained material which was in suspension till that time. In this pair of laminae, the coarser corresponds to the initial phases of the high-tide current, and the finer to the final phase of the low-tide current. After the emergence, the upper lamina is partly dried up and consolidated so that the following tide cannot wash it away, In this way, the tidal bedding develops between the mean high-and low-tide levels. As far as similar bedding originates under the low-tide level, it is due to a recurrent alternation of the intensity of tidal currents.

Second in importance is the bedding caused by storm tides (the so-called Sturmflutschichtung), developed as lenticular, finely rippled current bedding. Lenses of pelecypod shells or of echinoderm spicules, which are occasionally present, are virtually sections through minor aggraded ridges. The diagonal bedding can be of a large extent, occuring particularly in the vicinity of deeply cut discharge channels, and originates by their lateral shift. The shifting channels remove the sediment on one side depositing it on the other, parallel to the slope of the bank. The wedge-like current bedding occurs only in well-sorted sands in the vicinity of dune islands.

Reference is also made of the so-called internal unconformities which appear as an unconformable superposition of two sedimentary units. They are produced by the cutting of beds by channels and by the deposition of diagonal bedded sediments. This phenomenon, which in ancient sediments would appear as a great angular unconformity, developed virtually without any break of sedimentation.

Because of the presence of an extremely rich benthos on tidal flats, many papers are devoted to the relation of benthos to the stratification. The opinion prevails, as mentioned above, that a perfect preservation of stratification is directly contradictory to the presence of a major amount of benthos. However, in places of rapid deposition or of a continuous reworking of sediments, the benthos is not able to disturb the primary bedding at such a rate. The tidal flats afford the best example of this phenomenon. W. SCHÄFER (1956) wrote a comprehensive study on the mode of life of the common organisms of tidal flats and their influence on the upper layer of the unconsolidated sediment. On the basis of experiments both in tidal flats and in aquarium he came to the conclusion that each organic species disturbs the sediment in a particular way.

The most frequent and best recognized structure of tidal flats are the ripple-marks. They may be divided into several groups according to size or shape. On the basis of shape, the following principal types are distinguished: 1. Normal ripples of a wave length up to 20 cm, 2. Giant ripples of a wave-length of 20 cm to 10 m, 3. Submarine dunes of a still larger wave length (sometimes exceeding 100 m).

On the basis of shape, symmetrical, asymmetrical, and linguoid ripples are differentiated. The giant ripples are a very marked structural type of tidal flats. They develop only in sands, their shape is asymmetrical and the course of their crests irregular. Their wave length ranges from 90 to 1,800 cm and their amplitude from 10 to 70 cm. The windward side slopes at 4—8°, the leeward side at 14—33°. The material deposited between the crests is usually coarser than the capping of the crests. Ripples constitute the whole fields in tidal flats, usually of elliptical form, which represent parts of shallow-water sand flats. Their shape changes in short intervals, even after high and low tide. Sometimes, they are rapidly covered with finer-grained material, at another time they are removed by erosion during one high tide. From tidal flats, the existence of remarkable ripples developed in fine-grained sediments (silty clays) has also been described. No analogous forms are known from other sedimentary environments. Rounded crests of ripples, resulting from the activity of ebb-tide currents, have been regarded as characteristic of tidal flats (TH. WEGNER 1932). Recent investigation, however, has not corroborated that they cannot develop in any other environment. Characteristic of ripples in tidal flats are the abundant deformations due particularly to the flowing of material down the crests, so that some structures are already transitional to those originated by flow.

Mud cracks are a relatively frequent structure of tidal flats. They can develop also under a constant water cover by processes included in the collective term "synaeresis", meaning a rapid dewatering of sediments after their deposition, as a result of the ageing of colloids. The erosion of mud cracks produces clay balls, by the slight displacement of which intraformational conglomerates can arise. The formation of mud cracks leads to the breaking up of the upper layer of sediment just at the place of a bioglyph or mechanoglyph occurrence, which is a site if minimum resistance. Thus, the surfaces covered by various markings are, moreover, deformed by the formation of mud cracks.

The occurrence of clay galls and shreds have frequently been observed in the deposits of tidal flats. They are partially consolidated already by a mere drying of "Schlick". As described above, these rapidly consolidated, eroded, porous, strongly clayey sediments may be incorporated in contemporaneous or slightly younger sediments.

All the other structures known from beaches have also been found in tidal flats (e. g. flowage structures, rills, etc.). In channels even typical slump structures originate. Bubble-impressions left by the escaping air or gas are extremely frequent in places. In a fresh state, a channel of supply is discernible in these minute mounds and after piercing, traces of methan escape. Raindrop-impressions are rare and indistinct.

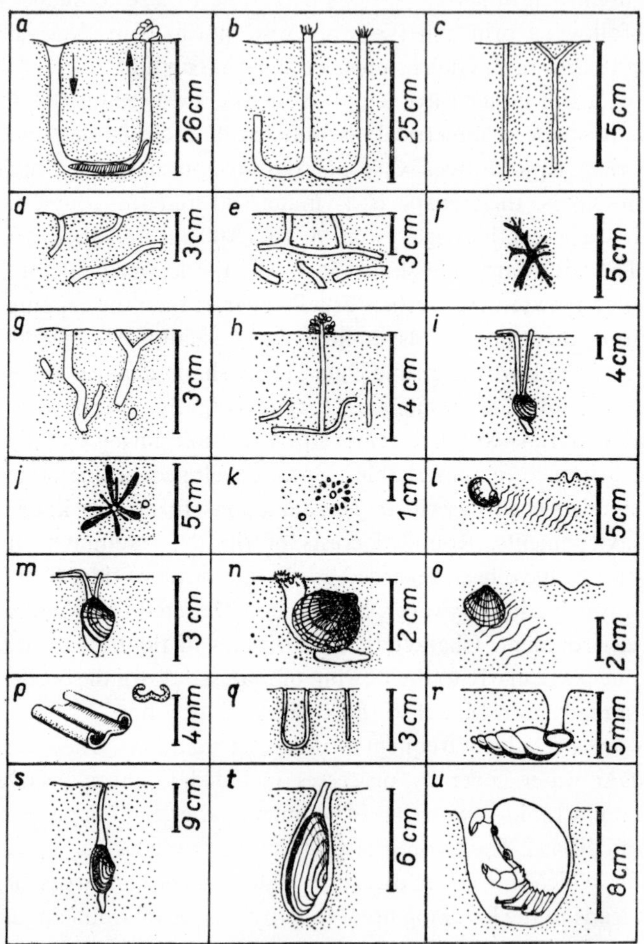

Fig. 92. Structures produced by organisms inhabiting the intertidal flats. a) *Arenicola marina*, b) *Lanice conchilega*, burrow, c) *Pygospis* sp., burrow, d) *Nephtys* sp., burrow, e) *Scoloplocs armiger*, burrow, f) *Nereis diversicolor*, surface trails, g) *Nereis diversicolor*, burrow, h) *Heteromastus* sp., burrow, i) *Scrobicularia plana*, surface trails, k) *Macoma baltica*, surface trail, l) *Littorina* sp., surface trail, m) *Macoma baltica*, burrow, n) *Cardium edule*, burrow, o) *Cardium edule*, surface trail, p) *Mytilus edulis*, burrow, q) *Corophium* sp., burrow. r) *Hydrobia ulvae*, burrow, s) *Lutraria elliptica*, burrow, t) *Pholas candida*, burrow, u) crab, burrowing. After G. Evans (1965).

Large surfaces of tidal flats are covered by numerous markings of the worm *Arenicola.* In fine-grained sandy sediments traces of the worm *Nereis*, similar to fossil chondrites, are also frequent. The imprints of jelly-fish, *Phialidium*, often resemble the structure which in fossil sediments are taken for raindrop imprints. The morphological and genetic variety of bioglyphs is immense. The multiform tracks of one species, as for instance, of eurypterid *Corophium volutator* or the Polychaeta species *Lanice conchilega*, prove the inadequacy of the division of biogenic markings on a systematic-taxonomic basis. The more so, that, on the other hand, similar and even identical tracks are produced by a series of various organisms, which widely differ in their systematic assignment. Thus, the star-shaped tracks occurring abundantly in tidal flat and other environments, can be produced by the amphipod genus *Corophium*, starfish *Asterias*, pelecypod *Scrobicularia plana* (by its siphon), fish of the genus *Gobius*, sponge *Spongia ottoi*, and worm of the genus *Nereis.* The comparatively frequent fossil star-shaped tracks on bedding planes are attributed to the sponge *Spongia ottoi.*

Biogenic reworking of tube dwelling annelids is remarkable and intensive process. According to D. C. Rhoads (1967) *Clymenella* is estimated to ingest 247 ml of sediment per worm per year. *Pectinaria* ingest 400 ml of sediment in the same time. The selective ingestion and transportation of sand and smaller particles leads to the formation of biogenically graded deposits.

The biology of tidal flat sediments

Palaeontological study carried out extensively in the waddens is extremely important for sedimentology. We have mentioned above the significance of diatom frustules for the deposition of fine-grained sediments and the immense amount of shells of pelecypods and other organisms in aggraded ridges. R. Richter (1927) also described the activity of worms, the so called "sand-corals" of the genus *Sabellaria* which in some places of tidal flats build colonies reminiscent of bioherms. He recorded that the worm built up tubes by sticking 2,000—2,500 sand grains daily. This worm lives in various places of tidal flats but always under the low-tide level. The worm *Lanice conchilega* builds similar tubes at a rate of 3 cm per day.

Of no less importance is the study of the orientation of pelecypod shells in the sediment. On the basis of R. Richter's results (1942) the following rule has been established as universally valid (the "Einkippungsregel"): If a valve of saucer-shape is found in a sediment it must be first found out whether it does not represent:

1. An occurrence in an aggraded ridge.
2. A preserved living position or a position which was enabled by the decay of soft parts of the body.
3. Sedimentation in a current shadow.
4. Sinking of the valve into a soft substratum.

When these four possibilities can be ruled out, it holds good universally that the valves are oriented with the convex side upwards.

We must not omit the remarkable role of some worm species in sedimentation. The worms *Arenicola* and *Nereis* can live with a tenth of normal oxygen content and for a certain period even without oxygen, drawing it form the glycogen of their own bodies. This explains the presence of worm-produced markings in deposits of reducing environments.

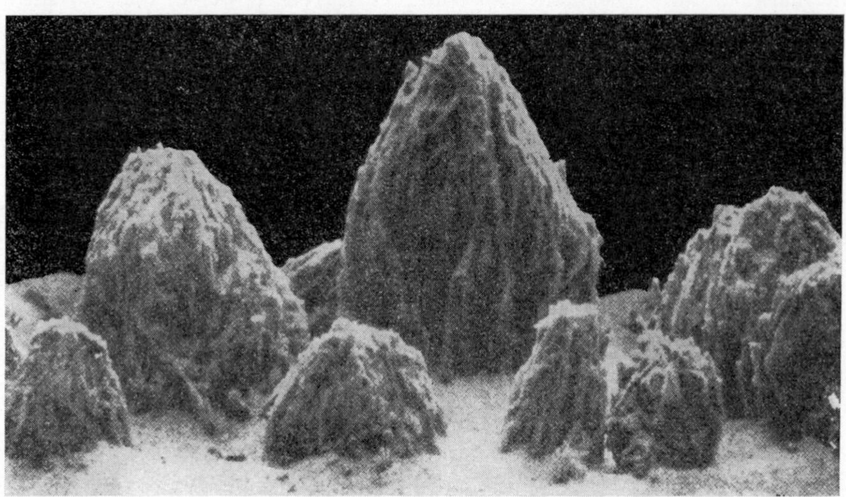

Photo 7. Sabellaria "reefs" from North German wades formed from worm tubes of genus Sabelaria. After R. Richter (1927).

The following palaeoecological and palaeontological criteria can serve for the determination of fossil sediments of tidal flats:

1. Flats of fine-grained sands with worms of the genus *Arenicola*.
2. Silty sand with worms of the genus *Lanice*.
3. Sand with tubes of *Lanis coheni*.
4. "Schlick" with abundant faecal pellets of the worm *Heteromastus filiformis*.
5. Coastal silt densely inhabited by the crustacean of the genus *Corophium*.
6. Clayey-silty sand with the pelecypod *Cardium edule*.
7. Sand flats with the pelecypod *May arenaria*, in living position.
8. Growths entrapping faeces and mud with the *Mytilus* pelecypod.
9. Oyster beds.
10. Silty clay with the pelecypod *Scrobicularia plana*.
11. Clayey-silty sand in shallow waters at the shoreline with the growths of sea grass and pelecypods *Littorina littorea*, *Hydrobia* and *Nassa reticulata*.
12. Echinoderms *Echinus miliaris* and worms *Echiurus* on the not very steep walls of discharge channels that do not emerge during low water level.

13. The echinoderm *Echinocardium cordatum* in the sands of outer tidal flat, 20—30 cm under the surface.

14. "Reefs" of tubes of the worm of *Sabellaria* genus in the outer tidal flat.

References

Alluvium. 1—424, Berlin, 1936.

Brockmann Ch. (1935): Diatomeen und Schlick im Jade-Gebiet. Abh. Senckenb. Naturf. Gess., Bd. 430; 1—64, Frankfurt a. M.

Ehrhardt H. (1937): Das Watt. P. 1—96, Hamburg.

Evans G. (1965): Intertidal flat sediments and their environments of deposition in the Wash. Quart. Jour. Geol. Soc., vol. 121: 208—245, London.

Het Waddenboek (1965): Nederlandsche Geol. Vereinigung. P. 1—223, Amsterdam.

Hummel K. (1930): Tierfährten am Tropenstrand. Natur u. Museum, Bd. 60; 81—89, Frankfurt a. M.

Klein DeVries G. - Sanders J. E. (1964): Comparison of sediments from Bay of Fundy and Dutch Wadden Sea tidal flats. Jour. Sedimentary Petrology, vol. 34: 18—24, Menasha.

Kuenen P. H. - Straaten L. M. J. U. van (1957): Accumulation of fine grained sediments in the Dutch Wadden sea. Geol. en Mijnbouw, Jg. 19: 329—354, Amsterdam.

— — (1958): Tidal action as a cause of clay accumulation. Jour. Sedimentary Petrology, vol. 28: 406—413, Menasha.

Kukal Z. (1958): Wades and their importance for the investigation of Recent and fossil sediments (in Czech). Časopis pro min. a geol., vol. 3: 73—96, Prague.

Lüders K. (1930): Entstehung der Gezeitenschichtung auf den Watten im Jadebusen. Senckenbergiana, vol. 12: 229—254, Frankfurt a. M.

Reineck H. E. (1955): Haftrippeln und Haftwarzen. Ablagerungsformen von Flugsand. Senckenbergiana lethaea, vol. 36: 299—304, Frankfurt a. M.

— (1969): Tidal flats. Bull. Am. Assoc. Petrol. Geol., vol. 53: 737, Tulsa.

Reineck H. E. - Singh I. B. (1967): Primary sedimentary structures in the Recent sediments of the Jade, North Sea. Marine Geology, vol. 5: 227—235, Amsterdam.

Rhoads D. C. (1967): Biogenic reworking of intertidal and subtidal sediments in Barnstable Harbor and Buzzard Bay, Massachusetts. Jour. Geology, vol. 75: 461—476, Chicago.

Richter R. (1927): Die fossilen Fährten und Bauten der Würmer. Zeitschr. Paleont., vol. 9: 231, Berlin.

— (1942): Die Einkippungsregel. Senckenbergiana, P. 215—244, Frankfurt a. M.

Schäfer W. (1956): Wirkungen der Benthos-Organismen auf den jungen Schichtverband. Senckenbergiana lethaea, vol. 37: 183—263, Frankfurt a. M.

— (1962): Aktuo-Paläontologie. P. 1—666, Frankfurt a. M.

Trusheim F. (1929 *a*): Eigenartige Entstehung von Tongallen. Natur u. Museum, vol. 59: 70—72, Frankfurt a. M.

— (1929 *b*): Wattenpapier. Natur u. Museum, vol. 59: 72—79, Frankfurt a. M.

— (1939): Rippeln in Schlick. Natur u. Volk, vol. 66: 103—106, Frankfurt a. M.

Valentin H. (1954): Die Küsten der Erde. P. 1—118, Gotha.

Wegner Th. (1932): Unter Gezeitenwirkung entstandene Wellenfurchen. Centralblatt f. Min., Abt. B, p. 31—34, Stuttgart.

Weyl F. (1953): Beiträge zur Geologie El Salvadors. N. Jhrb. Geol. Pal., Mh, p. 198—202, Stuttgart.

18. *Sediments of inland seas*

This group of basins includes all seas that have only a narrow connection with the ocean, as well as large lakes (e. g. the Caspian Sea and the Baikal) which show the same sedimentary conditions as the seas. We shall instance the sediments of inland seas by the deposits of the Black, Caspian, Baltic and Red Seas.

The beach sediments of the Black and Caspian Seas have the same composition as the deposits of oceanic beaches that are dealt with in a separate chapter. The deep-water parts of these basins and their sediments display a particular character.

Sedimentation is influenced to a great depth by the type of the material supplied, which in turn depends on the topography of the land surface. As the northern and western coasts of the Black Sea display small topographic differences, the eastern and southern ones being mountainous, the shallow-water and partly also deep-water sediments differ depending on the relief of the adjoining land. The fundamental differences are summed up in the Table 130 (the data concern the Black Sea, but can be applied broadly to all intracontinental seas).

Apart from the above shallow-water parts, the bottom of the Black Sea is covered by a mixture of calcareous and fine-grained terrigenous material.

The content of carbonate increases with depth so that in the offshore direction silty clays and clays grade into more or less pure calcareous clays. In the two central deepest depressions there are nearly pure calcareous muds. The calcareous sediments of very similar petrography bear local names. Sediments of central depressions are called abyssal calcareous clay, its shallower variety as Midia clay and Phaseolina clay (according to the content of organic remains). These principal sediments have the following chemical composition (Table 131).

Table 130

Sediments of two different parts of the Black Sea

Parts of the basin with a flat coast	Parts of the basin with a mountainous coast
Wide shallow-water zones with beach sands and abundant accumulations of pelecypods to a depth of a few tens of metres; strongly calcareous sediments, slow deposition, in places non-deposition	Narrow shallow-water zones, steeper slope of the bottom, narrow or no beaches; clastic sediments near the mouths of rivers; gravels and sands can reach a depth of as much as several hundreds of metres. In places, areas free of sediments, due to sliding of unconsolidated sediments

Table 131

Chemical composition of principal deep-water sediments of the Black Sea

	Deep-water clay %	Midia clay %	Phaseolina clay %
Insoluble res.	75·0	75·8	74·0
$CaCO_3$	14·3	16·3	19·2
Fe	4·83	4·76	5·25
Mn	0·066	0·04	0·055
P	0·11	0·10	0·10
Org. C	1·74	1·61	1·61

The increase in the amount of carbonate basinwards is not due to a greater rate of its deposition but to the decrease in the rate of supply of terrigenous material away from the shore. The rate of deposition both of terrigenous material and carbonate is highest near the shore, but the rate of sedimentation of terrigenous detritus decreases basinwards much more rapidly than that of carbonate. N. M. Strakhov (1954) considers the carbonate to be mainly of detrital origin and imported in river suspension. In shallow-water parts, the percentage of skeletal material is higher, particularly on shallow flats with sandy sediments and abundant calcareous tests. The question of biochemical, chiefly bacterial precipitation of carbonate remains open. This mode of origin was favoured by earlier authors. At present, we think that at least part of carbonate forms in this way, because quite abundant needles of calcite have been found in calcareous clays, which is suggestive of chemical precipitation.

In the sediments of the Black Sea there is a remarkable percentage of iron sulphides. Their origin should undoubtedly be connected with bacterial activity

Table 132

Forms of sulphur in the surface layer of sediments of the Black Sea (after I. I. Volkov - E. A. Ostroumov 1957)

	Sample No 1 %	Sample No 2 %	Sample No 3 %
FeS	60·48	60·07	60·12
FeS_2 (melnikovite	20·72	30·05	34·90
FeS_2 (pyrite, markasite)	0·17	0·25	0·11
S free		0·20—0·30	
$FeSO_4$		0·09—0·24	

by the reduction of sulphates of the sea water to a large extent and subordinately by the transformation of organic matter. According to I. I. VOLKOV and E. A. OSTROUMOV (1957) several forms of sulphur are found in sediments of the Black Sea. From analyses of three samples taken from the surface layer, the following percentages of individual forms have been calculated (Table 132).

Photo 8. Coarse-grained calcarenite from the continental shelf edge. Composed mostly of fragments of calcareous organic remains. X 6.5. Guyana shelf. After D. J. G. Nota (1959).

In the top sedimentary layer, sulphide sulphur is predominant. Bacterial processes produce colloidal clusters of hydrotroilite which dehydrate soon afterwards passing into concretions of anhydrous sulphide. Only ofter the annexation of a sulphur atom, metastable melnikovite originates which passes diagenetically into pyrite or marcasite.

The remarkable stratification of deep-water sediments of the Black Sea has already been described many times. The calcite-rich laminae of seasonal origin alternate with those richer in biological and terrigenous material. The thickness of carbonate laminae increases basinwards. In shallow-water sediments, deposits formed predominantly of calcareous organic tests alternate with calcareous clays.

In the deep-water sediments of the Black Sea there is an increased percentage of organic matter (often above 5% org. C). This is not a result of an increased organic production but of the dominantly reducing conditions and retardation in the decay of organic substances. They are not oxidized but only bacterially transformed into stable forms of organic matter, the so-called marine humus, and in part regenerated into inorganic compounds (mainly CO_2 and subordinately phosphates, nitrates and others).

The Caspian Sea

Sedimentologically, the Caspian Sea is very interesting. It is divided into two parts. The northern part is encompassed by a low coast, of a platform character, with small height differences between the source and depositional areas. The southern half of the Caspian Sea is encircled by mountains; the slopes of the bottom are substantially steeper and, according to some authors, this part has a typical geosynclinal character. Beach sediments of the Caspian Sea are developed only in places, particularly in the northern platform part. Gravel beaches are scarce. The greater part of the northern Caspian Sea in influenced by the supply of detritus by the river Volga and its sediments resemble in grain-size the material in suspension in that river. They are chiefly sands, clayey sands and silt-clayey sands, i. e. mostly sediments with a predominant sand fraction. The Md of grain-size and sorting of the sediments are as follows (Table 133).

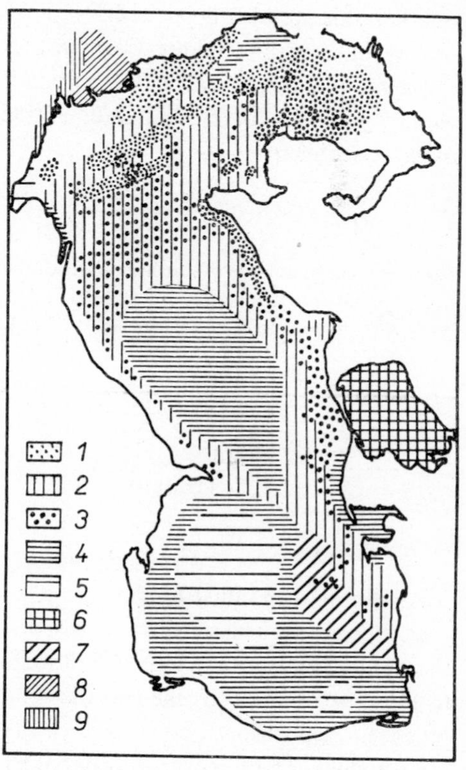

Fig. 93. The distribution of Recent sediments in the Caspian Sea. 1. Sands, 2. Sandy clays, 3. Shells, 4. Silty clays, 5. Calcareous clays, 6. Silts, 7. Carbonate muds, 8. Sediments of Volga delta, 9. Clayey sediments of delta bay. After M. V. Klenova (1959).

Table 133

Parameters of sediments near the Volga delta region in the Caspian Sea

	Range mm	Average mm
Md	0·057—0·80	0·09
So (Trask's)	1·410—2·57	2·07

At smaller depths (to about 20 m) and particularly on moderate elevations, finer-grained sediments are washed out and the sediments become enriched in the fine and medium-sandy fractions. The shallow and flat parts of the Caspian Sea are the site of wide-spread deposition of entire pelecypod shells. The accumulations develop to a depth of 8—10 m. The content of $CaCO_3$ varies between 81 and 95%. They pass basinwards into sands and clayey sands and shorewards into beach sediments. In these deposits, tests of the species *Cardium edule*, many species of the genera *Didacna, Dreissensia* and others are predominant. In addition, fragments of tests of *Monodacna caspia* and *Monodacna edentula* are found. At smaller depths, oolitic sands are also deposited.

The southern half of the Caspian Sea is supplied chiefly by the suspended load of the mountain river Kura. The suspended material is of silt grade to which the prevailing grain-size of basinal sediments corresponds. Table 134 lists the parameters of these sediments.

Table 134

Parameters of sediments of southern parts of the Caspian Sea

	Range mm	Average mm
Md	0·14—0·047	0·030
So (Trask's)	2·40—4·5	3·1

As can be seen, all sediments are of the silt and clay-silt grades.

Because of its character as a closed sea, the Caspian Sea is a typical example of two-fold sedimentation: on a shallow flat (in the N) and in a typical geosynclinal trough (in the S). In the northern, platform parts, synsedimentary wash-outs, breaks in deposition, and, thus, occurrences of sandy and even bimodal sediments are frequent. The deposition occurs at the lowest rate in the shallowest shelf; terrigenous material is carried over them and only carbonate skeletal material accumulates. On the basis of data from various sources, the following rates of sedimentation in individual environment and the composition of respective sediments have been established (Table 135).

Sediments of the Caspian Sea, like those of the Black Sea, show a regular bedding of two fundamental types: In the northern, shallower part of the basin, vertical alternation of beds composed of pelecypod shells with beds of sandy clays or clayey sands is frequent. In the central part of the sea, beds of non-carbonate clayey silt, or clayey-silty sand alternate with calcareous silty clays. Some silty clays enclose regular laminae of pelitomorphic carbonate.

Table 135

Composition of sediments and their rate of sedimentation in various environments of the Caspian Sea
(after M. V. Klenova 1956)

Environments	Rate of sedimentation cm/100 years	Composition of sediments
Southern and Western parts of the basin with abundant detritus carried by river Kura	30—100	silts, silty clays
Southern parts of geosynclinal character, far from Kura mouth	10— 20	silty clays up to clays
Central parts of basin with transitional character	1·2—2·5	clays, sandy clays, silty-sandy clays
Shallow-water flats in northern part of basin	0 up to several cm, in places nonsedimentation and erosion	oolites, accumulations of shells
Northern parts of basin near Volga delta	5—15	sands, silty-clayey sands
Northern parts of basin, far from Volga mouth	0—15	sands up to sandy clays

Lake Baikal

Lake Baikal differs from other inland basins by a large amount of diatomaceous sediments. The maximum content of authigenic SiO_2 (soluble in Na_2CO_3) in the deposits is up to 60%. Diatomaceous sediments are developed particularly in the central part of the basin, where the terrigenous sedimentation is slower. In the marginal parts, up to a depth of 10 m, fine-grained sands with the Md ranging from 0·07 to 0·13 mm occur. Their sorting coefficient usually exceeds 2·5, as a result of a considerable admixture of clay and silt. The sand fraction decreases basinwards, so that in the central belt the sediments contain no more than 2—4 per cent of the fraction above 0·1 mm. The diatom frustules are of silt grade and the Md of grain-size of diatomaceous clays varies between 0·013 and 0·025 mm. Distinctive of the sediments of the Baikal are also phosphate concretions composed of vivianite that are found particularly in the deeper parts.

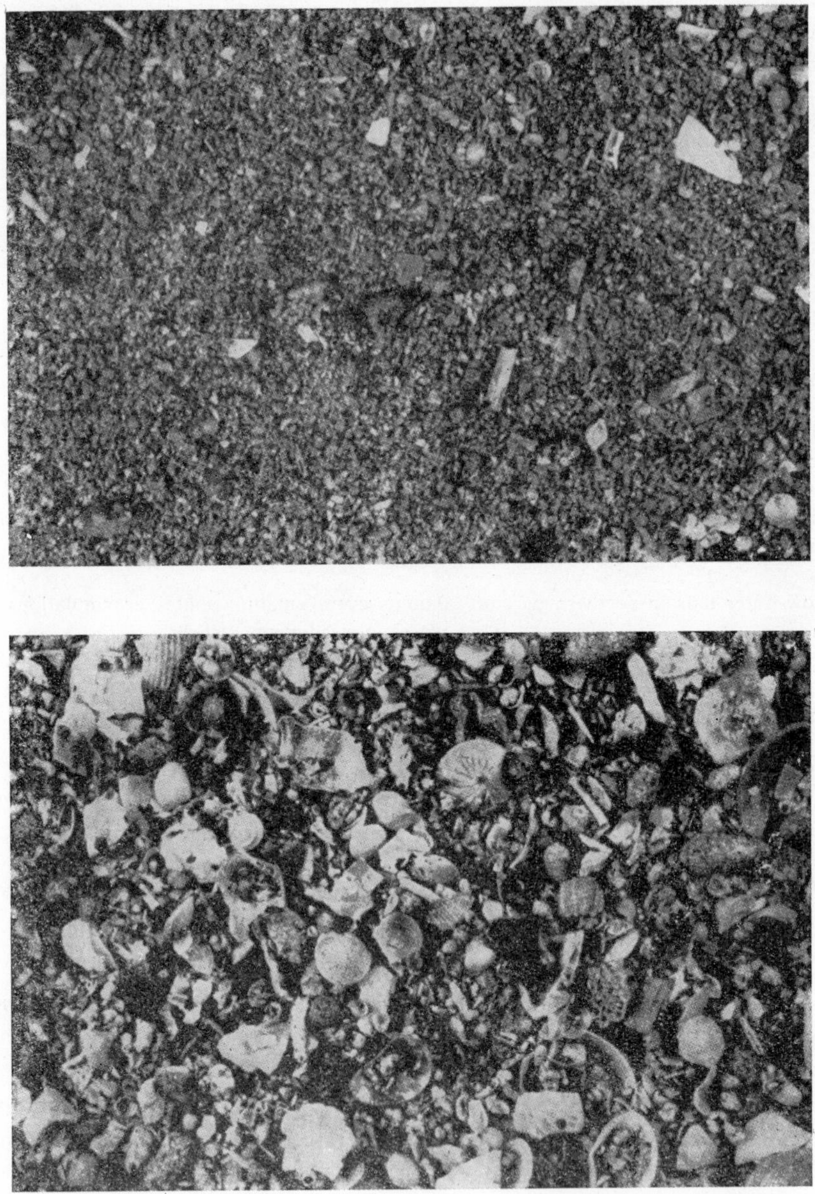

Photo 9. Sediments of Guyana continental shelf. Upper photo: Medium-grained calcarenite with terrigenous admixture (X 7). Lower photo: Fine-grained sand with fragments of shells (X 10). After D. J. G. Nota (1959).

Lake Aral

Lake Aral falls into two lithologically different parts, one of which is represented by the deltas of the Amu-Darya and the Syr-Darya, composed dominantly by clay-silt sediments. The Md of grain-size varies between 0·003—0·05 mm. In the other part of the basin, a quite broad zone of sandy sediments reaches to a depth of several tens of metres. The sands are fine-grained, poorly sorted, grading to silts. In the central part of the basin, unaffected either by an appreciable terrigenous supply or by shallow-water sedimentation, silty clays and clays predominate again.

Lake Balkhash

This lake is renowned for its Recent dolomitic sedimentation which will be described later in the text. Here, it should be mentioned that dolomitic sediments, like other chemical deposits are developed in the northern part of the lake, which is dissected in many bays. In the wider, rather shallow southern part of the lake, fine-grained sands pass gradually into calcareous clays which occur at a greater depths. A large amount of calcite is precipitated (similarly as dolomite) chemically or biochemically. The prevalently chemical sediments (calcareous clays) of Lake Balkhash have the following mineralogical composition (Table 136).

Table 136

Composition of chemical sediments of Lake Balkash
(after N. M. Strakhov 1954)

	Range %	Average %
Insoluble residue	18·16—34·75	25·41
Carbonates	49·83—66·21	58·45
Calcite	19·94—47·93	34·61
Dolomite	5·32—40·62	23·84
Fe total	0·97— 3·15	2·13
Org. C	1·24— 2·74	1·97

The Baltic Sea

The Baltic Sea is one of the basins which are characterized by a peculiar mineralogical and chemical composition of bottom deposits. The Baltic Sea bottom deposits are largely influenced by the following factors:

1. Small vertical differences in the relief of the surrounding landmass.

2. Considerable supply of clay material by rivers. An appreciable supply of organic matter, a strong inflow of fresh water and slow evaporation.

3. Poor aeration of the bulk of water masses and, consequently, impoverishment in oxygen, in some cases the presence of free hydrogen sulphide. The aeration is hindered by the so-called salinity-stratification (i. e. alternation of layers of different salinity, which prevents the vertical mixing of water and the introduction of oxygen into lower layers).

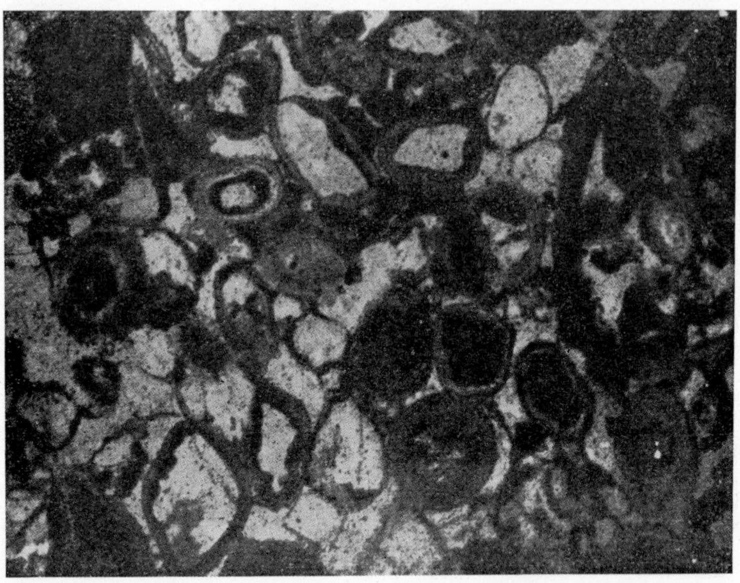

Photo 10. Calcareous ooids and encrusted quartz grains from shallow-water sediments of the Caspian Sea. (X 15). After M. V. Klenova (1960).

4. The presence of glacial sediments in the adjacent areas and on the bottom of the basin.

The individual factors produce the following effects on sedimentation:

Ad 1. On account of small topographical differences coarse-grained sediments are absent from the basin.

Ad 2. The same as ad 1; moreover, a large amount of organic matter and more acid reaction of waters which impairs the precipitation of carbonate. Therefore, the sediments of the Baltic are poor in carbonate (generally 1—2% $CaCO_3$, only sporadically more than 20%). The content of organic matter is frequently above 10% org. C.

Ad. 3. This factor is responsible for the peculiar character of sediments and their organic content. Gyttja sedimentation at the presence of a certain amount of free oxygen above the bottom prevails, but in the top layer there are already sediments of reducing environment. The reducing conditions in the sediment affect the particular mineral association:

Clay minerals (esp. of illite group) — organic matter with org. C/org. N ratio in excess of 10 — a small proportion of siderite in minute concretional forms. Sediments contain an increased percentage of Mn and U. Siderite, however, changes into pyrite already at a small depth under the top layer.

The clay sediments of the Baltic Sea are extremely fine-grained, Md varies around 0·5 microns. They contain only 3—6 per cent of fraction above 20 microns. Fine sand fraction is almost absent from these sediments and silt is very scarce. Sandy beach and shallow-water deposits reach to a depth of 3—6 m, where they pass into silty sands and sandy silts with a rapidly rising content of clay. Approximately at a depth of 20—30 m, the silt sediments pass into the above mentioned fine-grained clay deposits.

The Red Sea

The area of the sea is about 438,000 km^2, maximum depth is 2,920 m. The sea contains some of the hottest and saltiest ocean waters of the world (temperature up to 56 °C and salinity 7·5 × that of ordinary sea-water) in isolated deep depressions. Carbonate deposits prevail in the shallower parts of the sea, but remarkable deposits occur in the depressions. Seven types of sediments have been described:

1. Iron-montmorillonite with sphalerite admixture,
2. Goethite deposits,
3. Sulfide sediments rich in iron monosulfide, marmatite, sphalerite, chalcopyrite and pyrite,
4. Manganosiderite,
5. Manganite,
6. Anhydrite,
7. Ordinary sediment, consisting mostly of biological carbonate constituents.

References

BISCHOFF J. L. (1969): Red Sea geothermal brine deposits. Their mineralogy, chemistry and genesis. In: Hot brines and Recent heavy metal deposits in the Red Sea, 368—406. Berlin—Heidelberg—New York.

DEBYSER J. (1957): Contribution a l'étude des sédiments organiques de la mer Baltique. Rev. Inst. Franc. Pétrole, p. 3—13, Paris.

DEGENS E. T. - ROSS D. A. (Editors, 1969): Hot brines and Recent heavy metal deposits in the Red Sea. P. 1—600, Berlin—Heidelberg—New York.

GORSHKOVA T. I. (1963*a*): Bottom sediments of the Baltic sea and of its bays (in Russian). Deltaic and shallow-water littoral marine sediments, p. 14—21, Moscow.

— (1963*b*): Bottom deposits of the Baltic Sea (in Russian). Baltica, vol. 1: 189—210, Vilnius.

GRIPPENBERG S. (1934): A study of the sediments of the North Baltic and adjacent seas. Fennia, vol. 60: 1—331, Helsingfors.

— (1939): Sediments of the Baltic Sea. Recent Marine Sediments, p. 298—321, Tulsa.

HARTMAN M. - NIELSEN H. (1969): δ^{37}S — Werte in rezenten Meeressedimenten und ihre Deutung am Beispiel einiger Sedimentprofile aus der westlichen Ostsee. Geol. Rundschau, vol. 58 621—655, Stuttgart.

KLENOVA M. V. (1956): Recent sediments of Caspian Sea (in Russian), AN SSSR, p. 1—302, Moscow.

MAYEV E. G. (1960): Genesis of sediments of southern Caspian Sea. (in Russian). Doklady AN SSSR, vol. 120: 154—157, Moscow.

OSTROUMOV E. A. - VOLKOV I. I. (1957): About the relation of phosphorus, vanadium and organic matter in the sediments of Black Sea. (in Russian). Geochimija, vol. 6: 518—527, Moscow.

RAUPACH F. (1952): Die rezente Sedimentation im Schwarzen Meer, im Kaspi und im Aral und ihre Gesetzmässigkeiten. Geologie, vol. 1: 78—132, Berlin.

VOLKOV I. I. - OSTROUMOV E. A. (1957): Concretions of iron sulphide in the sediments of Black Sea (in Russian). Doklady AN, vol. 116: 645—648, Moscow.

19. Deep-sea sediments

The term deep-sea sediments is used here to designate sediments of the continental slope and abyssal oceanic regions, i. e. those which were deposited under the — 200 m isobath. Thus, the boundary between shallow and deep-sea sediments corresponds roughly to the boundary between the continental shelf and the continental slope. Where the boundary between them is diffuse, the circumstances are more complicated. However, it has recently been ascertained that the character of sediments depends rather on the distance from the land than on the absolute depth. Therefore, it was necessary to develop a new classification of deep-marine regions and sediments, based on their distance from the landmass.

Before proceeding to a systematic description of deep-sea sediments, we shall outline briefly some fundamental problems, such as deep-sea currents, turbidity currents, suspended matter in sea water and others.

Suspended matter in sea-water

Suspended material in sea-water represents sediments in the first stages of its origin. As mentioned above, the erosion and transport action can be inferred from the composition of suspended stream load. Analogously, the type of sedimentation at various places of world oceans can be determined to some extent from the composition of marine suspensions. The study of the quantity of suspended material at separate places, chiefly near the mouths of major streams, leads to the recognition of surface and bottom currents.

Suspended material can be either of inorganic or organic origin. In the surface layers there is the maximum percentage of inorganic suspension in the proximity

Table 137

Amount of suspension in sea water

	Amount of suspension g/m^3
Baltic Sea, during phytoplankton bloom	7— 10
Caspian Sea, far from coast	5
Caspian Sea, coastal regions	10— 25
Caspian Sea, near Volga mouth	100—600

of land, particularly in the mouths of major rivers. Biogenic suspension attains its maximum in places of ascending currents and of mixing of warm and cold waters.

One cubic metre of sea-water contains on the average 0·5—4 g of suspended material. Yet in places the content can exceed many times these values (see Table 137).

Suspended material of sea-water derives from the following sources (in order of importance):

Inorganic suspension	Organic suspension
Suspended river material	Phytoplankton
Eolian material	Zooplankton
Colloid or solid precipitated substances	
Cosmic material	

In a normal oceanic enviroment, where there is an average production of plankton, the inorganic and organic suspensions are of about the same importance. In places where phytoplankton thrive the significance of biogenic suspension obviously exceeds that of the inorganic suspension. The suspension of eolian origin increases considerably in the zones to which material from deserts is brought by wind. In warmer and more salty waters, the water can be rich in chemically precipitated crystals of calcite or aragonite.

The mechanical analyses of suspended material do not present very different results. All samples contain a predominance of clay particles, a variable admixture of silt fraction and little sand. Fine-grained sand and silt are but occasional components of suspended load, whereas clay is invariably present. Table 138 gives grain-size analyses of suspended matter from the Indian Ocean.

Table 138

Grain-size analyses of several samples of suspension from the Indian Ocean (after A. P. Lisicyn 1961)

Suspension concentration g/m^3	Fractions (mm)					
	>0·1	0·1—0·05	0·05—0·01	0·01—0·005	0·005—0·001	<0·001
0·279	0·33	—	11·67	1·81	2.89	83.23
0·571	0·67	—	4·67	94·66		
0·800	0·20	0·20	2·80	3·09	2·83	90·86
0·062	1·00	1·00	11·30	2·92	5·39	78·68
0·068	2·00	0·50	21·00	76·50		
0·022	—	—	—	—	—	—
0·036	1·67	0·33	17·33	80·67		

In all cases, the suspended material is of a finer grade then the underlying sea floor sediment and is greatly enriched in the clay fraction.

The study of depth-distribution of suspended matter along profiles drawn across large parts of the oceans has shown the existence of certain changes. Thus, for instance, anomalous concentrations of mineral particles in the Atlantic occur chiefly at a depth of between 3,000 and 5,000 m. There exists the opinion that they probably represent the remainder of the finest particles of volcanic origin, ejected into the ocean where they settle at a low rate. The investigation of the distribution of suspended material above the bottom has revealed a laminar stratification in some places and a quite irregular distribution in others. In the former case, minor turbulent movement of water near the bottom is responsible for the regular stratification. This occurs rather in the bottom depressions, where the irregular distribution tends to develop rather than on the elevations.

The amount of suspended material and the rate of its decrease was also used for the calculation of the rate of sedimentation. A. P. LISICYN (1961) computed the sedimentation of diatomaceous ooze in the southern part of the Indian Ocean at 0·13 cm (of dry sediment) in 1,000 years; the rate of sedimentation in Bay of Bengal at 1·000 km distance from the coast was 35 cm in 1,000 years.

Deep-water currents

The system of surface currents is fairly well known, and detailed data on its quantitative and qualitative characteristics are given in all textbooks on oceanography. Table 139 presents some values of velocities of major surface oceanic currents.

Table 139

Velocities of some major surface surrents

Current	Average velocity cm/sec
Gulf Stream in the Strait of Florida	193
Gulf Stream in Central Atlantic	120—140
Equatorial Pacific current	50
Kuro-shio current	20— 89

Most oceanic currents have a sufficient velocity for transporting even sand sized material over certain distances.

The problem of deep-water currents is more complicated and of greater geological importance and as yet little known. The expedition of the "Meteor" into the equatorial Atlantic showed that besides the world-oceanic deep circulation a series of local deep currents exist, called forth by the differences in physiochemical properties of the water masses or by local unevennesses. The present-day ocean circulation is caused by climatic zoning.

Table 140

Velocity of deep-water currents in Equatorial Atlantic

Depth m	Velocity cm/sec
2,000	13—25
2,500	16—19
3,000	8

In Table 140 some data on absolute velocities of deep-water currents are given. Maximum velocities have been ascertained on submarine elevations. All current velocities on the Mid-Atlantic Ridge exceed 20 cm/sec. The Antarctic deep-current shows the following velocities at respective depths (Table 141).

Table 141

Velocities of Antarctic deep-water current

Depth m	Velocity cm/sec
5,000	5·8
4,500	8·4
4·000	9·0
3,750	12·5

The oceans are divided by the oceanic deep current system into several belts by zones of divergence and convergence. In the zones of divergence, the deep currents ascend to the surface. In the equatorial Atlantic these are near the equator and at 10°S latitude. The oceanic regions along the entire western coast of Africa and South America represent other sites of the ascent of deep-water currents. The zones of convergence are the sites of the descent of surface waters. In the Atlantic a zone of convergence runs along the 5°S latitude.

According to some authors, the oceanic current system is decisive for the distribution of carbonate and non-carbonate sediments. G. SCHOTT (1944) records that the rout of the Antarctic deep-water current northwards can be followed across the western part of the Atlantic after the occurrences of non-carbonate sediments. Some other writers ascribe the difference in the content of carbonates in sediments of the Northern Atlantic and the Northern Pacific to the oceanic circulation; in this system carbonate-rich Pacific waters move across the northern polar region into the Atlantic to deposit there the carbonate in the form of organic calcareous sediments.

Recently, much attention has been paid to the local currents caused by the unevennesses in the bottom or by revolutionary sedimentary processes. Near the Californian coast the following velocities have been determined at a depth of 800 m (Table 142).

Table 142

Velocities of local deep-water current near the Californian coast at a depth of 800 m

Centimetres above the bottom	Velocity cm/sec
126	26·3
51	21·6
21	15·6

Table 143

Data on velocities of local deep-water currents (after various authors)

	Current velocity cm/sec
Submarine canyon, Californian coast	25
Submarine canyons off New England	11
Continental slope of South California	34·8—37·7
Submarine bank near Californian coast	> 50
Adjacent depressions	about 10
Continental slope of Norway, 250 m depth	< 10
600 m depth	< several cm

The above data indicate that everywhere on the sea bottom currents exist which are capable of transporting clay and silt, sometimes even sand. However, it is not yet known whether the local currents have a stable or periodical character. The-currents above elevations seem to be rather periodical. In addition to the above-mentioned results of direct measurements, the presence of deep currents has been proved by a number of phenomena observed directly in deep-sea sediments. We should mention at least the following:

1. Finds of Tertiary foraminifers on the surface of sediments at many places of the Pacific. Their presence can be most adequately explained by the supply from the surrounding elevations which were sites of intensive erosion.

2. Traces of submarine erosion and local unconformities in some cores taken from deep-sea sediments, or the probable presence of stratigraphic hiatuses of different length.

3. The presence of coarse-grained sediments on the deep-sea bottom, as far as they originated by the washing out of the finest fractions from the originally ill-sorted sediments. This is true particularly for elevations, in the summits of which even several thousands metres below the sea-level also very coarse-grained sediments may be preserved or formed as a result of the continuous current activity.

4. Changes in the distribution of suspended material above the bottom, and the above-described movement of marine suspension.

Turbidity currents

The question of turbidity currents and their influence on deep-sea sedimentation has appeared in geological literature only in the last twenty years. At present, they are believed to be of great importance for deep-sea sedimentation.

Compared with normal currents, turbidity currents possess a high density, a large amount of suspended material and a multiple competency, sorting capacity and lateral constancy. Turbidity currents can be regarded broadly as a transitional mode of transport between normal water currents and mudflows. The difference between these three kinds of transportation are summarized in Table 144.

Ph. H. Kuenen (1951) established experimentally the relation between the density of the current and the length of transport as follows:

		Experiments I.	Experiments II.
Density of turbidity currents at the place of formation		1·57	1·58
Density at a distance 350 cm from the place of formation			
cm above bottom	12·5	1·06	1·01
	10·0	1·10	1·02
	7·5	1·10	1·05
	5·0	1·18	1·12
	2·5	1·24	1·15

Table 144

Difference between normal water curents, turbidity currents and mud-flows

Normal water currents	Turbidity currents	Mud flows
Water bears solid suspension	Water together with solid suspension forms a homogeneous environment	Sediment in moving draws along water particles
Density only slightly > 1	Density commonly between 1·2 and 1·6	Density higher, even > 2
Competence corresponds to the velocity of the current	Competency many times greater	
Erosive power of the current depends on the velocity; it is considerable at high velocities	At identical velocities the erosive power substantially lower	Erosive power negligible
Distribution of solid particles in suspension non-uniform, content of suspended material increases bottomwards and towards the more rapid sectors of the current	Distribution of suspended material more uniform, content of larger and heavier particles somewhat increases into lower and anterior sectors of the current	Distribution of solid particles regular
Occur even at minimum slopes of the bottom	Minimum slope required for their origin presumably 2°	Minimum slope probably > 5°
The resulting sediment predominantly current-bedded	sediment mostly graded-bedded	Sediment unbedded or chaotically bedded

From the above it follows that turbidity current is densest in the lower parts. The density declines upwards. As to overall density it is directly proportionate to the velocity, the latter also decreases in these directions. In consequence of the difference in velocity, the coarsest particles accumulate in the anterior and lower parts of the current. The coarsest particles are deposited first and buried by finer-grained sediments settling from the upper and rear parts of the current. The result of this process is the so-called graded-bedding; its character was observed in the field and imitated in the laboratory.

Because of their higher density, turbidity currents have a far greater competence than normal current. It has been proved experimentally and by computation that a turbidity currents of velocity of 50 cm/sec can carry a boulder of 1 kg weight. From experiments the following values are inferable (Table 145). From the comparison of experimental results with the composition of the products of natural turbidity currents R. G. WALKER (1965, 1967) concluded that several types of turbidity currents can originate, giving rise to fairly different product. He differentiated: 1. Immature

Table 145

Competency of turbidity currents
(after Ph. H. Kuenen 1951)

Turbidity current velocity cm/sec	Density	Weight of boulders which can be rolled kg
100	2·0	50
200	2·0	3,200
400	2·0	205,500

Table 146

Characteristics of some types of turbidity currents
(after Walker 1965)

1. Immature "traditional" turbidity current	Bulk deposition to give ungraded mixture of everything in the current-forming pebbly mudstone or muddy sandstone; or non-bulk deposition of the grains too big to be transported-giving small patches of gravel
2. Semi-mature currents	Nose sediments followed by tail sediments — poorly or moderately graded sandstone with interstitial mud
3. Perfectly mature "Traditional" currents	Nose sediments followed by tail sediments — perfectly graded sandstones with no interstitial mud
4. "Traditional" currents with traction carpet	Low shear — gradual accumulation of sand, giving structureless bed, or simultaneous deposition and reworking into laminations. High shear — traction carpet dispersed into current. Below critical value of shear lower part of carpet "freezes" forming poorly graded or ungraded sandstone with little interstitial mud
5. Turbidity currents with autosuspension	Low velocity decay — sediment deposited from suspension on to bed, giving perfectly graded sandstone with interstitial mud but no current-formed sedimentary structures. High-velocity decay — deposition from suspension with reworking

"traditional" turbidity currents, 2. Semi-mature turbidity currents, 3. Perfectly mature "traditional" turbidity currents, 4. "Traditional" currents with traction carpet, and 5. Turbidity currents with autosuspension. Products of these currents show the following composition (Table 146).

Turbidity currents affect sedimentation in several ways, the most important of which are as follows:

1. They can carry an immense load of solid particles simultaneously.
2. They can transport particles of enormous size, many times larger than currents of the same velocity.
3. Because of their considerable lateral constancy they can disperse sediments over a large area.

According to the present state of knowledge, turbidity currents can develop in these ways:

1. By a rapid import of sediments into a basin with stagnant water. It is sufficient for the density of the current to be 0·0001 times greater that the density of stagnant water. From this it evidently follows that turbidity currents in lakes are much more abundant than in the seas. The origin of turbidity currents necessitate the presence of at least 0·0242 g sediments in 1 cm^3 of river water at the site of its flowing into the sea. At the river mouth into the lake, however, a fiftieth of this amount is sufficient.
2. By bringing a large mass of sediments into movement. This may be produced particularly by earthquake movements or by gravity (when sediments are quickly accumulated up to non-equilibrium position).

This latter cause gives rise to the largest and most important turbidity currents. Recent experiments have adequately proved that turbidity currents cannot practically be induced by minor factors, as can, for instance, avalanches. The formerly postulated origin by the accrement of masses brought into movement cannot be taken into consideration.

The first information on the effects of turbidity currents on sedimentation has been obtained from the Swiss lakes. In Lake Zurich, for instance, between the laminated calcareous and clayey deposits, beds and laminae of silty sands have been found, which, as was determined, correspond in time to the years when large slides of lake banks took place. Before the mechanism of deposition by turbidity currents had been recognized, it was impossible to explain how this material could have been distributed over the whole bottom. At present, it is self-evident that the material after mixing with water produced a turbidity current which flowed not only downslope but because of its inertia also up the opposite slope (Photo 16).

In the literature, the following turbidity currents of recent times are described (Table 147).

In this Table, currents which were observed and studied in action are described. Besides, there is naturally a number of phenomena that originated by the effects of former turbidity currents, many of which were previously totally inexplicable, as for instance:

Table 147

Recent turbidity currents (after various data)

Place	Origin	Description of current
Grand Banks, Newfoundland 1929	Earthquake and slumping of great mass of sediments	Enormous turbidity current 130—200 km wide and above 500 km long with maximal velocity about 440 cm/sec; replaced 220 billions m^3 of material
Mouth of Columbia River		Small turbidity current
Orléansville, Algeria, 1954	Earthquake and slumping of great mass of sediments	Great turbidity current, with maximal velocity about 250 cm/sec. Distributed sediments over the area 3,000 km^2 large
Mouths of Congo River, Magdalena River and Tajo River	Due to floods and sudden transport of great amount of material	
Suva, Fiji, 1953	Earthquake	Turbidity current of medium intensity
La Jolla, California	Artificial turbidity current	
Congo submarine canyon	During great floods	Occur in 50 year periods
Mediterranean submarine canyons	Masses of sediments were put in motion by bathyscaph	Small local turbidity currents

1. Finds of terrestrial flora in the vicinity of the Galapagos Islands, at a depth of 3,000 m.

2. The presence of the alga *Zostera* near the mouth of the Magdalena River at a depth of more than 1,000 m.

3. Finds of shallow-water pelecypods at 5,000 m in the vicinity of the Bahama Islands and in the sediments of the Hudson submarine canyon at about 3,700 m depth.

4. Finds of algae of the genus *Halimeda* in the area adjacent to the Bermuda Islands at a depth of 4,500 m.

5. Finds of graded bedded sands and silts in the cores of deep-water sediments, some of which contain perfectly sorted remains of shallow-water fauna of the same

grain-size as the terrigenous material. Moreover, the sediments have an extraordinarily high content of $CaCO_3$, whereas the over- and underlying clays bear a negligible amount of carbonate.

6. Very frequent interlayers of coarse-grained material in the cores from deep-sea troughs. The material is either of pyroclastic or terrigenous origin. Their conspicuous graded bedding suggests that in most cases they were deposited by turbidity currents.

7. The fact that out of 1,500 long cores of deep-sea sediments only 20% show an indisturbed stratigraphy. A further 20% bear doubtless signs of deposition by turbidity currents.

On the basis of their most frequent occurrence in deep-sea trenches, the turbidity currents are operative mainly in topographically varied and seismically active areas. Recently the turbidite accumulations have been found almost in all the trenches of the Indo-Pacific region (W. A. ANIKOUCHINE - HSIN-YI LING 1967). This seems to be true also for other geological periods. The present investigation does not corroborate the former opinion that turbidity currents can originate on relatively flat parts of the sea bottom in shallow-water regions.

Sediments of submarine canyons and of deep-water deltas

Submarine canyons are narrow, steep-sided valleys cut into the shelf and continental slope. The slope of the canyon walls is up to about 40°. The height difference between the canyon and its environments sometimes exceeds 1,000 m. The heads and upper parts of the canyons begin mostly at the shelf-continental slope boundary. The morphology and development of submarine canyons is dealt with in detail in F. P. SHEPARD's (1963) book. Therefore, I shall give here only a brief description of their sediments and of their influence on sedimentation in the surrounding area.

Because of the steepness of the canyon walls, the sediments are distributed rather irregularly. In the walls, older sediments of volcanic rocks crop out, and patches of silt and clay, occassionally also of impure gravel mixed with finer material are observed. Surprisingly, clay layers are preserved even on very steep walls. As it is seen, the sedimentary cover of the walls is varied and no regularity in its distribution has so far been ascertained. The bottom of the canyons is generally formed of sandy clays and silts or clayey and silty sands. Towards depths more than 2,000 m, the irregular patchy cover passes into more regular deposits. On the whole, the grain-size of sediments decreases with depth. Yet in cores taken from the canyon bottoms, the alternation of fine and coarser deposits, usually of light-coloured sandy silts and darker clayey silts, have almost invariably been observed. The sandy silts are generally graded and in the upper parts show frequent ripple bedding connected with lamination. This mode of bedding is most likely associated with the activity of turbidity currents which recurrently flow through the canyon. B. C. HEEZEN *et al.* (1964) estimate that approximately 50 turbidity currents flow through the Congo submarine canyon during a 100 years. At the mouth of almost every canyon, a deep-sea fan-shaped deltas of various sizes is developed. Its slopes are inclined at a few degrees and the length of its axis can exceed 100 km. Best known and most thoroughly investigated is the delta of the Hudson canyon, from which this type of sedimentation was first described and La Jolla submarine canyon and fan (F. P. SHEPARD *et al.* 1969). An enormous amount of sediments is transported through the submarine canyons into the deep-sea regions. Part of its is laid down in the deep-water delta and the finest material is dispersed to settle slowly from the suspension; it can also be transported by deep currents to large distances.

Classification of deep-sea sediments

The classification of deep-sea sediments is based at present on other petrographical criteria. The classification by J. MURRAY and A. F. RENARD (1891) is still used widely along with the minor changes suggested by R. REVELLE (1944), but is now subject to revision to bring it into line with recent discoveries. The two main drawbacks of the old classification were the term "red clay" and the disregard of coarse-grained deep-sea sediments (deposited either by turbidity currents or in another way). We still distinguish two large groups of deep-sea sediments — the pelagic and the terrigenous (or the eupelagic and the hemipelagic). The latter differ from the former by the presence of a larger amount of terrigenous detritus. It is generally recorded that eupelagic sediments must have less than 30 per cent of terrigenous material coarser than 0·01 mm, i. e. of silt and sand grade. The classification that corresponds best to the present state of knowledge was proposed by F. P. SHEPARD (1963):

Pelagic

- Brown clay (having less than 30% biogenous material)
- Diagenetic deposits (consisting dominantly of minerals crystallized in sea water, such as phillipsite and manganese nodules)
- Biogenous despoits (having more than 30% material derived from organisms).
  - Foraminiferal ooze (having more than 30% calcareous biogenous, largely *Foraminifera*, particulary common in cores penetrating Tertiary, usually called Globigerina ooze).
  - Diatom ooze (having more than 30% siliceous biogenous largely diatoms).
  - Radiolarian ooze (having more than 30% siliceous biogenous largely radiolarians)
  - Coral reef debris (derived from slumping around reefs)
    - Coral sands
    - Coral muds (white)

Terrigenous

- Terrigenous muds (having more than 30% of silt and sand of definite terrigenous origin)
  - Green muds
  - Black muds
  - Red muds

Turbidites (derived by turbidity currents from the lands or from submarine highs)
Slide deposits (carried to deep water by slumping)
Glacial marine (having a considerable percent of allochthonous particles derived from iceberg transportation)

According to Ph. H. Kuenen (1950), the distribution of hemipelagic and eupelagic sediments depends first on the distance from the landmass and the topographical differences, and only subordinately on the absolute depth of sedimentation. We may quote as an example the narrow deep basins encircled by land where hemipelagic sediments are deposited to as much as 5,000 m (e. g. the East Indian Archipelago, deep-sea trenches, the Mediterranean). On the other hand, in pelagic areas distant from the coast, typical eupelagic sediments are often deposited on elevations, several hundred metres under the sea level. In classifying the sedimentary deep-sea environments, Ph. H. Kuenen (1950) adds the terms hemipelagic and eupelagic, expressing the distance from the depth conditions. He divides the deep-sea environments into the following groups:

1. Pelagic-abyssal environment which is characterized by a considerable distance from the shore (more than several hundreds of kilometres) and a depth exceeding 1,000 m. Eupelagic sediments (calcareous ooze, siliceous ooze, brown clay, etc.) are typical of it.

2. Hemipelagic-abyssal environment. The distance from the shore is less than a few hundred metres (approx. 200—500 km, depending on the topographical differences), the depth less than 1,000 m. Sediments have prevalenty a sandy or silty admixture and a minor amount of planktonic organisms.

3. Bathyal-environment. Depth between 200 and 1,000 m, distance from the shore ten to several hundred metres. Hemipelagic sediments are typical.

Table 148

Areas covered by marine sediments
(after Ph. H. Kuenen 1950)

Type of sediment	Area, millions of km^2	Percentage of sea floor	Average depth m
Shelf sediments	30	8	100
Hemipelagic sediments	63	18	2,300
Pelagic sediments	268	74	4,300
Globigerina ooze	126	35	3,600
Pteropod ooze	2	1	2,000
Brown clay	102	28	5,400
Diatom ooze	31	9	3,900
Radiolarian ooze	7	2	5,300

From this outline it is apparent that the division of the deep-sea sediments differs from the classifications used for other deposits. It has the advantage that already macroscopic and common microscopic characteristics make it possible to range the sediments into the system, but its disadvantage is the smaller petrographic accuracy. Therefore, it is used chiefly by oceanographers, whereas sedimentologists use the classification based on grain-size distribution. The grain-size of deep-sea deposits reflects not only the conditions of deposition, but also the content of the tests of organisms.

The individual kinds of Recent deep-sea sediments occupy the following areas on the sea bottom (data after PH. H. KUENEN 1950) (Table 148).

Table 149

Distribution of the pelagic sediments in the oceans
(after H. Sverdrup *et al.* 1942)

Type of sediment	Indian Ocean %	Pacific Ocean %	Atlantic Ocean %
Calcareous ooze	54·3	36·2	67·5
Siliceous ooze	20·4	14·7	6·7
Brown clay	25·3	49·1	25·8
Ocean	**Calcareous ooze %**	**Siliceous ooze %**	**Brown clay %**
Indian	26·9	33·9	15·7
Pacific	40·6	55·3	68·7
Atlantic	32·5	10·8	15·6

Table 150

Bathymetric conditions of sedimentation of individual types of deep-sea deposits
(after H. Svedrup *et al.* 1942)

Type of sediment	Minimal depth m	Maximal depth m	Average depth m
Globigerina ooze	777	6,006	3,612
Pteropod ooze	713	3,516	2,372
Diatom ooze	1,097	5,733	3,900
Radiolarian ooze	4,298	8,184	5,292
Brown clay	4,060	8,282	5,407

Table 149 presents the distribution of deep-sea sediments in the different oceans and Table 150 shows bathymetric conditions of sedimentation of the individual types of deep-sea deposits (data after H. SVERDRUP *et al.* 1942).

The overlapping of depth ranges is considerable but, nevertheless, it is well seen that siliceous ooze and red clay are deposited mainly at a greater depth than 4,000 m, the calcareous and diatomaceous oozes being confined to smaller depths.

Hemipelagic (terrigenous) sediments

Hemipelagic sediments contain more than 20% of terrigenous material of silt and sand grade. From some data on the mean size composition the following parameters can be inferred.

	Range	Average
Md	0·005—0·04 mm	0·02 mm
So	2·200—4·10 mm	3·00 mm

The sand content is usually not very large (unless the sediments have an admixture of glacial material), so that the sediments consist almost solely of clay and silt fractions. In nearly all hemipelagic sediments three main components are distinguishable:

1. Terrigenous material, prevalently of clay-silt grade.
2. Organic, mostly calcareous, less frequently siliceous components.
3. Chemical-authigenic components, including glauconite, phosphates, Fe- and Mn- compounds, or alteration products of volcanic material.

According to the colour and composition, three fundamental types of hemipelagic sediments are generally differentiated, i. e. blue, red and green muds. Following J. MURRAY - A. F. RENARD (1891) the composition of these sediments is a follows (Table 151).

From the foregoing data it follows that the differences between these three types of hemipelagic deposits are not great. Only green muds have a somewhat higher amount of glauconite in the coarse fraction. According to other observations, however, their green and grey-green colour is produced by clay minerals (illite and chlorite), possibly also by the admixture of chlorophyll.

Blue muds are a typical hemipelagic sediment and form an almost continuous belt around the continents. They can pass laterally into calcareous muds and at higher latitudes gradually into glaciomarine sediments. With the increase in carbonate components they can grade to hemipelagic calcareous muds. Towards deeper regions they generally pass into calcareous Globigerina ooze as a result of the decrease of the terrigenous component and the increase of the calcareous organogenic component. Calcareous muds are a very widespread variety of blue muds; for instance, they cover the whole bottom of the Mediterranean and of the sea of the East Indian

Table 151

Composition of hemipelagic sediments (after J. Murray - A. F. Renard 1891)

	Blue mud %	Red mud %	Green mud %
Organic constituents			
$CaCO_3$, maximum	34·3	60·8	56·2
minimum	tr	5·8	tr
average	12·5	32·3	25·5
Pelagic foraminifera			
maximum	24·0	30·0	35·0
minimum	0	2·0	tr
average	7·8	13·4	14·6
Benthonic foraminifera			
maximum	10·0	7·0	15·0
minimum	0	1·0	tr
average	7·8	3·3	2·9
Other calcareous remains			
maximum	16·0	40·8	31·2
minimum	tr	1·8	tr
average	3·2	15·5	8·0
Siliceous remains			
maximum	15·0	—	50·0
minimum	tr	—	1·0
average	3·0	1·0	13·7
Inorganic constituents			
$>$0·05 mm, maximum	75·0	25·0	13·0
minimum	1·0	10·0	1·0
average	22·5	21·1	21·1
$<$0·05 mm, maximum	97·0	68·3	83·1
minimum	16·1	28·1	9·7
average	61·8	45·6	33·7

Archipelago. It is of interest that the biological component is there more frequently represented by coccoliths than by foraminiferal tests. Faecal pellets form a frequent admixture of blue muds which can even grade into the so-called pellet muds. Phosphates also occur abundantly. A remarkable feature of blue muds is a few mm or cm thick oxidation surface layer, coloured red-brown which is of the same composition as the red mud. It is the product of halmyrolitic oxidizing processes, being the thicker the slower the sedimentation. When covered by the next layer of sediments, reduction re-appears and the colour turns to bluish-grey.

The normal primary red mud is a variety of hemipelagic sediments deposited in those places, where an immense amount of redbrown suspension from lateritic material is brought by streams (red muds are most extensive near the mouth of the Amazon, along the coast of Brazil). Red muds have an appreciable admixture of particles of silt grade; they do not contain either sulphides, organic matter and glauconite or a substantial amount of phosphates.

Green muds passing into green sands are characteristic particularly of continental slopes with a slow sedimentation.

The not very widespread black and grey muds are also ranged to the hemipelagic sediments. They represent deposits of the environment with stagnant water, a reduced amount of oxygen and occasionally with free hydrogen sulphide in the water, such as, for instance, longitudinal trenches below the Californian shelf, Norwegian fjords, the depressions in continental shelves (the well-known Caraico trench near Venezuela). The characteristics of these typical hemipelagic sediments derived from scanty data available are as follows (Table 152).

Table 152

Composition of black and grey muds

	Range	Average
Md	0·0003—0·02 mm	0·001 mm
So (Trask's)	?	4
$CaCO_3$	0—22·5%	2%
Org. C	0·5—15%	5%
S	0·2— 3%	1·3%

The group of hemipelagic sediments also includes sediments of deep-sea trenches, i. e. elongated depressions more than 5,000 m deep in the sea-bottoms, which generally ly adjoin on one side the are of young volcanic islands. In spite of the enormous depth of their deposition, the sediments have a sufficient amount of coarser terrigenous material. The deposits of Kurile, Mariana, Riukiu, Puerto-Rico and Peru-Chile trenches have been most thoroughly studied. Characteristics of the sediments of deep-sea trenches are summed up in Table 153.

The special nature of these sediments is caused by large height-differences between the source area (island chains) and the site of deposition. At a relatively small distance (tens or a few hundred metres) the vertical difference amounts to more than 10 km. The slope of the trough walls varies between 8 and 30° but can also be steeper. The walls are commonly interrupted by a few terrace steps marked by a sudden flattening of the inclination. Some parts of the slopes are sediment-free, others bear patches of gravels, and at a greater depth, sandy and silty clays.

Graded bedding is a frequent sedimentary structure developed in the sediments of deep-sea trenches, particularly in beds with volcanic material. Otherwise, both

Table 153

Characteristics of the sediments of deep-sea trenches

Grain-size and bathymetric distribution of sediments	Sand and gravel may reach depths of up to 3—3·5 km; deeper silts, silty sands with coarser admixtures occur
Content of carbonate	Decreases with increasing depth, below 5,000 m only small amounts occur. Only in exceptional cases layers rich in $CaCO_3$ occur below 7,000 m. This material was transported from moderate depths by turbidity currents or mud flows
Content of authigenic silica	Some sediments contain more than 30 % of authigenic silica, mainly as diatom frustules
Volcanic material	In Pacific trenches very abundant. Lapilli and pumice fragments form layers often graded bedded
Colour of sediments	Grey to brown, sometimes greenish
Type of sediments	Mainly grey muds with pyroclastic admixture, or brown clay passing into diatom ooze
Content of organic matter	Varies strongly. High in trenches with limited aeration. In grey and black muds more than 3% org. C, in brown clay below 1% org. C

regular and chaotic bedding is present depending presumably on the alternation of the modes of deposition — by turbidity currents, slumps, mud flows or from suspension.

The content of biological particles is related to hydrological conditions. When the trenches occur at the sites of ascending currents (as e. g. in the Peru-Chile trench), the sediments are rich both in skeletal particles and organic matter. A thin oxidation layer is developed also on grey trench sediments. In the Kurile trench, the changes in its thickness have been examined in detail. In the central parts of the trench the thickness is several millimetres to a few centimetres, increasing away from the trenches towards normal abyssal. As mentioned above it corresponds in composition to the normal red clay. Its thickness is indirectly proportionate to the rate of sedimentation.

The typical composition and development of sediments as well as the tectonic position of deep-sea trenches show that these have characteristic features of geosynclines at initial development stages. It is universally stated that the vertical difference between the source and depositional areas steadily increases, as the intensity of present erosion is not sufficient for lowering the rising island chains or mountainous coasts, and the rate of sedimentation is too low to compensate the deepening of trenches.

The rate of sedimentation in deep-sea trenches is relatively great. The precise values are not known, because of the alternation of revolutionary, recurrent sedimentary processes (turbidity currents, submarine slumps) and slow settling from suspension. Moreover, all deep-sea trenches function as collectors of sediments, as they frequently lie in the way of the main supply of sediments into abyssal regions and trap huge amounts of sedimentary material. They let through only fine suspended load. Therefore, the rate of sedimentation in trenches can be estimated at 10 times the value attained in other abyssal areas, i. e. on the average at more than 10 cm in 1,000 years.

The geosynclinal character of trenches is also evident from the composition of sediments. After compaction, the present-day deposits will be red and black clayey and silty shales, with intercalations of siltstone and sandstone in places and with beds of basic effusives, tuffs and tuffites.

The increased rate of sedimentation is associated with the considerable thickness of Recent and Subrecent unconsolidated sediments. Table 154 lists the thicknesses established.

Table 154

Thickness of unconsolidated sediments in deep-sea trenches (after various data)

Trench	Thickness of unconsolidated sediments
Puerto Rico	8,000 m
Tonga	200 m unconsolidated sediments covering thick lava flows
Kurile trench	3,000 m
Hawaian trench	2,000 – 3,000 m

Eupelagic (pelagic) sediments

Unlike terrigenous (hemipelagic) deposits, the eupelagic sediments contain the finest terrigenous material of clay grade and, at most, a small admixture of silt fraction. The terrigenous detritus is mixed with biogenic material (which is predominant) and a variable amount of volcanic and extratelluric (cosmic) material. The above-mentioned division of deep sea sediments into calcareous and siliceous oozes and brown clay is rather artificial, as they pass one into another only very slowly and as a matter of fact represent one sediment with a variable amount of different biogenic components.

The grain-size composition of deep-sea sediments, without the removal of the biological component, is not a suitable diagnostic characteristic of hydrodynamic conditions. Yet, for correlation purposes, parameters of deep-sea sediments are plotted on the diagrams below (Fig. 94 a,b).

The presence of tests of foraminifers and pteropods is responsible for the larger grain-size than would correspond to the hydrological conditions. The prevalence of this calcareous material results also in a high grade of sorting. Skewness equals zero or is positive. Whereas foraminifers have a silt up to fine sand grade, diatoms and radiolaria show a fine silty and sponge spicules a coarse silty to fine-sandy grain-size.

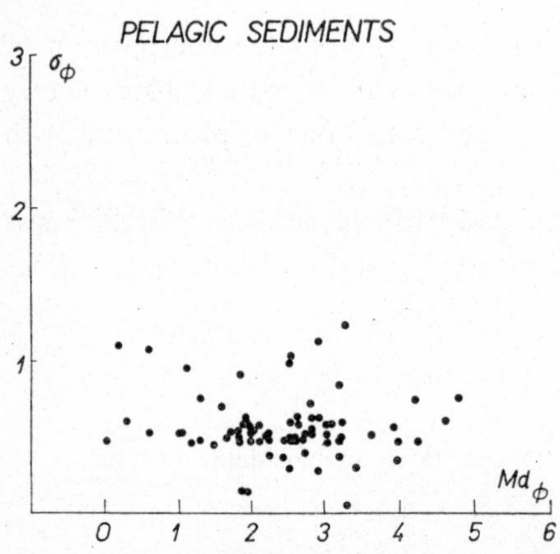

Fig. 94a. Grain-size parameters of pelagic sediments. Relation of Phi Median Diameter to Phi Deviation Measure. Data compiled from various authors.

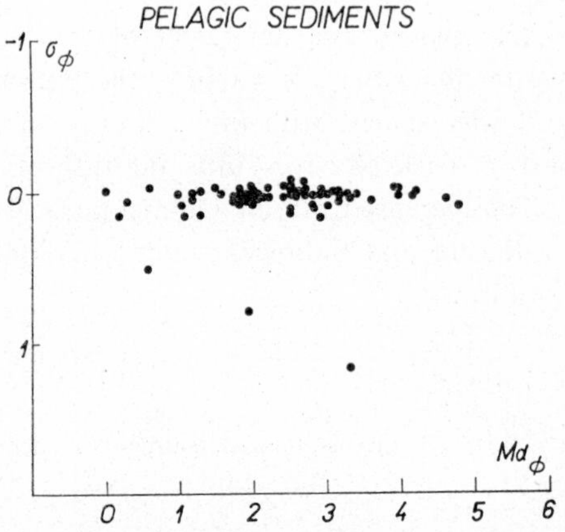

Fig. 94b. Grain-size parameters of pelagic sediments. Relation of Phi Median diameter to Phi Skewness Measure.

Foraminiferal (calcareous) oozes

Globigerina ooze is the most extensive calcareous ooze and probably the most widely developed of all deep-sea sediments. It must contain more than 30% $CaCO_3$, or more than 6,000 foraminiferal tests in 1 g of dry sediment (the number of foraminiferal tests in 1 g dry sediment is the so-called foraminiferal number; its maximum theoretical value may be 23,300 but the maximum number observed thus far has been only 12,800).

In the Globigerina ooze planktonic foraminifers are by far predominant. Although their quantity is enormous, the number of species is limited, whereas the less numerous benthonic foraminifers (generally about 1% of the total volume) occur in more species (see Table 155).

The question remains as to what extent the character and occurence of species in sediments on the sea floor correspond to their amount and occurrence in the water layers above. Most authors believe that generally the same foraminiferal species

Table 155

Number of species of foraminifers in Globigerina ooze
(after R. Revelle 1944)

	Total amount	Number of species
Planktonic foraminifera	about 60%	12– 23
Benthonic foraminifera	up to 1%	89–119

occur in sediments and in the overlying water column. In addition to foraminifers, calcareous oozes contain in places abundant tests of pteropods and heteropods. They form up to 30% of the total volume of sediments in the equatorial Atlantic. Deposits with an appreciable amount of these tests are called pteropod oozes and are regarded as a shallower variety of Globigerina ooze. They occur especially on the submarine elevations of the Atlantic Ocean (The Mid-Atlantic Ridge a. o.), at a depth of several hundred to a few thousand metres. The deposits termed coccolith oozes contain about 60% of tests of flagellates (coccolites). The best-known occurrences are in the Mediterranean (where they are the most abundant sediments) and in the tropical Atlantic.

In addition to the pronouncedly skeletal carbonate, the calcareous oozes contain fine-grained sediment – calcite – which does not bear signs of skeletal structure. Its chemical origin, postulated by some writers, has so far not been corroborated by investigation. It seems more probable that it represents the product of disintegration of foraminiferal and other tests. The disintegration of calcite crystals has been many times observed either directly or indirectly (e. g. according to the increase of the finest calcium carbonate fraction with depth in Globigerina ooze).

Table 156

Average chemical composition of Globigerina ooze
(after R. Revelle 1944)

	Maximum %	Minimum %	Average %
SiO_2	0·72	0·08	0·47
Al_2O_3	1·35	1·22	0·78
FeO, Fe_2O_3	1·08	0·51	1·08
MgO	0·16	0·10	0·13
CaO	54·52	53·12	53·82
CO_2	43·10	41·69	42·37

Globigerina oozes of lower geographical altitudes are coarser grained than those of higher latitudes, which is connected with the development of larger foraminiferal tests in warmer waters. In addition, differences in the grain-size of calcareous oozes from elevations and depressions have been observed; they are due to washing out of the finest fractions from the sediments of elevations and their accumulation in depressions. Coarse-grained Globigerina oozes are also richer in calcium carbonate because the insoluble residue is confined only to the finest fraction.

The average chemical composition of the Globigerina ooze is as follows (after R. REVELLE 1944, Table 156).

J. MURRAY-A. F. RENARD (1891) recorded the following composition of Recent Globigerina oozes (Table 157).

Towards higher geographical latitudes, the amount of authigenic SiO_2 in Globigerina oozes increases, owing to the increase in the number of diatom tests. On the average, the Globigerina ooze of the Pacific and Indian Oceans is richer in SiO_2 than that of the Atlantic Ocean.

Calcareous oozes occur even at a depth over 7,000 m (according to D. B. ERICSON *et al.* 1961) and have more than 70% $CaCO_3$. These oozes, however, represent undoubtedly redeposited material of shallow-water origin. Abundant traces of the dissolution of calcareous tests point to the instability of carbonate at this depth.

The presence of dolomite in the Globigerina ooze is still disputable. Dolomite in recent deep-sea calcareous oozes was recorded by three authors. J. MURRAY-A. F. RENARD (1891) found disseminated rhombohedra, which were optically similar

Table 157

Composition of Recent Globigerina ooze
(after F. Murray - A. F. Renard 1891)

Organogenic constituents	%	Inorganic constituents	%
$CaCO_3$, maximum	97·3	> 0·50 mm, maximum	50·0
minimum	30·2	minimum	1·2
average	64·5	average	3·3
Pelagic foraminifera		<0·05 mm, maximum	64·6
maximum	80·0	minimum	1·2
minimum	25·0	average	30·6
average	53·1		
Benthonic foraminifera			
average	2·1		
Other calcareous remains			
maximum	31·8		
minimum	1·2		
average	9·2		
Siliceous remains			
maximum	10·0		
minimum	1·2		
average	1·6		

to dolomite, but they did not examine them in detail. C. W. CORRENS (1939) found isolated rhombohedra of dolomite at small depths under the surface of sediments. E. A-ZEN (1960) ascertained a small admixture of dolomite in deep-sea calcareous sediments of the Peru-Chile trench.

There is considerable information available on the distribution of the Globigerina ooze. It forms a continuous cover over large areas of the sea bottom in the equatorial and northern Atlantic, in the western part of the Indian Ocean, and in the southern part of the Pacific. In other places, as for instance, in the northern Pacific, it forms isolated occurrences, particularly on the elevations, between continuous cover of clay or patches between areas covered by diatom ooze.

Diatom and radiolarian oozes

Diatom ooze, one of the two main varieties of siliceous oozes, is partly hemipelagic and partly eupelagic sediment. Being distributed in higher geographical latitudes where there is a large supply of coarse glacial material, it contains on the average more sandy and silty terrigenous detritus than all other deep-sea sediments. The mean composition of the diatom ooze is as follows (Table 158).

A variable, but generally quite high percentage of terrigenous clay and silt, tests of Radiolaria, spicules of sponges and tests of agglutinated foraminifers are common admixtures of the diatom ooze. Diatom tests are easily dissolved on account of their rather low chemical stability. The corrosion of freshly deposited tests has frequently been described. Inside the sediment, they are completely dissolved and transformed

Table 158

Composition of Recent diatom ooze
(after J. Murray - A. F. Renard 1891)

Organogenic constituents		%	Inorganic constituents		%
$CaCO_3$	maximum	36·3	>0·05 mm	maximum	25·0
	minimum	2·6		minimum	3·0
	average	23·0		average	15·6
Pelagic foraminifera			<0·05 mm	maximum	27·9
	average	18·2		minimum	12·5
Benthonic foraminifera				average	20·4
	average	1·6			
Other calcareous remains					
	average	3·2			
Siliceous remains					
	maximum	60·0			
	minimum	20·0			
	average	41·0			

into a pseudo-amorphous gel mass of SiO_2. As mentioned above, the diatom ooze is prevalently of silt grade, and shows the following mean content of individual fractions:

Sand	6·00%
Silt	88·00%
Clay	6·00%

Diatom oozes are deposited in places, where the production of phytoplankton predominates over that of zooplankton and where the supply of terrigenous detritus is not so great as to mask the biogenic sedimentation. The first condition can be fulfilled not only in higher geographical latitudes but also in many coastal areas which, however, usually do not comply with the second requirement. Both conditions are safisfactorily fulfilled mainly in the polar regions, in a continuous belt round the Antarctic Ocean areas, where glaciomarine sediments are laid down. In the Pacific, this belt of diatom oozes is developed also round the Arctic region. The polar areas of diatom oozes pass towards the poles into glaciomarine deposits and towards the Equator into Globigerina oozes and red clays. Diatom sediments also occur in isolated basins near the western coast of North America, where the production of phytoplankton is supported by ascending, nutrient-bearing currents, and which are at least partly isolated from the import of a substantial amount of terrigenous detritus. The production of diatoms can also be supported by submarine volcanic activity and submarine emanations the products of which supply suitable material for the building of diatom tests. The lamination of diatomaceous sediments reflects the seasonal variation in the production of phytoplankton.

Table 159

Composition of radiolarian ooze
(after J. Murray - A. F. Renard 1891)

Organic constituents	%	Inorganic constituents	%
$CaCO_3$ maximum	20	$<0·05$ mm maximum	5·0
minimum	tr	minimum	1·0
average	3·1	average	1·7
Pelagic foraminifera		$<0·05$ mm maximum	67·0
average	3·1	minimum	17·0
Benthonic foraminifera		average	39·0
average	0·1		
Other calcareous remains			
average	0·1		
Siliceous remains			
maximum	80·0		
minimum	30·0		
average	54·5		

Radiolarian ooze is a variety of red clay containing more than 20% of radiolarian tests and an admixture of diatoms and sponge spicules. Its mean composition is given in Table 159.

The conditions of deposition of radiolarian oozes resemble those postulated for the sedimentation of red clay. A large production of zooplankton prerequisite for their development means their restriction to the equatorial regions, in particular. The largest areas of radiolarian oozes are in the equatorial Pacific; smaller occurrences have been established in the Indian Ocean, especially SW of the East Indian Archipelago.

Brown clay (red clay)

Brown clay, one of the most widespread and most interesting deep-sea sediments, has recently been assigned by P. L. BEZRUKOV *et al.* (1961) to the so-called polygenous sediments. This opinion is well substantiated, because brown clay can originate in several ways, as summarized below:

1. By dissolution of calcareous oozes, leaving behind an insoluble residue.
2. From the so-called marine hydrogenic substance, which is a constant suspension of marine water, partly of detrital and partly of chemical and colloidal origin.

Table 160

Composition of brown (red) clay
(after J. Murray - A. F. Renard 1891)

Organic constituents		%	Inorganic constituents		%
$CaCO_3$	maximum	29·0	>0·05 mm	maximum	60·0
	minimum	0		minimum	tr
	average	10·4		average	2·4
Pelagic foraminifera					
	maximum	3·0	<0·05 mm	maximum	100·0
	minimum	0		minimum	31·0
	average	2·0		average	87·5
Benthonic foraminifera					
	maximum	3·0			
	minimum	0			
	average	0·6			
Other calcareous remains					
	maximum	6·3			
	minimum	0			
	average	1·0			
Siliceous remains					
	maximum	5·0			
	minimum	0			
	average	0·7			

3. From decomposed volcanic material.
4. From cosmic material.
5. By oxidation and halmyrolysis (i. e. alternation on the sea floor) of all above materials.

As a result of the many possible origins, the mineralogical and chemical composition of brown clay is also variable.

The term brown (red) clay is derived from its characteristic colouration which, depending on the content of manganese, changes from brick-red in the Atlantic to chocolate-brown in the Pacific.

The mean composition of brown clay, according to J. MURRAY-J. CHUMLEY (1924) is given in Table 160.

H. U. SVERDRUP *et al.* (1942) recorded the following mean contents of individual size fractions:

Sand	0% (or traces)
Silt	17·3%
Clay	82·7%

Brown clay is excessively rich in various accessories owing to a slow sedimentation. It contains grains of allothigenic terrigenous or volcanic origin (feldspars, augites, epidote, garnets, magnetite, zircon, tourmaline, a. o.) and authigenic components (sporadic glauconite. abundant grains of manganese minerals, palagonite and phillipsite). In addition, it bears an increased amount of particles of cosmic origin.

In the areas adjacent to volcanic islands and on extensive parts of the ocean bed, as for instance, in the north-western Pacific, or in the vicinity of deep-sea trenches, brown clay contains many volcanic particles of silt grade, suggesting its undoubted volcanic origin at least to some extent. The different genesis is reflected in the different mineralogical composition, although the differences are not so great as might be expected. It is because the halmyrolytic processes bring very close the resulting compositions of brown clay. Generally, in brown clays illite predominates over montmorrillonite; only in those of volcanic origin does montmorillonite occur in excess of illite and chlorite. From the brown clay of the Atlantic beidellite and occassionally a substantial amount of kaolinite has been described. Table 161 presents the semiquantitative composition of the average samples of brown clay from the Pacific, Atlantic and Indian Oceans (after S. K. ELWAKEEL - J. P. RILEY 1961).

K. SUZUKI - W. KITAZAKI (1954) ascertained that brown clay of the Pacific is formed particularly of the amorphous mineral of the alophane group and of newly formed illite and quartz. The question of the authigenesis of clay particles remains open. Some investigators (see above) would suggest an appreciable influence of the authigenesis on the final composition of brown clay and on the formation of clay minerals. Yet, it seems that the main component of brown clay is the finest allogenic terrigenous detritus with a variable admixture of altered volcanic material.

Table 161

Semiquantitative composition of the average samples of brown clay from the Pacific, Atlantic and Indian Oceans
(after S. E. ElWakeel - J. P. Riley 1961)

Brown clay	Atlantic ocean	Pacific Ocean	Indian Ocean
Silty fraction			
Quartz	5	5	5
Chlorite	2	1	2
Illite (mica)	1	1	2
Feldspar	1	2	1
Clay fraction			
Chlorite	2	2	2
Montmorillonite	1	1	1
Mixed-layer montmorillonite-chlorite	1	tr	tr
Illite	2	3	2
Quartz	2	2	2
Feldspar	1	1	1
Mixed layer illite-chlorite	tr	tr	1
Total sample			
Quartz		4	4
Feldspar	1	1	1
Chlorite	2	2	2
Illite	1	2	2
Montmorillonite	1	2	2
Other	tr	tr	tr

Brown clay can be deposited only in those places where no appreciable amount of calcareous and siliceous tests can be laid down. These conditions are fulfilled in cold waters and at a great depth, where calcareous tests are dissolved. As a matter of fact, brown clay can be taken for the most stable insoluble residue of all other components of deep-sea deposits. As a product of halmyrolysis, it largely corresponds in composition to the well-known oxidation layers at the surface of other deposits.

In some places in the Pacific, consolidated sediments have been found on the bottom surface. They proved to be common brown clays consolidated by phillipsite cement into a crust, with a great amount of volcanic material, glass and other more or less halmyrolized components. They are thought to be the relics of a continuous volcanic cover of certain parts of the Pacific.

The question of the origin of brown clay is closely connected with that of halmyrolysis. Although this term is widely used, it is not precisely stated which processes should be included in it. It seems, however, that the terms subaqueous oxidation and halmyrolysis are nearly equivalent. Halmyrolysis is in fact oxidation occuring in waters bearing free oxygen. The first information was derived from the finds of weathering rings around pieces of basic glass and the red-brown oxidation layers on blue and grey clays. In oxidation products, the content of trivalent iron increases greatly at the expense of bivalent iron, and so does the amount of manganese, alkalis and water. On the other hand, the percentage of Ca, Mg, slightly also SiO_2 and

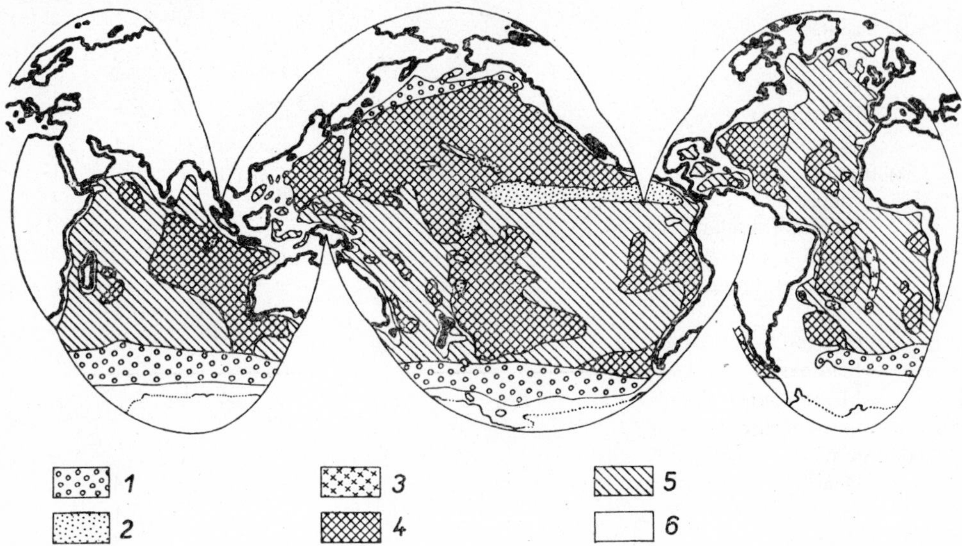

Fig. 95. Regional distribution of individual types of pelagic sediments. 1. Diatom ooze, 2. Radiolarian ooze, 3. Pteropod ooze, 4. Brown (red) clay, 5. Globigerina ooze, 6. Terrigenous deposits and continent. After H. U. Sverdrup *et al.* (1942).

occasionally Al_2O_3 decreases. The following alterations of marine sediment have so far been proved:

1. Pure oxidation, consisting in the development of free Fe-hydroxides at the expense of bivalent Fe of clay minerals, mafic minerals and occasionally also siderite.
2. Alteration of acid volcanic glasses into a mixture of quartz, kaolinite, illite and analcite (in ratio 1 : 2 : 2 : 0·5). In the alteration products opal seems to be present as well.
3. Alteration of basic volcanic glasses into palagonite, i. e. presumably a mixture of chlorite and illite minerals, compounds of trivalent Fe and phillipsite.
4. Migration of elements and enrichment of the upper layers in iron and manganese. In the superficial weathering crust of a basalt fragment, for instance, the percentage of Fe_2O_3 increases from 10·71 to 17·57% and that of MnO from 0·44 to 2·69%.
5. Special alterations, such as the transformation of plagioclases to orthoclases, or the generation of illite and muscovite from volcanic glasses.

The intensity of halmyrolytic processes is affected by the rate of sedimentation, the size of particles and the content of free oxygen in sea water. Slow sedimentation, the presence of finest particles of colloidal dimensions in the sediment and the normal content of free oxygen in water, that all contribute to the intensity of halmyrolytic alterations. Brown clay complies best with these requirements. It is unquestionable that, whether of terrigenous, volcanic or authigenic origin, its resulting composition is homogenized by these very halmyrolytic processes. The oxidation state of iron and the red or brown colour corresponding to it, the small amounts of carbonate and the increased total percentage of Fe and Mn always suggest the activity of submarine weathering processes. It has been found out, for instance, that similarly as the intensity of oxidizing conditions increases from the deep-sea trenches towards the open ocean, so grows also the thickness of the upper oxidation layer — brown clay — from several centimetres to a few metres (E. A. OSTROUMOV - V. M. SHILOV 1956).

Glaciomarine sediments

The substantial effects of glacial processes on the formation of deep-sea deposits was mentioned above. Glaciomarine sediments are widely distributed in the areas of higher geographical latitudes, on the continental shelf and slope as well as in abyssal regions. They are a mixture of common deep-sea sediments and material brought by ice or icebergs. This accounts for their, on the average, coarser grain-size. The size distribution corresponds generally to the intensity of glacial conditions, or the proximity of the glaciated area.

The effect of cold water contributes to the preservation of some unstable organic substances, as for instance, faecal pellets which may even be in the stage of transformation into glauconite. The percentage of $CaCO_3$ is variable, ranging usually from 1 to 10%. H. W. MENARD (1953) writes that the clay fraction of these sediments is often formed of undecomposed extremely fine rock flour. Other data, however, do not corroborate this presumption; some authors found abundant illite with and admixture of chlorite in these sediments.

In large parts of the oceans, glaciomarine sediments of the Last Glacial are still on the surface. On elevations, there are deposited gravels with striated pebbles, which, being evidently brought by glaciers, have not yet been covered by fine-grained Recent deposits.

Pyroclastic sediments

Pyroclastic deposits include two types differing in composition and genesis:

1. Products of subaerial volcanism, transported by wind and deposited on the sea bottom as fine silt and clay particles.
2. Products accompanying submarine volcanism.

The former type of pyroclastic deposits have been far more commonly recognized by deep-sea researchers. Analyses of volcanic ash falling on the sea surface have revealed that it is predominantly of silt grade (P. L. BEZRUKOV 1959). The composition of deep-sea pyroclasts is in full accordance with this finding. Their distribution is controlled not only by the position of volcanic centres but also by the predominating direction of wind. The transport of pyroclastic products has been traced directly. A. F. RICHARDS (1958) records that the mean velocity of the transport of pumice pieces across the whole of the Pacific was 18 cm/sec. Pumice ejected by the eruption on the Krakatoa Island was transported at a rate of 23 cm/sec. A minor amount of volcanic material can be borne on the surface of water up to a distance of 12,000 km.

Table 162

Review of volcanic ash and sediments in deep-sea deposits corresponding to the known volcanic eruption
(after various data)

Volcanic eruption	Distribution and characteristics of pyroclastic material
Hecla (Iceland)	Volcanic sand and ash layer surrounding the island. Admixtures of ash in pelagic deposits of North Atlantic
Katmai (Alaska)	Surface layer of pyroclastic material in NW Pacific. Asymetrical distribution of pyroclastic sediments due to wind action
Santorini (Mediterranean Sea)	Several layers of volcanic ash in calcareous sediments
Volcanoes on Kuriles and Kamchatka	Abundant pyroclastic sediments in adjacent trench. Pumice and lapilli occur together with ash and fragments of lava
Volcanoes of Java	Ashes in all marine sediments of East Indian Archipelago
Volcanoes of the Azores (Pleistocene volcanic activity)	Pyroclastic material disseminated in sediment of the whole North Atlantic. Several individualized layers of volcanic ash
Vesuvius (Italy)	Several layers of pyroclastic sediments in Tyrrhenian Sea. Tuffites with 10% of terrigenous material prevail

Table 162 gives a review of volcanic ash and sediments in deep-sea deposits corresponding to the known volcanic eruptions.

Besides these examples, numerous beds of volcanics of unknown provenance have also been found, e. g. in the Bay of California, Peru-Chile trench, etc.

Products accompanying submarine volcanism are less well known, but their wide distribution can, nevertheless, be postulated. H. W. MENARD (1953) estimates that lava sheets occupy 8% of the Pacific bottom, apart from sea mountains and guyots which are likewise of volcanic origin. Therefore, it cannot be presumed that such immense masses of lava would not be accompanied by volcanic ejecta, or at least by the products of lava granulation. The abundant fragments of basic glass and basaltic rocks, which were found in brown clay by earlier expeditions and described under the collective term palagonite undoubtedly are partly of this origin.

Innumerable traces of volcanic activity have been ascertained in deep-sea trenches, including coarse pyroclastic material, such as pieces of lava and several decimetres-large fragments of pumice. This material could have been redeposited from the slopes of the trenches, or was generated by volcanic activity right at the bottom.

Phosphorites

Concretions and nodules of phosphorites occur generally on the continental slope, at the margin of the shelf and on the isolated submarine elevations. The Agullas Bank not far from the southernmost tip of Africa is the best-known site of their occurrence. Otherwise, they are developed on the eastern continental slope of Japan, near the coast of Chile, Florida and virtually everywhere to a minor extent. The size of concretions ranges from a few milimetres to half a metre, and the surface is smooth or nodular. Their regular oval shape can be formed by a moderate transport and abrasion, but usually originates from a regular accretion of laminae. The main minerals of concretions are collophanite and fluor-apatite. In minute phosphate ooids interspersed among larger concretions, concentric shells of francolite and dahlite have also been determined. It has been ascertained that phosphates occur wherever slow sedimentation or non-sedimentation take place. They are precipitated chemically under suitable hydrological and hydrochemical conditions or as a result of a sudden supply of phosphates from the landmass by volcanic activity or by organisms perishing at the contact of cold and warm currents.

In places, sea water is saturated or oversaturated with phosphates so that it can produce metasomatic phosphatization. The phosphatization of calcareous foraminiferal tests has often been observed. Of interest is the reported phosphatization of a piece of wood only a part of which, embedded in the sediment, preserved its original composition, whereas the part protruding above the sediment has been altered.

Other admixtures in abyssal sediments

Barite concretions have been found in red clay and calcareous ooze at some places, as for instance, near Ceylon, in the East Indian Archipelago, in the basins near the Californian coast. The analysis of one of these concretions has shown the following chemical composition (Table 163).

Concretions near the Californian coast contain between 62 and 77% $BaSO_4$. They originate most probably by the precipitation of barium, supplied by warm waters outflowing on the fault zones. Besides, this accumulation of barium and other elements (Sr, Pb) is influenced by marine planktonic organisms. Comparatively high concentrations of barium, strontium, and lead are found in some marine plank-

Table 163

Chemical composition of barite concretion
(after K. Andrée 1920)

	%		%
BaO	53·85	Fe_2O_3	1·67
SO_3	28·56	CaO	2·01
SiO_2	6·46	MgO	0·42
Al_2O_3	2·32	Ign. loss	2·94

tonic organisms, which also contain considerable quantities of other heavy-metal ions. This suggests that biological extraction from surface sea water and subsequent sinking is important mechanism in accreting these elements in the sediment. Among the organisms notable in this respect are some species of Foraminifera, pteropods and heteropods.

Table 164

Grain-size distribution of phillipsite in brown clay
(after G. Arrhenius 1954)

Fraction μ	Amount of phillipsite %
> 32	63
10—32	89
3·2—10	51
1·0— 3·2	4
$< 1·0$	?

The zeolite phillipsite was early recognized as quantitavely important in slowly accumulating pelagic sediments. Recently, it has been found that other members of the phillipsite-harmotome series are widespread. Zeolite concentrations higher than 50% are not infrequent, and zeolitite consequently is a common pelagic sediment type. According to the experiments the chemical precipitation of zeolite from sea water is possible. Apart from that, microcrystalline zeolite is found within dissolved skeletons of siliceous organisms. This suggests that the high concentrations of dissolved silica from biogenic sources at the sediment-water interface induce the formation of zeolites in these cases. The frequent close association of zeolite with pyroclastic sediments points to their direct crystallization from those products. Recently, indication of the crystallization of zeolite on the interaction between sea waters and basaltic lava effusions have been recorded. According to G. ARRHENIUS (1954), the individual fractions of brown clays contain the following amount of phillipsite (Table 164).

The mean size of phillipsite crystalls varies between 0·05 and 0·027 mm. It has also been recognized that there is a direct relationship between the radioactivity of abyssal sediments and their content of phillipsite. The analysis of phillipsite-rich brown clay (about 20%) from the Pacific Ocean gave the following chemical composition (Table 165).

Table 165

Chemical composition of phillipsite-rich brown clay
(after G. Arrhenius 1954)

	%		%
Ign. loss	7·35	CaO	1·38
SiO_2	49·88	MgO	1·20
Al_2O_3	16·52	K_2O	5·10
Fe_2O_3	5·54	Na_2O	4·59
MnO	0·44	H_2O	9·33

Deep-sea sands and gravels

The most widespread abyssal sediment of an anomalous grain-size are undoubtedly the above-described glaciomarine sediments. However, abundant sands and gravels have been found on the sea bottom (or a few centimetres under its surface) for which a glacial origin must be excluded, so that another explanation of their genesis should be looked for. All these occurrences can be divided into local and regionally distributed ones. The characteristics of both groups are given in Table 166.

Table 167 lists some most important occurrences of coarse grained deep-sea deposits, their mode of deposition, structure, texture and their postulated origin.

From this survey it follows that deep-sea gravels and sands are widespread and at present one of the main abyssal deposits. On the basis of their distribution and

Table 166

Characteristics of local and regionally distributed deep-sea coarse-grained sediments

Local occurrences	Regionally distributed coarse-grained sediments
Mainly on elevations or in their surroundings	Form regular layers within the sediments or on their surface
Mostly gravels or coarse sands without clay, or as pebbles disseminated in fine-grained pelagic sediments	Better sorted and finer-grained sandy sediments (usually medium-grained and fine-grained sands)
Relics of Pleistocene sediments occurring at places of Recent non-sedimentation; pebbles carried by kelp or in plant roots, fragments originated by crushing during submarine tectonic processes; shallow water sediments transported towards deep-sea environment by mudflows or slumps	Transport and sedimentation mostly by turbidity currents, mud flows or normal deep-water currents
Variable petrographic composition, mostly non-quartzose sediments	Mostly quartzose sands, with feldspars and other common terrigenous admixtures

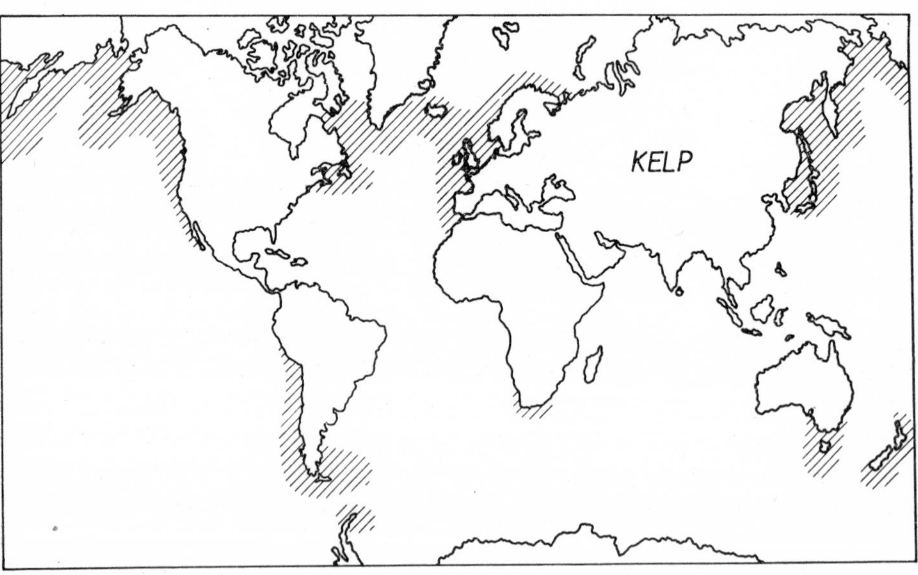

Fig. 96. Area of greatest importance for transporting agents of coarse sediments. Cross-hatching shows areas likely to be reached by kelp rafting from coastal zones where kelp lives. After McGill (1958) and K. O. Emery (1963). (In M. N. Hiil 1963.)

Table 167

Characteristics of important occurrences of coarse-grained deap-sea deposits

Localization	Composition	Suggested origin
Brown clay, Atlantic Ocean, 29°NL, depth 5,450 m	Gabbro pebbles disseminated in fine-grained pelagic sediments	Transport and sedimentation by slumps at the mouth of submarine canyon
Atlantic Ocean, surroundings of Madeira and Canary Islands	Coarse-grained badly-sorted sands with clayey admixture. Formed of fragments of young volcanites and rocks occurring on adjacent islands	Local origin by land abrasion and transport by slumps into adjacent troughs
Romanche trench (Eastern Atlantic) depth up to 7,500 m	Grain-size composition of sediment layer μ — upper parts % — lower parts % <6 — 6·4 — 4·7 6—62 — 12·6 — 0·9 62—125 — 46·6 — 11·3 125—250 — 22·9 — 33·3 250—500 — 9·4 — 34·1 <500 — 2·1 — 15·7 Maximum grain-size is 4 mm. Mixture of minerals and rock fragments, angular grains bear traces of crushing	Tectonic crushing and sliding towards the bottom of the trench
Western equatorial Atlantic	Medium and coarse-grained sands, mainly of quartz and feldspar. Th'ckness of layers up to 160 cm; increases towards South American coast. Contains remains of shallow-water fauna and plant remains. Graded bedding. Great areal extent	Sedimentation by turbidity currents flowing from American coast towards central Atlantic
Western parts of North Atlantic	Fine-grained sands and silts, graded bedding and lamination, great areal extent. Thickness increases towards American coast	Sedimentation by turbidity currents flowing through submarine canyons of North American continental slope
North-eastern Pacific	Layers of fine-grained sand and silt, graded bedded. Great lateral extent	Sedimentation by turbidity currents

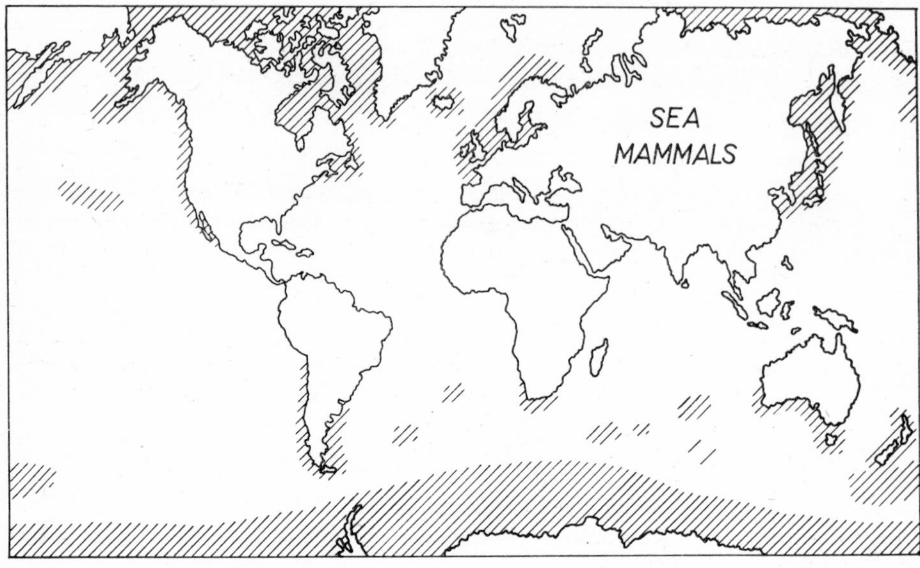

Fig. 97. Areas where transportation by sea-lions and most other sea mammals may occur. After K. O. Emery (in M. N. Hill 1963).

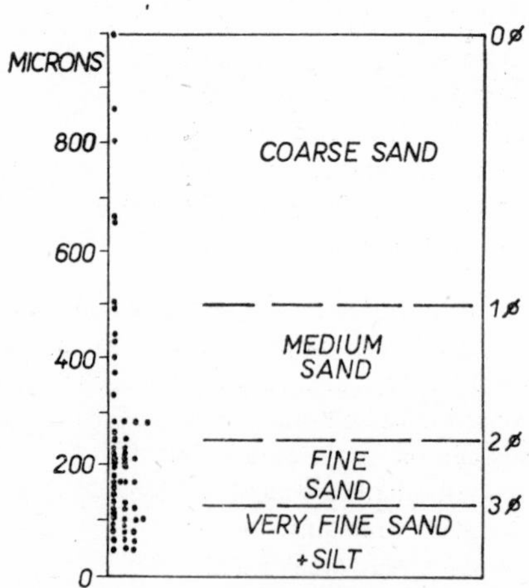

Fig. 98. Median diameters of base of deep-sea layers and of thin sand layers from various parts of the world. After F. P. Shepard (1963).

their relationship to the landmass, most authors believe that the regionally distributed deep-sea sands were deposited by turbidity currents. J. F. Hubert (1964), however, came to the conclusion that deep-sea sands of the Northwest Atlantic were laid down by normal deep bottom currents. He bases his opinion on detailed analyses of the composition and structures of the deposits and points out the following reasons: Deep-sea sand and silt layers are commonly current-laminated throughout their thickness. They contain consistently less than 5% clay, and none exceeds 12%. The sinusoidal trends present on scatter diagrams between mean size and size sorting, and between mean size and skewness, are similar to the trends known for sediments deposited in continental, littoral and shallow marine environments. PH. H. KUENEN (1967) advocates the turbidite origin of deep-sea sands.

Local occurrences of deep-sea gravels and sands can represent relics of Pleistocene deposits or can originate by tectonic crushing or anomalous ways of transport (i.e. by kelp, sea mammals, flotation, driftwood). The most widespread occurrences of deep-sea sands have been found in the cores of sediments flanking the mid-Atlantic ridge, which suggests that they were deposited when major parts of it rose above the water level.

Deposition of carbonates in deep-sea regions

The main source of carbonates in deep-sea sediments is the plankton. Chemical precipitation of calcium carbonate is highly improbable, although some scientists seem to think it possible, and terrigenous supply can be considered only for hemipelagic deposits. Moreover, an increased terrigenous supply does not produce a larger content of $CaCO_3$ in sediments, because it is masked by the even quicker deposition of silicate detritus. Thus, it seems most likely that the carbonate of deep-sea sediments is fully of biological origin. Its distribution is controlled by the equilibrium between the organic production, the possibility of sedimentation of biological material and the forces contributing to the removal of carbonate from sea water and sediments.

The following factors exert influence on the distribution of carbonate sediments in the oceans:

1. Dissolution of tests during sinking and during diagenesis in the sediment.
2. Deep-sea currents and their influence on the dissolution of $CaCO_3$.
3. Absolute depth, because the solubility of CO_2 and thus also of $CaCO_3$ rises with increasing pressure.
4. Volcanic processes, as the emanations of an acid character induce the dissolution of calcium carbonate.
5. Salinity of water.

All these factors influence to a certain extent the deposition and distribution of Recent calcareous deep-sea sediments. It remains to establish their relative significance and interrelationship. As at least several of them combine in their effects, it is evident that the overall picture of carbonate sedimentation cannot be explained by any one of these assumptions alone. The regional changes in the distribution of carbonate must be interpreted in terms of regional causes, such as deep-sea currents, volcanic activity and depth conditions (which limit the carbonate sedimentation bathymetrically). Local changes are then explained by local factors, e. g. topography, local currents, local organic production, mixing with terrigenous detritus and partially volcanic activity.

The dissolution of tests during sinking has been observed directly and established experimentally. In sinking by 300 m, 20% of foraminiferal tests is dissolved; this value increases to 50 per cent during sinking by 3,600 m.

The influence of deep-sea currents can be well illustrated by a comparison of the $CaCO_3$ content in sediments of the eastern and western parts of the equatorial Atlantic at the same depths. Carbonate in sediments of the Brazil basin is dissolved by cold Antarctic current flowing S—N through it (Table 168).

Table 168

Average content of carbonate in sediments of eastern and western parts of equatorial Pacific (after G. Schott 1944)

	Depth m	$CaCO_3$ %
Brasil Basin	5,150	4
	5,900	2
Congo Basin	5,200	80
	5,600	40

Table 169

Mean content of carbonate at various depths of the Pacific (after R. Revelle 1944)

Depth m	$CaCO_3$ %
0–1,000	46
1,000–2,000	62·1
2,000–3,000	60·2
3,000–4,000	48·7
4,000–5,000	26·0
$<$ 5,000	5·1

The effect of the absolute depth is strikingly apparent from the correlation of the mean contents of carbonate at various depths (R. Revelle's data from the Pacific, Table 169). The correlation between the $CaCO_3$ percentage and depth improves only after excluding the nearshore data to remove the effects of clastic influx. According to S. V. SMITH *et al.* (1968) water depth predicts about 40% of the variability in $CaCO_3$ percentages of pelagic sediments.

The influence of volcanic processes on the sedimentation of carbonate has been clearly proved in the Indian Ocean. In the western part not even traces of volcanic processes have been found, and the bottom is covered universally by carbonate oozes; on the other hand, the eastern part is rich in volcanic products and is covered by noncarbonate brown clay. The bathymetric limit of carbonate sedimentation lies much higher in minor seas with manifestations of volcanic processes than in seas free of volcanic activity.

The influence of salinity on the content of calcium carbonate in sediments was proved first by P. D. TRASK (1936), who found that the higher the salinity, be it only in superficial layers of ocean waters, the greater the content of carbonate in sediments. This relationship is connected with higher temperature and greater organic production.

Organic matter in deep-sea sediments

The average annual production of organic matter in the oceans amounts to 1,000 g/m^2, but only 0·02% of this amount becomes sediments. Approximately one-fourth of organic matter is destroyed before the sediment is covered by 30 cm of further deposits. During diagenesis, about 40% of the original content of organic matter is lost. Table 170 present a survey of the content of org. C in various sediments.

Diatomaceous sediments are richest in organic matter, because the phytoplankton contains the greatest amount of most stable organic substances. Besides, in cold waters not even the unstable components of organic matter are decomposed so readily as elsewhere.

Table 170

Content of organic matter in various deep-sea sediments

Sediment	Content of organic carbon
Grey hemipelagic clays	1—2% org. C and sometimes much more
Black clays of basins with stagnant water	3—10% org. C
Blue and red muds	0·2—0·8% org. C
Globigerina ooze	0·2—1·0% org. C
Diatom ooze	0·5—1·2% org. C
Brown clay and radiolarian ooze	traces

The amount of organic matter in Recent deep-sea deposits is affected by these factors:

1. Organic production in the upper part of the water masses. Organic production is most effective at the sites of ascending currents, at the higher geographical latitudes and other places of an impetuous development of phytoplankton. The decomposition of organic matter and its decrease are, however, so great that they can reduce these originally high amounts to a minimum.

2. Decomposition of organic matter in the water and the sediment. The decomposition is most rapid in warm water and oxidizing environment. This factor is of great importance because, as mentioned above, only a negligible percentage of primarily produced organic matter is preserved in the sediment.

3. Dilution of organic matter by terrigenous detritus and carbonate. This factor is relevant for the creation of regional differences in the content of organic matter. Thus, for instance, parts of the oceans with Globigerina ooze have generally a lower content of organic matter than the areas with brown clay. The influence of terrigenous supply is perceptible mainly in the variable content of organic matter in hemipelagic sediments.

4. The character of the sediment. This factor is connected with the above-mentioned possibility of the decomposition of organic matter. Fine-grained sediments are usually richer in organic matter than the coarser-grained ones, as a result of their slower decomposition in clays and the concomitant deposition of organic matter and clay.

5. The rate of sedimentation. This factor has two contradictory effects. When sedimentation occurs slowly, the organic matter can concentrate but is exposed to decomposition for a longer time, so that a sediment poor in organic matter usually develops. On the other hand, during rapid sedimentation organic matter is masked by terrigenous material to such a degree that the resulting deposits in it are poor. Therefore, the most suitable rate of sedimentation for the preservation of as large an amount of organic matter as possible seems to be about 5 mm in 1,000 years.

Sedimentation and preservation of organic matter are affected by the sum of all above factors, each of which can be fundamental in some areas and subordinate in others.

Sedimentation of iron and manganese in deep – sea sediments

The increased concentration of iron and manganese in deep-sea sediments was known already in the last century. Whereas the explanation of the concentration of iron does not present any difficulties, the interpretation of the anomalous concentrations of manganese is not unanimously accepted.

Iron requires mainly terrigenous supply, being concentrated also by halmyrolytic processes in sediments deposited at a low rate. Manganese concentration in deep-sea sediments reaches seven times the value of iron concentration. For their interpretation four main theories have been developed:

1. Theory of a volcanic source of manganese.

2. The assumption of an increase of manganese concentration by the biogenic adsorption on foraminiferal tests. After the dissolution of calcite, concentrations of manganese can increase many times.

3. The assumption of the concentration of manganese by the adsorption on clay particles. This theory has a theoretical substantiation and this mode of concentration was experimentally imitated.

4. The assumption of the enrichment during halmyrolytic processes under a perfect oxidation of deep-sea sediments.

A number of modern authors (E. Bonatti - Y. R. Nayudu 1965, G. Arrhenius 1965) is of the opinion that volcanic processes are in fact the sources of manganese, both submarine exhalations and decay products of volcanic material on the sea bottom. Local differences in the content of manganese can be interpreted in terms of a different dilution of manganese compounds by terrigenous material. The precipitation of Mn-compounds was explained either by inorganic processes or by organic, especially bacterial activity. G. Arrhenius (1965) suggests specifically that manganese is removed from the bottom waters by catalytic oxydation of manganous ion by colloidal ferric hydroxide at the sediment-water interface. In support of the biotic transfer it has been demonstrated the presence of organic matter in the nodules. It has also been proved that the nodules contain bacteria capable of reducing manganese; it is difficult at the present time to evaluate the biotic hypothesis against an inorganic one.

Individualized concentrations of manganese, so-called manganese nodules, are plentiful in all deep-sea sediments deposited at a low rate. They form either typical

Table 171

Amount of manganese nodules in various kinds of deep-sea sediments (after R. S. Dietz 1955)

Type of sediment	Number of samples	Average depth m	Number of samples with Mn nodules
Globigerina ooze	772	3,700	43
Blue mud	342	2,250	12
Brown clay	126	5,410	54
Volcanic mud	102	1,490	1
Pteropod ooze	40	2,610	6
Green mud	17	903	0
Coral mud	9	2,020	0
Diatom ooze	8	3,840	0
Red mud	8	1,170	0
Calcareous sand	2	775	0

nodules from micronodules to 10 cm large, or irregular crusts and concentrations. According to H. W. MENARD (in G. ARRHENIUS 1963), approximately 10% of the bottom surface of the pelagic Pacific is covered by manganese nodules. New data from the Atlantic corroborate this wide distribution. R. S. DIETZ (1955) records the following amounts of manganese nodules in various kinds of deep-sea sediments (Table 171).

In places, the concentration of manganese nodules is amazing. Thus, for instance, at a depth of 6,000 m at some places in the Pacific, up to 200 nodules were ascertained in an area of 2 m^2. The nodules originated obviously as films round small cores, whose substance was later replaced metasomatically. Some nodules contain plenty of organic remains, frequently of Tertiary age. The rate of growth of manganese nodules is very low, approximately 1 mm in 1,000 years, which shows clearly an indirect relation between the number of nodules and the rate of sedimentation of the adjacent deposits (e. g. D. S. CRONAN - TOOMS J. S. 1968).

Recent investigation has shown that the mineral composition of manganese nodules is very complicated. The nodules consist of intimately intergrown crystallites of different minerals, particularly opal, goethite, rutile, barite and nontronite. The manganese components are represented by MnO_2 and manganite. The manganite layers are composed of MnO_2 lamina and a lamina formed of hydrated $Mn(OH)_2$ or $Fe(OH)_2$. The Fe/Mn ratio varies between 0·26 and 1·04 depending on the amount of goethite. Manganese nodules are enriched in a number of trace elements. According to L. H. AHRENS *et al.* (1967) the average composition of manganese nodules is as follows:

Mn	21·6 %	Zn	0·06%
Fe	11·6 %	Pb	0·10%
Co	0·3 %	Cd	9 p.p.m.
Ni	0·62%	Bi	7 p.p.m.
Cu	0·20%	Tl	140 p.p.m.

Sn < 2 p.p.m.

Distinct relationship between the contents of Cd and Zn has been established.

Chemical composition of deep-sea sediments

Deep-sea sediments lend themselves well for the study of some chemical processes active during sedimentation and of the behaviour of individual elements in the course of the sedimentary process. It is true both of essential and trace elements.

F. W. CLARKE (1924) presents the average chemical composition of the main kinds of deep-sea sediments as follows (Table 172). On the basis of data collected from different sources (e. g. L. H. AHRENS 1968) Table 173 has been compiled to give the

Table 172

Average chemical composition of main types of deep-sea sediments (after F. W. Clarke 1924)

	Red clay %	Radiolarian ooze %	Diatom ooze %	Globigerina ooze %	Pteropod ooze %
SiO_2	62·10	56·00	67·92	31·71	3·65
Al_2O_3	16·06	10·52	0·55	11·10	0·80
Fe_2O_3	11·83	14·99	0·39	7·03	3·06
MnO_2	0·55	3·23		tr	
CaO	0·28	0·39			
MgO	0·50	0·25			
$Ca_3P_2O_5$	0·19	1·39	0·41	2·80	2·44

Table 173

Content of trace elements in deep-sea sediments

	Hemipelagic sediments	Globigerina ooze	Radiolarian ooze	Mn nodules	Diatom ooze	Brown clay
	p.p.m.					
Li				55		78
Rb						391
Ca						13
Sr						60
Ba		180	180			200
B	16—155	155				
Sc		4·6	3	3		4
Zr						140
Mn						1,770
Ni						253
Cu				3,000		160
Y	0	8	0	8		8
Ag				0·2		
Ga_2O_3		50	100	50	50	1,000
GeO_2						50
Se				19		
Cd		4	4	51—84	4	4

average content of trace elements in deep-sea sediments. D. S. CRONAN (1969) found some inter-element correlations. In ordinary pelagic muds these are between V—Cr—Ti and Mn—Ni—Mo. In manganese nodules between Mn—Cu—Co, Co—Pb, Fe—Ti and Cr-detrital constituents.

According to K. H. WEDEPOHL (1958), the average content of individual elements in all deep-sea sediments is as follows (Table 174).

Table 174

Average content of elements in all deep-sea sediments (after K. Wedepohl 1958)

	%		%
Si	23·0	Ni	0·032
Al	9·2	Cu	0·074
Fe	6·5	Cr	0·0093
Ti	0·73	V	0·045
Mg	2·1	Pb	0·015
Ca	2·9	Mo	0·0045
Na	4·0	Zr	0·018
K	2·5	Yb	0·0021
Sr	0·071	Y	0·015
Ba	0·39	La	0·014
B	0·03	Sc	0·025
Mn	1·25	Co	0·016
Ga	0·0019		

The behaviour of elements in deep-sea sediments is of fundamental importance for the solution of their geochemical equilibrium. Therefore, we shall mention here briefly the régime of a few of the most important elements.

Mg, Ca and Sr — the Mg concentration increases with the diminishing of grain size, because this element is mostly bonded to clay minerals. In slowly settling deposits, large amounts of Ca are concentrated in phillipsite and biogenic apatite. In biogenic sediments Ca concentrates at a greater rate than Mg and, consequently, the Ca/Mg ratio rapidly increases. Strontium is concentrated mainly in the apatite phase of fish scales, attaining up to 0·2% of Sr. High concentrations of this element are also in phillipsite and in Fe- and Mn-oxides. In pure clay fraction there is roughly only 0·05% of strontium.

Ba — In some sediments of minor regional distribution, as for instance in the equatorial Pacific, this element is strongly concentrated. A positive correlation established between Ba and Mg suggests that Ba was probably precipitated also as a carbonate. The amount of barium is likely to increase proportionately with organic production. Surprisingly enough, a post-sedimentary migration of Ba has been evidenced. It is concentrated together with Ra in excrements of benthonic organisms.

B — forms anomalously high concentrations in all marine sediments. Approximately 10 per cent of the total amount of boron can be removed with the exchange of sorbed ions. The remaining boron is firmly bonded to authigenic minerals and probably replaces part of Si in the tetrahedra of clay minerals.

Y, Sc, Th and rare earths — are concentrated in biogenic apatite and in mafic Fe and Mn minerals.

Table 175

Decrease in the amount of certain elements away from the source of volcanic material (Hawaiian Islands)
(after G. Arrhenius 1954)

Growing distance away from Hawaiian Islands in miles	1,000 Ti/Al	1,000 V/Al	1,000 Cr/Al
26	270	11	4·5
100	91	4·2	1·0
137	77	3·9	1·1
188	66	3·3	0·87
1,820	50	4·3	0·01

Ti — is represented in deep-sea sediments in roughly the same amount as in igneous rocks. Higher concentrations are always caused by the presence of basic volcanic material. A considerable decrease in the amount of Ti (and other elements) is observable away from the source of volcanic material, in this case from the Hawaiian Islands (Table 175). In samples lacking pyroclastic material there is less than 0·8 TiO_2. G. Arrhenius (1954) attempted to determine the rate of sedimentation from the TiO_2 content in deep-sea sediments.

Different concentrations of titanium indicate different rates of sedimentation.

Zr — is scarcer in deep-sea sediments than in igneous rocks. The stability of this element is often overestimated; a fairly large amount of it passes into the hydrosphere. It occurs partly in authigenic Mn and Fe minerals, partly in the finest colloidal clay fraction.

V — attains 2·5 times the amount contained in igneous rocks. The major part of it occurs in biogenic apatite and phillipsite, and in the finest fractions of sediments. The enrichment of coarser fractions by vanadium is due to its presence in mangetite and ilmenite. The amount of V drops with the decrease of basic volcanic material.

Cr — occurs in a variable amount (from 35 to 530 p.p.m.). It is frequent in sediments rich in basic pyroclastic material. Similarly as Ti, it is largely concentrated in coarser fractions, owing to its presence in mafic Fe and Mn minerals. In places, increased concentrations of Cr have been observed in montmorillonites originated obviously by a decomposition of basic glasses. The Cr content above 0·01% is a good indicator of the presence of unaltered volcanic material.

Ni — deep-sea sediments are appreciably enriched in nickel; anomalously high concentrations are particularly in manganese nodules. Besides, it is also confined to the finest fractions, which reveals a direct relationship between Fe and Ni, and Mn and Ni. The opinion prevails that Ni derives from the following sources: 1. Supply from basic rocks of the dry land, 2. Precipitation from sea water, 3. Submarine volcanism, 4. Cosmic material. The greatest concentration occurs most likely as a result of adsorption on manganese gels. The increased content of Ni may also occur in sediments with a larger amount of organic matter.

Co — as a direct relation between Fe and Co concentration has been established, it is likely that cobalt is absorbed on Fe gels and deposited along with themi.

Cu — from a direct relationship between Mn and Cu concentration copper is inferred to be appreciably concentrated in manganese concretions. As it also constitutes abundant organic complexes, it also occurs in sediments with an increased amount of organic matter.

Cd — the content of Cd increases with the rising percentage of organic matter in the sediment. It has been proved experimentally that considerable amounts of Cd are precipitated simultaneously with calcium carbonate. A relatively large amount of cadmium concentrates in manganese nodules.

The content of a number of trace elements in Recent deep-sea sediments is largely dependent on the distribution of manganese nodules, phosphate sediments and deposits rich in organic matter. In phosphate concretions mainly Zn, Cd, In and Bi are concentrated, whereas V, Mo, Cu and Ni accumulate in bituminous deposits. The amount of trace elements is additionally affected by the mixing with terrigenous material and by the deposition of calcareous and siliceous components of sediments.

On the basis of their source and their history in deep-sea sediments the elements can be divided into three groups:

1. Elements imported into the sea from the weathering products of the landmass.
2. Elements of primarily marine origin, concentrated mainly by organisms.
3. Elements of mixed origin.

The study of elements in deep-sea sediments is of special importance for the establishment of their geochemical cycle and equilibrium. It is because deep-sea sediments distributed over an immense surface can contribute to the explanation of many anomalies in the concentration of elements in igneous and sedimentary rocks, as well as in the relation to their mean content in the upper lithosphere. The elements which concentrate in deep-sea sediments are mainly Mn and Pb and, according to the results of recent investigations, also Ni, Co and several other elements.

The question of the equilibrium of Ca between igneous and sedimentary rocks is extremely interesting and so far unsolved. It has been found that in deep-sea sediments calcium is so abundant that it is considerably in excess of continental sources — igneous rocks. In the opinion of some authors, this anomaly can be explained only by assuming that the earlier deep-sea sediments were substantially poorer in calcium.

Differences between the present and earlier deep-sea sedimentation are the object of intensive study. Thus, for example, the ratio of elements indicates the differences between the content of certain components in Tertiary and Quarternary deep-sea deposits (Table 176, after G. ARRHENIUS 1959).

Table 176

	Ratio average content of element in Tertiary deep-sea sediments/ average content in Recent deep-sea sediments
SiO_2	0·97
Al_2O_3	0·83
Fe_2O_3	1·3
TiO_2	0·69
MgO	0·96
CaO	1·41
Na_2O	1·20
K_2O	0·89
SrO	1·6
BaO	1·5
B_2O_3	0·93
MnO	2·9
Ga_2O_3	0·70
NiO	1·2
CuO	1·1
Co_2O_3	0·50
Cr_2O_3	0·44
V_2O_5	1·11
PbO	1·3

Regional distribution of deep-sea sediments

The distribution of the principal deep-sea sediments is comparatively well known. The differences in the maps published are due to the use of various classification schemes. The distribution of deep-sea sediments in the oceans can to a certain extent be inferred from the laws controlling deep-sea sedimentation. The decisive factors are as follows:

1. The supply of terrigenous material by rivers, turbidity currents, traction bottom-currents, floating ice and wind.
2. Organic production of calcareous and siliceous material,
3. Dissolution of carbonate in water and sediment.
4. Supply of pyroclastic and extratelluric material.

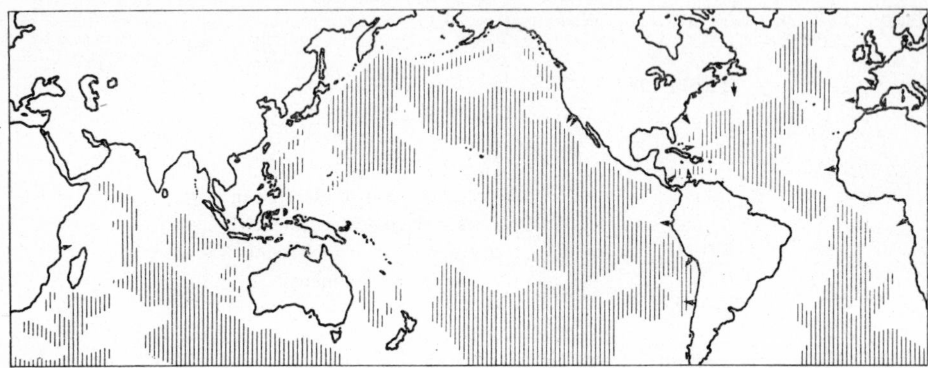

Fig. 99. Map illustrating areas inaccessible to shallow-water turbidity currents. Arrows denote one or more historical turbidity currents (since 1880). With the exception of two off southern California, all are based on submarine cable breaks. After B. C. Heezen (in M. N. Hill 1963).

As all these factors are not equivalent, their interrelationship must be taken into consideration.

The division of deep-sea sediments into terrigenous (hemipelagic) and pelagic (eupelagic) is based on the supply of coarser-grained terrigenous detritus.

On the continental slopes the deposition generally occurs at a maximum rate. In most of these areas the production of phytoplankton or zooplankton is also quite voluminous. Although neither intensive dissolution of carbonate nor other reduction of biological content has been observed, hemipelagic sediments usually contain a lower percentage of $CaCO_3$. This can only be interpreted in terms of a more rapid deposition of terrigenous detritus than of the biological components which in this way becomes masked. Towards the open oceans the biological components gradually increase at the expense of the terrigenous components. In sections drawn across the entire basins the increasing rate of sedimentation is traceable towards the dry lands (e. g. in the section across the Northern Atlantic a several centimetres thick layer of eupelagic sediments corresponds to a more than 1 m thick bed of hemipelagic sediments — blue mud). It has also been proved that, for instance, Globigerina ooze with a terrigenous or volcanic admixture is deposited at a greater rate than pure ooze without any admixture. These facts unquestionably prove that the presence or absence of large terrigenous supply is decisive for deep-sea sedimentation.

A continuous belt of hemipelagic sediments encircles all continents. In deep-sea troughs they reach to depths below 5,000 m and some inland seas, such as the Mediterranean, are filled completely with them. The width of the band of hemipelagic deposits depends on many factors, such as the climate (in the higher geographical altitudes the belt is wider) and the topography of the continental slope (as a result of particular topography of the continental slope on the western coast of South and North America, terrigenous material cannot be transported farther into the basin). The main factors seems to be cold climate, the presence of glaciers and floating ice which take clastic material into abyssal regions.

The rate of deposition of terrigenous material is comparable only with the organogenic sedimentation conditioned by an immense organic production, such as is the population of diatoms and other phytoplankton in places of ascending currents and cold waters. Therefore, in some places the depositional rate of diatom oozes can equal the rate of sedimentation of hemipelagic deposits, and diatom oozes with an almost invariably coarser terrigenous admixture can develop (indicating that the deposition of terrigenous detritus was also rapid). This has been observed not only in the belt round polar regions but also in basins near the Pacific coast of North America. The rate of accumulation of other plankton cannot keep up with that of the sedimentation of terrigenous detritus. Therefore, it constitutes sediments only in abyssal regions that are out of reach of terrigenous supply. The development of individual kinds of sediments is then connected with the relationships among the individual plankton groups and with the possibility of dissolution of carbonates in sediments.

Zooplankton is widespread particularly in upper-water layers of normal salinity and higher temperature. As foraminifera are nearly everywhere more abundant than radiolaria, it must be presumed that, should all the carbonate be preserved in sediments, Globigerina ooze or its varieties would develop. As a result of dissolution of carbonate, brown clay or Radiolarian ooze are formed. The sedimentation of Radiolarian ooze in the equatorial Pacific and elsewhere seems to be caused both by slower deposition of clays and a higher production of siliceous zooplankton.

From this short review, the distribution of individual kinds of eupelagic sediments in world oceans and the relative rates of their sedimentation can be deduced. Hemipelagic (terrigenous) sediments are laid down at the greatest rate, being followed by diatom oozes. The rate of Globigerina ooze sedimentation is much lower and that of Radiolarian ooze and brown clay is lowest; the last named are in fact insoluble residue of calcareous oozes.

References

AHRENS L. H. (Editor, 1968): Origin and distribution of elements. P. 1—1178, Oxford.

AHRENS L. H. - WILLIS J. P. - OOSTHUIZEN C. O. (1967): Further observations on the composition of manganese nodules. Geochim. Cosmochim. Acta, vol. 31: 2169—2180, London.

ARRHENIUS G. (1954): Origin and accumulation of alumosilicates in the ocean. Tellus, vol. 3: 215—220, Stockholm.

— (1959): Sedimentation on the ocean floor. Research in Geochemistry, p. 1—24, New York.

— (1963): Pelagic sediments. In "The Sea". Vol. 3: 655—728, London.

A-ZEN E. (1960): Carbonate equilibria in the open ocean and their bearing on the interpretation of ancient carbonate rocks. Geochim. Cosmochim. Acta, col. 18: 57—71, London.

BERRIT C. R. (1955): Etude des teneurs en manganese et en carbonates de quelques carottes sédimentaires Atlantiques et Pacifiques. Med. Oceanogr. Inst. Göteborg, vol. 23-61, Göteborg.

BEZRUKOV P. (1959): Sediments of trenches in the North-Western Pacific. Eclog. Geol. Helv., vol. 51: 500—505, Basel.

BEZRUKOV P. L. - LISICYN A. P. - SKORNIAKOVA N. S. (1961): The map of distribution of recent sediments in oceans (in Russian). In Recent Sediments of Sea and Oceans, 73—85, Moscow.

BONATTI E. - NAYUDU Y. R. (1965): The origin of manganese nodules of the ocean floor. Am. Jour. Sci., vol. 263: 17—39, New Haven.

BOURCART J. (1964): Les sables profonds de la Méditerranée occidentale. In "Turbidites". P. 148—155, Amsterdam.

BRUUN A. F. (1957): Deep sea and abyssal depths. Geol. Soc. Am. Mem. 67; 641—672, Baltimore.

CHAVE K. E. - MACKENZIE E. T. (1961): A statistical technique applied to the geochemistry of pelagic muds. Jour. Geology, vol. 69: 572—582, Chicago.

CHOW T. J. - PATTERSON C. C. (1962): The occurrence and significance of lead isotopes in pelagic sediments. Geochim. Cosmochim. Acta, vol. 26: 263—308, London.

CLARKE F. W. (1924): The data of geochemistry. US Geol. Surv. Bull. 770: 1—871, Washington.

CONOLLY J. R. - EWING M. (1967): Sedimentation in the Puerto Rico trench. Jour. Sedimentary Petrology, vol. 37: 44—59, Menasha.

CORRENS C. W. (1939): Pelagic sediments in North Atlantic Ocean. In Recent Marine Sediments. P. 48—135, Tulsa.

CRONAN D. S. (1969): Inter-element associations in some pelagic deposits. Chemical Geology, vol. 5: 99—106, Amsterdam.

CRONAN D. S. - TOOMS J. S. (1968): A microscopic and electron-probe study of manganese nodules from the northwest Indian Ocean. Deep Sea Res., vol. 15: 215—233, London.

DAVIES D. K. (1968): Carbonate turbidites, Gulf of Mexico. Jour. Sedimentary Petrology, vol. 38: 1100—1109, Menasha.

DIETZ R. S. (1955): Manganeous nodules on ocean floor. California Jour. Geol. Mines, vol. 33: 218—228, Berkeley.

DILL R. F. (1964): Features in the heads of submarine canyons. Narrative of underwater film. Deltaic and shallow marine deposits, p. 101—104, Amsterdam.

— (1965): Bathyscaph observations in the La Jolla submarine fan valley. Am. Bull. Assoc. Petrol. Geol., vol. 49: 338, Abstr., Tulsa.

ELWAKEEL S. K. - RILEY J. P. (1961): Chemical and mineralogical studies of deep-sea sediments. Geochim. Cosmochim. Acta, vol. 25: 110—146, London.

ERICSON D. B. - EWING G. - WOLLIN G. - HEEZEN B. (1961): Atlantic deep-sea sediment cores. Bull. Geol. Soc. Am., vol. 72: 193—286, Baltimore.

EWING M. - EWING J. (1965): Distribution of sediments in the world ocean. Bull. Am. Assoc. Petrol. Geol., vol. 49: 339 Abstr., Tulsa.

GOLDBERG E. D. - ARRHENIUS G. (1958): Chemistry of Pacific pelagic sediments. Geochim. Acta, vol. 13: 153—212, London.

GOLDBERG E. D. - KOIDE M. (1962): Geochronological studies of deep-sea sediments by ionium-thorium method. Geochim. Cosmochim. Acta, vol, 26: 417—450, London.

GOLDBERG E. D. - PARKER R. (1960): Phosphatized wood from the Pacific Sea floor. Bull. Geol. Soc. Am., vol. 71: 631—632, Baltimore.

HEEZEN B. C. (1959): Modern turbidity currents. Eclog. Geol. Helv., vol. 51: 521—522. Basel.

— (1965): Abyssal basin sedimentation. Bull. Am. Assoc. Petrol. Geol., vol. 49: 344, Abstr., Tulsa.

HEEZEN B. C. - HOLLISTER C. (1964): Turbidity currents and glaciation. In "Problems in Palaeclimatology". P. 99—109, London.

HEEZEN B. C. - MENZIES R. J. - SCHNEIDER E. D. - EWING W. M. - GRANELLI N. C. (1964): Congo submarine canyon. Bull. Am. Assoc. Petrol. Geol., vol. 48: 1126—1149, Tulsa.

HUBERT J. F. (1964): Textural evidence for deposition of many western North Atlantic deep-sea sands by ocean bottom currents rather than turbidity currents. Jour. Geology, vol. 72: 757—785, Chicago.

KRASINCEVA V. V. - SHISHKINA O. V. (1959): Problem of boron distribution in marine sediments (in Russian). Doklady AN SSSR, vol. 128: 572—582, Moscow.

KUENEN PH. H. (1950): Marine geology. 1—568, New York.

— (1951): Properties of turbidity currents of high density. Soc. Econ. Pal. and Min., Spec. Publ. 2: 14—33, Tulsa.

— (1964): Deep-sea sands and ancient turbidites. In "Turbidites". P. 3—33, Amsterdam.

— (1967): Emplacement of flysch-type sand beds. Sedimentology, vol. 9: 203—243, Amsterdam.

KUKAL Z. (1960): Deep-sea sediments in the light of modern researches (in Czech). Knihovna ÚÚG, vol. 35: 1—180, Prague.

LISICYN A. P. (1961): Distribution of suspended sediments in seas and oceans (in Russian). In "Recent Sediments of Seas and Oceans". P. 124—174, Moscow.

MELLIS O. (1960): Gesteinsfragmente im Roten Tone des Atlantischen Ozeans. Med. Oceanografiska Inst. Göteborg, vol. 28: 1—17.

MENARD H. W. (1953): Pleistocene and Recent sediments from the floor of the Northeastern Pacific Ocean. Bull. Geol. Soc. Am., vol. 64: 1278—1294, Baltimore.

— (1964): Marine geology of the Pacific. P. 1—217, New York.

MERO J. L. (1962): Ocean floor manganese nodules. Economic Geology, vol. 57: 747—767, Lancaster.

— (1965): The mineral resources of the sea. P. 1—312, Amsterdam.

MOORE D. J. (1961): Submarine slumps. Jour. Sedimentary Petrology, vol. 31: 343—357, Menasha.

MURDMAA I. O. (1961): Recent sediments and composition of suspended material in the region of Kurile Archipelago (in Russian). In "Recent Sediments of Seas and Oceans". P. 404—418, Moscow.

MURRAY J. - CHUMLEY J. (1924): The deep sea deposits of Atlantic Ocean. Trans. Roy. Soc. Edinburgh, vol. 54: 1—252, Edinburgh.

MURRAY J. - RENARD A. F. (1891): Report on deep sea deposits based on specimens collected during the voyage of HMS Challenger in the years 1872—1876. Challenger Reports, p. 1—525, London.

NAYUDU Y. R. (1962): Origin and distribution of deep-sea sand-silt layers in Northeastern Pacific. Bull. Am. Assoc. Petrol. Geol., vol. 46: 273—274, Tulsa.

NESTEROFF W. D. - HEEZEN B. C. (1962): Essais de comparaison entre les turbidites modernes et le flysch. Revue de Géographie Physique et Géologie Dynamique, vol. 5; 115—127, Paris.

NORIN E. (1958): Die Sedimente des zentralen Tyrrhenischen Meeres. Geol. Rundschau, Bd. 47: 207—217, Stuttgart.

OSTROUMOV E. A. - SHILOV V. M. (1956): Sulphuric iron and hydrogen sulphide in the bottom deposits of Northwestern Pacific Ocean (in Russian). Doklady AN SSSR, vol. 106: 501—504. Moscow.

PETTERSON M. N. - GRIFFIN J. J. (1964): Volcanism and clay minerals in the Southeastern Pacific. Jour. Marine Research, vol. 22: 13—21.

PRATT R. N. (1968): Atlantic continental shelf and slope of the United States. US Geol. Surv. Prof. Pap. 529 B: 1—44, Washington.

RAD U. von (1968): Comparison of sedimentation in the Bavarian flysch (Cretaceous) and Recent San Diego Trough (California). Jour. Sedimentary Petrology, vol. 38: 1120—1154, Menasha.

REVELLE R. (1944): Marine bottom samples collected in the Pacific ocean by the Carnegie on its Seventh cruise. Carneg. Inst. Publ. 556: 1—179, Washington.

RICHARDS A. F. (1958): Transpacific detection of Moyoyn Volcano. Bull. Geol. Soc. Am., vol. 39: 818—831, Baltimore.

RODOLFO K. S. (1967): Sedimentation in Andaman Basin, Northeastern Indian ocean. Bull. Am. Assoc. Petrol. Geol., vol. 51: 479, Tulsa.

SCHOTT G. (1944): Geographie des Atlantischen Ozeans. P. 1—438, Hamburg.

SEARS M. (1961): Oceanography. P. 345—366, Washington.

SHEPARD F. P. - DILL R. F. - RAD U. von (1969): Physiography and sedimentary processes of La Jolla submarine fan and fan-valley. Bull. Am. Assoc. Petrol. Geol., vol. 53: 390—420, Tulsa.

SMITH S. V. - DYGAS J. A. - CHAYES K. E. (1968): Distribution of calcium carbonate in pelagic sediments. Marine Geology, vol. 6: 391—400, Amsterdam.

SUZUKI K. - KITAZAKI W. (1954): Mineralogical studies on the red clays from the Western Pacific Ocean, East of Bonin Island. Japan. Jour. Geol. Geograph., vol. 24: 171—180, Tokyo.

STRAATEN L. M. J. U. VAN (1964): Turbidite sediments in the southeastern Adriatic Sea. In "Turbidites". P. 142—147, Amsterdam.

SVERDRUP H. U. - JOHNSON M. W. - FLEMING R. H. (1942): The oceans. P. 1—1060, New York.

TRASK P. D. (1936): The relationship of salinity to the calcium carbonate content of marine sediments. Prof. Papers, 186 N, p. 273—300, Washington.

— (1961): Sedimentation in a modern geosyncline off the arid coast of Peru and Northern Chile. Internat. Geol. Congr., 21. Sess., P. 23: 103—118, København.

WALKER R. G. (1967): Turbidite sedimentary structures and their relationship to proximal and distal depositional environments. Jour. Sedimentary Petrology, vol. 37: 25—43, Menasha.

WEDEPOHL K. H. (1958): Comparison of the deep sea and nearshore clays with reference to some minor elements. Geochim. Cosmochim. Acta, vol. 14: 166, Abstr., London.

— (1960): Spurenanalytische Untersuchungen an Tiefseetonen aus dem Atlantik. Geochim. Cosmochim. Acta, vol. 18: 200—232, London.

20. Shallow-water carbonate sediments

Shallow-water marine carbonate sedimentation occurs 1. on coasts 2. on shelves and 3. on coral reefs.

Littoral carbonate sediments

Some kinds of carbonate beach sediments, especially accumulations of tests on beaches, are described in the foregoing chapters. There are three prerequisities for the accumulation of organic remains on beaches and other parts of the shores: a small supply of terrigenous sand, suitable conditions for the growth of organisms in the environs, and favourable hydrodynamical conditions. Carbonate beach sediments occur in abundance in tropical, subtropical and temperate climatic zones, in places even above 60°N geographical latitude. They consist mostly of the shells of the pelecypods, gastropods and some crustaceans. Many occurrences of sediments containing more than 50% carbonate minerals indicate that carbonate sediments can be deposited on almost any continental shelf, regardless of latitude. Organisms which deposit carbonate skeletons are present in shallow water almost everywhere and these will form carbonate sediments where terrigenous clastics do not dilute them too much (K. E. Chave 1967).

The group of carbonate beach sediments are sometimes cemented to form the so-called beachrock, i. e. solid, mostly calcareous sand-grade rock or terrigenous sediment, perfectly cemented by calcium-carbonate. It occurs near the water surface, most frequently between the mean high- and low- tide levels, and is often found associated with coral reefs of the marginal parts of shallow-water areas. Beachrock is widely distributed: it extends for a maximum distance of 1,600 km along the Atlantic coast of Brazil. There it is composed of detrital particles of sand — gravel grade, covered with crust of radial aragonite. The mean porosity of this sediment is 34%. There are different opinions as to the origin of the beachrock, but the genesis as given below seems most plausible. The alternation of photosynthesis and respiration of phytoplankton results in the enrichment of upper-water layers by carbon dioxide and, consequently, in the changes of pH and solubility of calcium carbonate. The alternation of these conditions also induces an intensive dissolution and precipitation of $CaCO_3$ in the intergranular interstices, so that in a short time the rock becomes well cemented. An analysis of O^{18} isotopes has shown that $CaCO_3$ is precipitated only from saline waters. Thus, the development of "beachrock" requires increased temperature, higher salinity and the presence of a source of carbonate material, mainly fragments of older limestones. This mode of genesis is

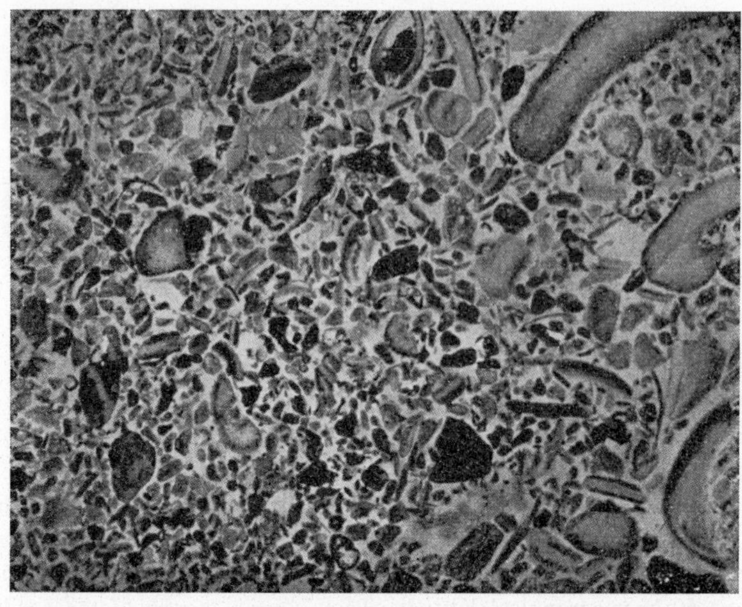

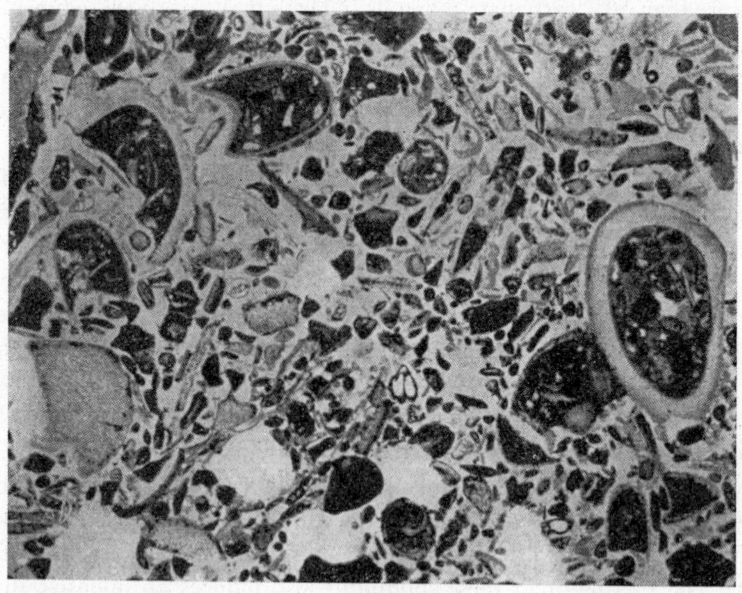

Photo 11. Shallow-water carbonate sediments of the Persian Gulf. Upper photo: Calcarenite composed of rounded fragments of corals, bryozoans and pelecypod shells. Lower photo: Calcarenite composed of angular fragments of shells. (X 7). After J. H. C. Houbolt (1959).

also proved by the observation that during the day (as a result of photosynthesis) water in the pores and in the environs of calcareous sediments is oversaturated with calcium carbonate up to 800%. The beachrock often develops from very coarse detritus and its cavities of irregular shape are filled with calcite of concentric structure. Such cavities are usually regarded as a typical feature of reef sediments and some fossil beachrocks, after their structure has been somewhat affected, can be mistaken for part of a coral reef.

Carbonate sediments of shallow basins and shallow-water areas

Shallow-water detrital carbonate sediments are very widespread on some shelves (Bahama, Florida), in shallow bays and small basins (Persian Gulf, Red Sea and others) and around coral reefs. Their origin is predominantly biological and only in part inorganic; they are virtually the detritus of free grains, the grain-size reflecting the predominating hydrodynamical conditions. On the margins of shelves and at shallow depths they are coarser than in the depressions of flats, in current shadows or a greater depths. In shallow-water detrital carbonate the following components can be differentiated:

1. Organic remains or their fragments.
2. Grains generally originated from several minor intergrown aragonite individuals.
3. Aragonite ooids.
4. Aragonite mud, of biological (algal) or inorganic origin.
5. Faecal pellets.
6. Fragments of older, but Recent or maximally Subrecent carbonate rocks.

Detailed data on the relation between the depth and distribution of individual kinds of carbonate sediments have been gathered from the Persian Gulf (J. H. C. Houbolt 1959). Calcareous mud is deposited only at depths below 35—50 m. Moreover, two different kinds of calcareous sands (calcarenites) confined to definite depth zones are distinguishable. Calcarenites with a prevalence of rounded grains occur at depths smaller than 18 m. They are composed of rounded fragments of shells, bryozoans and corals. Calcarenites made up of unrounded shell fragments are common between 15 and 30—50 m. These values give the maximum depths, because in sheltered areas the transition from rounded to non-rounded calcarenites occur already at a depth of a few metres. Many special types of carbonate sediments have been recently described by various authors (e. g. H. Füchtbauer, Editor, 1969).

The typical and best investigated shallow-water carbonate area is the Great Bahama Bank covering more than 100,000 km^2. It is separated from the neighbouring continent by deep trenches, that trap any terrigenous material, so that this material does not contaminate sediments of the bank. The Great Bahama Bank is of subdued relief; huge submarine dunes of calcareous sand are the only major unevennesses.

Photo 12. Aerial view on a part of the shallow-water Bahama platform. After N. D. Newell *et al.* (1956).

The overall bank is several centimetres to a few under the low tide level, except for some parts which rise above the water. Because of considerable speed of tidal currents sweeping over the large flat area the deposition of carbonate mud is confined to depressions and current shadows. The major part of this area is covered with sediments of sand and silt grade. On the basis of numerous mechanical analyses the following parameters have been determined:

	Range	Average
Md	0·02—0·70 mm	0·08 mm
So	1·20—3·20 mm	1·50 mm

The coarsest sediments of the Md 0·5—0·7 mm occur at the margin of the banks in association with a plenty of eolian sands. The amount of biological particles in the sediments is variable; fragments of calcareous algae (*Lithophyllum*, *Goniolithon*, *Halimeda*), larger fragments of pelecypod shells, a minor amount of radiolarian tests, fairly abundant pieces of bryozoans, corals, crustaceans and calcareous tubes of worms are the main components. Grains in various stages of aggregation represent mineral particles of sand and silt fractions. The irregular grains, sometimes even in semiplastic state, are formed by the binding together of minute aragonite granules. After a grain has attained a certain size, further aragonite needles or grains adhere to its surface. In addition to aragonite, organic matter plays an important role. When

Photo 13. Microphotos of some Bahamian sediments. 1. Faecal pellet sediment, 2. Grapestone, i. e. carbonate sediment formed by accretion of minute aragonite grains, 3. Oolite. X 10. After N. D. Newell *et al.* (1956).

the size of grains corresponds to the hydrodynamical conditions of the environment, they can settle, be laid down one upon the other, and become cemented into the above-described beachrock. That is the fundamental difference between the components of Bahamian sediments and other carbonate deposits of detrital origin.

The fine fractions of the sediment are composed of minute aragonite needles, the presence of which has long been regarded as an evidence of chemical origin. New investigation (particularly H. A. LOWESTAM and S. EPSTEIN 1957) and analysis of O^{18} have shown that the needles may be decay product of calcification of calcareous algae. Soon after deposition and after being covered with younger sediments, they loose their acicular texture as a result of a slight compression. Algae can produce fine aragonite mud in great quantity, as proved by K. W. STOCKMAN *et al.* (1967). The rate of deposition of fine mud produced by alga *Penicillus* sp. is 1·0—2·0 cm/1,000 years.

Aragonite oolites are a frequent shallow-water sediment found in the Bahama Islands, near Florida, the Texas coast, in the Mediterranean near the Egyptian coast, in the Persian Gulf, the Red and Caspian Seas, etc. They are abundant on the margins of the Bahama platform, where they form submarine dunes, as mentioned above.

Table 177

Chemical composition of aragonite ooids of the Bahama Platform (after H. Illing 1954)

	%		%
CaO	53·8	SO_4	0·30
MgO	0·34	P_2O_5	0·014
SiO_2	0·08	K_2O	0·04
Al_2O_3	0	NaCl	0·19
Fe_2O_3	0·07	Na_2O	0·34

Depths from 1 to 3 m are most suitable for their development. The mean grain-size is 0·4—0·7 mm, depending on the intensity of currents; an oval shape is not always necessary. Table 177 gives their chemical composition. They also contain a considerable amount of organic matter, by the elementary analysis of which it has been established:

Ash	11·97%
C	42·42%
H	6·44%

The question of the influence of organic material on carbonate sedimentation still remains open. The organic matter is termed collectively peloglea and forms a sticky film on some objects in water environment. Calcarenite grains are bound together by a gelatinous substance of organic origin. According to R. G. C. BATHURST (1967) this gelatinous mat is produced either by blue-green algae or by diatoms, because both groups are known to they secrete mucilages. This mat is destroyed rapidly if buried. It has probably greater significance for carbonate sedimentation than has been presumed so far.

According to O^{18}/O^{16} isotope analyses, oolites are undoubtedly of inorganic, physico-chemical origin.

The mineral composition of some shallow-water carbonate sediments from the Bahamas is as follows (Table 178).

Table 178

Mineral composition of 8 different shallow-water carbonate sediments from the Bahamas

	1. %	2. %	3. %	4. %	5. %	6. %	7. %	8. %
Aragonite	80	76	71	78	94	85	99	91
Mg-high calcite	15	17	21	16	5	11	1	7
Calcite	5	7	8	6	1	4	0	0
Dolomite	0	0	0	0	0	0	0	0

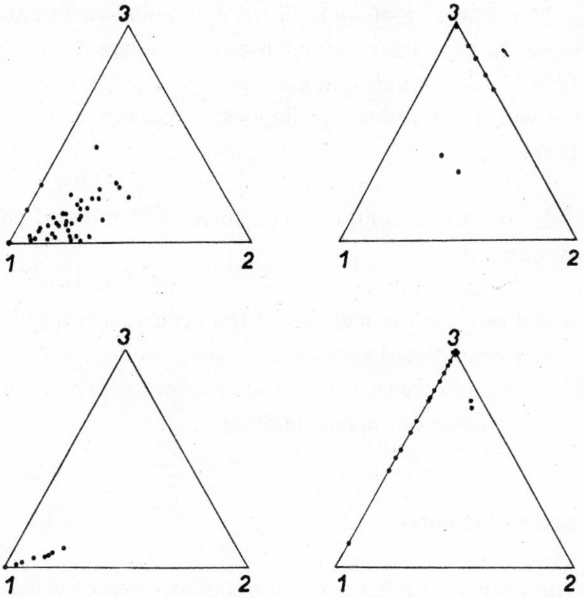

Fig. 100. Mineralogical composition of some shallow water Recent carbonate sediments: Florida (upper left triangle), China Sea and Persian Gulf (upper right), Andros Lobe of the Bahama Banks (lower left), and of Pleistocene carbonate rocks of Florida (lower right triangle). 1. Aragonite, 2. High-Mg calcite, 3. Calcite. After F. G. Stehli - J. Hower (1961).

It is remarkable that the content of aragonite rises on the average with the content of fine fractions. Calcite with Mg in solid solution is more abundant in Recent sediments than pure calcite.

The Bahamian shallow-water carbonate sediments differ from other carbonate shallow-water deposits chiefly by their non-detrital genesis. The differences are summed up in several points in Table 179.

Table 179

Differences between Bahamian shallow-water carbonate sediments and other carbonate shallow-water deposits

Bahamian sediments	Other deposits
Aragonite prevails, Mg-high calcite also abundant, dolomite absent	Mg-high and Mg-low calcites usually prevail
Many carbonate grains originated from accretion of several smaller ones	Carbonate grains originated from fragmentation of larger grains
Enormous purity of carbonates, terrigenous material almost absent	Small admixture of terrigenous material generally occurs
Sediments composed mostly of faecal pellets are present	

The character of some of the Bahamian carbonate sediments is so distinctive that some authors have coined for them a new term — the bahamites (E. W. BEALES 1958). It is probable that analogous sediments will be found among ancient deposits, for which the following diagnostic features can be postulated (after E. W. BEALES 1958):

1. High purity, carbonate content above 99%.
2. All grains conform in chemical and mineral composition, i. e. the sediment is monomictic.
3. A small amount of distinguishable tests and their fragments, which can be found only on the margins of the occurrences along with oolitic sediments.
4. Consolidated rocks are virtually non-porous.
5. Occurrence of isolated intercalations of anhydrite or gypsum.
6. Absence of current bedding.

Recent dolomites

The presence of Recent dolomites has attracted the attention of geologists, because it presented a suitable opportunity to study the genesis of these rocks. According to earlier data, only a few classic dolomites occurrences in Recent sediments have been known which, moreover, have not been safely established owing to imperfect analysis, but in the course of time and with the improvement of research methods there is a continuously growing amount of knowledge of their occurrence. Many occurrences of Recent dolomites have been registered in the course of the last decade. They are summarized by G. M. FRIEDMAN - J. E. SANDERS (1967), L. C. PRAY - R. C.

Table 180

Occurrences, composition and genesis of Recent dolomites

Localization	Composition	Suggested origin
Lake Balkash	Up to 40% of dolomite in calcareous clays, in form of fine-grained mud	Precipitation influenced by high salinity and pH (8·8—8·9), alkalic reserve and probably by greater amount of organic matter as well
Great Salt Lake	Small rhombohedras of dolomite in aragonite muds. Several cm below the surface layer of unconsolidated dolomite 11,000 years old has been found	Precipitation caused by higher salinity. In the second case diagenetic origin is probable
Atlantic Ocean	Small disseminated dolomite crystals occur in calcareous clays, generally several cm below the surface	Diagenetic origin is suggested
Peru-Chile trench	Dolomite occurs in clayey-calcareous sediments in the form of grey, mostly indurated muds	Early diagenetic origin is possible
Australian lagoons and lakes	Fine-grained dolomite in calcareous clay; mixture of Mg-calcite and Ca-dolomite $(Ca_{0\cdot77}Mg_{0\cdot23})CO_3$	Precipitation from high saline waters (pH 8·2—8·4). Dolomite was ascertained also in suspension
Small salt lakes on Danube floodplain, e. g. Krivan	Mixture of calcite and dolomite mud	?
Florida Bay	5% dolomite in aragonite mud, grain-size of dolomite 1—60 μ	Detritic origin is highly probable
Coral reefs (e. g. Eniwetok atoll)	Rhombohedra in coralline algae, in corals or other organic remains. Only in deeper parts of reefs	Late diagenetic or epigenetic origin
Deep Spring Lake, California	Calcium rich dolomite, with crystals less than 1 μ.	Early diagenetic origin
Persian Gulf	On the surface of sediments with gypsum	Primary precipitation during evaporation

Table 180 (*continued*)

Localization	Composition	Suggested origin
Lake Neusiedler (Austria)	Calcareous and dolomitic clays having: CaO 4·5–21·1% MgO 4·1–11·0% In upper layer protodolomite was ascertained, in lower layers dolomite	Diagenetic origin with origin of unstable protodolomite
Etocha Basin (South Africa)	Calcareous and dolomite mud in ephemeral lakes	Precipitation from highly saline waters
Bahamas	Dolomite originating during salt production (artificial) by evaporation. Dolomite crystals (0·14 mm) appear after 8 years together with aragonite and calcite	Diagenetic origin was proved

MURRAY (1965) and H. FÜCHTBAUER (1969). The most suitable conditions for the deposition of dolomites are as follows: 1. Increased pH value, 2. Increased alkaline reserve. 3. Increased partial pressure of CO_2, 4. Presence of a major amount of organisms with Mg-calcite in tests, 5. Predominant sedimentation of fine-grained material, 6. Probably also an absence of a larger amount of organic matter. 7. Excess of Mg ions due to gypsum precipitation. In Table 180, the known occurrences of Recent dolomites are listed, along with a concise characterization of their composition, mode of deposition, and postulated genesis.

It can be assumed that Recent dolomites will be found also in other carbonate and clay-carbonate sediments. The recognized dolomite occurrences are generally of diagenetic origin. The primary (penecontemporary) origin of dolomite has, however, been proved in hypersaline, periodically evanescent lakes and lagoons. The origin due to the excess of Mg ions caused by gypsum precipitation is mostly suggested. Although under normal marine conditions the primarily precipitation of dolomite is extremely rare, it cannot be excluded to have taken place in other geological periods.

Coral reefs

Coral reefs represent a particular sedimentary environment that has been the subject of an extensive study. Particularly in the last twenty years many detailed data on the composition, distribution and structure of coral reefs have been gathered. Less attention has been paid to the ecological and biological aspects of their origin. Works contributing substantially to the solution of this problem have appeared only recently. Many misunderstandings and erroneous interpretations of reef structures in fossil sediments call for a perfect recognition of the structure of Recent reefs. The discrepancies are largely due to the differences in the conception of reefs, bioherms and biostromes in Recent and ancient sediments. Fossil reefs and bioherms are defined first morphologically, as their size, shape and relation to the adjacent sediments is often easier traced. On the other hand, Recent reefs are defined mainly ecologically, on the basis of their relation to the sea level, and according to the amount and species of organisms. It would be most convenient if the definition of a reef implied both morphological and ecological features. H. A. LOWESTAM's (1950) definition supplemented by several other writers, seems to comply best with these conditions: *A coral reef is a body of sediments built by organisms, the upper part of which occurs near the sea level and is wave resistant.* It should fulfil two conditions: a) the skeleton of the reef should be formed of reef-building organisms which by intergrowing and outgrowing would form a solid structure resistant to strong wave agitation, b) the reef as an active organic structure should constitute in itself a particular environment which would influence the sedimentation in wide surroundings.

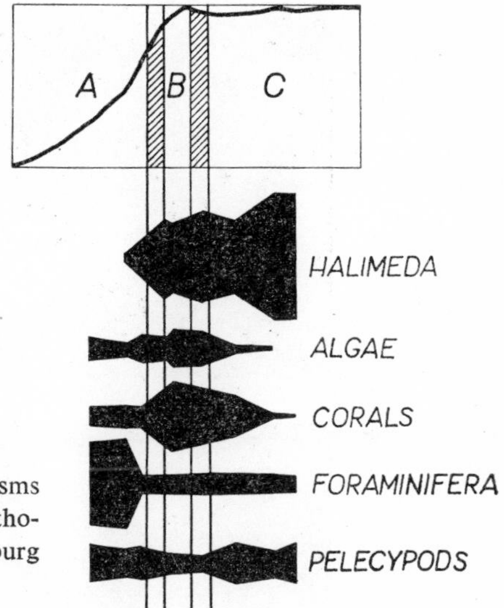

Fig. 101. Distribution of sediment-forming organisms in fringing reef of Florida. A. Outer slope, B. Lithothamnium ridge, C. Reef flat. After R. N. Ginsburg (1956).

As recorded in all textbooks, three main types are differentiated on the morphological basis: 1. Fringing reefs, 2. Barrier reefs, 3. Atolls. In the course of time, further, morphologically not very different types were added: Patch reefs, which are of a homogeneous structure and simple cupola shape. Table reefs — flat mounds devoid of lagoons. This division does not take into account genetic conditions; it is merely morphological, based on the relation of reef to the neighbouring land. Genetic classifications, insofar as they exist, are extremely complicated and are beyond the scope of this book. One of such modern classifications has been suggested by W. G. H. MAXWELL (1968).

According to the terminology used at present, a coral reef is always a bioherm, but not every bioherm is a coral reef. Some bioherms lack a solid framework and are

Photo 14. Origin and development of Pacific atolls illustrated by the morphology of some contemporaneous reefs. Fringing reef → barrier reef with narrow lagoon → atolls. The arrows indicate prevailing wind directions. After N. D. Newell *et al.* (1956).

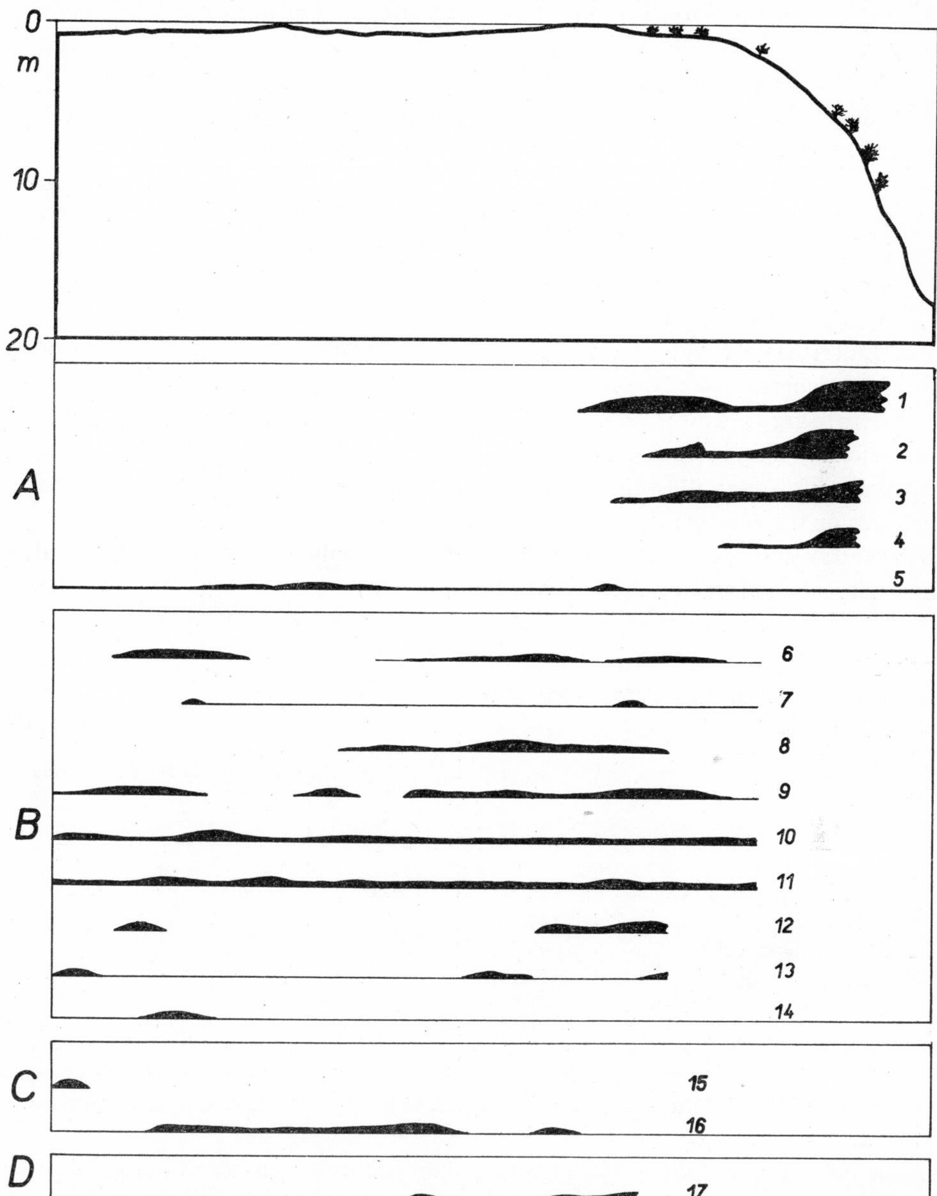

Fig. 102. Areal distribution of sediment-forming organisms and various structures in the Alacran Reef in the Gulf of Mexico. A. Skeletal frame-building organisms: 1. Massive corals, 2. *Acropora palmata*, 3. *Millepora*, 4. *Acropora cervicornis*, 5. *Porites porites*. — B. Sedimentary fill: 6. *Halimeda*, 7. *Sponges*, 8. *Ophiuroidea*, 9. *Echinoidea*, 10. *Gastropoda*, 11. *Pelecypoda*, 12. Alcyonarian corals, 13. Fishes, 14. Chitons. — C. Cement-forming organic compounds: 15. *Thalassia*, 16. *Sargassum*, 17. Coralline algae. After L. S. Kornicker *et al.* (1962).

not wave-resistant. Therefore, they can be formed also at a greater depth or near the water-level in places of weaker hydrodynamical activity, as in bays, gulfs, etc.

Although it has been stated many times that other organisms than corals generally predominate in reefs, we use the term coral reefs because the corals are responsible for the solid skeleton of reef structures. The skeleton is built up by the growing of young colonies over the old ones. The incrustation activity of calcareous algae, which cement the fragments to one another and join them to the coral skeleton is complementary.

An ideal reef is made up of three structural units:

1. Skeleton formed of corals and coraline algae.
2. Detrital filling formed of fragments of corals, coraline algae, pelecypods and foraminifers.
3. Biological and chemical cement, formed of bryozoans, foraminifers, occassionally also inorganic calcite.

Coral reefs are an illustrative example of organic activity which on the one hand depends on the environment, but on the other affects it strongly and controls the sedimentary conditions in the neighbourhood in supplying the basin with detritus and nutrients, influencing the water circulation, etc.

Table 181

Amount of individual organic groups on some reefs

Reef	Algae %	Reef-forming corals %	Foraminifera %	Pelecypoda %
Isle Murray (Torres Strait)	42·5	34·6	4·1	15·2
Honolulu	44	24		
Pearls and Hermes Reef	48·5	16·6	6·3	17·8
Florida	25·1	9·3	9·0	17·5
Tahiti	45·96	48·47	0·1	10·45

In Table 181 the representation of individual groups of organisms on some reefs is given in percentages.

The distribution of individual organic groups in the Saipan Island in the Mariana Archipelago was studied in detail by P. E. Cloud (1959) (Table 182).

The data quoted represent an average for the whole reef. On every reef there is a conspicuous zoning of organisms so that the composition of rocks from arbitrary places cannot be correlated.

Ecological conditions of the origin of reefs and living possibilities of reef-building organisms have not yet been fully solved and some deep-rooted and currently quoted views had to be revised. Ecological conditions are summed up in Table 183.

Table 182

Summary of the identified marine organisms of Saipan (after P. E. Cloud 1959)

Larger systematic group	Families	Genera	Species and subspecies
Alage	10	18	28
Flowering plants	2	3	3
Foraminifera	27	91	200
Sponges	9	9	10–11
Corals	19	33	63–75
Scleractinians (true stony corals)	12	26	56–68
Alcyonarians (spicular corals)	4	4	4
Actinarians and zoanthids (soft corals)	3	3	3
Hydrozoans	1	1	1
Flatworms	2	4	4
Brachiopods	1	1	1
Bryozoans	1	1	1
Sipuncoloids	1	4	9
Echiuroids	1	1	1
Polychaete annelids	20	35	38
Molluscs	58	98	164–174
Amphineurans	1	2	2
Gastropods	35	60	116–122
Pelecypods	20	34	44– 48
Scapophods	1	1	1
Cephalopods	1	1	1
Crustaceans	27	27	140
Ostracods	3	14	17
Cirripedes	2	2	2
Malacostracans	22	71	121
Stomatopods	1	2	4
Amphipods	1	4	4
Decapods	20	65	113
Brachyurans	10	38	61
Anomurans	2	5	6
Macrurans	8	22	46
Echinoderms	13	24	30
Crinoids	1	2	3
Echinoids	2	6	6
Asteroids	3	5	7
Ophiuroids	5	8	11
Holothurians	2	3	3
Total of invertebrates	179	389	662–685
Fish	19	30	32– 33
Total animals	198	419	694–718

Table 183

Ecological conditions of reef-building corals

Conditions	Old views	Revised views
Substratum	Only firm fundament	On rock and on unconsolidated mud as well. Some bioherms sink into rigid calcareous mud
Content of suspension in sea water	Only clear water, coral reefs are absent in front of river mouths, etc.	Some exceptions were ascertained. *Porites limosus* lives even in turbid water as well as Madrepora. With the increasing amount of suspension it changes the form of its colonies
Salinity	Corals are stenohaline animals and live only in sea-water of normal salinity	Several species such as *Madrepora cribripora* live also in brackish water

Table 184

Differences in ecology of hermatypic and ahermatypic corals
(after C. Teichert 1958)

Hermatypic corals	Ahermatypic corals
They have symbiotic zooxanthellae	They have no zooxanthellae
Occur only in waters warmer than 18·5 °C and only in depths less than 100 m	Occur at all depths down to 7,500 m and in temperatures as low as −1·1 °C
They are restricted to the tropical belts where many of them form coral reefs	Their distribution is world-wide. The majority of them are single (or solitary types) but the group includes a fair proportion of species that build ramified, colonial skeletons

In discussion the influence of temperature and penetration of water by light, two large groups of coral are sometimes distinguished, i. e. the hermatypic and the ahermatypic. Their definition and their differences in ecology are in Table 184.

The description of Recent coral sediments nearly always concerns only hermatypic corals, whose properties are commonly extended to all corals. It is necessary to deal with the qualities of ahermatypic corals in more detail, because the correlations of fossil and recent structures may lead to grave errors if they are neglected.

Recent occurrences of bioherms with ahermatypic corals are known from the continental slope of Norway and from the Mediterranean; they develop to a depth below 1,000 m and as far as 71°N latitude.

They are formed mainly of skeletal coral *Lophohelia prolifera* which build ramified colonies commonly several decimetres in size, growing one over the other or deposited side by side. In addition to the above-mentioned species there are another skeletal species of *Madrepora ramea*, the genera of *Stylaster* or *Allopora* genera. The further 190 recognized species include:

Coelenterata	32	Pelecypods	32
Worms	18	Spongiae	6
Echinodermata	48	Bryozoa	7
Crustacean	32	Brachiopods	4

On the continental slope of Norway their bioherms are distributed at the following depths (Table 185).

Table 185

Bathymetric distribution of ahermatypic coral patches on the continental slope of Norway (after C. Teichert 1958)

Depth m	Number of coral patches recorded	Depth m	Number of coral patches recorded
100–200	12	600– 700	3
200–300	24	700– 800	6
300–400	19	800– 900	5
400–500	8	900–1,000	5
500–600	4	1,000–1,300	23

Ahermatypic corals live undoubtedly at greater depths also in the tropical oceans but they have not yet been studied there. The maximum depths of their occurrence are presumed to rise towards the lower geographical latitudes.

The interpretation of fossil coral reefs and bioherms should be made with regard to the amount of reef-building organisms in the body and to the character of sediments in which it is enclosed. Provided that these conditions are known, the sedimentary environment of the body is relatively easy to determine on the basis of the following instructions (Table 186).

The morphology of reefs is extremely varied. All types, however, display several zones of a relatively stable position. The classic zonal structure of a fringing or barrier reef is as follows:

← Sea		Land →	
Slope with terrace	Elevated Lithothamnium ridge	Reef flat formed of various subzones	Coral, *Heliopora* and *Porites* zones passing into beach sediments

Table 186

Interpretation of sedimentary environments of bioherms and reefs (after C. Teichert 1958)

Biota	Environments
Few species of frame-building coelenterates, rich invertebrate association, no calcareous algae	Deep and cool waters, shelf or bay environments; temperate to polar belts, or in dysphotic zone of tropical belt
Calcareous algae, no corals, normal invertebrate association in places	Shallow and cool waters near-shore or bay environments, temperate to polar belts
Few species of frame-building coelenterates, associated with calcareous algae and rich invertebrate fauna	Moderately deep to shallow, warm waters, near-shore or bay environment, subtropical to warm temperate zones
Rich assemblage of frame-building coelenterates and calcareous algae in association with rich invertebrate fauna	Shallow, warm waters; open-shelf to near-shore environment; tropical belt.

The zonal structure of all reefs differs according to the zonal distribution of coral species. On the Saipan fringing reef in the Mariana Islands the following zones can be differentiated on this basis:

← Sea		Land →	
Algal ridge	Reef flat		
Pocillopora *Acropora* *Goniastrites*	outer *Acropora*	transitional *Heliopora* *Seriatopora*	inner *Porites*

The zones in atolls with lagoons display the same distribution. The inner, lagoonal part of the reef has a structure with a reverse sequence of zones. That means that nearest to the island (from the side of lagoon) the zone with the *Porites* genus occurs and farthest, lagoon-wards, is the zone with the *Acropora* genus. Because of a lower hydrodynamical energy on the lagoonal side, the individual zones occur there at shallower depths than on the marine side of the reef. Table 187 (after E. D. McKee 1958, from Kapingaramangi atoll in the Caroline Islands) shows the depths of coral occurrences on the lagoon and the sea sides of the atoll.

Table 187

Bathymetric distribution of corals on the lagoonal side and marine side of the Kapingaramangi Atoll (after E. D. McKee 1958)

Species	Bathymetric distribution	
	on lagoonal side m	on oceanic side m
Acropora rayneri	25—35	20— 60
A. rambleri	25—50	38— 60
A. reticulata	10—24	34— 38
A. vaughani	7—35	34— 38
Astreopora tabulata	24—35	35—120
Montipora marshallensis	50	28— 60
Leptoseris incrustans	14—35	20— 78

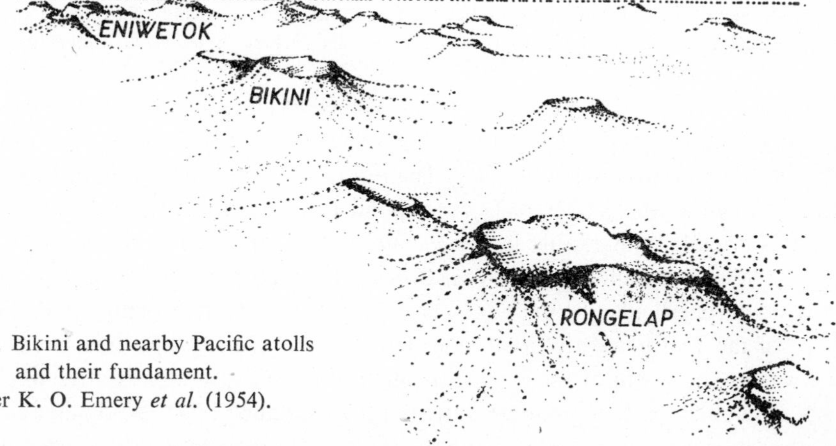

Fig. 103. Bikini and nearby Pacific atolls and their fundament. After K. O. Emery *et al.* (1954).

The zonal structure also occurs in patch reefs and reef knolls. The marginal *Lithothamnium* zone with abundant calcareous algae is followed by a moderately sloping reef flats (most frequently *Acropora* zone passing into *Porites* zone).

The following zones are developed either perfectly or rudimentarily, in fringing, barrier reefs or atolls.

1. A moderately elevated front of the reef — usually a zone of most intensive organogenic growth with an abundance of calcareous algae. It is sometimes called the *Lithothamnium* or Algal ridge.

2. A wide and irregularly dissected reef flat (platform) is divided into seawards or external, transitional and landwards or internal parts.

3. Foreshore zone of various appearance depending on the form of the reef.

Fig. 104. Sketch of sea reef off Tufa Island, Rongelap Atoll, showing relationship of marginal zone of reef and offshore algal ridge or spurs to older surface. After K. O. Emery *et al.* (1954).

The outer seaward slope of the reef is inclined at an angle of 30—40°. The coarser the fragments of talus the steeper the slope. The slopes of reefs at the protected shores or in the lagoon are lower (10—20°). The terrace levels occurring on the outer sides of atolls, fringing and barrier reefs correspond to standstill in the rise of sea-level in the Postglacial. Remarkable in the Pacific atolls is the 18 m-terrace which is developed both on sea and lagoon sides. Broad and several metres deep gullies running roughly perpendicular to the edge of reef reach to this depth (Fig. 104). The algal ridge of the reef is mostly somewhat elevated and during low tide it rises above the water. The width of the reef platform varies from several tens of metres to a few tens of kilometres. It is predominantly flat with subdued relief, exposed continuously to wave erosion, particularly during the storms. On the surface there are only socles of corallites, and in places many-ton boulders torn off and ejected by a rough sea. In depressions, growths of ramified corals, algae and occasionally coral detritus occur. In places, groups of micro-atolls or small eroded patch reefs are developed.

The lagoonal part of reefs is narrower and morphologically more differentiated, because it is not so strongly exposed to surf and wave action. Therefore, especially

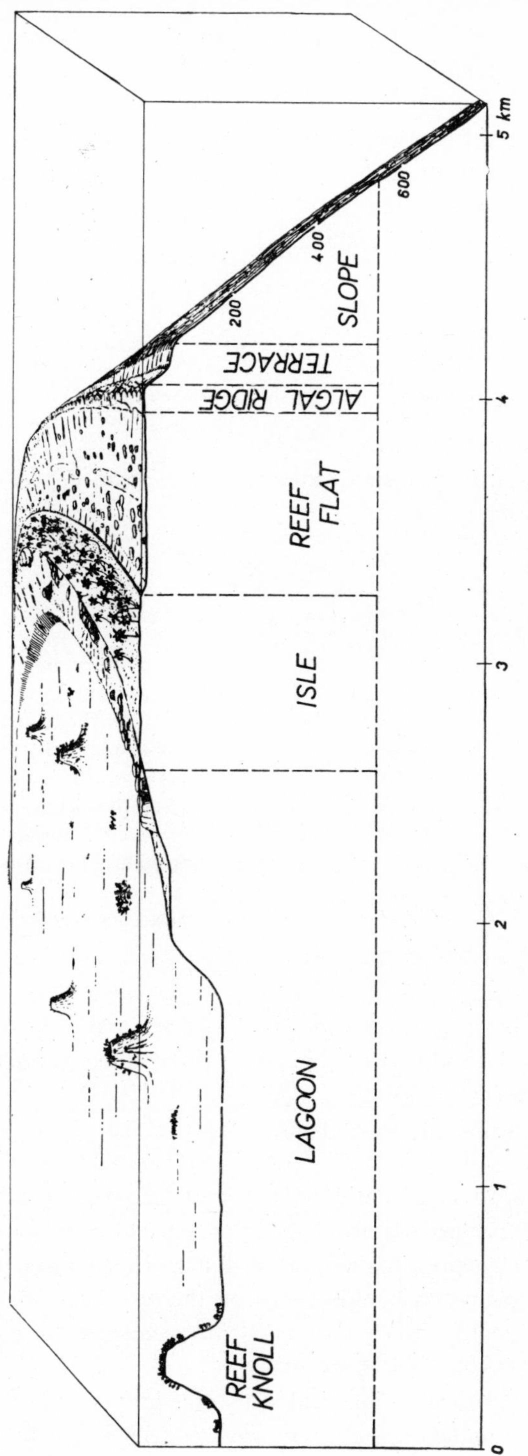

Fig. 105. Ideal schematic section through coral atoll showing the distribution of main zones.

in lagoons, minor patch and knoll reefs also develcped. The size of a lagoon varies (it is larger in atolls then in barrier reefs); its average diameter is estimated at 3—5 km. The mean depth of a lagoon is 20—40 m, the maximum depth is a little less than 100 m; on the whole, the depth is directly proportionate to the size. Being an accumulation part of the reef protected from intense wave action, the lagoon is the site of deposition of fine-grained sediments; plentiful knoll reefs, several tens of metres high with steep sides (up to 60°) rise from them, emerging above the water level at low tide.

Sediments of coral reefs

Sediments of coral reefs can be divided into three groups:

1. Solid biological sediments.
2. Detritus derived from disintegrated solid biological deposits.
3. Primarily detrital sediments.

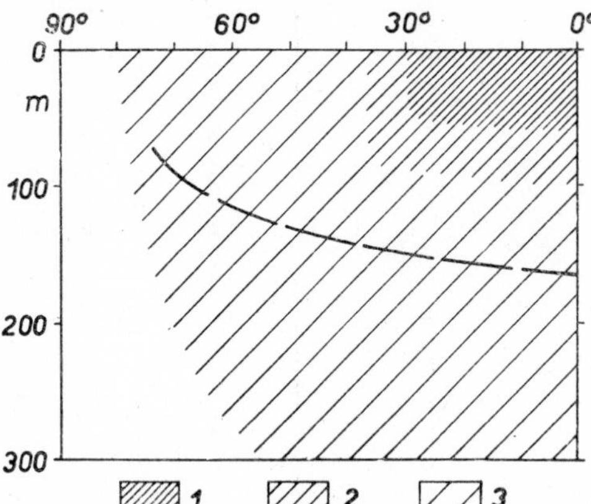

Fig. 106. The bathymetric distribution of hermatypic and ahermatypic corals and coralline algae in relation to climatic zones up to the depth of 300 m. 1. Tropical coral reefs, 2. Hermatypic not frame-building corals, 3. Ahermatypic corals. Dashed line indicates depth limit of occurrene of calcareous algae. After C. Teichert (1958).

The third group includes only sediments of lagoons and deeper outer parts of the reef: Halimeda sand, sand of Coralline algae and sediments built up from tests of foraminifers and pelecypod shells.

Solid biological sediments occur from a depth of 10—20 m on the oceanic side of the reef right up to the island. This part, being continuously affected by wave and surf action, is eroded and considerably smoothed at the surface. Organogenic gravels and sands, particularly from the fragments of corals and Coralline algae can develop only in depressions on the reef flat. The sea agitated during storms throws on the flat many-ton boulders torn off the reef faces. These together with a larger amount of smaller stones can be rapidly cemented by processes similar to those that give rise to the "beachrock".

The disintegration of solid biological structures into unconsolidated (loose) sediment occurs on an immense scale. The resulting loose products on recent reefs

Photo 15. Unconsolidated material dredged from the outer slopes of Bikini Atoll at a depth 190–265 m. Composed mainly of Foraminifera: 2 large species of *Cycloclypeus* and an encrusting type. Delicate pelagic shells are present but the algae, corals, and mollusks of the surface reef are absent. Natural size. After K. O. Emery *et al.* (1954).

Table 188

Coral reef environments and sediments

Environment	Characteristics of detrital sediments	Md mm	So (Trask's)
Reef front grooves	Coarse-grained detritus, fragments of corals and algae, unrounded	> 2	about 1·5
Coral sands encircling barrier reefs	Medium-grained or coarse-grained well-rounded coral sands; coral and algae fragments prevail	0·23	1·21
Reef front detritus, exposed to wave action	Well-sorted sand, grain-size corresponding to beach sands. Various constituents	0·21	1·29
Sand in depressions on reef flat	Badly sorted sand with admixture of gravel. Fragments of corals, genus *Acropora* prevailing	0·63	1·58
Coarser sediments in inlets	Coarse-grained coral sand, frequently bimodal (maximum about 2 mm and about 0·5 mm)	0·96	1·74
Coral gravel and sand in shallow parts of lagoon	Coarse-grained sand and gravel of coral fragments, mixed with Halimeda and foraminiferal sand	1·12	2·02

have been estimated at 2—3 times the volume of the solid skeleton. They form not only talus at the foot of the reef but also fill the numerous voids and pores. Disintegration is performed jointly by mechanical and biological (pelecypods, holothurians, fishes, algae, bacteria) factors. Biological destruction predominates on the lagoonal side, preparing and facilitating mechanical erosion. On the outer reef slopes, frequent gravity slides take place which may pass into turbidity current and transport coral detritus to depths attaining thousands of metres. Coarser deposits are preserved mainly on gently sloping terraces.

On the basis of grain-size and sorting, reef sediments are divisible into several groups which in places correspond to clearcut sedimentary environment (Table 188).

P. E. Cloud (1959) records 0·63 mm as an average Md of size and 1·7 as the mean So for all detrital reef sediments.

The most abundant accumulation takes place in lagoons, where it is of a detrital, chemical and also biological origin. Some authors maintain that one half of the

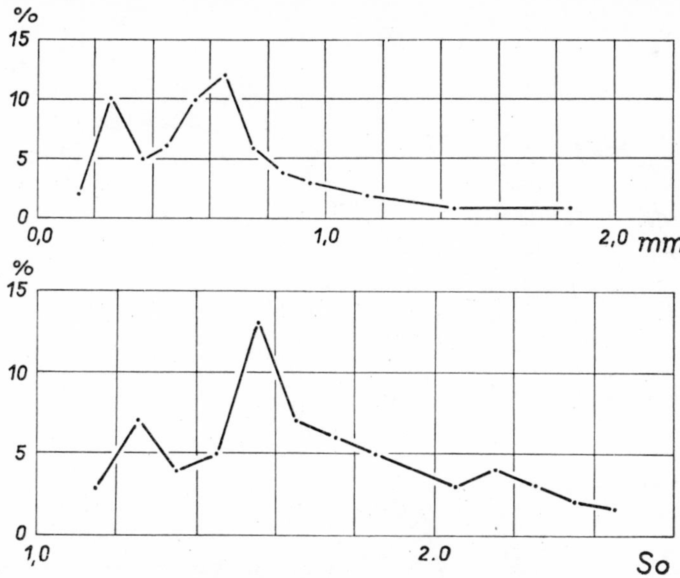

Fig. 107. Median-frequency distribution of 63 shoal sands from Saipan (upper diagram). Sorting-frequency distribution of 64 shoal sediments from Saipan. After P. E. Cloud (1959).

Table 189

Characteristics of sedimentary belts in lagoon of the Kapingamarangi Atoll (after E. D. McKee - E. B. Leopold 1959)

Depth range m	Diagnostic element	Textural feature	Dominant mineral composition
up to 15 mostly 0—8	*Amphistegina madagascariensis, Marginopora vertebralis*	Foraminiferal sand	Calcite
up to 15 mostly 8—15	Clastic molluscs shell fragments	Shell fragment sand	Mostly aragonite
10—35	Debris of dead branch corals	Gravel and sand rubble	Aragonite
30—55	*Halimeda macroloba* fragments	Irregular shape, granule size	Aragonite
40—70	*Amphistegina lessoni*	Foraminiferal sand	Calcite
50—80	Anhedral grains and needles of aragonite; some microfossils	Calcium carbonate mud	Mostly aragonite

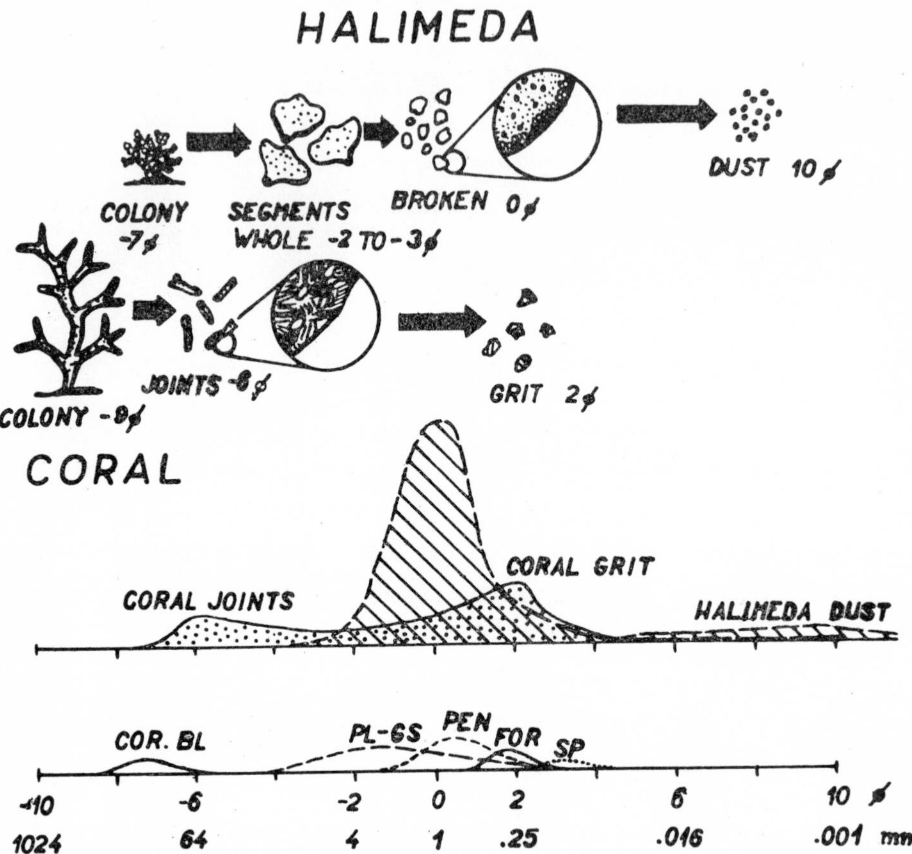

Fig. 108. Attempt to estimate the total quantity of each organic constituent in each grain-size for all the beach sediments of Isla Perez. It is subject to considerable error but is probably not too inaccurate. The relative volume of each organic constituent is proportional to the area of its frequency curve, and the probable grain-size distribution for each constituent is also shown. Minor constituents are shown on the lower scale: COR, B1, brain coral blocks and elkhorn coral blades, also includes large conch shells; PL-GS, pelecypods and gastropods and their fragments; PEN, peneroplid foraminifera; FOR, smaller foraminifera and SP, gorgonian spicules, coralline algal stems, etc. The upper scale shows the two most abundant constituents, each of which breaks down to two selective sizes because of structural factors. Staghorn coral forms branching colonies, and these on impact break into -6_{Φ} sticks that represent the length of the joints or branches; on abrasion, the coral breaks down further to 2_{Φ} grains because this is the approximate size of the fibrous crystalline packets of which the coral is ultimately composed. There is little coral material of intermediate size. Halimeda "bushes" first come apart into individual leaflike segments, but these are so thin and fragile that they tend to break into a dozen or so discoidal pieces averaging about 0_{Φ}. On abrasion, the Halimeda segments eventually disintegrate into 10_{Φ} dust, the size of the individual aragonite microcrystals that make up the calcareous skeleton. None of this *Halimeda* dust occurs in beach sediments, but small quantities are present in submerged areas, particularly grass flats, and more is present in deaper lagoonal areas. This diagram attempts to show why gravel on Isla Perez is chiefly made of coral, 0_{Φ} sand is mostly Halimeda, and 2_{Φ} sand is again mostly coral. After R. L. Folk (1964).

Table 190

Amount of microfossils in lagoonal sediments of the Kapingamarangi Atoll (after E. D. McKee - E. B. Leopold 1959)

(Number of microfossils/g of bottom sediments; upper number—calculation based on 1g of total sediment; lower number—calculation based on 1 g of sediment fraction finer than ½ mm)

	Depth					
	0 m	0·5 m	3 m	5 m	18 m	40 m
Sponge spicules	0	127 2,500	191 1,100	228 1,100	1,600 6,400	2,400 6,100
Diatoms	2	103 2,100	549 3,400	306 1,400	6,000 24,000	5,400 13,800
Micro-foraminifera	4	475 7,500	89 550	77 350	183 730	97 250
Pollen	14	24 480	11 68	28 130	17 68	— —
% fines below 0·5 mm	94	5	16	22	25	39

	Depth				
	84 m	156 m	186 m	226 m	240 m
Sponge spicules	7,970 80,000	7,040 70,000	5,700 19,000	30,800 30,800	60,600 60,600
Diatoms	2,700 27,000	19,700 197,000	17,200 59,000	23,600 23,600	20,200 20,200
Micro-foraminifera	77 770	127 1,270	71 350	2,600 2,600	2,260 2,260
Pollen	— —	— —	36 120	73 73	90 90
Dinoflagellate algae	— —	— —	36 120	950 950	1,000 1,000
% fines below 0·5 mm	10	10	29	100	100

sediments is of detrital and biological origin and the other of chemical origin. According to other data, biological and detrital deposition predominates.

The depth zoning of deposits is traceable also in lagoonal sediments. The lagoonal sediments of the Kapingamarangi atoll are given here as an example (after E. D. McKee *et al.* 1959) (Table 189).

The bathymetric zonal distribution reveals an interesting fact, i. e. that mechanical processes have a negligible effect on the origin of calcareous oozes in the deepest parts of lagoons. The *Amphistegina* zone, which is one of the deepest zones is calcitic, whereas the underlying muddy sediments are aragonitic. This shows clearly that the ooze could not have originated by the mechanical disintegration of the sediments of shallower parts.

In addition to these sediments, thousands of reef knolls of different size are scattered. They are encircled by slightly coarser-grained detritus, particularly coral and Halimeda sands. On their surface pieces of dead corals are mixed with foraminiferal sand.

Lagoonal sediments are unquestionably the most voluminous deposit of atolls and of some barrier reefs. Therefore, their identification can be of primary importance for the recognition of coral reefs. Detailed study has shown that lagoonal sediments contain a large amount of various microfossils (Table 190).

Total amounts are given in the upper column and fractions below 0·5 mm in the lower column. The Table shows clearly the proportion of biological particles in lagoonal sediments. It cannot be excluded that also structurless carbonate particles occurring in this environment originated by disintegration. This fact can be likewise applied to other dense limestones, even though they do not bear so many discernible biological particles.

The distribution of reef sediments is greatly influenced by storms, which may even be the decisive factor in the open, well-exposed parts of the reef. Additionally, large masses of loosened detritus may also be transported over the islands into lagoons, constituting there interlayers of coarse, predominantly coral and algal detritus in finer sediments.

The mineral composition of sediments depends mainly on their biological constituents. Generally, it can be said that aragonite is strongly predominant in Recent sediments of coral reefs. Pure calcite occurs only in foraminiferal sand of lagoons. Table 191 gives some data on the presence of three principal components of carbonate sediments (aragonite, Mg-calcite and calcite) in various deposits of coral reefs.

In recent papers by Ch. M. Hoskin (1968), G. M. Friedman (1968), Ch. E. Weaver (1968) and O. H. Pilkey - B. W. Blackwerder (1968) the relation between the mineralogy and Mg, Sr and Ba contents is described. In addition to high-magnesium and low-magnesium aragonites high-strontium and low-strontium aragonites have been described.

In Alacran Reef (Mexico) the amounts of Mg and Sr are different in various environments. Maximum Mg (mean $1{\cdot}38 \times 10^4$ p.p.m.) has been found in places adjacent to Mg rich coralline algae sediment, whereas maximum Sr ($0{\cdot}58 \times 10^4$

Table 191

Mean standard deviation of the carbonate mineralogy of sediments from Florida Bay, South Florida Reef and Back Reef, Bermuda, the Campeche Bank and three size fractions of the carbonate fraction of the United States South Atlantic shelf
(after O. H. Pilkey 1964)

Localization	Aragonite %	High—Mg calcite %	$MgCO_3$ %
Florida Bay	66 ± 6	19 ± 9	11·5 ± 2·5
Florida Reef	75 ± 9	20 ± 7	14·0 ± 1·5
Bermuda	56 ± 16	64 ± 15	
South Atlantic shelf (coarse sand)	49 ± 14	18 ± 12	10·5 ± 1·5
South Atlantic shelf (fine sand)	59 ± 20	13 ± 12	10·5 ± 1·5
South Atlantic shelf (silty sand)	66 ± 20	9 ± 6	10·5 ± 1·5

p.p.m.) has been found in cellular reef deeps adjacent to rich stony corals and codiacean algae.

It is likely that in earlier geological periods and in ancient sediments patch reefs passing into the table reefs are most widespread. Viewed from above they are mostly of circular shape or moderately elongated down the current. Their length can attain even several hundred metres.

The thickness and mode of deposition of reef sediments

The thickness of fringing and barrier reefs varies from tens to hundreds of metres depending on the length and the intensity of the positive movement of the sea level. Numerous borings have been sunk particularly in the fringing reefs of the Saipan Island in the Marianas, in the barrier reefs of Florida, in the Great Barrier Reef of Australia, in the shelf of Borneo, etc. In oceanic atolls many shallow and several deep borings down to the reef substratum were carried out (an earlier bore hole on Funafuti atoll and a recent boring on Kita-Daito-Zima, Eniwetok and Bikini atolls). The logs of shallow boring are illustrative of the vertical changes of reef environment. For this reason we give here a characteristic profile of the boring on the Bikini atoll accompanied by a description of sediments and an interpretation of the environments (Table 193).

From this profile the prevalence of lagoonal sediments (in volume) is clearly seen, which is common for all atolls and in part also for barrier reefs. The opinion prevails that in these reefs lagoonal sediments make up as much as 90% of the total volume of biological sediments. The borings, however, rarely encountered the true core of the reef which appears as corallites in the growth position, with abundant cavities filled by zonal granular calcite.

Table 192

Part of the profile of the boring on the Bikini Atoll with the interpretation of environment (after K. O. Emery *et al.* 1954)

Thickness m	Sediment	Environment
4	Unconsolidated, coarse foraminiferal sand containing several thin streaks of gravel	Shallow part of lagoon
2	Consolidated foraminiferal sand with layers of algal sand	Lagoon edge
8	Foraminiferal sand with great fragments of corals and Lithothamnion. Unworn tests of *Calcarina spengleri, Marginopora*	Lagoon edge
3	Broken pieces of coral and cemented detritus. The coral is fresh and unworn specimen of *Acropóra* cf. *palifera*. Some fragments are coated with a thin crust of Lithothamnion. The detritus consists of moderately well-cemented tests of *Calcarina spengleri, Marginopora*, and segments of Halimeda	Reef flat edge
7	Coralliferous detrital limestone. Recovered core consists largely of coral colonies, some of them in position of growth. *Acropora, Stylophora*, and *Astreopora* are the most common genera. Detrital material contains abundant Halimeda segments and gastropod shells	Shallow water, close to the lagoon edge of a reef
3	Colonies and fragments of corals, especially *Asteropora* and *Acropora*. The detrital material includes Halimeda segments. Well-preserved operculum of *Turbo* and few large cross-sections of echinoid spines	Near-reef environment of lagoon, part of the reef-knoll
2	Halimeda sand. 50 % Halimeda segments, 10 % coral and mollusc fragments from 1 mm to 1 cm, 30 % foraminiferal tests and 10 % fine detrital material	Lagoon

Deep borings on the Pacific atolls struck in the substratum of Recent sediments, coral deposits of Pleistocene, Pliocene, Miocene and Eocene ages. It is of interest that in all borings a stratigraphical hiatus corresponding to Oligocene occurs. The substratum was in all cases formed of basalt, the same rock from which all guyots in the Pacific are built. Limestones were consolidated mostly from a depth of several tens of metres, but loose sediments were also found at greater depths. According to prevailing views, they are the product of secondary decementation. Aragonite was present in Recent and Pleistocene sediments, being rare in the Tertiary recrystallized deposits. Dolomite occurs in the lower part of the boring on Funafuti atoll, at a depth below 200 m and in the upper part of the profile on Kita-Daito-Zima. In shallow bores on various Pacific atolls, described already at the beginning of this century by W. Skeates, dolomitic sediments with 1—15% MgO and dolomite crystals were found mostly in the upper parts of reefs. The dolomite was obviously of epigenetic or late diagenetic origin, being generated at the redistribution of magnesium liberated from Mg-rich calcites during the recrystallization of sediments. On other reefs, the enrichment of the upper parts by magnesium is not due to the presence of dolomite but to the presence of unstable Mg-rich calcite.

The total thickness of coral sediments of oceanic atolls may exceed 1,000 m. Thanks to recent geological findings and geophysical data the origin of atolls is on the whole clear. In the Pacific and other oceans there are many guyots, i. e. submarine table mountains whose summits were levelled by waves and surfs when they were near the sea level. The levelling was succeeded by a rapid subsidence to depths of several thousand metres, either a result of the general rise of the world ocean, or rather by an isolated subsidence of individual guyots which was caused by the concentration of large masses in one place. As long as the guyots were near the sea level, reef-building corals began to grow on their surface. If the subsidence of guyots occurred at a rate that could be compensated by the growth of corals, they have continued growing up to the present. If the subsidence was rapid, coral perished and were preserved as relics at 1000 m. In the Mid-Pacific mountain ridge, primordial development stages of coral reefs of Cenomanian to Aptian age have been ascertained on the summits of guyots. These finds corroborated fully Darwin's subsidence theory of the development of coral atolls.

Subrecent primordial reefs are found in many places of the continental slope. They are initial stages of cupole-shaped bioherms or reefs, with a diameter of several dm to a few m, which originated at a time when the sea level was considerably lower. In some places they are already covered by fine-grained sediments, thus forming a transition to fossil bioherm.

References

ADAMS J. E. - RHODES M. L. (1960): Dolomitization by seepage refluxion. Bull. Am. Assoc. Petrol. Geol., vol. 44: 1912—1920, Tulsa.

ALDERMANN A. R. (1958): Aspects of carbonate sedimentation. Jour. Geol. Soc. Australia, vol. 6: 1—10, Adelaide.

BATHURST R. G. C. (1967): Subtidal gelatinous mat, sand stabilizer and food, Great Bahama Bank. Jour. Geology, vol. 75: 736—738, Chicago.

BEALES E. W. (1958): Ancient sediments of Bahaman type. Bull. Am. Assoc. Petrol. Geol., vol. 42: 1815—1830, Tulsa.

BLACK M. (1933): The precipitation of calcium carbonate on the Great Bahama Bank. Geol. Magazine, vol. 70: 455—466, London.

CHAVE K. E. (1967): Recent carbonate sediments- an unconventional view. Counc. of Education in the Geol. Sciences, Short Rev. 7; Jour. Geol. Educ., vol. 15: 200—204, Washington.

CHILINGAR G. V. - BISSELL H. J. (1963): Formation of dolomite in sulfate-chloride solution. Jour Sedimentary Petrology, vol. 33: 801—803, Menasha.

CLOUD P. E. (1959): Geology of Saipan, Mariana Islands. Prof. Pap. K, p. 361—445, Washington.

FAIRBRIDGE R. (1950): Recent and pleistocene coral reefs of Australia. Jour. Geology, vol. 58: 330—401, Chicago.

FOLK R. L. - ROBLES R. (1964): Carbonate sands of Isla Perez, Alacran Reef, comples Yucatán.

FRIEDMAN G. M. (1968): Geology and geochemistry of reefs, carbonate sediments and waters, Gulf of Aqaba, Red. Sea. Jour. Sedimentary Petrology, vol. 38: 895—919, Menasha.

FRIEDMAN G. M. - SANDERS J. E. (1967): Origin and occurrence of dolostones. In: Development in Sedimentology, vol. 9A: 268—348, Amsterdam.

FÜCHTBAUER H. (Editor, 1969): Lithification of carbonate sediments. Vol. 1. Spec. issue of Sedimentology, vol. 12, Amsterdam.

GINSBURG R. N. (1956): Environmental relationship of grain size and constituent particles in some of South Florida sediments. Bull. Am. Assoc. Petrol. Geol., vol. 40: 2384—2427, Tulsa.

GINSBURG R. N. - LOWESTAM H. A. (1958): The influence of marine bottom communities of the depositional environment of sediments. Jour. Geology, vol. 63: 219—223, Chicago.

GRAF D. L. - EARDLEY A. J. - SHIMP N. F. (1961): A preliminary report on magnesium carbonate formation in glacial Lake Boneville. Jour. Geology, vol. 69: 219—223, Chicago.

HAMILTON E. L. (1956): Sunken islands of the Mid-Pacific Mountains. Geol. Soc. Am. Mem. 64: 1—97, Baltimore.

HOSKIN Ch. M. (1968): Magnesium and strontium in mud fraction of Recent carbonate sediment, Alacran Reef, Mexico. Bull. Am. Assoc. Petrol. Geol., vol. 52: 2170—2177, Tulsa.

HOUBOLT J. H. C. (1959): Surface sediments of the Persian Gulf near the Quatar penninsula. Diss., p. 1—113, Utrecht.

HUBBS C. L. - BIEN G. S. - SUESS H. E. (1963): La Jolla natural radiocarbon measurements III. Radiocarbon, vol. 5: 254—272.

ILLING L. V. (1954): Bahamian calcareous sand. Bull. Am. Assoc. Petrol. Geol., p. 1—95, Tulsa.

ILLING L. V. - WELLS A. J. - TAYLOR J. C. M. (1965): Penecontemporary dolomite in the Persian Gulf. SEPM, Spec. Pap. 13: 89—111, Tulsa.

IMBRIE J. - PURDY E. G. (1962): Classification of modern Bahamian carbonate sediments. In "Classification of carbonate rocks". P. 253—272.

KORNICKER L. S. - BOYD D. W. (1962): Shallow water geology and environments of Alacran reef complex, Campeche Bank, Mexico. Bull. Am. Assoc. Petrol. Geol., vol. 34: 203—214, Tulsa.

LADD H. S. - SCHLANGER S. O. (1960): Drilling operation on Eniwetok atoll. Geol. Surv. Prof. Pap., p. 260-Y, Washington.

LOWESTAM H. A. (1950): Niagaran reefs of the Great Lakes area. Jour. Geology, vol. 58: 430 — 487, Chicago.

LOWESTAM H. A. - EPSTEIN S. (1957): On the origin of sedimentary aragonite needles of the Great Bahama Bank, Jour. Geology, vol. 65: 364—375, Chicago.

MAXWELL W. G. H. (1968): Atlas of the Great Barrier Reef. P. 1—258, Amsterdam.

MAXWELL W. G. H. - JELL J. S. - MCKELLER R. C. (1964): Differentiation of carbonate sediments in the Heron Island reef. Jour. Sedimentary Petrology, vol. 34: 294—308, Menasha.

McKee E. D. (1958): Geology of Kapingaramangi atoll, Caroline islands. Bull. Geol. Soc. Am., vol. 69: 241—278, Baltimore.

McKee E. D. - Chronic J. - Leopold E. B. (1959): Sedimentary belt in lagoon on Kapingaramangi atoll. Bull. Am. Assoc. Petrol. Geol., vol. 43: 501—562, Tulsa.

Miller D. N. (1961): Early diagenetic dolomite associated with salt extraction process, Inagua, Bahamas. Jour. Sedimentary Petrology, vol. 31: 473—476, Tulsa.

Newell N. D. (1956): Geological reconnaisance of Raroia (Kon Tiki) atoll, Tuamotu archipelago. Bull. Am. Mus. Nat. Hist., vol. 109, Art. 3: 317—372, New York.

Peterson M. N. - Bien G. S. - Berner R. A. (1963): Radiocarbon studies of Recent dolomite from Deep Spring Lake. Jour. Geophys. Res., vol. 68, No 24.

Pilkey O. H. (1964): The size distribution and mineralogy of the carbonate fraction of United States South Atlantic Shelf and Upper slope sediments. Marine Geology, vol. 2: 121—136, Amsterdam.

Pilkey O. H. - Blackwerder B. W. (1968): Mineralogy of the sand size carbonate fraction of some Recent marine sediments. Jour. Sedimentary Petrology, vol. 38: 799—810, Menasha.

Pray L. C. - Murray R. C. (1965): Dolomitization and limestone diagenesis. A Symposium. Soc. Econ. Pal. Min., Spec. Publ. No 13: 1—180, Tulsa.

Revelle R. (1958): Chemical erosion of beach rock and exposed reef rock. US Geol. Survey, Prof. Pap. 260-T; p. 699—709, Washington.

Revelle R. - Emery K. O. (1957): Bikini and nearby atolls, Marshall Islands. Geol. Surv. Prof. Pap., vol. 260 A-T, p. 1—709, Washington.

Rigby J. K. (1969): Reefs and reef environments. Bull. Am. Assoc. Petrol. Geol., vol. 53: 738, Tulsa.

Seibold E. (1962): Untersuchungen zur Kalkfällung und Kalklösung am Westrand der Great Bahama Bank. Sedimentology, vol. 1: 50—74, Amsterdam.

Shinn E. (1963): Spur and groove formation on the Florida reef tract. Jour. Sedimentary Petrology, vol. 33: 291—303, Menasha.

Shinn E. A. - Ginsburg R. N. - Lloyd R. M. (1965): Recent supratidal dolomite from Andros Island SEPM, Spec. Publ. 18: 112—123, Tulsa.

Siegel F. R. (1961): Variations of Sr-Ca rations and Mg contents in Recent carbonate sediments of the northern Florida Keys area. Jour. Sedimentary Petrology, vol. 31: 336—342, Menasha.

Skinner H. C. (1960): Formation of modern dolomitic sediments in South Australian lagoons. Bull. Geol. Soc. Am., vol. 71: 1976, Baltimore.

Stehli F. G. - Hower J. (1961): Mineralogy and early diagenesis of carbonate sediments. Jour. Sedimentary Petrology, vol. 31: 358—371, Menasha.

Stockman K. W. - Ginsburg R. N. - Shinn E. A. (1967): The production of lime mud by algae in South Florida. Jour. Sedimentary Petrology, vol. 37: 633—648, Menasha.

Storr J. F. (1964): Ecology and oceanography of the coral reef tract, Abaco Island, Bahamas. Geol. Soc. Am., Spec. Pap. 79: 1—98.

Taft W. H. (1962): Dolomite in modern carbonate sediments, Southern Florida. Bull. Am. Assoc. Petrol. Geol., vol. 46: 281, Abstr., Tulsa.

Teichert C. (1958): Cool- and deep-water coral banks. Bull. Am. Assoc. Petrol. Geol., vol. 42: 1064—1082, Tulsa.

Weaver Ch. E. (1968): Geochemical study of a reef complex. Bull. Am. Assoc. Petrol. Geol., vol. 52: 2153—2169, Tulsa.

Wells A. J. (1962): Recent dolomite in the Persian Gulf. Nature, vol. 194, No 4825: 274—275, London.

Williams M. - Barghoorn E. S. (1959): The genesis of marine carbonates. General Petroleum Geochemistry Symposium, p. 64—70, New York.

21. *Changes in sedimentation in the Sub–Recent and their causes*

The study of changes in sedimentation in the Sub-Recent and of their causes gives a clue to the recognition of the laws controlling the sedimentation in earlier geological periods. There are objective proofs of the causes of changes in sedimentation in the near past (particularly in the Pleistocene), and it may be presumed that changes in earlier periods had been produced by the same or similar factors.

The changes in sedimentation during recent geological periods were influenced by the following changes of a geological of geographical character:

1. Changes in the basinal environment induced by climatic fluctuations; they include changes of the current system, the physico-chemical properties of waters and the biological properties of the basin.

2. Changes in the depth of the basin of eustatic or tectonic origin; the tectonic cause did not produce any deformation of the hinterland.

Table 193

Stratification of Recent lacustrine sediments and factors producing them

Types of sediment stratification	Factors causing the changes in sedimentation
Change of gyttja into sapropel	Normal development of the lake without external influence; in some cases increased humidity of climate
Alternation of calcareous and clayey sediments rich in organic matter	Seasonal climatic changes during which the growth and death of plankton alternate
Alternation of clay sedimentation into carbonate one, eventually sedimentation of Fe ores and then into gyttja and sapropel deposition	Normal hydrochemical and hydrobiological development of the lake caused by internal changes of its regime
Alternation of silt-sand laminae with the clayey ones, richer in organic matter. The more sandy laminae are richer in carbonate	Seasonal alternation of sedimentation in glacial lakes; the more sandy laminae can develop during the spring thaw and the carbonate sediments in the warmer, usually summer period; clay laminae are the product of the cold period
The alternation of diatomite, rich in organic matter, with calcareous and clayey-calcareous sediments	Seasonal changes in the growths and death of phytoplankton

Table 194

Stratification of Recent bay and shelf sediments and factors producing them

Types of sediment stratification	Factors causing the changes in sedimentation
Alternation of carbonate-rich and carbonate-deficient laminae. Alternation of laminae rich and poor in organic matter	Seasonal changes reflected in the changes of the biochemical carbonate regime
Alternation of beds rich and poor(er) in organic matter without changes in carbonate content	Seasonal changes affecting the summer bloom and abrupt autumn death of phytoplankton
Beds of silt or sand in fine-grained clay sediments	Sediments of stormy periods, when more sand was distributed over the basin area by flotation
In lagoons or bays intercalations of coarser, mainly organogenic sediments in finer-grained ones	The effects of periodical storms that brought coarser material into the depositional area of finer sediments

3. Changes of the neighbouring relief due to tectonic processes or to changes in the velocity of denudation.

4. Catastrophic changes in the environs of the basin or in the basin itself, such as earthquake, large-scale gravitational phenomena, or the swift displacement of the stream channels, etc.

Climatic changes are probably the most frequent phenomena affecting the character of sedimentation during the recent geological periods. They manifest themselves wherever they are not masked by other factors, e. g. tectonic or catastrophic. Closed basins, lakes or inland seas are most sensitive to climatic changes. Some changes in lacustrine sedimentation and climatic or other factors producing them are given in Table 193.

In shallow-water sediments of open coasts seasonal climatic changes are mostly difficult to trace, because they are masked by other factors. In bays, shallow and deep inland seas, on the other hand, both seasonal and climatic changes are easily discernible. Examples are given in Table 194.

On the shelves this kind of stratification is concealed by other effects. The following stratification can be discerned only in some depressions:

The alternation of green clay (with benthos) and grey clay rich in organic matter, free hydrogen sulphide and pyrite	Changes in circulation, that caused the change of an aerated basin into that with stagnant water. These changes are often combined with tectonic processes.

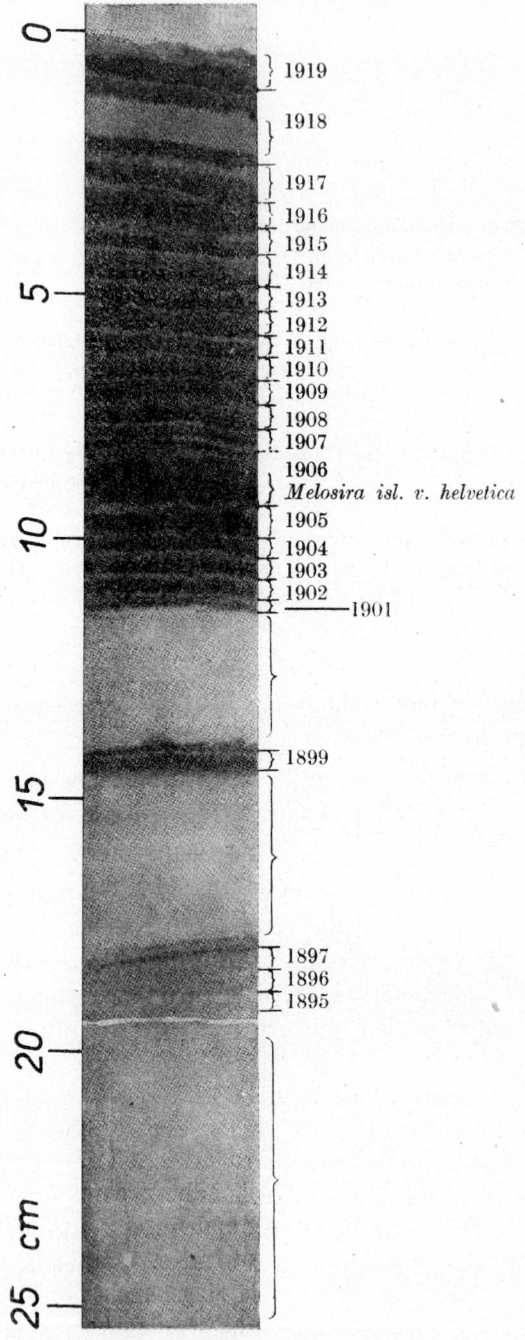

Photo 16. Core from sediments of Lake Zurich. Laminated seasonal sediments interrupted at several places by intercalations of fine-grained sands and silts. These sands and silts are products of turbidite sedimentation, because they correspond exactly to the time range, when great slides of shore material occurred. After L. Minder (1939).

On the continental slope and shelf of glaciated areas stratification caused by climatic conditions occurs in abundance.

Alternation of coarse-grained detrital sediments, most frequently gravel with unsorted sands or with laminated silts and clays	The alternating intensity of glacial conditions. The cooler was the climate, the more frequent were ice floes and icebergs and the coarser sediment could settle.

On some continental slopes stratification of a very interesting genesis has been observed:

Deposition of calcareous ooze changing into sedimentation of green clay rich in organic matter and free hydrogen sulphide	The change of normal conditions into hypertrophic conditions in the top water layers. Owing to a considerably import of nutrients an immense production of Dinnoflagelatae took place whose toxic excrements killed nearly all other organisms.

Climatic effects are easily traceable in deep-sea sediments. Their stratification presents a picture of climatic changes during the latest geological period, which show themselves above all by different rates of production of plankton and different rates of supply of terrigenous material. Differences in the current system can influence sedimentation to a large extent. The primary differences in the production of calcareous and siliceous plankton are modified secondarily by the dissolution of calcareous particles in the deep cold currents. A number of methods have been used for the study of the stratigraphy of the youngest deep-sea sediments:

1. The determination of absolute age by current and C^{14} methods.
2. The determination of palaeotemperatures on the basis of O^{18}/O^{16} isotopes.
3. Geochemical method consisting in the study of the variability of Ti- and carbonate contents.
4. Biological, ecological and palaeontological methods. In this respect, the ratio of warm and cold-living species of foraminifers and sometimes also the proportion of right- and left-coiling forms of some species have been studied.

By means of these methods the stratigraphy of deep-sea sediments has been studied mainly in the equatorial Atlantic, the Caribbean Sea and in some parts of the Pacific and the Indian Ocean, and the sequences of the youngest sediments have been correlated there.

The sedimentation of the Last Glacial differed from that of the Recent mainly by the following phenomena:

1. Glaciomarine and hemipelagic sediments were far more extensive.
2. The deposition both of terrigenous and carbonate material over the whole of the deep-sea floor occurred at a greater rate. It can be determined from several measurements that during the Last Glacial the deposition of terrigenous material was 2·5 times greater than it is today and the sedimentation of carbonate 1·3 times the present one.

3. During the Last Glacial the average production of calcareous plankton was probably greater, because the increased circulation of atmosphere and water masses induced a more intensive water exchange and a more perfect restoration of nutrients.

4. Siliceous, particularly diatom, oozes were more widespread owing to the immense production of phytoplankton in cool near-surface waters.

5. The association of foraminifers in sediments altered. The individual planktonic species indicate the presence of definite climatic zones (see Table 195).

Table 195

Distribution of planktonic foraminifera in various climatic zones

Zone	Species	Stratigraphic distribution
Arctic and Antarctic	*Globigerina deutertrei*	Eocene-Recent
	Globigerina pachyderma	Miocene-Recent
Temperate	*Globigerina bulloides*	Eocene-Recent
	Globigerina inflata	Eocene-Recent
	Globorotalia crassula	Miocene-Recent
	Globorotalia canariensis	Miocene-Recent
	Globorotalia truncatulinoides	Miocene-Recent
	Globorotalia hirsuta	Pliocene-Recent
Tropical	*Orbiculina universa*	Miocene-Recent
	Globigerina dubia	Eocene-Recent
	Globigerinella aequilateralis	Oligocene-Recent
	Globigerinoides rubra	Oligocene-Recent
	Globigerinoides sacculifera	Oligocene-Recent
	Globigerinoides conglobata	Pliocene-Recent
	Globorotaria menardii	Miocene-Recent
	Globorotalia tumida	Pliocene-Recent
	Globorotalia scitula	Miocene-Recent
	Sphaerodinella defiscens	Miocene-Recent
	Pulleniatina obliquiloculata	Pliocene-Recent

The principal associations corresponding to warm climate include *Globorotalia m. menardii*, *G. flexuosa* and *G. tumida*. Associations with *G. punctulata* are characteristic of cool climate.

On the basis of the ratio of warm- and cold-water foraminifers, a correlation was made of deep-sea cores along a line more than 6,000 km long across the Caribbean Sea and the Equatorial Atlantic Ocean. The study of O^{18}/O^{16} isotopes, used as a complementary method, has yielded the same results. It has also been ascertained that the main turning point from the Glacial to Post-Glacial conditions was 11,000 years ago. The maximum of the Last Glacial was not, of course, simultaneous in all geographical latitudes. In the areas of higher latitudes the changing intensity of glacial conditions is traceable according to the grain-size of deposits. As mentioned above, coarser-grained sediments correspond to a more intense glaciation. Thus, for instance, the following stratigraphical succession has been ascertained on the continental slope of the Antarctics:

Recent—	6,000 years —	fine-grained glaciomarine deposits
6,000—	12,000 years —	fine-grained glaciomarine sediments with clay prevailing
12,000—	29,000 years —	coarse-grained glaciomarine sediments
29,000—	40,000 years —	medium-grained glaciomarine sediments with prevailing sand
40,000—	133,000 years —	coarse-grained glaciomarine sediments with several intercalations of fine grained deposits
133,000—	173,000 years —	fine-grained — clayey-silty, laminated glaciomarine sediments, with scarce pebbles
173.000—	350,000 years —	coarse-grained gravelly glaciomarine sediments
350,000—	420,000 years —	fine-grained sediments with laminated horizons.

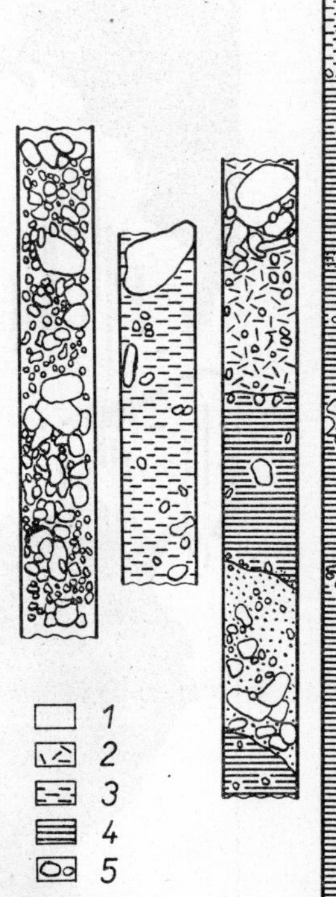

Fig. 109. The cores of glaciomarine sediments from the Sea of Japan shelf. 1. Sand, 2. Silt, 3. Silty clay, 4. Clay, 5. Boulders. After N. M. Strakhov (1961).

For geological periods analogous to the last transition of glacial into non-glacial conditions the following changes of sedimentation can be postulated:

1. Reduction of areas in which coarse glaciomarine sediments were deposited.
2. Lowered rate of sedimentation, manifested, for instance, by an increased amount of benthonic organisms or the presence of more frequent secondary structures.
3. Decreased amount of siliceous sediments and a more or less increased amount of organogenic calcareous sediments. In some places the change can be reverse.
4. Change of cold-water planktonic associations into warm-water ones.

The changes in the depth of the sedimentary basin are best apparent from the Holocene transgression when the sea level rose about 90—100 m all over the world. Analogies to the changes that then took place in shallow sedimentary environments can be found also in past geological periods.

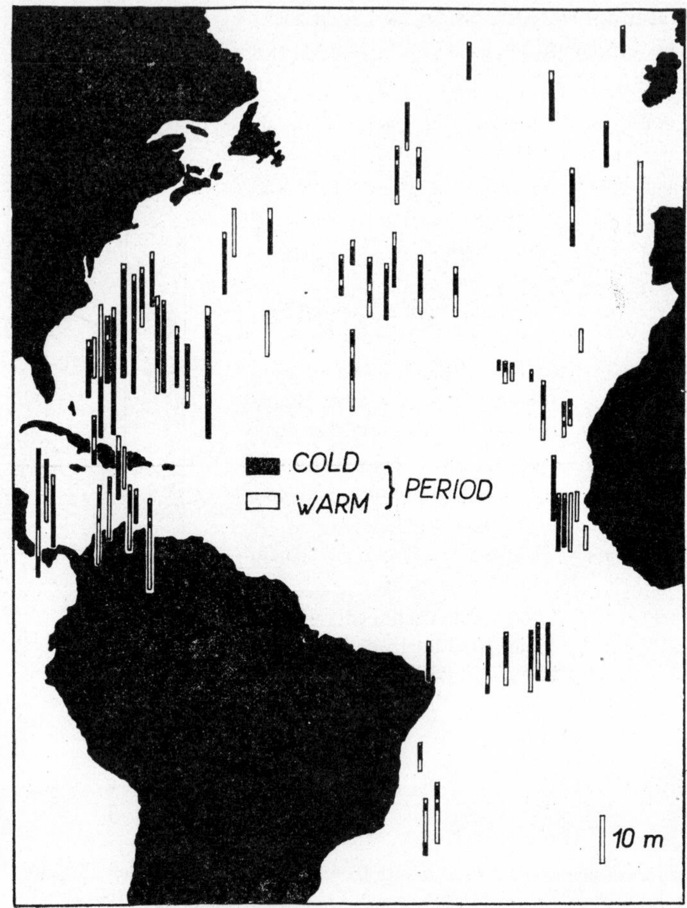

Fig. 110. Distribution of the climatic stages of the deep-water cores from the Atlantic Ocean. The top of the column is the approximate location of the core station. Beginning from the top of the columns the stages indicated are: the Postglacial (open), the Last-Glacial 2—3 (solid); the interstadial of the Last-Glaciation (open); the Last-Glacial 1 (solid); the Last-Interglacial (open). Interpretations are based on investigations of the planktonic foraminifers and lithology of all the cores and on radiocarbon dates, isotopic temperatures and grain-size analyses of selected samples of selected cores. After D. B. Ericson *et al.* (1961).

In the chapter on deltaic sediments, changes in sedimentation produced by the Post-Glacial rise of sea level have been referred to. Shelf sediments likewise bear numerous signs of a Holocene transgression. Not very deep under Recent fine-grained sediments there appear beds of shells dated by C^{14} analyses as 15,000—20,000 years old, that were deposited at a depth of a few metres under low-water conditions. During transgression, they were covered at first by calcareous, later by quartzeous sandy deposits, and finally by silty clay and clay. Coarser-grained sediments, such as sorted gravels and sands corresponding to the relics of Pleistocene beaches, are at present buried by finer-grained, usually calcareous-clayey shelf sediments.

Sediments of lagoons closed fully or partly by sand barriers migrate landwards, being followed by the movement of sand banks and bars. In the vertical sequence, the basal continental sediments, usually sand dunes, are followed mostly by clayey sediments of lagoons or bays with in turn are overlain by sands or sand banks. If the transgression had a continuous course, the deposition of sands could persist for a long time, because the barriers sufficed to grow up to and above the water level. When the rate of the rise of sea level increased, as happened in many places, the barriers could be overlain by strongly calcareous, mostly sandy shelf sediments.

In some coastal basins or fjords, the deposition of black clays laid down in stagnant non-aerated environment was replaced, as a result of transgression, by the sedimentation of green calcareous clays laid down in normal, well-aerated environment.

All the above examples are a manifestation of the Post-Glacial eustatic transgression occurring all over the world. Besides, a number of local transgressions provoked by other factors also took place. A classic example is provided by the Caspian Sea, where during the last several decades the fluctuation of the sea-level attained several metres. At the critical depth between 5—10 m (of the present state) the fluctuation showed itself by the alternating sedimentation of calcareous shells (during shallowing) and clayey sediment (during deepening). From this follows that even relatively small changes of depth can induce widespread changes in sedimentation over extensive shallow parts of the basin.

The change of calcareous biological to terrigenous, predominantly clayey sedimentation is of utmost importance for the solution of some problems of ancient sedimentation. For this reason, the Recent and Sub-Recent stratifications from the Taman Bay (Azov Sea) deserve to be mentioned (B. I. KOSHECHKIN 1959). There, carbonate and clayey sedimentation alternate in such a way that calcareous sediment sets sharply on the underlying clayey sediment, passing upwards again into clays. During the sinking of the sea level, major parts of the basin become shallower so that sediments could be disturbed by wave and current action. In these places, terrigenous sedimentation is interrupted and the bottom is covered by the shells. When the sea level rises and the major part of the basin deepens, the clay component can be deposited over a large area, benthonic organic production decreases and the sedimentation passes into clay through the reduction of shells. This process also occurs in many other environments and it should be realized that such an explanation of analogous phenomena would be the simplest and most probable also for ancient sediments. This phenomenon can be interpreted wrongly by the current aggradation of shells and its origin can be likened to that of graded bedding.

In half-closed bays and lagoons, changes in the water level produce an alternation of normal marine conditions, represented most frequently by sandy-calcareous sediments, and conditions of poorly aerated basins, represented by black clay sediments rich in organic matter. Shallow bays are particularly sensitive to these changes and their sediments are the best indicators of them. The stratification of Recent and Sub-Recent sediments is very well developed in the Kara-Bougas-Gol Bay, where, on the basis of various kinds of sediments Sub-Recent transgressions and regressions could be discerned (the conditions depend not only on the absolute height of the water level, but also on the altitude of the barrier which had been recurrently disturbed and rebuilt). During transgression, clay sedimentation with silt and fine sand intercalations always takes place. During regression and the closing of the bay, calcareous deposits, gypsum, mirabilite, halite, epsomite and astrachanite are progressively deposited.

In the Great Salt Lake a regressive sequence of sediments can well be traced. In the course of shallowing and final evanescence of the lake the following succession of sediments developed: clay—calcareous clay —calcareous-dolomitic clay—skeletal calcareous sediments—oolites—calcareous and aragonitic crusts—soils.

Changes of the adjacent relief

Changes of the adjacent relief can be produced by tectonic agencies or the fluctuations in the rate of denudation of the landmass. Unfortunately, well substantiated changes in sedimentation brought about by these causes are very few in Recent and Sub-Recent deposits. On the one hand, they have not yet been studied in detail, and several thousands of years are a rather short period for the registration of these processes. One of the classic examples is the increase of silt towards the top beds of the Baltic Sea deposits (S. Gripenberg 1934), which is interpreted in terms of the uplift of the Scandinavian Shield. During tectonic rest, in minor basins, particularly mountainous lakes, the grain-size diminishes upwards, depending on the reduction of the relief energy. These effects are generally masked by other factors.

Abrupt to catastrophic changes due to different causes

The characteristic phenomena of this kind influencing sedimentation are the sudden changes in the position of stream channels, subaqueous slumps, turbidity currents, and mud flows.

Changes in sedimentation caused by a sudden shift of stream channels have been described above. Channel deposits form only a small part of river sediments, while fine-grained deposits of the flood-plains and lateral accretion predominate. Therefore, a shift of the channel usually manifests itself only by an alternation of coarse and fine-grained sediments. The ideal vertical succession of sediments after an abrupt displacement of the river channel coincides with the lateral sequence across the river valley. Channel deposits cover sharply the old flood-plain, passing gradually upwards into the lateral accretion sediments, which are then covered by the young flood-plain.

In continental sediments these catastrophic phenomena are reflected by an alternation of alluvial (or eolian and lacustrine) and widely different proluvial deposits (alluvial cones, fanglomerates). Catastrophic phenomena and the sharp alternation of river deposits and mud flows or transitional sediments occur even in alluvial cones alone.

The characteristics of sediments and of bedding produced by revolutionary sedimentary processes, such as turbidity currents and mudflows, were given above (see p. 297, 298).

Breaks in sedimentation

One of the important and most difficult tasks of the study of ancient sediments is the restoration of stratigraphic hiatuses — breaks in sedimentation. It is necessary to draw information from Recent sediments where the break of sedimentation shows itself by the following phenomena:

1. Accumulation of organic remains, chiefly tests of benthonic organisms, *in situ* or redeposited. In places of non-sedimentation the floor on the shelf and continental slope is covered by pelecypod shells. Roughly speaking, the percentage of shells is indirectly proportionate to the rate of sedimentation.

2. The occurrence of relic coarse sediments. In Recent sediments, relics of earlier coarse-grained, mainly glacial sediments are developed in many places of non-sedimentation on shelves, continental slope and isolated elevations. Therefore, some intercalations of coarse-grained sediments in ancient deposits could be interpreted as a feature of long non-sedimentation following their deposition.

3. Occurrence of bimodal clastic sediments. They occur in places of direct erosion: in river channel, shallow-water sediments, or deep-water sediments when strong currents exist near the bottom. Therefore, they mostly designate a retardation of sedimentation, non-sedimentation or erosion.

4. The occurrence of horizons enriched in Fe-hydroxides or Mn-oxides. In fairly well-aerated (subaqueous) environments these deposits are concentrated in the upper, several mm to a few cm-thick layer, the thickness of which is indirectly proportionate to the rate of sedimentation. M. V. Klenova (1960) described an illustrative example from the Barentz Sea. Where the rate of sedimentation exceeds 30 cm/1,000 years, an oxidation bed does not develop. Where the sedimentation rate ranges from 10—30 cm in 1,000 years, a full superficial oxidation and concentration of Fe and Mn occurs. The preservation of this oxidation bed is another problem. It can be preserved when rapid burial and lithification take place; when sedimentation is slow

the Fe and Mn-compounds are fully reduced. A fossil red bed in which concretions of Fe and Mn-oxides occur within sediments of another coloration, originated probably as a result of slow sedimentation or a break of sedimentation.

5. The occurrence of phosphate, ferrous and manganese concretions and nodules. Phosphate concretions are accumulated mainly in places of slow or non-sedimentation. They are often associated with coarse-grained relict deposits. The materials (pieces of wood, calcareous tests) become phosphatized when they are long in contact with sea water. Manganese nodules are distinctly linked with anomalously slow sedimentation. An indirect proportion between the content of manganese and the rate of sedimentation has been established. The extremely slow deposition of layers in manganese nodules (1 mm in 1,000 years) proves that they can actually occur only in places of slowest sedimentation.

6. The occurrence of glauconitic sediments, insofar as the the glauconite is authigenic.

7. The occurrence of crusts of calcareous algae on pebbles, calcareous crusts of unknown origin or finds of the so-called hard-ground.* These highly special features require, of course, a very detailed study.

The possibilities of the preservation of sediments of various environments

The presence of sediments of various environments in geological history depends on a) Their distribution; b) The possibilities of their preservation; and c) Geotectonic factors (such as the stability and permanence of the oceans, large-scale vertical movements of the Earth's crust, etc.).

The estimated percentage of sediments of various environments on the present Earth's surface is given below:

Deep-sea sediments	
abyssal	53%
bathyal	20%
Sediments of open shelf	8%
Sediments of sheltered shelf	4%
Delta sediments	7%
Sediments of inland seas	5%
Sediments of river valleys and alluvial plains	2%
Beach sediments	<1%

Sediments of other environments are not considered because of the difficulties connected with the calculation or estimation of their areal extent.

* The term "hard ground" is used for the bedding planes of limestones that has been long exposed under water and consolidated before the deposition of further sediments. They contain accumulations of organic remains and abundant traces of the activity of benthos.

Deep-sea sediments are undoubtedly most prone to preservation. Having been buried by a thick complex of sediments they become part of fossil series on the Earth's surface after subsequent movement of the crust. Continental sediments have the slightest opportunity of being preserved. Sediments of subsiding intermontane depressions with the deposits of periodical lakes, alluvial cones and mudflows, further sediments of the lower courses of streams, alluvial plains and of all tectonically sinking parts of the continent with rapid sedimentation have comparatively good possibilities for preservation. Eolian and glacial sediments, as well as deposits of the middle reaches of streams when covered by other deposits, may occasionally become preserved. Lacustrine sediments are frequently preserved, especially in sinking parts of the continent, in the proximity of sea. Littoral, particularly beach sediments, are preserved only to a minimum extent because beaches are disturbed both during the positive and the negative movement of the sea. They can be protected from removal only when covered by deltaic deposits, alluvial cones or mud flows. In no case can they be presumed to constitute a major part of fossil complexes. The deposits of large deltas are preserved more frequently than all other shallow-sea sediments, because deltas are often places of a long-term sedimentation at the sites of a relatively intensive tectonic subsidence. Sediments of lagoons and bays, that are also characteristic of sinking coasts, are frequently preserved. The relative possibilities of preservation for sediments of various environments can be estimated as follows (100 — full preservation, O — preservation impossible).

Continental sediments	
Upper and middle courses of rivers	0–10
Lower courses and alluvial plains	50
Great lakes	40
Small lakes	20
Eolian sediments of dune fields	5
Loesses	5
Sediments of transitional environments	
Deltas of great rivers	80
Deltas of small rivers	25
Lagoons and bays	50
Marine sediments	
Open shelf	50
Deep-sea sediments	90

An evaluation of the possibilities of preservation for Recent sediments can be based on the frequency of their transition into the Subrecent state. The amount of the above-mentioned sediments of the individual environments passing into Subrecent state is of the same order as given above.

Geological and geotectonic conditions of sediment preservation can be defined as the transformation into such a fossil state that they can be found on or near the Earth's surface. The best prerequisites in this respect exist for continental sediments; they are worse for shallow-water and are worst for deep-sea deposits.

Thus, in geological history the deposits that are met with are those that satisfy all three conditions. From an evaluation of these, we can range roughly the sediments of various environments according to their possibilities of preservation as follows:

1. Sediments of sheltered shelf, deltas of major streams, and large inland basins.
2. Sediments of open shelf, deposits analogous to present sediments of continental slope, sediments of lower courses of streams, of alluvial plains, lakes and intermontane troughs.
3. Sediments analogous to present abyssal deposits, beach sediments and the major part of other continental sediments.

References

GRIPENBERG S. (1934): A study of the sediments of the North Baltic and adjoining seas. Fennia, vol. 60: 1–231, Helsingfors.

KLENOVA M. V. (1960): Geology of Barents Sea (in Russian). P. 1–366, Moscow.

KOSHECHKIN B. I. (1959): Stratigraphy of the bottom sediments of Taman Bay and its relation to the changes of climate (in Russian). Doklady AN SS SR, vol. 127: 846–848, Moscow.

OLAUSSON E. - OLSSON I. U. (1969): Varve stratigraphy in a core from the Gulf of Aden. Palaeogeography, Palaeoclimatolog., Palaeoecolog., vol. 6: 87–103, Amsterdam.

22. *Changes in sediments following their deposition – transition of Recent sediments into ancient deposits*

After deposition all sediments undergo a series of alterations before they are converted into sedimentary rocks. Ancient deposits have some features in common with Recent deposits and other features which are different. Some of the latter features are the result of diagenesis and lithification. In discussing the origin of ancient sedimentary rocks, attention must be first paid to the ascertainment of alternations that have affected the respective deposit since its deposition.

The approach to the study of diagenetic processes in Recent and Subrecent sediments may be diverse.

1. Direct study of the changes of mineralogical, structural and chemical composition in the drilling cores of Recent deposits.

2. The study of changes of the environment at various levels of the sediment. The pH and Eh conditions under the surface differ from those existing on the surface, and a transformation of the original mineral associations into other takes place.

3. The study of differences between Recent sediments and fossil sedimentary rocks.

Changes within Recent sediments

Changes in the physical state must be differentiated from those in the mineral and chemical composition. The former do not show the same character in all kinds of sediments.

In clays several stages of mechanical alterations are observable. In recently deposited sediments adsorbed films of water envelop all particles; they are squeezed out by pressure until the grains touch one another. At a 45% pore volume that first stage ends, and the stage of the reconstruction and the formation of denser structures begins, lasting until 37% pore volume is attained. Proceeding compression is accompanied by a deformation of clay minerals that are squeezed out into the interstices between harder grains. At the end of this stage, the pore volume is about 10%. During the final stage the harder grains, too, are deformed, the pore volume is negligible and the rock occupies only 20—45% of the original volume. The dewatering of sediments occurs intensively immediately after deposition, particularly within the few upper centimetres. The consolidation by the pressure of overlying sediments takes place only at a greater depth, the pressure attaining several hundred to a few thousand atmospheres.

In the calcareous sediments of biological and mechanical origin no substantial changes in the physical state are observable. Their consolidation requires a greater thickness of overlying sediments than in clay deposits. Loose organodetrital calcareous deposits were found even several hundred metres under the surface.

Calcareous, aragonite and calcite muds. For the mechanical compaction of aragonite muds a pressure of several tens of atm. is sufficient, the rock becoming fully recrystallized.

The changes in chemical, mineral and structural composition are much more varied and occur under the surface of all sediments.

Sands. In Recent sands, aureoles of secondary quartz are perceptible already several centimetres under the surface. Calcareous tests are dissolved and under favourable conditions calcareous cement is precipitated. In fine-grained sand rich in benthos, dark reducing zones with organic matter and the acid reaction of environment occur several centimetres under the surface.

Table 196

Composition of organic matter in Bering Sea sediments (Core No 619) (after O. K. Bordovskyi 1965)

Level cm	Org. C %	Total N %	Humid acids %	Bitumen content %
20— 40	1·57	0·191	0·78	0·070
125— 145	1·05	0·121	0·82	0·049
215— 230	1·12	0·114	0·74	0·106
280— 304	1·00	0·079	0·38	0·465
440— 460	0·61	0·066	0·27	0·079
670— 690	0·71	0·069	0·27	0·069
865— 885	0·68	0·071	0·22	0·074
1,250—1,260	0·61	0·067	0·29	0·059
1,340—1,350	0·71	0·069	0·24	0·059

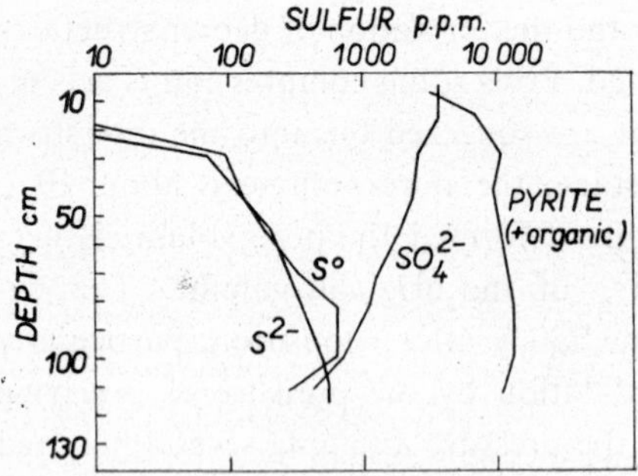

Fig. 111. Distribution of forms of sulphur with depth; core L-62 in the Gulf of California. After R. A. Berner (1964).

Clay sediments. In some Recent sediments mineral changes below the sediment surface depth have been observed, particularly the origin of illite or chlorite from montmorillonite. Simultaneously, K_2O increases in clays and a diagenetic loss of organic matter and a change of its substance take place. According to average data,

Table 197

Forms of iron in Bering Sea sediments (after O. K. Bordovskyi 1965)

Core No	Level cm	Quantity (% of dry sediment)				
		Fe total	Fe detrital	Fe pyritic	Fe^{2+} mobile	Fe^{3+} mobile
1532	5— 15	3·39	0·89	0·37	0·32	1·81
	50— 62	3·28	0·78	0·46	1·22	0·81
	120— 130	4·37	1·99	0·19	1·46	0·72
	130— 198	3·01	0·99	0·39	1·10	0·92
540	4— 10	2·84	0·87	0·18	0·87	0·92
	60— 70	2·91	0·81	0·30	0·94	0·86
	545— 560	4·03	1·15	0·29	1·57	1·25
	740— 755	3·50	0·51	0·34	1·69	0·95
	923— 943	2·90	0·61	0·47	1·43	0·39
	1,620—1,645	3·61	0·74	0·23	1·69	0·99
619	20— 40	2·49	0·12	0·12	1·11	1·15
	215— 230	3·50	0·20	0·38	1·48	1·43
	280— 304	3·53	0·31	0·29	1·34	1·60
	440— 460	3·61	0·51	0·16	1·12	1·68
	865— 885	3·62	0·52	0·16	1·09	1·84
	1,340—1,350	3·51	0·07	0·32	1·25	1·88

30—40% of organic matter gets lost during diagenesis, as a result of chemical and bacterial oxidation and regeneration. Humic and bituminous substances decrease with depth (see Table 196) and the sulphur régime alters. The percentage of pyrite-sulphur, elementary and bivalent sulphur increases with proceeding diagenesis, whereas the percentage of sulphate decreases. It is due to an intensive reduction of trivalent Fe and bacterial reduction of sulphates from sea water. The forms of iron change likewise. The percentage of pyrite-iron and mostly also trivalent Fe increases at the expense of bivalent particularly mobile Fe (Table 197).

The intensity of diagenetic changes and the resulting composition depend largely on the rate of sedimentation. At a rapid deposition, no equilibrium association can develop before consolidation, and rocks with organic matter and free Fe-hydroxides are formed. The study of Eh and pH conditions during the diagenesis of sediments suggests that the bulk of pyrite in sediments is of diagenetic origin. Under the surface of Recent sediments pyrite forms minute pellets, and aggregates of large crystals. Because of the moderately acid reaction, which frequently exists at a small depth under the surface, calcareous tests are dissolved.

In carbonate sediments diagenetic changes are extremely varied. The loss of Mg from Mg-rich calcites, the transformation of aragonite into calcite and of Mg-high calcites into dolomites and Mg-low calcites progressively increases with depth and time. In addition, other trace elements, such as Mn, Sr and Ba are segregated. Calcareous oozes soon after burial loose all the aragonite which is converted to calcite. Aragonite of the aragonite shells is generally more stable, sometimes surviving all diagenetic processes.

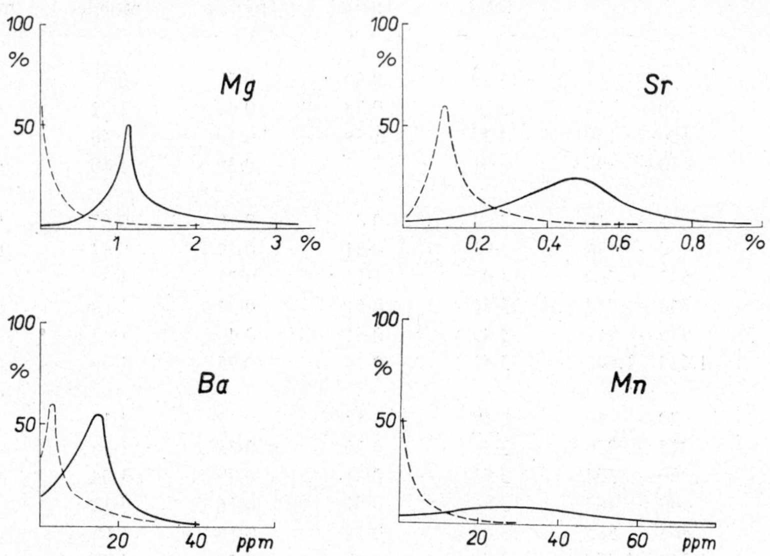

Fig. 112. Magnesium, strontium, barium and manganese concentrations in Recent and Pleistocene carbonate sediments from Florida. Full-line — Recent sediments, Dashed line — Pleistocene rocks. After F. G. Stehli - J. Hower (1961).

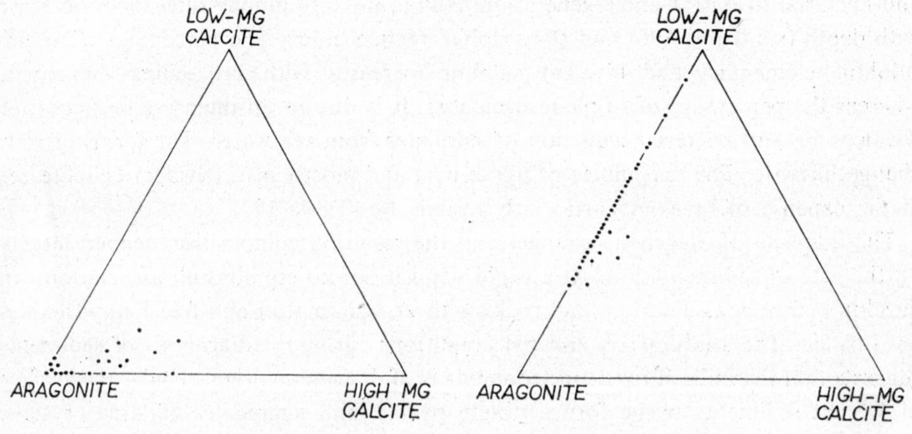

Fig. 113. Mineralogy of Recent oolite from the Great Bahama Bank (left triangle) and of Pleistocene oolite from the same area. After G. M. Friedman (1964).

The changing physico-chemical conditions produce frequent structural changes also on the surface (e. g. the formation of beach-rock). As a result of the dissolution and precipitation of carbonate in pores, chemical or biological cementation takes place under the surface. During the latter stages of diagenesis, the calcareous material

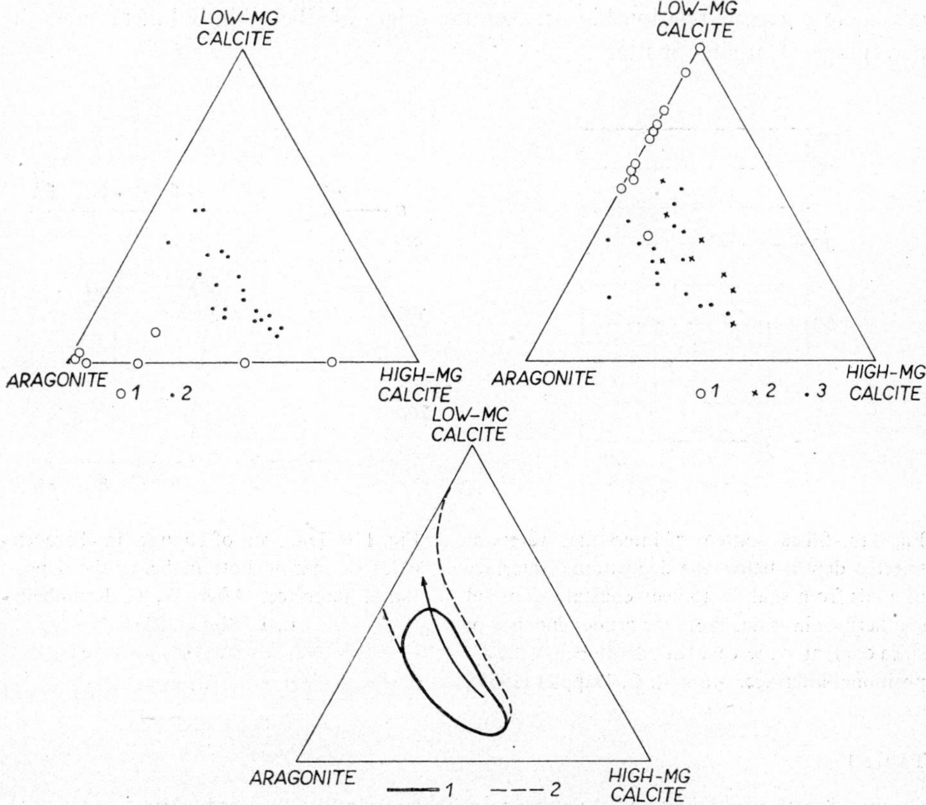

Fig. 114. Mineralogy of Recent carbonate sediments from Bermuda (upper left triangle) 1. Reef sediment, 2. Skeletal sediment. Mineralogy of Pleistocene skeletal sands from the same area (upper right triangle). 1. Beyond reach of marine spray, 2. From marine spray zone, 3. Submerged in sea water.
Lower triangle: Mineralogy of Recent and Pleistocene skeletal sands from Bermuda. The arrow indicates the direction of diagenetic alternation of those carbonate sands exposed to fresh water leaching. After G. M. Friedman (1964).

of impure fine-grained calcareous sediments differentiates and accumulates into nodules and concretions. At the transition to epigenesis, the selective leaching of carbonate from clayey-calcareous sediments leads to a vertical alternation of beds rich and poorer in carbonate.

Siliceous sediments. Diagenetic silicification was observed in Recent and Subrecent sediments. In the clayey sediments of tidal flats and deltas exsolution of pseudo-amorphous gel of silica were observed close under the surface. The accumulation of silica occurred during early diagenesis. In cores recovered from Recent clayey sediments the content of SiO_2 increases with depth. It also rises in layers richer in organic matter. In the seas of the Far East the content of authigenic SiO_2 universally increases with the increasing percentage of organic matter. These findings suggest a close genetic relationship between the origin of silicites in sediments and the percentage of organic matter.

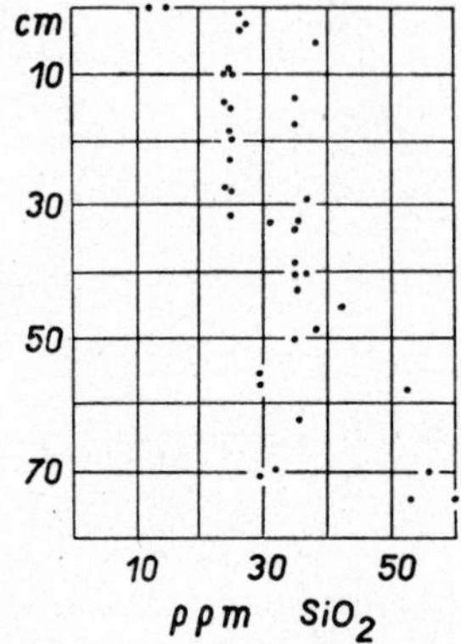

Fig. 115. Silica content of interstitial waters at selected depths below the depositional interface of muds from shallow to deep coastal waters off southern California. Note the crude tendency of silica content to increase with depth below the depositional interface. After E. C. Dapples (1959).

Fig. 116. Diagram of changes in characteristics of marine bottom below the depositional interface. After W. C. Krumbein - L. L. Sloss (1951).

Table 198

Composition of pore waters and their changes with depth in Black Sea sediments (after O. Shishkina 1961)

Depth in sediment cm	Br	Cl	SO_4	K	NH_4	Ca	Mg	pH
	mg/eq/kg							
0– 20	0·49	327	9·3	7·5	2·1	6·6	53	8·2
30–100	0·47	319	5·9	6·5	3·3	4·7	52	8·6
220–230	0·45	299	5·1	5·7	2·4	7·8	47	8·1
310–320	—	284	4·7	4·3	3·8	9·2	46	8·05
480–490	0·41	259	6·8	3·9	1·7	14·5	38	8·05
595–605	0·39	242	3·8	2·3	2·7	19·4	33	8·0
710–721	0·36	224	4·6	2·0	3·4	21·2	30	7·9

Salts are sediments most affected by diagenetic processes which often produce and absolute mineral transformation. The change leads from complex, strongly hydrated salts to simpler, less hydrated salts. The changes of conditions within the sediment after deposition and the changes in the composition of pore solutions have been studied in detail. Table 198 is an illustrative example of these changes in the sediments of the Black Sea (after O. SHISHKINA 1961).

Table 199

Reconstructions of original state of sediment by elimination of changes produced by diagenetic processes

Present state of fossil sediment	Diagenetic processes	Original composition of fresh sediment
Black clayey shale with 1% org. C, great amount of pyrite and slight carbonate admixture	Several cm below sediment surface pyritization caused by Eh changes, great decrease of organic carbon content, solution of organic remains and migration of carbonate	Black clay with 3–5% org. C, with siderite admixture and scarce calcareous shells
Dense limestone a little dolomitic with disseminated dolomite rhombohedra, with small amount of pyrite and organic matter	Compaction of aragonite mud and slight recrystallization, diagenetic dolomitization, several cm below sediment surface pyritization and loss of organic matter	Calcareous or aragonite mud with small admixture of organic remains; org.C about 1%
Nodular limestone with 70–80 % $CaCO_3$ in nodules and 30–50 % $CaCO_3$ in clayey matrix; in clayey streaks 2% pyrite and 2% org. C; admixture of authigenic silica	During diagenesis mobilization and migration of carbonate, differentiation of originally homogeneous carbonate mud into carbonate-rich nodules and carbonate-deficient matrix; pyritization of clayey streaks, concentration and preservation of organic matter in parts richer in silica	Carbonate mud with 50 to 60% $CaCO_3$, 3% org. C and probably slight admixture of sponge spicules or other siliceous organic remains
Clay shale with mixed layers clay minerals (illite-chlorite); kaolinite admixture	Diagenetic changes in clay mineral composition. Change of montmorillonite into illite-chlorite is probable	Pure clay with illite-chlorite ML and montmorillonite
Siliceous shale with scarce diatom frustules and sponge spicules, with 30% SiO_2, grains of glauconite and 1% org. C	Silicification by dissolution of diatom frustules, migration and precipitation of authigenic silica; decomposition of organic matter	Diatomaceous clay with 3% org. C

In all cores of Recent sediments the content of anions (particularly Cl′) decreases with depth and the amount of cations (especially Ca^{2+}) increases. In interstitial water of some sediments horizons richer in Cl′ anion have been ascertained. This increased concentration can be a genuine relic of the higher salinity of sea-water or the chlorine can be joined to the horizons with a higher content of organic matter. There is evidence available for both theories.

At a depth of several centimetres under the surface of the sediment a strong reduction of sulphates takes place; its high intensity is connected with the maximum number of bacteria in these layers. The pH value is usually lowered, but in the lower layers of the sediment it rises again roughly to the degree existing on the surface. The final composition of the sediment depends largely on the velocity at which it passes through the individual zones. When the sediment remains several centimetres under the surface for a long time and then passes swiftly through the lower zones and is lithified readily, it may preserve numerous characteristics of this distinct subsurface layer.

N. M. STRAKHOV (1961) proved that diagenetic processes were decisive for the regime of iron in sediments. Trivalent Fe transported in suspension is not reduced in sea water but under the surface of the sediment.

From what has been said above the importance of diagenesis as the controlling factor of sedimentation is obvious.

References

BERNER R. A. (1964): Distribution and diagenesis of sulfur in some sediments from the Gulf of California. Marine Geology, vol. 1: 117—140, Amsterdam.

BORDOVSKYI O. K. (1965): Transformation of organic matter in bottom sediments and its early diagenesis. Marine Geology, vol. 3: 83—114, Amsterdam.

BURST J. F. (1969): Diagenesis of Gulf Coast clayey sediments and its possible relation to petroleum migration. Bull. Am. Assoc. Petrol. Geol., vol. 53: 73—93, Tulsa.

DAPPLES E. C. (1959): The behavoiur of silica in diagenesis. Silica in sediments, p. 36—54, Tulsa.

EARDLEY A. J. - GVOSDETSKY V. (1960): Analysis of Pleistocene core from Great Salt Lake, Utah, Bull. Geol. Soc. Am., vol. 71: 1323—1344, Baltimore.

FISCHER A. G. (1961): Stratigraphic record of transgressing seas in light of sedimentation of Atlantic coast of New Jersey, Bull. Am. Assoc. Petrol. Geol., vol. 45: 1656—1666, Tulsa.

FRIEDMAN G. M. (1964): Early diagenesis and lithification in carbonate sediments. Jour. Sedimentary Petrology, vol. 34: 777—813, Menasha.

GORSHKOVA T. I. (1961): Sediments of the Norwegian Sea. Internat. Geol. Congr., 21. sess., vol. 19: 83, København.

HATHAWAY J. C. - ROBERTSON E. C. (1960): Microtexture of artificially consolidated mud. Bull. Geol. Soc. Am., vol. 71: 1883, Baltimore.

KAPLAN I. R. - RITTENBERG S. C. (1963): Basin sedimentation and diagenesis. In "The Sea". P. 583—619, London.

KRUMBEIN W. C. - SLOSS L. L. (1951): Stratigraphy and sedimentation. P. 1—489, San Francisco.

KULLENBERG B. (1952): On the salinity of the water contained in marine sediments. Medd. Oceanogr. Inst., vol. 21: 1—37, Göteborg.

LARSEN G. - CHILINGAR G. V. (Editors, 1967): Diagenesis in sediments. Developments in Sedimentology, vol. 8: 1—551, Amsterdam.

LOVE L. G. (1967): Early diagenetic iron sulphide in Recent sediments of the Wash (England). Sedimentology, vol. 9: 327—352, Amsterdam.

MAKEDONOV A. V. (1957): Some laws governing the geographical distribution of recent concretions in sediments and soils (in Russian). Izvestija AN SSSR, p. 43—58, Moscow.

PRAY I. C. (1946): Compaction in calcilutites. Bull. Geol. Soc. Am., vol. 71: 1946, Abstr., Baltimore.

RICHARDSON S. H. - BRAY E. E. (1963): Biogeochemistry of sediments in experimental Mohole. Jour. Sedimentary Petrology, vol. 33: 140—172, Menasha.

RICHTER V. G. (1961): Bottom deposits of Kara-Bougas-Gol Bay as the indicator of water level changes of Caspian Sea (in Russian). Bull. MOIP, otd. geol., vol. 36: 115—126, Moscow.

SHISHKINA O. V. (1961): Types of waters originating within marine sediments during diagenesis (in Russian). Recent sediments of seas and oceans. P. 548—559, Moscow.

SIEVER R. - GARELS R. (1962): Early diagenesis: composition of interstitial water of recent marine muds. Bull. Am. Assoc. Petrol. Geol., vol. 46: 279—280, Tulsa.

STEHLI F. G. - HOWER J. (1961): Mineralogy and early diagenesis of carbonate sediments. Jour. Sedimentary Petrology, vol. 31: 358—371, Menasha.

STRAKHOV N. M. (1961): Fundaments of the theory of lithogenesis. vol. 2: 1—516, Moscow.

THOMAS L. A. - BIGGS D. L. (1960): Rate and trend of carbonate diagenesis. Bull. Geol. Soc. Am., vol. 71: 1992; Abstr., Baltimore.

WELLER J. M. (1959): Compaction of sediments. Bull. Am. Assoc. Petrol. Geol., vol. 43: 273 — 310, Tulsa.

— (1960): Stratigraphic principles and practice. P. 1—725, New York.

WHITEHOUSE I. G. - MCCARTER R. S. (1958): Diagenetic modification of clay mineral types in artificial sea water. Clay and Clay Minerals, Nat. Acad. of Sci. Publ. 566; 1—81, Washington.

23. The application of knowledge of Recent sediments to ancient sediments

This chapter summarizes the methods and informations which make it possible to identify the sedimentary environment of ancient deposits. The correlation of Recent and ancient sedimentation, however, is not so easy and simple as can be presumed. The very approach to their investigation should be somewhat different as will be shown below (Table 200).

Table 200

Differences in the sedimentological investigation of Recent and ancient sediments

Recent sediments	Fossil sediments
Mainly thin surface layer over large areas	Mainly vertical section, horizontal distribution restricted
Environmental studies in horizontal sense	Facies variation in horizontal sense interrupted by size of outcrops
Time correlation of deposits well known	Time correlation of layers practically impossible

Although the composition and origin of ancient and Recent sediments have many features in common, some quantitative and qualitative differences are perceptible at first sight (Table 201).

There are the commonest and most striking differences between Recent and ancient sediments determinable by a rough comparison.

During the history of sedimentological and geological studies these differences have been emphasized and the possibility of the actualistic approach to the study of fossil sediments has been denied. An absolutely mechanical application of the knowledge of Recent sedimentation to the fossil deposits would lead to a harmful uniformitarian trend. Physical and chemical laws are invariable, but under different conditions they change their field of activity. The following examples will show to which errors the uniformitarian standpoint may lead:

Example 1. At present, coarse sand with gravel is found on continental slopes to a depth of several hundred metres. The mere application of this fact would lead to the conclusion that gravels and sands can be deposited under analogous conditions. This would be erroneous for most environments, because the occurrences of coarse sands are of Pleistocene age and do not correspond to contemporaneous sedimentary conditions.

Example 2. In deep-sea trenches sand occurs also deeper than 8,000 m. The mechanical application of this fact would be incorrect because the present deposition of sand at such depths was made possible by the steepness of slopes of deep-sea trenches and the enormous topographical differences which in many geological eras could not have existed.

Table 201

Differences between Recent and ancient sedimentation

Recent sediments	Ancient sediments
Calcareous oolitic sediments, shallow-water silicites, marine iron ores, phosphates, comparatively scarce	Oolitic limestones, iron ores, shallow-water silicites and phosphates very abundant in some geological epochs
Shallow-water limestones, with exception of coral reefs do not form thick and widespread deposits	Shallow-water limestones are widely distributed
Recent dolomites are rare	Ancient dolomites are very abundant
Sandy deltaic sediments are badly developed. Deltas with clayey-silty sediments prevail	Sand deltas are very frequent in shallow water sediments
Evaporites are comparatively rare	Evaporites are abundant in some geological epochs
Secondary sedimentary structures prevail (as result of disturbance of primary bedding by the action of organic and inorganic factors)	Primary structures are frequently preserved

Example 3. At present, 95% of the continental shelf area is covered by clastic sediments, in particular sands, silts and clays. From this it could be inferred that the shallow parts of the sea along continents were also the sites of terrigenous sedimentation. However, this was not the case in geological periods when small topographical differences caused only slight mechanical denudation and a voluminous biochemical sedimentation on extensive shallow water flats. On the other hand, in the present postorogenic and postglacial periods, great topographical differences between the source and sedimentary areas and intensive denudation cause extensive terrigenous sedimentation on continental shelves.

Thus, the correlation of Recent and ancient sediments is hampered by two basic factors:

1. The present postorogenic and postglacial stage is connected with sedimentation of a particular type; in consequence of the intensive mechanical denudation of the continent and the reworking of a great amount of glacial deposits terrigenous material is predominantly deposited.

2. As a result of the rapid rise of the sea level in the last 15,000 years, a large part of the sea floor has become in a state of dis-equilibrium, as sedimentation has not yet compensated for the change in depositional conditions. Therefore, in many places of world oceans and seas there are sediments corresponding to Pleistocene conditions of sedimentation (e. g. 30% of the continental shelves and part of the continental slopes).

Whole series of methods may be used for the reconstruction of ancient sedimentary environments. They can be divided into the following groups:

1. Petrographical, mineralogical and chemical methods.
2. Structural and lithological methods.
3. Biological, palaeontological and palaeoecological methods.
4. Facial methods.
5. Geological, palaeogeological and palaeogeographical methods.

A succesful reconstruction requires the use of as many criteria as possible.

Petrographical methods

Petrographical criteria are of primary importance for the reconstruction of ancient sedimentary environments, because the knowledge of the composition of sediments is a prerequisite for the recognition of their origin. In clastic sediments, which make up the prevalent part of present-day deposits, the determination of size-distribution is the first task. The basic factors affecting the grain-size of sediments are:

a) Size-distribution of parent rock and their weathering products and the mode of weathering;

b) Length and mode of transport;

c) Hydrodynamical conditions on the site of deposition.

In the last case, eolian and glacial sediments, whose mode of transportation is rather special, are not considered.

Material transported in individual environments differs in grain-size, as shown in Table 202 (data after F. Hjulström 1935 and Ch. Nevin 1946).

Glaciers, turbidity currents and mud flows have an unlimited capacity of transport. In discussing the relation between the environment and size distribution, the interference of the so-called special modes of transport must be taken into consideration. This concerns, for instance, transport by flotation, in tree roots, ice floes and ice

Table 202

Current velocities and corresponding maximal grains transported in various environments

Environment	Current velocity cm/sec	Maximal dimension of transported grains mm
Rivers: Upper course	1,000	1,000
Middle course	300	250
Lower course	200	150
Tidal currents in near-shore environments	150–160	120 (in suspension)
Oceanic currents	up to 30	5
Marine currents	up to 20	3
Currents in inland seas and great lakes	10	0·8
Longshore currents	up to 100	100
Deep-water currents in oceans	up to 25	3·5
Deep-water currents in seas	up to 10	0·8
Deep-water currents on shelf	up to 20	up to 3

blocks (which can carry boulders to a distance of several thousand km) and by brown algae.

Near the shores, the intensity of transport is modified by wave action that increases with the increase of basin dimension (cf. p. 12). The wave intensity is so immense that at the points where the waves break even a very huge body can be set into motion.

From these data we can already make some conclusion about the sedimentary environments of certain deposits. We take into consideration such environments, where the given grain-size occurs most frequently and do not consider other ones, where these grain-sizes do not occur.

A more detailed estimate of the genesis of sediments can be made on the basis of their total size distribution. Recently, a number of research workers have attempted to determine the sedimentary environment from the character of size distribution curves (their course or parameters), proceeding from the following considerations: In every environment one or more modes of deposition are predominant, each of which is typical by its more or less defined grain-size distribution. For convenience these modes can be divided into the following groups:

1. Sedimentation from unidirectional water stream (fluviatile processes).
2. Sedimentation under wave influence.
3. Tranquil sedimentation from suspension.
4. Sedimentation from turbidity currents.
5. Sedimentation from mud flows.
6. Glacial sedimentation.
7. Eolian sedimentation.

In individual environments the following modes of sedimentation are active (arranged in descending order):

Continental environments	
River channel	1, 2
Flood-plain	1, 2, 3
Alluvial lakes, marshes	3, 1
Small lakes	3, 2, 1
Great lakes and inland seas	3, 2, 1
Marine environments	
Deltas of great rivers	1, 2, 3, 4
Sheltered shelf	2, 3, 1
Beaches	2, 1
Inner open shelf	1, 2, 3
Continental slope	3, 1, 4
Abyssal regions	3, 4
Deep-sea trenches	3, 4, 5

According to the present state of study of size-distribution in sediments, the correlation of parameters (Md, Sorting coefficient, and Skewness, cf. p. 21) seems to be the best method available. In the text, parameters of sediments of different

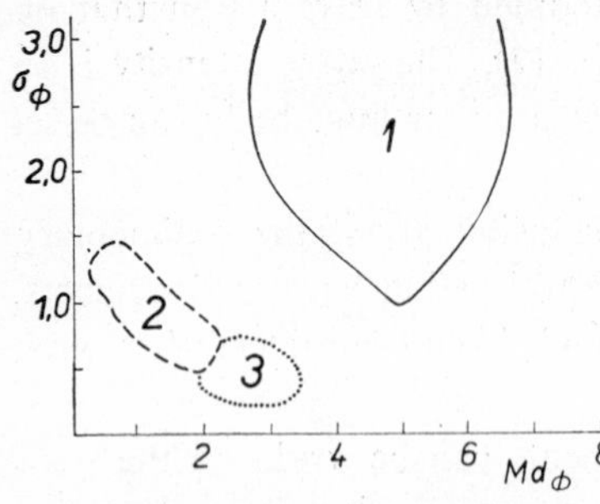

Fig. 117. Grain-size parameters of shallow-water sediments of San Miguel lagoon in the form of a diagram illustrating the relation between Phi Median Diameter and Phi Deviation Measure. 1. Sediments quietly deposited from suspension, 2. Sediments deposited by unidirectional currents, 3. Sediments deposited under wave action. After J. H. Stewart (1958).

environments are given. In plotting the points of parameters into J. H. STEWART's (1958) diagram (Fig. 117), the products of different types of sedimentation can be distinguished. Sediments deposited from suspension in tranquil water have Md below 0·05 mm, Phi Deviation above 2·0 and the Phi Skewness varies in value.

Sediments deposited under influence of wave action have Md between 0·23 and 0·120 mm. Phi Deviation is low, almost invariably below 0·5, and Phi Skewness is either positive or negative. Recently many data proved that due to the presence of greater amount of coarser fractions beach sediments are mostly negatively skewed.

The distinctive feature of beach sands in their small content of clay and silt fraction (up to 5%) which differentiates them from sands of eolian origin. All beach sands contain a clay admixture, because they develop from heterogeneous material by a constant outwashing of finer fractions, a small part of which is entrapped among the grains or adhere to grains of a coarser grade. Dune sands, however, never contain finer material, because they originated by the winnowing of grains of a given size from the surrounding beach, fluviatile, glacial and other sediments or soils. As will be shown below, water-laid sediments can be distinguished from the eolian ones by the light/heavy fraction ratio.

The differentiation of beach sediments (deposited under the influence of wave action) from fluviatile sediments (deposited by unidirectional current) is not always unequivocal either, as fluviatile sediments have also a prevalence of coarser fractions and, consequently, a negative Phi Skewness, yet their sorting is mostly poorer. In 95% of cases Phi Deviation (σ_Φ) of beach sediments varies between 0·20 to 0·40, while in fluviatile sediments this values are generally higher. In 30% of cases σ_Φ of river sands exceeds 0·5. The Sorting coefficient (Trask's) is 1·1—1·5 for beach sands and in approximately 20% of cases it is higher than 1·5 in river sands. Recently some authors prefer to use moment measures for distinguishing beach sands from river and dune sands. According to S. K. MISHRA (1968) the dune and beach sands can be separated on the basis of scatter plots of first and third moment plot. The river sands may be distinguished from beach and dune sands by the first and second moment plot.

The deposition from turbidity currents does not show any pronounced size distribution. Most of these deposits have Md of silt and fine sand grade (from 0·03 to 0·068 mm). The Phi Deviation usually oscillates between 0·5 and 1·0; Trask's So varies between 1·7 and 2·1.

The sediments of mud flows have a particular size distribution differing from that of other sediments by poor sorting. The values of σ_Φ are generally above 2·0 and So is higher than 5.

The size distribution of glacial sediments is most varied. Their Md corresponds to the values of gravel, sand and clay fractions. Sorting is generally poor (σ_Φ above 3, So above 5), but there also exist sediments (especially glaciofluvial) with σ_Φ below 0·8 and So below 1·5.

For the determination of sedimentary environments based on size distribution the relation of the Md of grain size and the Phi standard deviation (Sorting coefficient) have proved convenient. Besides J. H. STEWART (1958), mentioned above, L. B. RUKHIN (1947), D. L. INMAN and T. K. CHAMBERLAIN (1956) and R. A. CADI-

GAN (1961) were the first to use this method. Later, it was applied by many others (e. g. W. R. DICKINSON 1968).

The solution of the genesis of sediments based on the shape of grain-size curves is usually unsuccesful, as for the derivation of sedimentary conditions not one type of analysis but the general tendency of parameter distribution must be applied. The CM method of R. PASSEGA (1957, 1962) seems to be promising; it uses the ratio of Md/first percentile (virtually the maximum grain-size) as the genetic parameter. The points of individual analyses form characteristic patterns on the diagram,

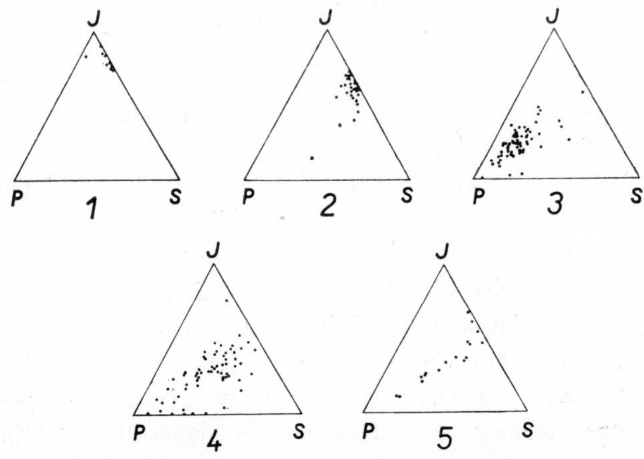

Fig. 118. Grain-size distribution of marine sediments, expressed in triangular diagrams, with end-members: clay (J), silt (S), sand (P). 1. Central parts of the Mediterranean Sea, 2. Northern parts of the Red Sea, 3. West coast of North America, 4. Barataria Bay, 5. Gulf of Mexico, 6. Equatorial Atlantic Ocean. After G. Müller (1961).

depending on the mode of transport and sedimentation. Although thousands of analyses have already been evaluated with the help of this method, it does not yet seem to provide a sufficiently good basis for interpretation of the environments of sedimentation.

Besides the above-mentioned methods, many others have been proposed, of which the Dutch (D. J. DOEGLAS, TJ. VAN ANDEL and H. POSTMA, D. G. J. NOTA), French (A. RIVIÉRE) and German (K. H. SINDOWSKI) should be mentioned. All these methods, however, appear to be very generalized and are not easily applicable owing to the differences in the mode of illustrating the size distribution. Some of these methods are based on the presumption that the resulting size composition in various places is produced by the mixing of modes (fractions of a definite grain-size range), which, according to the hydrodynamical régime, behave quite independently and are separately introduced into the basin, mixing in some parts and differentiating in others.

In some cases the plotting of sand - silt and clay fractions may prove useful. Diagrams of this type are shown in Figs 118 and 119. In sediments deposited from unidirectional current all three fractions are in equilibrium, or part of silt can be missing. Sediments deposited under the influence of wave action are richest in sand, whereas deposits settled from suspension are richest in clay and have a somewhat variable admixture of silt and fine sand. A general rule is valid: sediments deposited rapidly in places where no reworking occurred contain a fairly large amount of silt or the contents of all three fractions are in equilibrium. Where erosion of whatever type took place, a tendency to the formation of uni- or bimodal sediments appears. Sand and clay fraction may be enriched at the expense of the silt fraction (e. g. on shallow-water flats, in straits, etc.).

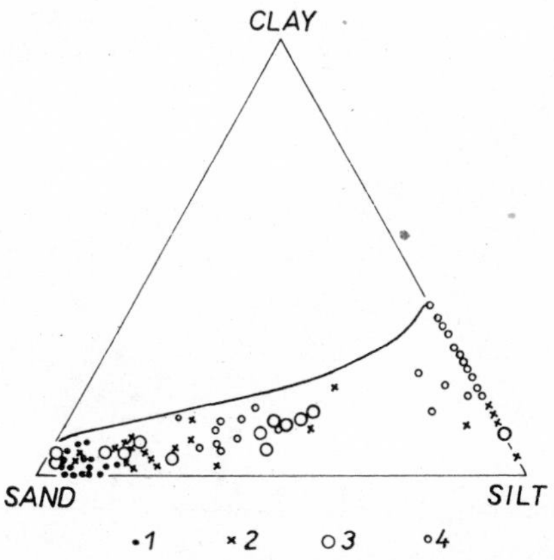

Fig. 119. Relationship between particle size of Burton Lake sediments and water depth. Sample depth: 1. 0—1·5 m, 2. 2—3 m, 3. 3·5—8 m, 4. 8—16 m. After J. R. Reid (1961).

These questions are already closely connected with the bimodality of sediments, which can contribute largely to the solution of the sedimentary environment of ancient deposits. This bimodality (shown by a two-peak histogram) may originate in several ways.

1. Mixing of several fractions widely differing in grain-size, as for instance, at the transition from barrier sands into lagoonal clays or from eolian sand dunes into bay clays. These products of mixing, however, are of no diagnostic significance for the determination of environments.

2. Erosion. In Recent deposits, bimodal sediments have been found to exist mainly in shallow-water conditions, in places of occasional non-sedimentation or erosion, in streams, in shallow-water environment affected by wave and current action, straits, etc. Coarse-grained river sediments are distinguished by the typical bimodality type: gravel — medium and fine-grained sand, with the hiatus in coarse-sand fraction. Bimodal sediments of sand-clay (without silt) composition occur wherever concomitant erosion of Recent sediments takes place. Thus, for instance, in the Caspian Sea 90% of bimodal sediment occurrences are at smaller depth than 40 m, their origin at greater depths being likewise affected by deep currents of greater velocity

that erode the developing sediments. Pronouncedly bimodal sediments of sand-clay composition occur in straits connecting lagoons with the shelf sea or island seas with ocean. Bimodality may also appear as a mixture of larger tests with finer inorganic sediment. According to R. J. RUSSELL (1968) the bimodality in fluviatile and similar sediments can originate by means of the removal of "hydrodynamically unstable" fraction 1—6 mm which is sorted out of bed loads of rivers and moved shoreward to accumulate in beaches.

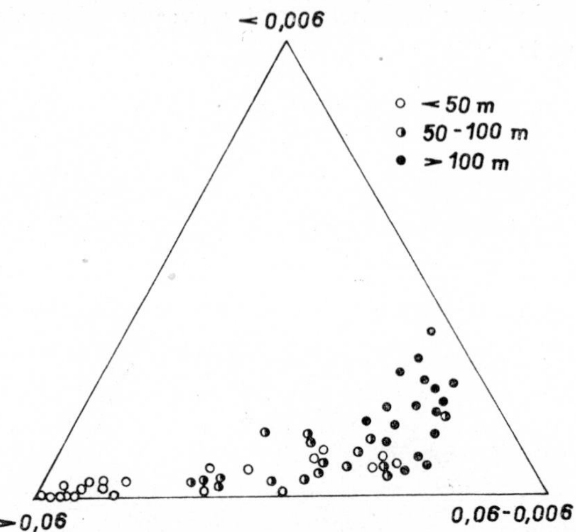

Fig. 120. Relationship between particle size of the sediments of the Gulf of Naples and water depth. After H. Sindowski (1958).

The occurrence of bimodal sediments can be generally evaluated as follows: The gravel – fine-grained sand composition suggests a river or, in the case of marine sediments, a strait or beach environment of deposition. From the sand-clay composition with more than 10% of sand, the deposition in shallow-water environment of slow sedimentation (possibly also erosion) or at boundary between the sandy and clayey deposition (e. g. between sand barriers and lagoons) may be deduced. The admixture of less than 5% of sand grains, especially when rounded or with a dull surface, points to the presence of eolian sand.

The depth distribution of individual clastic sediments, which is one of the most important questions to be solved, depends on the following basic factors:

1. Difference in the altitude of the source and depositional areas. The greater the difference and the smaller the distance between the respective areas, the deeper the coarse-grained sediments reach.

2. Hydrodynamical properties of the basin. The depth of the occurrence of coarse-grained sediments is directly proportionate to the energy of wave and current action. As this energy increases with the size of the basin, coarse-grained sediments are found in large basins, provided that other conditions remain constant.

3. The slope of the basin floor and the seismicity within its area. These two factors generally act together, because the steeper are the slopes of the basin, the greater their seismicity. The increase of these factors results in the spread of coarse sediments to greater depths.

Table 203

Bathymetric conditions of gravel and sand deposition in some basins together with the intensity of various factors

Basin	Maxim. depth of gravel occurrences	Maxim. depth of sand occurrences m	Morphological differences	Hydro-dynamical energy	Seismicity and basin slopes
North German lakes	several cm	1	1	1	1
Caspian Sea (north. part)	2 m	8	2	2	2
Caspian Sea (south. part)	5 m	17	4	2	3
Lake Michigan	1 m		1	2	1
Norwegian shelf and slope	30 m	1,000	4	3	3
Lake Issyk-Kul	34 m	100	4	1	3
Pozzuoli Bay	1 m	6	2	1	1
Mediterranean Sea (Egypt)		7	1	3	1
Kurile trench	800 m	3,000	5	3	5
Santa Barbara Basin (California)		2,000	4	4	5

Table 203 presents bathymetric conditions of gravels and sands in some basins together with the intensity of individual factors (in ascending order, from 1 to 5).

The review shows clearly that the bathymetric distribution of coarse sediments is affected chiefly by the slope of the basin floor, seismicity and topographical differences. The influence of hydrodynamic energy and other factors is subordinate.

Table 204

Interaction of basic factors influencing the bathymetric distribution of gravels (upper value) and sands (lower value)

		Morphological differences		Hydrodynamical energy		Seismicity and bottom slopes	
		small	great	small	great	small	great
Morphological differences	small	—	—	few cm 2 m	1— 2 m 5—10 m	2 m 10 m	3,000 m unlimited
	great	—	—	50 m 100 m	100 m 1,000 m	10 m 100 m	3,000 m unlimited
Hydro-dynamical energy	small	—	—	—	—	few cm 2 m	3,000 m unlimited
	great	—	—	—	—	1 m 10 m	3,000 m unlimited

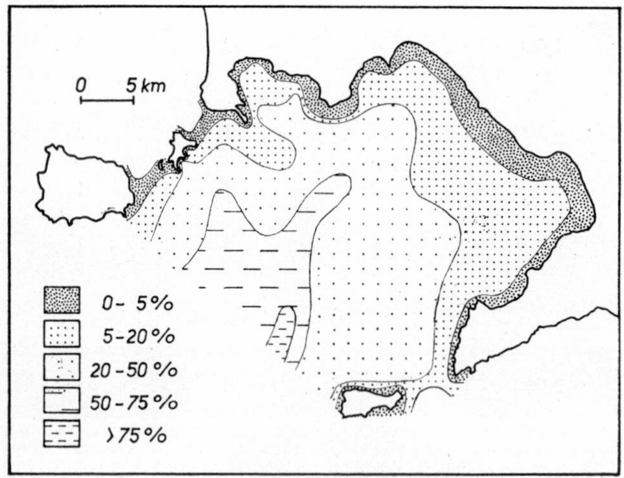

Fig. 121. Regular distribution of grain-size fraction <0·002 mm in the sediments of the Gulf of Naples. The amount of this fine fraction regularly grows with increasing depth. After G. Müller (1963).

The interaction of the basic factors and the estimations of bathymetric conditions for coarser sediments are tabulated in Table 204 (after existing conditions in Recent basins; the upper line refers to gravels and the lower for sands).

Individual environments, as already recorded in the descriptions of their sediments, occassionally show a characteric enrichment in some fractions. Figs 123 and 124 express diagrammatically these conditions; the size-distribution of individual environments may, of course, differ in details from this diagram.

In solving the conditions of the whole sedimentary area, we are confronted with many diverse ancient environments, particularly when the margins of the basins are preserved. In this case, the comparison of size distributions (either parameters, or the contents of individual fractions) is of upmost importance. An instruction was published, for instance, by K. O. Emery (1955); his Table lists all basic parameters of all environments of one sedimentary area — the Californian coast, up to deep-water region (Table 205).

Table 205

Md and So of sediments of the California coast region (after K. O. Emery 1955)

Environments	Md μ	So (Trask's)
Coastal soils	80	3·1
Mud-flow sediments	840	3·1
Cave sediments	500	11·0
Eolian dust	31	1·5
Bottom of playas	4	5·0
River channel	610	2·0
River suspension	46	2·2
Dunes	260	1·3
Marshes	32	4·2
Beaches	350	1·3
Continental shelf	140	1·6
Insular shelf	150	1·8
Bank surface	270	2·3
Straits and inlets	1,000	1·9
Isolated deep basins	3·5	3·8
Continental slope	9·2	4·2
Abyssal ocean (Globigerina ooze and brown clay)	1,3	2·7

The environment of gravel deposition can, to a large extent, be determined on the basis of grain-size parameters. According to K. O. Emery (1955), the gravels of various environments show the following difference (Table 206).

The best diagnostic feature of the origin of conglomerates and gravels is the degree of sorting. Whereas the Md varies irregularly, the values of Sorting coefficient may be fairly characteristic of some environments.

The findings bearing on clastic sediments can be applied to limestone of mechanical origin. Illustrative examples are recorded from the Persian Gulf (cf. p. 239); the following relations established there can be applied to fossil limestones. The depth

Table 206

Differences in grain-size and sorting of gravels of various environments (after K. O. Emery 1955)

Gravels	Md (mm)		So (Trask's)	
	mean	range	mean	average
Marine beach	56·0	10·8—750·0	1·25	1·13—2·14
Lacustrine beach	47·0	13·0—125·0	1·15	1·09—1·21
Fluviatile	19·0	10·4—355·0	3·18	1·34—5·49
Alluvial fans	16·5	10·0— 64·0	5·33	2·50—8·95

boundary between well-sorted bioclastic limestones with rounded grains and those with unsorted, unrounded grains ranges from 10 to 18 m. The boundary between the deposition of skeletal and fine grained dense limestones is drawn between 35 and 50 m.

Roundness and morphology of grains

The roundness of clastic grains is the result of transport and of the processes that functioned during deposition. In the comprehensive literature, several contradictory opinions on the development of roundness and its significance for the solution of sedimentary environments have been expressed. We shall mention here only those factors which can help us to throw more light on ancient sedimentary areas. First it should be noted that a detailed study of roundness by the various methods that are available is extremely time-consuming; so far, the comparative study, particularly after Krumbein standard scale, has proved most reliable.

As shown by a number of authors, the rounding of quartz grains is a very slow process and no substantial rounding of grains by river transportation can be achieved during one sedimentary cycle. Eolian, and to a minor extent also littoral processes can produce an appreciable roundness. Ph. H. Kuenen (1959) came to the conclusion that nearly all quartz grains had been rounded by wind action. Although this opinion is rather exaggerated, it draws attention to the predominant importance of eolian processes for the rounding of grains and to the minimum influence of fluvial processes. In evaluating the roundness two important factors must not be omitted:

1. Rounded grains are multicycle products; they derive from earlier sediments and were rounded during one or several foregoing cycles.

2. The rounding of grains depends, to a high degree, on their size. Therefore, only strictly defined fractions should be compared. The fractions 0·15—0·25 mm is most convenient for a genetic evaluation.

The highest degree of roundness is shown by eolian sands, particularly in coastal dunes and wind-blown parts of sand banks and shallow-water flats. It has been proved statistically that eolian sands show a somewhat higher values of roundness coefficient than beach sands (70% of the cases investigated corresponded to this presumption). The differences, however, are so small that the roundness can

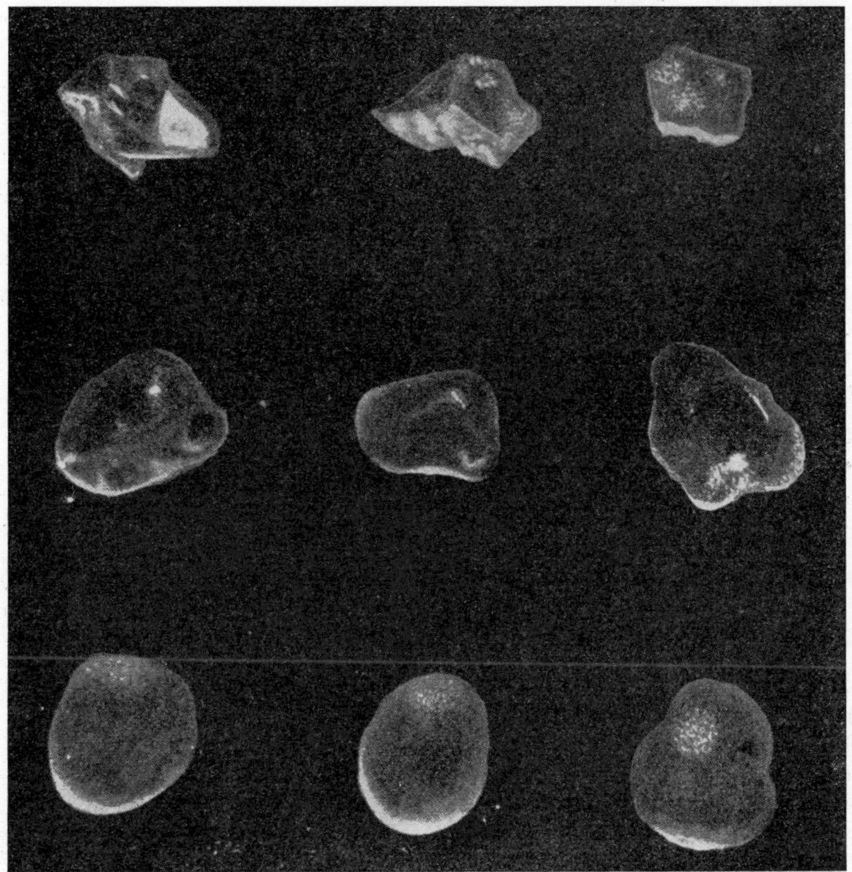

Photo 17. Morphometric types of quartz grains. Angular grains (upper row), rounded grains with polished surface (middle row) and rounded grains with dull surface (lower row).

not serve as a key for the differentiation of these two kinds of sands. Yet, they are easily distinguishable (when they are first-cycle sediments) from sands which have originated in other environments.

Generally, the sphericity of grains is not a suitable diagnostic feature, because of its dependence on the original shape and the mechanical properties of grains.

A method frequently used for the reconstruction of an environments is the study of grain surface, so-called morphoscopy. One of the simplest procedures is the examination of the dull grain surface. The problem has already been discussed above (p. 126, 127); in connection with the problem dealt with in this chapter, it should be repeated only that there are two types of dull surfaces:

1. The dullness is produced by the presence of minute irregular pits with rounded edges. These grains originate only in eolian sediments and their dull surface was produced by the impact of other grains.

2. The dullness resulting from fairly regular, several microns large etch figures, that are formed by solution by aggresive waters.

It is obvious that the determination of the type of dull surface is the first step in the reconstruction of a sedimentary environment from this characteristic.

Previously, A. CAILLEUX's method (1943, 1945), based on the grain size-roundness-grain surface relation, was frequently used. This author distinguishes three kinds of grains: angular, rounded with polished surface, and rounded with dull surface. The grains of the second group are generally of marine origin and those of the third group are deposited by wind. A. CAILLEUX's Table 207 summarizing the results concerning the Sahara sediments is given below.

Table 207

Characteristics of quartz grains of sediments of different ages in the Sahara (after A. Cailleux 1945)

Age	Environments	Angular grains %	Rounded polished grains %	Rounded dull grains %
Quarternary	continental	15	15	70
Pliocene	continental	60	5	35
Oligocene	continental	35	45	20
Cretaceous-Eocene	marine	60	35	5
Permian-Triassic-Jurassic	continental	35	10	55
Devonian	marine	?	?	present
Cambrian-Ordovician	continental passing to marine	30	0	7

The results show clearly that rounded dull grains are mostly of eolian origin. The situation is rather complicated in littoral sediments, where there is a constant exchange between the beach and the eolian sediments.

A. CAILLEUX (1945) expressed the following rule: If the fraction above 0·3 mm contains more than 30% of rounded and polished grains, the sediment is of marine origin. As mentioned above, the validity of this method is limited by a number of subsidiary effects and can be applied only to unconsolidated sediments, lacking any sign of silicification. The same author introduced a method for the differentiation of sedimentary environments for gravels, using two parameters — the Flatness coefficient and Dissymmetry coefficient. The parameters for pebbles of different origin are as follows (Table 208).

Table 208

Average parameters for pebbles of different origin (after A. Cailleux 1943)

Parameter	Fluviatile	Marine
Coefficient of flatness of limestone pebbles (5 cm diameter)	1·600—3·400	2·400—3·000
Coefficient of dissymmetry of limestone pebbles (diameter 3—3·5 cm)	0·559—0·585	0·545—0·558

This method is best applicable when a large number of analyses is available and the tendency in the distribution of coefficient of flatness and dissymmetry is traceable. The disadvantage of this method is the necessity to compare pebbles of the same composition and size. It has been experimentally evidenced that pebbles of various fractions and rocks frequently behave quite differently during transport. Surf action is the most effective factor in the rounding ot pebbles (PH. H. KUENEN 1964). Under surf conditions produced in the laboratory, limestone pebbles become rounded within an hour. Quartzite pebbles need a three times and chert pebbles up to thirteen times longer period to get rounded. Natural surf, of course, does not affect pebbles continuously and the process is estimated to last 9,000 times longer than under experimental conditions.

Some authors are of the opinion that flat pebbles are more frequent in beach deposits than in any other environments. Unfortunately, the difference is not great enough to be used for reconstruction purposes. It is of interest that PH. H. KUENEN (1964) does not ascribe the prevalence of flat pebbles in beach deposits to surf abrasion but to selection according to shape (shape sorting).

The striation of pebbles is not such a decisive proof of glacial origin as presumed so far. It is much scarcer even in true glacial sediments and morainic deposits have been found from which all striated pebbles were absent. The striation of pebbles appears most frequently on slightly resistant rocks, chiefly on limestones. It is not the most abundant structure of glacial (p. 149). deposits; various pressure phenomena, a pitted surface or traces of abrasion are far more frequent but these do not point unmistakably to glacial origin. The wedge-like shape of pebbles is also regarded as indicative of glacial origin, but this diagnostic feature does not seem to be very exact. Investigation has revealed that glacial pebbles are very difficult to differentiate from pebbles of other origin, because their roundness does not differ much from that of river gravels transported to a small distance; more than 10% of glacial pebbles have a value of roundness above 0·4.

Reconstruction of environments on the basis of heavy mineral content

The quantitative content of heavy minerals may be an important diagnostic feature of a ancient sedimentary environment. The theoretical study of this question and the direct examination of sediments have shown that the differences in the grain-size of light and heavy fractions are the most reliable basis for achieving the best results.

Table 209

Ratio of light/heavy fractions in water-laid and eolian sediments

	Ratio r of quartz/r of heavy mineral in water	Ratio r of quartz/r of heavy mineral in air
Magnetite	2·56	1·96
Pyrite	2·39	1·87
Garnet	1·94	1·58
Zircon	2·24	1·78
Tourmaline	1·26	1·16

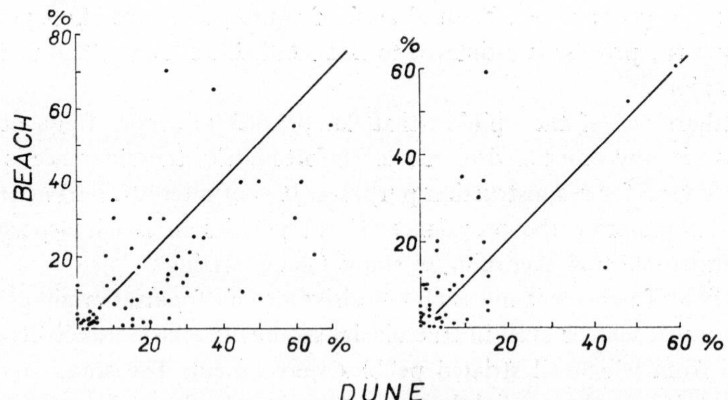

Fig. 122. Left diagram: Comparison of black mineral percentages in the silt fraction of the beach (foreshore) and dune pairs. The higher content of black minerals in the dunes along coast is indicated. — Right diagram: Comparison of percentages of shells in the 0·25—0·5 mm fraction in the paired beach (foreshore) and dune samples. Here a greater quantity of shells is found on the beaches. After F. P. Shepard *et al.* (1961).

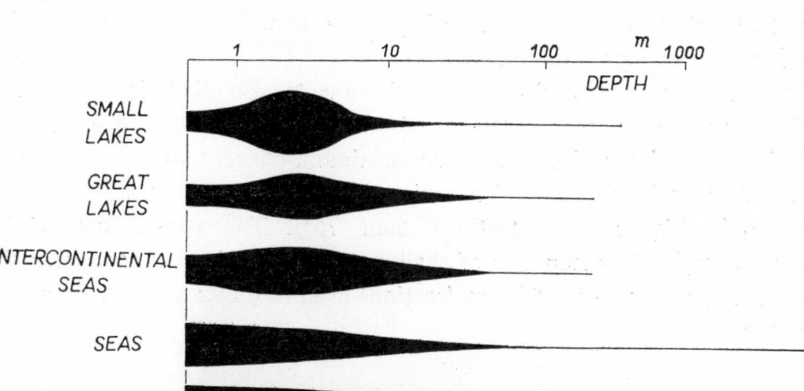

Fig. 123. Bathymetric distribution of Recent sands in the individual sedimentary environments.

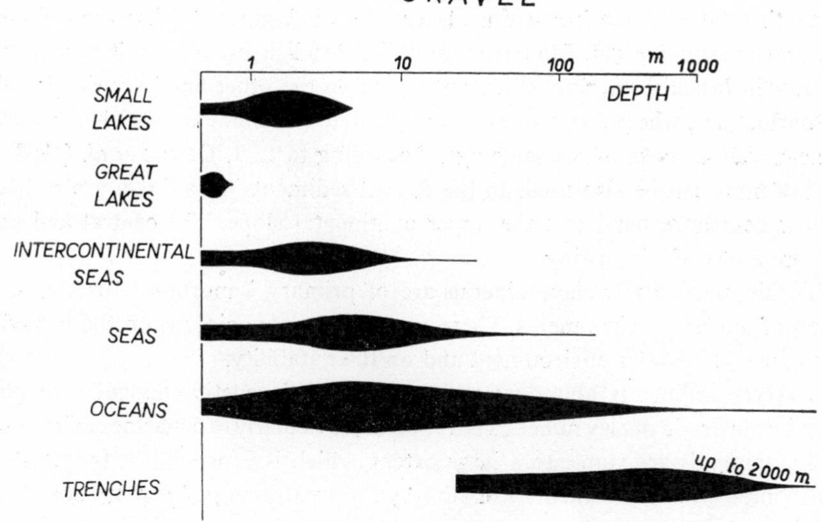

Fig. 124. Bathymetric distribution of Recent gravel in the individual sedimentary environments.

The lesser is the density of environments of transport and sedimentation, the lesser are the differences in the grain-size of these two fractions. In eolian sediments the grades of the two fractions would be closer to each other than in aqueous deposits. Several relations theoretically derived are given in Table 209.

Although this method has not yet been fully utilized, the existing experiences indicate that sufficient data will make it possible to differentiate between the two above-mentioned groups of sediments.

Shallow-water sediments of various environments also differ in the content of heavy fractions. Natural concentrations of heavy minerals exist both in marine and in fluviatile and eolian sediments, yet the maximum concentrations are confined to marine, and occassionally to eolian sediments. Fluviatile heavy mineral fractions do not attain higher concentrations than 75%. The comparison of heavy mineral contents in the complexes of shallow-water deposits has very often shown that the maximum amounts of heavy fractions exist in dunes and in the upper part of sand barriers.

The reconstruction of environment on the basis of mineral composition

Clastic minerals reflect the source of material rather than the sedimentary environment. Therefore, a detailed study of clastic minerals in Recent sediments has not revealed any likelihood of their applicability for determining sedimentary environments. A study of sediments of the Tertiary Molasse in the Alps shows that in some cases continental and marine sediments can be distinguished after the colour of biotite and tourmaline (H. FÜCHTBAUER 1963). Reddish brown biotites have been found only in brackish-marine sediments, which on the other hand are rich in olive-green tourmalines, whereas brown and blue-green tourmalines are enriched relatively in limno-fluviatile rocks of the same age. According to L. J. DOYLE *et al.* (1968) the amount of mica can be also used. In the Recent sediments mica flakes are restricted to narrow nearshore band and the upper continental slope. The central and outer shelves are areas of winnowing.

Authigenic, particularly clay, minerals are of primary importance for the reconstruction of ancient environments. There are two different opinions on the behaviour of clay minerals in water environment and on their stability:

1. In clayey sediments abundant syngenetic and diagenetic changes take place, and authigenic origin of clay minerals is frequent, particularly in a marine environment.

2. Clay minerals are stable to a large extent, which is expressed by the constancy of their composition from their formation by the weathering of parent rocks to the lithification of sediments.

The stability of clay minerals has been demonstrated mainly by fact that in some basins their composition corresponds to that of clay minerals present in the suspended matter of rivers feeding the basins or in the superficial mantle of the adjacent landmass.

The syngenetic and diagenetic formations of clay minerals has shown to exist by the following evidence (Table 210).

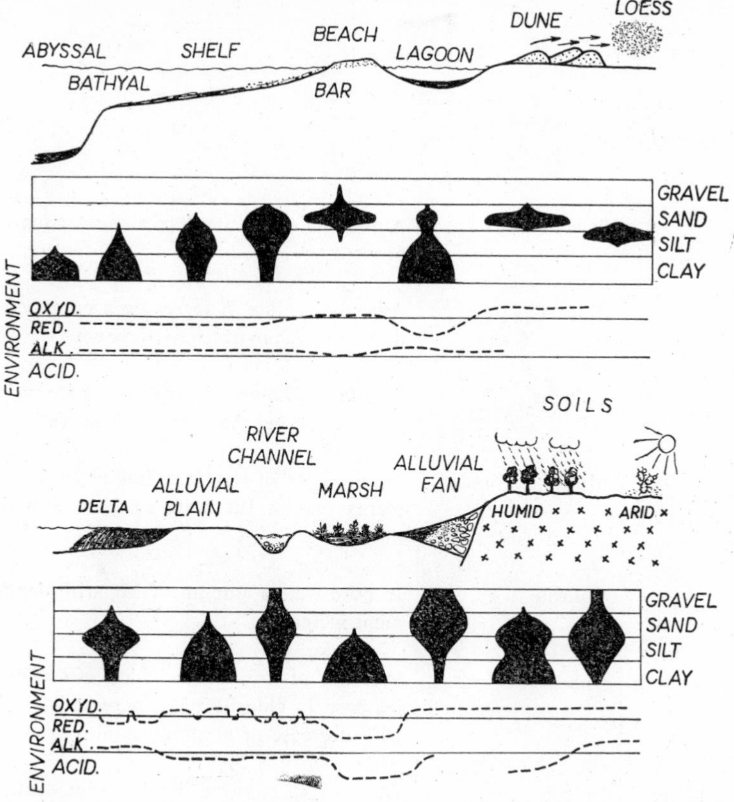

Fig. 125. Distribution of gravel, sand, silt and clay in continental and marine sedimentary environment. The curves illustrating Eh and pH conditions are added. After R. L. Folk (1961).

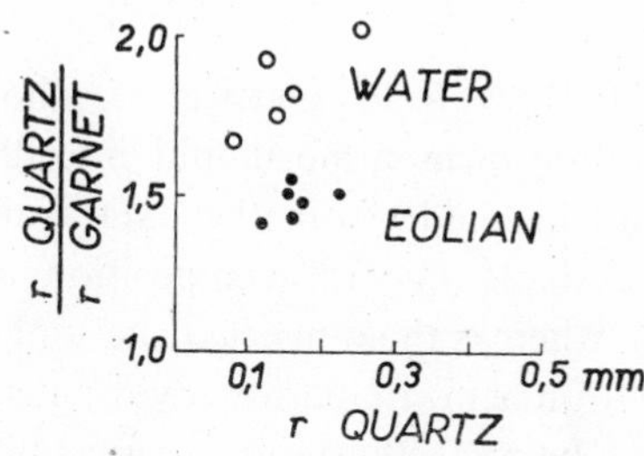

Fig. 126. Distinction between water- and wind-laid sands. From G. M. Friedman (1961) (modified Engelhard's diagram).

The alternations of clay minerals in the aqueous environment has been proved also by other observations. H. H. Murray - A. S. Sayyab (1955) found that in deep-sea sediments the crystallinity of all clay minerals increased towards the greater depths below the surface of Recent sediments. Mineral changes inside clay sediments

Table 210

Syngenetic and diagenetic changes of clay minerals

Original mineral	Alternate product	Note
Kaolinite	Decomposed	Amount of kaolinite decreases near marine environment
Kaolinite	Illite	Amount of illite increases on account of kaolinite decrease near marine environment
Kaolinite	Chlorite	Amount of chlorite increases on account of kaolinite decrease near marine environment
Illite	Montmorillonite	Occurrences of montmorillonite in deep-sea sediments just in the state of transformation from illite
Illite	Chlorite	In pore waters amount of Mg with depth in sediment increases
Montmorillonite	Chlorite	Presence of mixed-layer chlorite-montmorillonite; increase in chlorite with increase of salinity and with decrease of montmorillonite
Montmorillonite	Illite	Decrease in montmorillonite coincidental with increase in illite and increase of salinity; decrease in montmorillonite seaward; Na/K ratio of interstitial water greater than in sea water

were described already in 1942 by R. E. GRIM *et al.* who established the increase of K-amount and the alteration of montmorillonite into illite in vertical and lateral direction. The inference can be made from the data available that about 50% of clay sediments can have a stable mineral composition and reflect the mode of weathering of parent rocks, whereas the other half succumbed to authigenic changes, reflecting a sedimentary environment. In marine environment, illite minerals are most stable, montmorillonite may be perfectly stable or susceptible to alterations, chlorite shows a neutral behaviour and kaolinite is unstable. On the other hand, it is stable in acid waters of lagoons, in swamps and lakes. The beidellitization of clay minerals from a marine environment has also been described.

The differences in mineral composition are accompanied by the differences in chemical composition. The compounds MgO, K_2O, Al_2O_3 and Fe_2O_3 are most sensitive indicators showing to the distance from the adjacent shore. The summarized data on the chemical composition of near-shore clays and those more distant from

Table 211

Differences in chemical composition of nearshore and offshore clays (after M. C. Powers 1957)

	1 %	2 %	3 %	4 %
MgO				
Pelagic clays	2·02	2·45	2·90	3·38
Nearshore clays	9·04	3·37	4·16	2·11
K_2O				
Pelagic clays	0·84	2·38	2·51	3·43
Nearshore clays	0·84	1·51	2·36	2·52
Al_2O_3				
Pelagic clays		19·02	18·14	19·98
Nearshore clays		18·05	22·16	22·31
Fe_2O_3				
Pelagic clays		7·07	9·89	5·56
Nearshore clays		6·35	8·79	6·70

	nearshore	pelagic
1. Chesapeake Bay	salinity 1/1,000	salinity 22—25/1,000
2. Gulf of Mexico	delta	pelagic region
3. Pacific Ocean	bay	continental slope
4. California Bay	delta	transition from bay environment to pelagic environment

dry land reveal several differences that could be used for the determination of the distance of clay sedimentation from the shore. Examples are given in Table 211.

The percentage of K_2O mostly increases towards the pelagic sediments. The behaviour of MgO is variable: both a considerable increase and a moderate decrease have been observed. The content of Al_2O_3 and Fe_2O_3 depends also on other factors and thus is not indicative of the changes in composition only.

According to J. M. Mabecone (1963) the Ca/Na ratio in the clays of estuaries decreases landwards (see Table 212).

The observation of clay minerals under experimental conditions (in artificial sea-water) has yielded a number of interesting details (Table 213). It has been safely

Table 212

Landward decrease of Ca/Na ratio in estuarine clays of River Guadelete (Spain) (after J. M. Mabecone 1963)

Distance landwards from river mouth km	Ratio Ca/Na
48	0·026
25	0·033
16	0·035
14	0·038
10	0·040

Table 213

Alternation of chemical composition of clay minerals under the influence of artificial sea water (after D. Carroll - H. C. Starkey 1960)

Illite	Original composition %	Composition after action of sea water %
MgO	2·29	3·16
CaO	0·32	0·10
Na_2O	amount slightly decreasing	
K_2O	5·71	6·48
Kaolinite		
MgO	0·16	0·50
K_2O	0·42	0·27

proven that a prolonged action of sea-water on clays can produce the alteration of montmorillonite into illite and chlorite. Kaolinite and illite did not show any change except for a variation in the content of some oxides.

In montmorillonite MgO increases appreciably, K_2O rises moderately, and the percentage of Na_2O considerably decreases.

The total chemical composition of clay sediments which is occasionally quite characteristic, might serve as a basis for the differentiation of marine and fresh-water, and possibly also brackish environments. Carboniferous sediments are particularly suitable for these studies, as we have already some objective information on the environments of their origin. M. L. KEITH - E. T. DEGENS (1959) computed the mean chemical composition of marine and fresh-water claystones of Carboniferous age (Table 214).

Table 214

Mean chemical composition of marine and fresh-water claystones of Carboniferous are (after M. L. Keith - E. T. Degens 1959)

	Marine claystones %	Fresh-water claystones %
SiO_2	54·53	57·29
TiO_2	0·92	0·94
Al_2O_3	20·89	21·84
MnO	0·80	0·12
CaO	0·54	0·31
MgO	1·65	1·74
Na_2O	0·22	0·21
K_2O	3·71	3·53
P_2O_5	0·23	0·17
Fe	6·67	5·82
S	0·92	0·15
	p.p.m.	p.p.m.
B	115	44
Ga	8	17
Li	159	92
F	817	642

These analyses of ancient sediments conform roughly to the results obtained on Recent sediments, particularly as regards the increase of K_2O and mostly also MgO towards the marine environment. Of other components, phosphorous increases nearly always in the same direction. The data on the behaviour of Fe_2O_3 and Al_2O_3 vary.

In the opinion of some authors (A. B. RONOV - Z. V. KHLEBNIKOVA 1957) in continental environments, clays of cold and temperate zone can be differentiated from those of the tropical zone on chemical grounds. The former are chemically analogous to igneous rocks, they have more SiO_2, Na_2O and K_2O and less Al_2O_3, whereas the latter show a prevalence of Al_2O_3 and mostly also of Fe_2O_3.

In recent years, trace elements have been used more or less successfully, for the differentiation of fresh-water and marine sediments. It should be borne in mind that the concentration of trace elements can be affected not only by palaeosalinity but also by other factors, such as the content of organic components, mineralogical composition, grain-size, etc. Therefore, the concentrations of trace elements are comparable only in rocks of approximately the same mineral composition and an identical grain-size. According to investigations by S. LANDERGREN - F. T. MANHEIM (1963) P. E. POTTER *et al.* (1963) and Ch. T. Walker (1968), marineclay sediments are on the average richer in B, Cr, Cu, Ni, and V than fresh-water deposits. Of these elements, boron proved to be the best indicator of palaeosalinity. With regard to the changes in B content with the changing content of clay minerals, it has been recommended to correlate chiefly the illitic clays or to recalculate the content of boron to the percentage of illite in the rock. In Recent sediments, the direct relation between the salinity of waters and the content of boron in sediments has been proved to exist. In the Tyrrhenian Sea the following contents have been ascertained:

B	Salinity
0·014%	3·5%
0·010%	1·8%
0·003%	0·58%

In the illitic clays of the Dovey estuary the relation between the salinity of water and the content of boron is as follows (T. D. ADAMS *et al.* 1965, Table 215).

Table 215

Equivalent boron of illites and depositional salinity of sediments from the Dovey Estuary (after T. D. Adams *et al.* 1965)

Physiographic environment	Arithmetic mean salinity %	Equivalent boron (p. p. m.)
Marsh	25·5	355
Marsh	25·5	345
Marsh/flats	20·5	270
Channel/flats	18	260
Estuarine	21·5	275
Channel/flats	18	280
Channel/flats	18	260
Marsh	25·5	355

The last named authors tried to derive an equation for the computation of palaeosalinity:

The regression equation of average salinity (y) on equivalent boron (x) computed from product moment is:

$$y = 0{\cdot}0977x - 7{\cdot}043\,.$$

For y = 0 equivalent boron cotent is 72 p.p.m.

Table 216

Equivalent boron and depositional salinity of lithified rocks (after T. D. Adams *et al.* 1965)

	Average equivalent boron (p. p. m.)	(‰)
Dolomites below Middle Permian Gypsiferous marl, southern Yorkshire	1,000	92·5
Swinden Reef knoll	550	47·5
Average Yoredale limestone	394	31·5
Average Yoredale Shale, mainly marine but in part non-marine	373	29·5
Average Yoredale sandstone, mainly non-marine but in part marine	338	26
Coal Measure non-marine lamellibranch beds	190	12

On the basis of this formula the salinity of some ancient basins has been calculated (Table 216).

According to W. Ernst - H. Werner (1960) the following contents of boron are regarded as limits between the individual environments in Carboniferous deposits:

	B_2O_3
Marine claystones	above 0·035%
Brackish claystones	0·025—0·035%
Fresh-water claystones	below 0·025%

The bounding of boron in clay sediments is firm and it seems to replace Si in tetrahedra. The only drawback of the B—method is the rather difficult analytical determination of boron. In addition, attention must be paid to the possible presence of fresh-water clays transported into the marine environment.

For the differentiation of fresh-water and marine sediments other elements are also used:

1. Ga — Some authors record that this element is more abundant in fresh-water than in marine claystones (e. g. after M. L. Keith and E. T. Degens 1959, the former contain 25—120 p.p.m. Ga, and the latter only 7—40 p.p.m. Ga). According to other writers, this finding is not correct, reference is made even to higher contents of Ga in marine deposits than in corresponding fresh-water ones.

2. Rb — Some data record higher contents in marine sediments (100—1100 p.p.m. compared with 70—650 p.p.m. in fresh-water deposits).

3. Li — Analogously to Rb, its content is higher in marine sediments (above 100 p.p.m.) than in fresh-water sediments (under 100 p.p.m.).

4. Sr and Ba — This pair of elements and their relation to salinity has been much discussed. It is generally believed that the Ba/Sr ratio increases with salinity. In continental sediments, the Ba/Sr ratio in sandstones is 0·17, in siltstones 0·33, and in claystones 1·05. In some cases the absolute amounts of Ba can be applied, which also increases with salinity. The Sr content is affected mainly by the aragonite content.

Table 217

Distribution of trace elements in various sediments and their relation to salinity (in per cent) (after S. Landergren - F. T. Manheim 1963)

	1.	2.	3.	4.	5.	6.	7.	8.
Org. C	0·30	0·30	0·30	1·8	3·6	6·3	1·8	17·1
S				0·28	1·7	1·5	0·08	6·14
F				0·08	0·07	0·10	0·10	0·06
Ti	0·42	0·45	0·49	0·44	0·47	0·43	0·45	0·43
V	0·011	0·011	0·012	0·011	0·014	0·0060	0·0073	0·15
Cr	0·006	0·012	0·005	0·009	0·010	0·008	0·0053	0·007
Mn	0·30—1·98	0·49	0·30	0·044	0·47	0·057	0·13	0·013
Fe	5·1	3·6	5·0	3·0	4·9	4·2	3·6	6·7
Co	0·015	0·003	0·003	0·0007	0·0015	0·0012	0·0014	0·0035
Ni	0·021	0·010	0·006	0·0038	0·0050	0·0036	0·0033	0·020
Ga	0·002	0·0093	0·0024	0·0018	0·0033	0·0024	0·0230	0·0035
Cu	0·056	0·0072	0·0098	0·0020	0·0088	0·0054	0·0033	0·017
Zn	0·0085	0·0037	0·013	0·0059	0·011	0·013	0·0068	0·0080
Pb	0·0063	0·0018	0·0035	0·0033	0·0050	0·0031	0·0027	0·0038
Mo	0·0027	0·0001	0·0004	0·0003	0·0014	0·0007	0·0003	0·0089
Th				0·0010	0·0009	0·0012		0·0007
U	0·00026		0·00026	0·0006	0·0008	0·0011	0·0007	0·016
Sr	0·013	0·15	0·011	0·024	0·017	0·018	0·0090	0·0091
Ba	0·12	0·11	0·038	0·038	0·062	0·055	0·042	0·050
Sr/Ca	0·025	0·0044	0·025	0·0067	0·017	0·017	0·009	0·14
Ba/Ca	0·24	0·0032	0·087	0·011	0·062	0·051	0·042	0·078
Salinity	34—35	34—35	34—35	32—35	7—15	4—7	0	—
B	0·0135	0·012	0·013	0·0115	0·0105	0·0062	0·0015	0·0145

1. Deep-sea ooze, 2. Calcareous deep-sea ooze (Pacific), 3. Deep-sea ooze from Atlantic, 4. Clay (Skagerak), 5. Clay (Baltic Sea), 6. Clay (Fjord Kyrkfjörden in Baltic Sea), 7. Clay (Swedish lakes), 8. Alum shale of Cambro-ordovician age.

5. Mg — In Recent sediments the Mg content generally increases with salinity, but is simultaneously affected by so many other factors that it cannot serve as an adequate indicator. G. V. CHILINGAR (1963) notes that in carbonate sediments the Ca/Mg ratio increases seawards with depth.

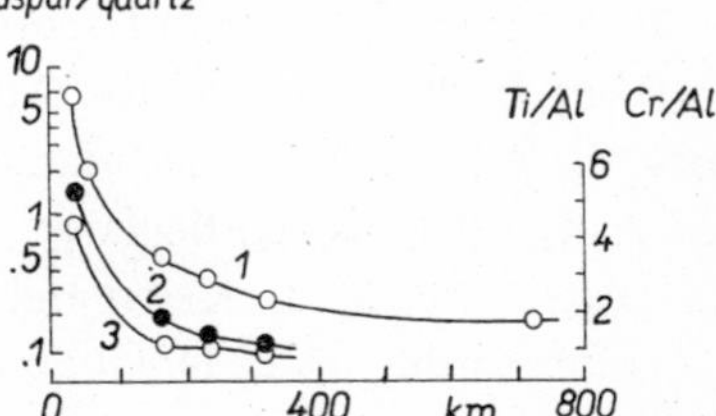

Fig. 127. Relative ratio feldspar/quartz in the North Pacific off Hawaii, showing the rapidly vanishing influence of the basaltic province. This is also reflected by the decrease in the relative concentration of Ti and Cr, mainly occurring in pyroxene crystallites. After Rex (1958) and M. N. Hill (1963).

6. The Cr/V and Cr/Ni ratios mostly drop with the distance from the shore. This decrease is linked with the reduction of coarser terrigenous admixtures. On the other hand the value of Nb/Ta ratio increases with the distance from the shore.

7. According to earlier opinions, the Th/U ratio can also be indicative of the environment. The Th content in continental clay sediments was recorded as attaining up to 24 p.p.m., and only 2—4·8 p.p.m. in marine deposits. Recent investigations have not corroborated these data.

8. Halogenes, Cl, Br and I — Chlorine seems to be the most promising of all halogenes. The necessary conditions of adequate results is again to correlate sediments of the same mineral composition, preferably those with chlorine content. W. D. JOHNS (1963) records that fresh-water chlorites contain less than 100 g/t Cl, whereas marine chlorites have about 1,000 g/t Cl (Fig. 135). In some cases, the

Table 218

Analyses of organic matter from marine and fresh-water Carboniferous claystones (after M. L. Keith *et al.* 1959)

	Fresh-water claystones			Marine claystones			
	p. p. m.						
Pb	400	150	100	40	50	40	50
Ni	25	25	20	150	100	80	70
Sn	30	40	20	5	5	5	10
Cu	300	1,500	1,000	150	100	150	800
Zn	400	600	1,500	400	300	500	1,000
Ag	2	5	2	2	8		2
Co	40	40	20	20	20		40
Mo	4	4	3	6	2		2
V	100	100	80	150	150	100	200

same relation has been ascertained for other halogenes, being more marked with I than with Br. As far as F is concerned, the effects of salinity are concealed by other factors, so that it gives no indication as to marine or non-marine origin.

9. Pyrite of fresh-water clayey shales is richer in Co, As, Ag and Cu than pyrites of marine clayey shales. No more details are known.

10. Organic fractions of sediments are also sensitive to geochemical differences between fresh and salt water. After separation from marine and fresh-water claystones, the following content of trace elements has been determined in them (averages of many analyses, particularly of Carboniferous sediments from various localities are given, after M. L. KEITH and DEGENS E. T. 1959, Table 218).

11. According to E. D. GOLDBERG *et al.* (1969) marine barites can be distinguished from continental ones by means of isotopic analysis of uranium and thorium.

The composition of organic matter and its relation to the enrichment in certain trace elements indicate sedimentary environments. It is important to distinguish the sapropel from the gyttja organic substances. Sapropel substances originated, as mentioned above, without the access of air, in peat bogs, eutrophic and dystrophic lakes, possibly also in coastal swamps, whereas gyttja sediments develop in places with a limited access of oxygen, particularly in oligotrophic lakes or slightly eutrophic lakes, inland basins, lakes and bays. Sapropelites are enriched particularly in V and Mo and gyttja sediments in Pb, Zn, Cu and Ni. Thus, the V/Ni ratio may be indicative of the character of organic matter. The percentage of phosphorous is higher in gyttja than in sapropel. The V/Cr ratio in gyttjas is about 1 and about 10 in sapropel. The V/Mo ratio varies between 5—10 in sapropel and is lower than 3 in gyttjas.

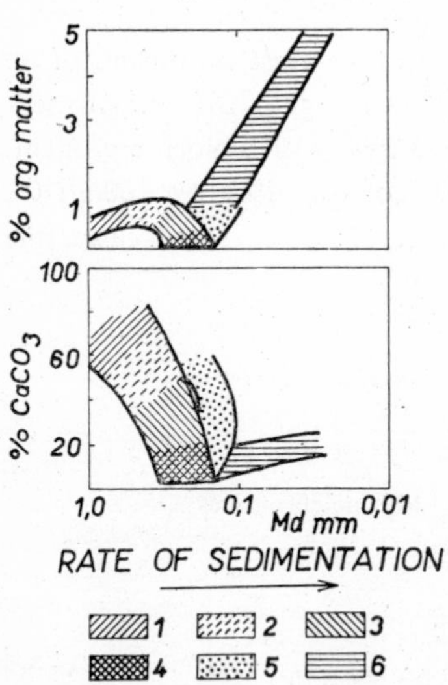

Fig. 128. Amount of organic matter and calcium carbonate in various environments of sheltered shelf. 1. Straits, 2. Bank tops, 3. Elevations on shelf, 4. Beach, 5. Open shelf, 6. Sheltered bays and depressions on shelf. From K. O. Emery *et al.* (1957).

Palaeosalinity, i. e. the salinity of waters in ancient basins, has recently been determined also by the conductivity of water extract, which in fact represents the content of adsorbed cations and anions in clay sediments. In clayey sediments of the Ostrava—Karviná coal-basin, the boundary between the occurrences of marine and brackish-water fauna was established by this method at $11{\cdot}5 \times 10^{-4}\ cm^{-1} \times \Omega^{-1}$, and the boundary between fresh-water and brackish environment at $8{\cdot}5 \times 10^{-4}\ cm^{-1}\Omega^{-1}$

(R. KÜHNEL 1960, J. TOMŠÍK 1960). Additionally, the content of Cl′, SO_4'' and/or Br′ and I′ was determined. However, the results achieved do not justify as yet a generalization of conclusions. Data on pH of sediments derived from various environments show that in Carboniferous clayey rocks, marine sediments have a higher pH (8·5) than the fresh-water ones (7·9—8·2). Attempts have been made to reconstruct palaeosalinity from the ratio of anions Cl′/HCO_3' but the method has not yet been perfectly elaborated.

G. D. NICHOLLS (1967) believes that the total trace element population of a sediment may afford some guidance to the depositional environment. High contents of Mo (>5 p.p.m.), Co (>40 p.p.m.), Cu (>90 p.p.m.), Ba (>1,000 p.p.m.), Ce (>100 p.p.m.), Pr (>10 p.p.m.), Nd (>50 p.p.m.), Sm (>15 p.p.m.), Gd (>15 p.p.m), and other rare earth except La and Y (>5 p.p.m.), Ag (>2 p.p.m.), Cd (>0·6 p.p.m.), Ni (>150 p.p.m.), and Pb (>40 p.p.m.), especially if associated with low contents of U (<1 p.p.m.) and Sn (<3 p.p.m.) in sedimentary rocks would suggest that the possibility of original formation under water deeper than 250 m might be considered.

Reconstruction of environments from the content and composition of organic matter

Organic matter can be a good quantitative and qualitative diagnostic key for some sedimentary environments. The total amount of organic matter is generally determined as percentage of organic carbon.

Both in shallow and deep-water environments, the presence of Recent sediments containing more than 2% org. C indicates sedimentation in a poorly-aerated environment, i. e. with a decreased amount of free oxygen above the bottom, or in water devoid of oxygen or even with free hydrogen sulphide. Shallow-water environments of this type include eutrophic and distrophic lakes, freshwater and saline swamps, bays, lagoons and tidal flats; deep-sea trenches, depressions in continental slope and shelf, fjords, and some large inland basins belong to deep-water environments of this kind. The amount of organic matter in fossil black shales is not a satisfactory key to distinguish whether they originated in shallow-water or deep-water environments. Therefore, other diagnostic features must be used. These are listed in Table 219.

The presence of benthos remains may serve as a basic distinctive characteristic between shallow- and deep-water black clays. Whereas black clays of even non-aerated bays and tidal flats contain some benthos remains, the deposits of deep-water basins carry them quite exceptionally. This difference is due to the presence of free hydrogen sulphide which is frequently present above the bottom of non-aerated deep-water environments. As is well-known, in this poisoned medium only some worms which take oxygen from the glycogene of their own bodies can live. In shallow-water non-aerated environments impoverished in oxygen, numerous species of benthos can still live permanently or periodically.

Table 219

Diagnostic features of shallow-water and deep-water black clays

	Shallow-water clays	Deep-water clays
Amount of coarse-grained admixtures and their grain-size	Variable, clays with various amounts of coarse-grained admixtures occur, sometimes well developed bimodality	Usually small amount of coarser admixtures, mainly silt; bimodality is scarce
Amount of carbonate	Generally small, up to 5 % $CaCO_3$	Variable, generally up to 30 % $CaCO_3$
Amount of plankton remains	Sometimes enormous, mainly diatom frustules	Variable; from minimum up to considerable amount (mainly foraminifers and diatoms)
Amount of benthos remains	Always present	Generally absent
Total amount of org. C	Between 1 and 10 %, sometimes more, mostly between 1 and 3 %	Generally above 5 %
Regional extent	Variable	Only great
Composition of organic matter	Humic and bituminous; generally higher values of ratio org. C/org. N	Mostly bituminous; lower values of ratio org. C/org. N

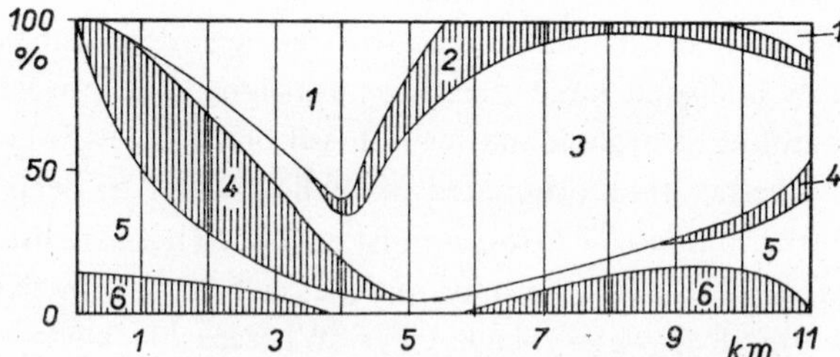

Fig. 129. Ideal presentation of the bathymetric distribution of Recent sediments. 1. Calcareous oozes, 2. Siliceous oozes, 3. Clays, oxidation environments, 4. Clays, reduction environment, 5. Sands and silts, 6. Rock without sediments cover. After J. M. Weller (1960).

Another suitable diagnostic feature is the character of organic matter which, from the genetic point of view, is divided into humic and bituminous substances. Humic matter is dominantly of plant origin and is composed mainly of cellulose, lignin and a subsidiary amount of nitrogen compounds and carbohydrates. Bituminous substances derive from faunal organic remains, being characterized by an increased amount of carbohydrates and nitrogen, proteins, fats and similar compounds. The composition of organic matter can be determined by a number of special analyses which are, however, too exacting for current petrographic practice. In Table 220 several petrographic methods are listed.

Table 220

Diagnostic features of humic and bituminous organic matter

	Clays with humic organic matter	Clays with bituminous organic matter
Ratio org. C/org. N	> 10	< 10
Following trace elements concentrate	Ge, Bi, U, Mo, As, Sb	V, Mo, Ni, Zn, Pb, Co, Cu, U
Amount of sulphur	Small, generally < 1 %	Generally higher (with exceptions)
Total amount of organic C in sediment	Generally higher > 2 %	Generally lower < 2 %
Organic remains	Mainly plant remains; animal remains, fresh water and marine	Animal remains, marine
pH value of environment during sedimentation	Mostly 3—7; strongly to slighty acid	Mostly 7—10, neutral to strongly alkaline

Reconstruction based on the content and composition of carbonates

The factors affecting carbonate sedimentation are so manifold that it is very difficult to make a generalization on the distribution of carbonate in every environment and to find decisive criteria for distinguishing various environments. With the exception of the shallow-water environment with pure carbonate sedimentation, the laws expressed on the diagram in Fig. 128 are generally valid. In shallow-water sediments, the local increase of carbonate content, especially bioclastic and detrital, mostly indicates the boundary environment between bay and open sea, i. e. the straits and the bay mouths.

The interrelationship between the content of carbonate and size distribution shows that in places of more rapid sedimentation the carbonate amount increases to the finer fractions (in fluviatile, deltaic and analogous sediments). In marine environments, in places of slower sedimentation or non-sedimentation, the carbonate content increases to the coarser fractions (A. SAADALLAH - Z. KUKAL 1969).

Some conclusions on sedimentary environments can also be drawn from the mineralogical composition of carbonates. The presence of an increased amount of aragonite indicates a higher temperature of water, i. e. tropical climate. Aragonite had long been regarded as unstable but recently it has been found as a primary compound not only in Mesozoic and Palaeozoic sediments, but probably also in Pre-Cambrian rocks. Finds of different forms of aragonite in ancient sediments point to the following environments (Table 221).

Table 221

Environmental interpretation of aragonite finds in ancient sediments

Type of rock	Sedimentary environment
Limestone with cloddy texture of accretion origin, eventually cloddy limestone of algal origin with clear algal incrustations	Shallow-water carbonate flat, or fossil analogy of Bahamian sediments
Pelitomorphic limestone	Lagoons, or recent analogies of sediments formed from aragonite needles
Organogenic limestone with relics of frame-building organisms	Bioherms or reefs

Three elements accumulated in organic shells show differences in concentration between fresh and marine environments: Ba, Fe and Mn (G. M. FRIEDMAN 1967). A study of marine, lagoonal, and fresh-water gastropods and pelecypods shows that the shells of nonmarine and lagoonal molluscs contain greater abundance of barium, iron and manganese than do marine molluscs.

Of dolomites, only the primary, i. e. homogeneous, dense or pelitomorphic, and fauna-free varieties can serve as indicators of sedimentary environment. In addition, they should have a large lateral distribution and/or be associated with gypsum or anhydrite. Those dolomites are typical sediments of saline bays, lagoons in tropical and subtropical climate or sediments of periodical lakes.

The reconstruction of environments based on the composition of carbonate concretions is not derived from findings on Recent sediments, but from a comparison of marine, freshwater and/or brackish Carboniferous sediments. This approach to the problem has brought some generally valid conclusions, as for instance: Marine sediments contain concretions rich in CaO, concretions of lacustrine and lagoonal deposits are rich in FeO, and concretions of saline lagoons and bays have a large MgO component. The formation of siderite during sedimentation may prove especially useful for environmental determination (Table 222, after J. N. WEBER 1965).

Table 222

Composition of siderite nodules from fresh-water, brackish and marine environments. Mean (in per cent) and standard deviations (after J. N. Weber 1965)

	SiO_2	Al_2O_3	MgO	CaO	Ba	V
FW	30·75 (12·8)	13·42 (7·25)	2·22 (0·674)	2·55 (2·17)	0·0360 (0·0118)	0·0094 (0·0039)
B	13·67 (7·44)	5·77 (3·95)	2·47 (1·32)	2·97 (0·824)	0·0180 (0·0110)	0·0110 (0·0042)
M	11·56 (3·25)	4·84 (1·12)	3·58 (1·02)	5·81 (1·66)	0·0140 (0·0033)	0·0140 (0·0043)

Care must be taken, however, not to overrate the conclusions of the composition of concretions, because this also depends on the surrounding sediments. In calcareous sediments there is a tendency to the formation of calcareous concretions, whereas in clastic sediments to the development of siderite concretions.

The study of the composition of Recent sediments has provided some information that coincides, to a certain degree, with the findings concerning ancient carbonate concretions. Siderite deposits have, in fact, been ascertained only in more or less brackish seas and bays (the Baltic Sea, Barentz Sea); in seas of normal or increased salinity only calcite concretions develop (apart from a sporadic finds of dolomite concretions described above). Thus, an increased content of iron in carbonates signifies a transition of marine conditions to an environment of brackish lagoons, bays, or inland seas and continental conditions themselves.

In Recent sediments, the content of Sr in carbonate sediments appears to be a reliable indicator of the salinity and the temperature of the environment. It increases both with salinity and temperature. The following values concerning this relationship have been established:

	Number of Sr atoms in 1,000 Ca atoms
Deep-water carbonates oozes	2·38— 2·61
Aragonite sediments of shallow-water tropical regions	4·00—10·80 and more
Lagoonal dolomites	11·4

Table 223

Differences in trace element distribution in reef and non-reef limestones (mean and range) (after R. Chester 1965)

	Average total rock trace element content			
	Reef limestone		Non-reef limestone	
	p. p. m.			
Ni	24	(1·400—169)	62	(15·00—224)
Co	5·1	(1·900—47)	18	(1·00—126)
Cr	7·5	(0·700—49)	128	(9·50—499)
V	4·0	(1·80—15)	49	(11·00—11)
Ba	44	(9·500—137)	252	(53·00—903)
Sr	67	(2·400—1,000)	533	(133·00—1,350)
Cr/Ni	0·31	(0·003—6·1)	2·1	(0·59—9·7)
Ni/V	6·0	(0·340—114)	1·3	(0·65—3·6)
Ba/Sr	0·65	(0·070—9·1)	0·48	(0·04—1·1)
	Average non-detrital (acid soluble) trace element content			
	Reef limestone		Non-reef limestone	
	p. p. m.			
Ni	22	(0·41—100)	34	(4·90—140)
Co	0·29	(0·03—4·4)	4·7	(0·18—18)
Cr	3·1	(0·31—29)	5·4	(0·11—16)
V	2·7	(0·30—18)	8·1	(0·60—36)
Cu	20	(2·60—50)	20	(4·00—88)
Pb	13	(0·58—68)	22	(1·10—96)
Ga	12	(0·50—54)	13	(0·35—55)
Ni/Co	37	(2·70—544)	7·2	(0·66—209)
Cr/V	1·1	(0·12—17)	0·67	(0·01—4·8)
Pb/Ge	1·1	(0·30—3·5)	1·7	(0·33—88)
Cu/Co	34	(1·20—136)	4·3	(0·45—37)

Unfortunately, this interdependence cannot be applied to ancient sediments, because Sr disappears already during diagenesis. Nevertheless, higher contents of Sr in limestones point to their origin at a higher temperature and possibly also to a higher salinity. The content of Sr can also be indicative of inorganic or biological origin. The organic fraction (e. g. in the tests of pelecypods, brachiopods) contains more Sr than the inorganic parts of limestones, as shown below.

	Number of Sr atoms in 1,000 Ca atoms
Average content in shells of Recent sediments	4·64
Average content in total Recent carbonate sediments	2·68

It is frequently difficult to decide whether the bulk of dense or pelitomorphic limestones (calcitutites) is of mechanical, chemical or biological origin. In many cases, the above-mentioned method (provided that suitable comparative material is available) could settle the question, as the increased percentage of Sr indicates the presence of biological detritus.

Glauconite and chamosite

The presence of glauconite in sediments can also be used as an indication of the sedimentary environment, but with due regard to the possibility of its frequent mechanical displacement. Recently analogous minerals to glauconite have been found in continental sediments and even in soils, so that glauconite can no longer be regarded as characteristic of marine sediments. It has been observed that normal glauconite grains are winnowed from shelf and beach deposits into eolian dunes. In marine sediments its bathymetric distribution is as follows (Table 224).

Table 224

Bathymetric distribution of glauconite

Author	Maximal depth of occurrence m	Occurs mainly in depths m
J. Murray - A. F. Renard (1891)	2,000	200—300
F. W. Clarke (from Kuenen 1950)	scarce < 900	50— 70
H. Sverdrup *et al.* (1942)	1,350	outer shelf
Ph. H. Kuenen (1950)	2,450	130—400

In Recent sediments, the major part of glauconite is confined to the central and outer parts of the continental shelf, i. e. to places of slow deposition or non-sedimentation. There, glauconite may make up as much as 90% of all deposits. According to present knowledge, it can be generated in neutral to slightly reducing environment and rather in cooler than in warm waters. Glauconite can be used best as an indicator of depositional conditions in complex shallow-water environments. Considering glauconite as an authigenic mineral, its amount is indirectly proportionate to the content of Fe-sulphides (at least in the superficial layer of sediments). According to D. H. PORRENGA (1967) temperatures lower than 15°C tend to be favourable for the formation of glauconite.

On the other hand, recent occurrences of chamosite are restricted to shallow marine environments in the tropics and this may indicate that warm (over 20 °C) bottom water is essential for its formation.

Phosphates

The presence of phosphates cannot be unequivocally accepted as suggestive of environmental conditions because in Recent sediments phosphates are found both in oxidizing and in reducing environments. Oxidizing phosphates occur in association with gravels on elevations and parts of the continental shelf and continental slope. Reducing phosphates are abundant in black clays of non-aerated basins. According to R. G. BROMLEY (1967) most favourable conditions occur in shallow, warm water, particularly between 30 and 300 m. For a large accumulation of organic phosphorus to build up on the sea floor, the sea must be considerably less than 1,000 m deep. A separate phosphatizing environment may also exist in estuaries.

In continental environments, phosphates always indicate intensive biological sedimentation.

Evaporites

Evaporites are, on the whole, an unquestionable indicator of sedimentary environments. They are either of continental or lagoonal origin. There is plenty of gypsum and anhydrite that originated diagenetically in various environments.

Environments reconstructed from the results of isotope studies

Isotopes are used as modern indicators of sedimentary environment, making it possible to differentiate fresh-water and marine sediments, in some cases also organic and inorganic deposits, and to establish the temperature of environments. The determination of O^{18}/O^{16} ratio is most widely used. The enrichment in heavier isotope indicates increased temperature, occasionally also increased salinity. Moreover, this relationship is affected by biological factors because some organic groups concentrate heavier isotopes. On the basis of oxygen isotopes in belemnite rostra and in oyster shells, curves of palaetemperatures were plotted for the whole of the

European and American Mesozoic and part of the Tertiary. Besides, this method, based on the ratio of isotopes in foraminiferal tests, is used for the stratigraphical study of Recent and Pleistocene deep-sea sediments.

M. L. KEITH - R. H. PARKER (1965) have ascertained that C^{13}/C^{12} ratios in mollusc shells are much more sensitive to the change in salinity (Fig. 136). Attempts were made to use also Si isotopes for the differentiation of environments (J. H. REYNOLDS - J. VERHOOGEN 1953); it was found that marine sediments are enriched in the heavier Si^{30} isotope. The study of the differences between marine and fresh-water diatomites yielded more promising results.

Lithological criteria for the differentiation of sedimentary environments

Different sedimentary structures correspond to different modes of deposition. As referred to above, the grain-size distribution of sediment depends on it directly. From this it follows that a close relation must exist between the grain-size distribution of sediments and the presence of certain sedimentary structures.

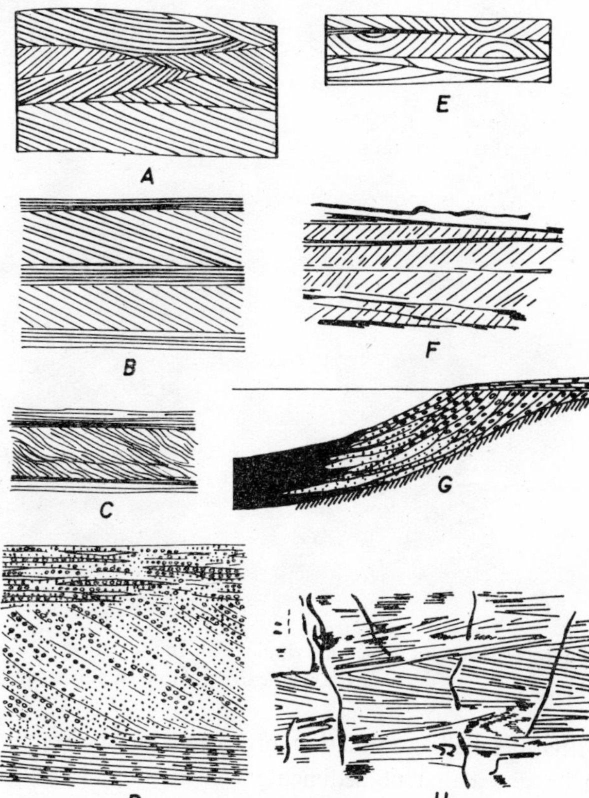

Fig. 130. Some typical examples of sediment stratification in various environments. A. Eolian sediments, B, C, D, F. River channel sediments, 3. Eolian dunes, G. Lake delta, H. Marine sediments. After L. N. Botvinkina (1962).

Photo 18. Form of preserving ripple field. In the course of high tide ripple field is covered by a sand carpet. After H. Ehrhardt (1937).

After detailed examination of various structures in individual environments it can be stated that no sedimentary structure exists which would clearly indicate a definite environment. All of them have several modes of occurrence and can be inter-

Table 225

Survey of sedimentary environments and their most characteristic structures

Sedimentary environment	Sedimentary structures
River channel	Current bedding dependent on grain-size of sediment, current ripple bedding, trough cross-stratification, sometimes flat horizontal stratification
Flood-plain	Parallel lamination and bedding, in coarse-grained layers graded bedding, sometimes ripple bedding occurs
Lateral accretion of river deposits	Flat, horizontal stratification, ripple stratification, homogeneous sediment with plant remains and structures
Alluvial plains	Regular tabular bedding with wide lateral extent, ripple bedding, tabular-planar cross-stratification with intermediate angle, plant rootlets in fine-grained deposits
Alluvial fans	Transition from structures typical for river channel deposits to those characteristic for mud-flow deposits. Trough cross-stratification, low-angle simple or planar cross-stratification occur most frequently
Eolian dunes	Wedge-planar cross-stratification (high-angle) trough cross-stratification, and also flat horizontal stratification, layers of clay balls
Eolian silts (loesses)	Homogeneous structure, indistinct lamination, plant rootlets
Small oligotrophic lakes	Fine regular lamination in middle parts
Small eutrophic and dystrophic lakes	Fine lamination passing to homogeneous structure
Salt marshes	Undulating, nodular lamination which shows usually considerably lateral continuity. Graded bedding in laminae. Little or no disturbance by burrowing. Abundance of very thin plant rootlets
Salt-water lagoons	Mostly little or no lamination, secondary biogenic structures very abundant, mud cracks in temporarily emerging parts of lagoon bottom
Brackish lagoons	Usually very thin lamination
Hypersaline lagoons	Lamination or stratification with subordinate burrowing structures, locally penetrated by mud crack structures

Table 225 (*continued*)

Sedimentary environment	Sedimentary structures
Beach	Structures dependent on grain-size, in foreshore low-angle simple or planar cross-stratification, in backshore trough cross-stratification, in off-shore low-angle planar cross-stratification and tabular-planar cross-stratification with intermediate angles. Varieties of ripple marks, swash marks and rill marks, bioglyphs. Various kinds of structures due to entrapment of air in the sand
High parts of tidal flats	Secondary structures prevail due to abundance of burrowing organisms, irregular lamination, coarser layers ripple bedded, sometimes mud cracks, enormous quantity of bioglyphs
Lower parts of tidal flats	Irregular lamination, lenticular bedding, ripple bedding, shell and clay pebble beds with imbricated structure, relatively few burrows, sometimes mud cracks. Bioglyphs
Tidal creeks (Prielen)	Coarse lenticular lamination, ripple bedding, sometimes small slump structures in silty clays, shell and pebble beds with imbricated structure, in coarse-grained layers trough cross-bedding
Upper parts of delta slopes	Well developed regular bedding and lamination, ripple bedding, in coarse-grained layers the same structures as in river channel, ripple marks, few load casts, some horizontal burrows of organisms
Lower parts of delta slopes	Lamination and regular bedding, not very distinct, abundance of horizontal organic burrows
Open shelf with active deposition	Lamination usually almost completely disturbed by burrowing organisms, in some cases sediment homogeneous
Open shelf without sedimentation	Mottled structures or homogeneous sediment
Fjords with stagnant bottom waters	Fine lamination or primary homogeneous sediments. No burrowing
Continental slope	Various sorts of bedding, most frequently parallel stratification, with ripple bedding in coarser layers. Regular parallel stratification passing into irregular one and in mottled structures. Vertical and horizontal burrowing
Floor of abyssal ocean	Primary homogeneous structures, fine lamination, or secondary homogeneous structures
Deep-water trenches	Parallel lamination and bedding. Graded bedding in coarser-grained layers, fine-grained layers with current ripple bedding

preted in several ways. An objective picture of a sedimentary environment can be obtained only when a structure is considered in relation to other criteria or combined with other structures as a whole. Two factors can be distinguished which influence the structure of Recent and ancient sediments:

1. Primary factors producing various kinds of bedding, such as changes in the intensity of currents, in the supply of material, in the velocity of sinking of different fractions, in the reworking on the bottom by the current and wave action.

2. Secondary factors that modify and disturb the primary factors or prevent the primary structures from developing. They are either biological, e. g. the action of benthonic organisms, or inorganic, or mechanical, e. g. formation of mud cracks, load casts, slump structures and convolute structures.

The interrelationship of these two factors depends on the rate of sedimentation or of the reworking of the sediment, on the amount of benthonic organisms and on the effects of secondary inorganic factors. Homogeneous sediments can be the product of an absolute alternation produced by secondary factors or, on the other hand, the result of the absence of alterations in primary factors. The former case is frequent in shallow-water sediments and the latter is rather typical of deep-water sediments.

In some environments, as for instance, beaches, parts of tidal flats, parts of deltas, the displacement is so rapid that primary structures cannot be disturbed by benthos. According to some opinions, the displacement of sediments (i. e. deposition and removal) must occur at an average rate at least 1,000 cm in 1,000 years, should primary structures be preserved at the presence of benthos. New direct measurements of the reworking of sediments by *Yoldia limatula* have shown, however, that this rate must in fact be much greater. This pelecypod was observed in Buzzard Bay to rework 5—6 lit. per sq. m per year (after D. C. RHOADS 1963). From this it follows that in places where primary structures are preserved either the reworking benthos was absent or scarce or the mechanical displacement occurred at a much greater rate. Table 225 gives a survey of structures characteristic of various environments and which can help in their identification.

In Recent sediments fewer primary structures are preserved than in ancient sediments. Moreover, many apparently homogeneous fossil sediments, after a reasonable treatment (e. g. saturation with oil, irradiation by X-rays) show a number of primary structures, such as various kinds of current bedding, etc. The difference in the preservation of primary structures in Recent and ancient sediments is considerable, but it is difficult to explain. Two possibilities come into consideration:

1. In ancient sediments, there is a predominance of environments of a rapid sedimentation and displacement, i. e. those in which secondary factors cannot affect to a greater extent the formation of secondary structures.

2. Benthonic organisms were not so widely developed and, consequently, their activity was not so intensive as it is in Recent sediments.

Table 226

Occurrences of individual bedding types in different environments
(after E. D. McKee 1964)

	1.	2.	3.	4.	5.	6.	7.	8.	9.	10.	11.
Flat, horizontal stratification			X		X	X		X			
Ripple stratification			X			X					
Trough cross-stratification	X			XO	X					X	
Low-angle, simple or planar cross-stratification	X								XO		XO
Tabular-planar cross-stratif. intermed. angle		X					X				XO
High-angle, wedge-planar cross-stratification					XO		X				
Graded bedding											
Recumbent fold structures							O				
Contorted bedding											
Convolute structures											
Irregular structures											
Mottled structures								X			
Structureless homogeneous bodies						X					

1. Alluvial fans, 2. Alluvial plains, 3. Flood-plains, 4. Point-bars, 5. Dunes, 6. Delta, upper part, 7. Delta cones, 8. Tidal flats, lower, 9. Beach, foreshore, 10. Beach, backshore, X - present O - typical occurrence.

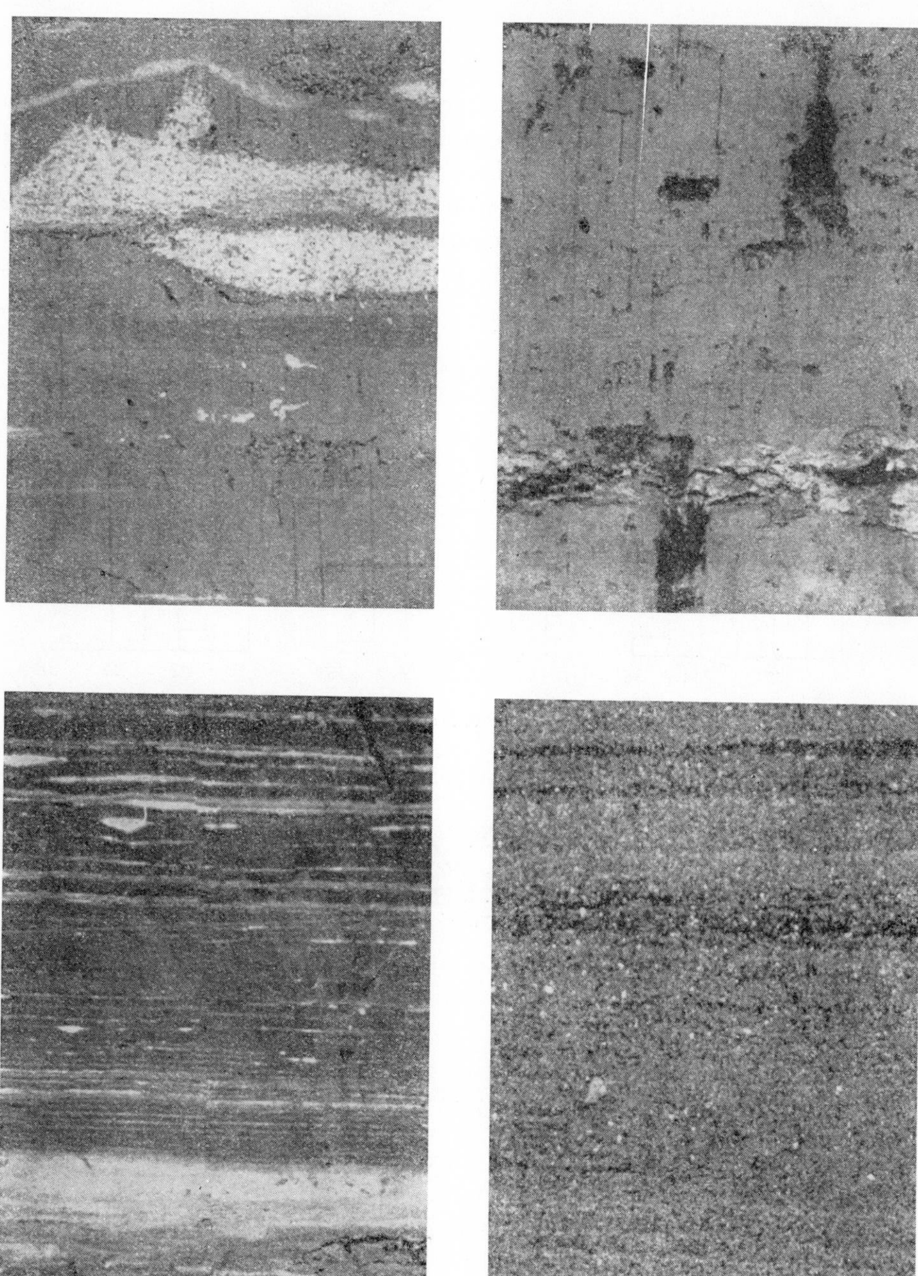

Photo 19. Some structures from the cores of shelf sediments. Mottled structure (upper left core), homogeneous structure (upper right core), laminated sediment with sand laminae (lower left core), and laminated sediment with heavy minerals laminae (lower right core). After H. N. Fisk *et al.* (1959).

Current and cross stratification

Current stratification is one of the most frequent primary sedimentary structures. In the literature, many attemps at its genetic interpretation are recorded. Unfortunately, none of these interpretations have been successful, because no type has so far been found that would be characteristic of a certain environment only. E. D. McKee (1964) presents a table summarizing occurrences of individual bedding types in different environments; the data are based on field observation and experimental results (Table 226).

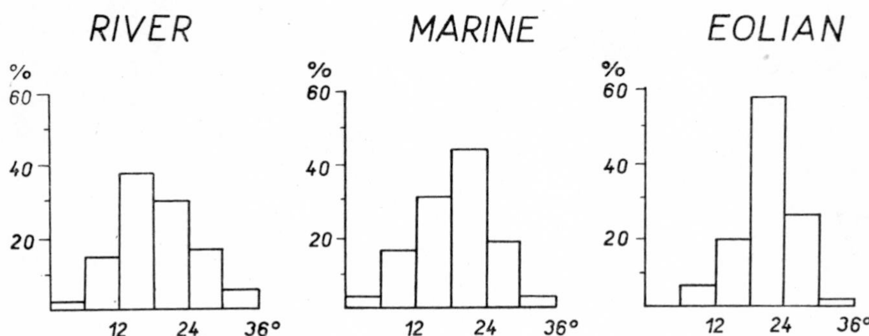

Fig. 131. Fluvial, marine and eolian cross-bedding, inclination. After F. J. Pettijohn (1962).

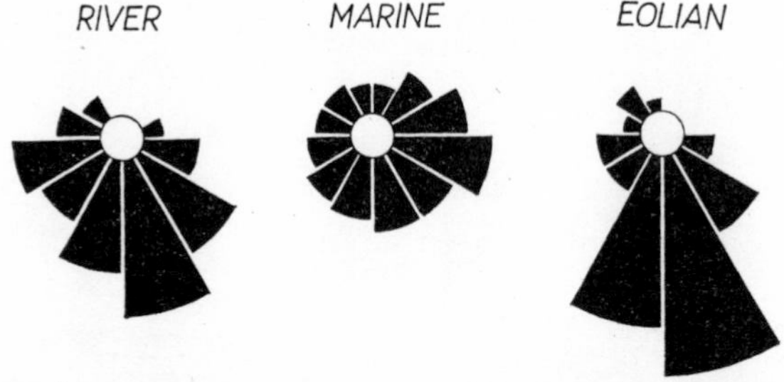

Fig. 132. Fluvial (Lafagette), marine (Framconia) and eolian (Coconino) cross-bedding, azimuths. After F. J. Pettijohn (1962).

From the observation it follows that probably only high-angle wedge planar cross stratification with large dips of laminae (above 30°) can be regarded as characteristic of eolian dune sediments (this type of stratification can, however, occur also in delta cones). Trough cross-stratification, which has been formerly thought to be typical of river sediments (point bars), has been found many times also in beach sediments on the backshore and even in eolian dune sediments. Flat, horizontal stratification and ripple stratification are virtually meaningless for the interpretation of environment. Abundant occurrences of low angle, simple or planar cross-stratification may be indicative of beach sedimentation on the foreshore.

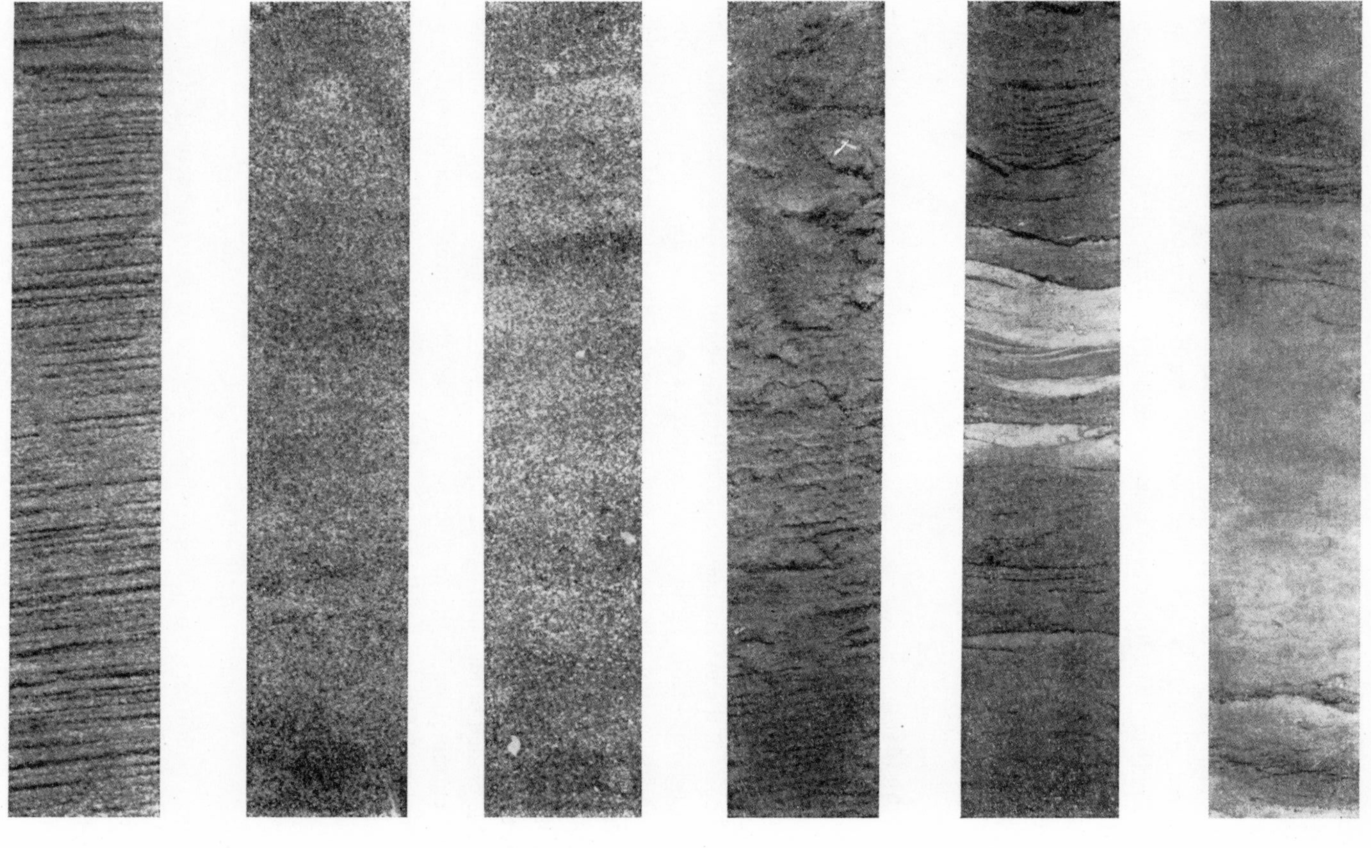

Photo 20. Sedimentary structures of some shallow-water sediments. Beach (three cores from the left), marsh (fourth profile from the left) and tidal flat (two profiles on the right). Netherland tidal flats. After L. M. J. U. van Straaten (1959).

The environmental interpretation of cross-bedding is made still more difficult by the dependence of bedding types on the grain-size and sorting of sediments.

On the other hand, the study of orientation of cross-beds and laminae can contribute to the identification of environments. A sufficient number of measurements taken in many outcrops revealed that eolian sediments showed a maximum tendency to the preferred orientation of inclined laminae and beach sediments tend least to its development (Fig. 131). It has been ascertained that the mean dip of eolian sediments is 21°, that of fluviatile sediments 19·8° and marine deposits 18° (Fig. 132). The scatter of dips is maximum in fluviatile sediments and minimum in the eolian deposits. The conditions in various Recent environments as to the sediment dispersal pattern have been summarized by G. DE VRIES KLEIN (1968).

Graded bedding

Graded bedding is more frequent and of a different character in ancient than in Recent sediments. Characteristic occurrences of graded bedding and its properties are given in Table 227.

On the basis of the present knowledge of Recent sediments, the environments can be ranged as follows according to the abundance of the occurrences of graded bedding: flood-plain, delta and deep-water delta, deep-water trenches and abyssal plains, lakes, sheltered parts of shelf, open shelf, continental slope and eolian sediments.

Graded bedding occurs commonly in sediments and in no way can be regarded as exceptional. It is produced by all processes that start abruptly and die out gradually, such as turbidity currents, normal traction currents, floods, storms, tidal currents, etc. The whole graded bedded unit can be deposited at once (e. g. in deep-sea sediments of turbidity currents), or be the product of a gradual slackening of the current (e. g. in river sediments). Graded bedding produced as a result of the life habits of organisms was described from intertidal flats (e. g. J. E. WARME 1967).

Ripple marks

The earlier opinion on the origin of asymmetric ripple marks by the activity of currents and of symmetrical ripples by wave action has been fully disproved both experimentally and by direct observation. Both types of ripple marks can originate by the action of both these agents. As a result of this finding, the importance of ripple marks as an indicator of palaeogeographical and palaeogeological conditions greatly decreased.

The rule stating that the ripple index (i. e. wave length: amplitude) of eolian ripples is greater than 10 and that of waterlaid ripples smaller than 10 (generally about 5) has proved valid.

Table 227

Characteristic occurrences of graded bedding and its properties

Environments	Characteristics of graded bedding
River channel	Irregular decreasing of grain-size from gravels to clays; thickness of graded bedded units is from several cm up to several m, middle and upper parts of these units are usually ripple and diagonal bedded; sharp basal contact, eventually great petrographical changes within units
Flood-plains	Grading from medium or fine-grained sands to clays. Thickness of units about several cm to several dm; recurrent grading occurs frequently
Alluvial plains	Like the graded bedding in flood-plain deposits; thicker graded units occur with coarser sediments on their bases. Thickness of graded bedded units is dependent on the grain-size of basal member
Glacial lakes	Glacilacustrine varves are often graded bedded in coarser parts. Grading is from fine sands into silts or clays
Bays	Grading occurs in sandy or silty layers of laminated sediments. Grading usually from fine-grained sands or silts into clays
Delta	In upper and middle parts the same type of graded bedding as in alluvial plains occurs. Grading is usually from medium-grained sands into silty clays
Deep-water delta	Graded beds generally several cm thick. Recurrent grading occurs frequently
Abyssal plains	Well developed and very frequent graded bedding. Grading from fine sand or silt into clay, thickness of units from centimetres up to decimetres, exceptionally up to 50 cm. Sharp basal contact. Recurrent gradation abundant
Deep-water trenches	Gradation abundant. Thick graded bedded units (up to metres). Grading from sand or gravel to sity or silts clays (coarser fraction usually formed from pyroclastic particles)

On the basis of shape, several types of ripple marks are distinguished, some of which are important for the reconstruction of sedimentary environment. The most significant are:

1. Rhomboid ripple marks indicate the deposition on beaches, most frequently on the foreshores of beaches. The orientation of rhombs indicate the trend of backswash and their shape the slope of beach (see p. 221).

2. Small symmetric ripple marks with a wave length below 1 cm, occurring in the profile in several successive laminae with crests and troughs immediately one above another. Their presence invariably suggests shallow-water sedimentation, generally in bays.

3. Giant ripple marks with wave length of several tens or hundreds of metres. They indicate sedimentation on shallow-water flats. They are not easy to identify in ancient sediments.

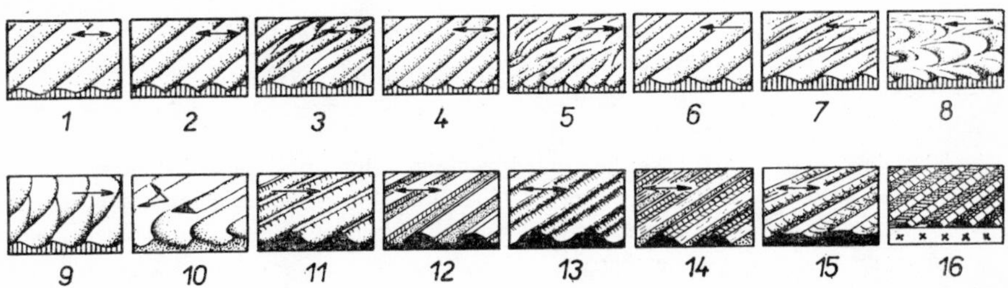

Fig. 133. Genetic classification of ripples according to their profile crest form and distribution of coarse material.

1, 2, 3. Coarse-grained material in troughs — shallow water, up to 200 m.
4. Shallowest environments, up to several metres.
5. Up to several decimetres.
6. Occur in various depths, mostly in shallow water.
7. Various depths, up to 2,500 m.
8. Coarse-grained material randomly distributed — shallow water high-velocity currents (2 to 2·5 m/sec).
9. Beach swash zone.
10. Beach-origin by combination of longshore current and swash motion.
11. Shallow water, lee side of sand waves.
12, 13, 14, 15. Shallow water. Rapid changes in water depth in the course of ripple formation.
16. Lack of loose material.

After V. I. Popov *et al.* (1963).

Ripple-marks with a regular linear course of crests are usually larger and develop in open-shelf environments with a fairly intense action of waves and currents. Ripple-marks with bent crests, frequently bifurcating and rejoining are characteristic of closed or sheltered environments, such as bays, lagoon beaches, etc. Interference ripple-marks occur commonly only in sheltered shallow-water environments. Ripple-marks with truncated crests are sometimes believed to be confined to tidal flats.

The differences in the size-distribution of material forming the crests and troughs of eolian and aqueous ripple-marks have been the subject of controversy. It is likely that aqueous ripples can have coarse material both on crests and in troughs. The distribution of coarse material depends on the amount of sediment; if it is sufficient coarser material occurs on the crests of ripples (after C. I. KING 1959). Eolian ripples bear mostly coarser material on crests.

V. I. POPOV - V. A. BABADAGLY (1963) have attempted to compile a general genetic classification of ripple marks (Fig. 133).

W. F. TANNER (1967) introduced five new parameters in order to distinguish the agents producing the ripple marks. If his theory will prove correct the determination of following types will be enabled:

1. Ripples parallel with current — longitudinal,
2. Transverse to current, or parallel with wave crests:
   A. Symmetric: a. Swash zone ripple marks,
      b. Wave formed ripple marks,
   B. Asymmetric: a. Wind types,
      b. Water types: α. Water formed ripple marks,
         β. Wind formed ripple marks,
3. Modifications and special cases.

Excellent summary about the origin of current formed ripples appeared recently in the book by J. R. L. ALLEN (1968).

It is much easier to interprete the whole ripple fields than the individual ripple marks. Ripple fields cover fine sandy beaches and shallow-water tidal flats. Their occurrences in deeper-water environments need not be taken into consideration.

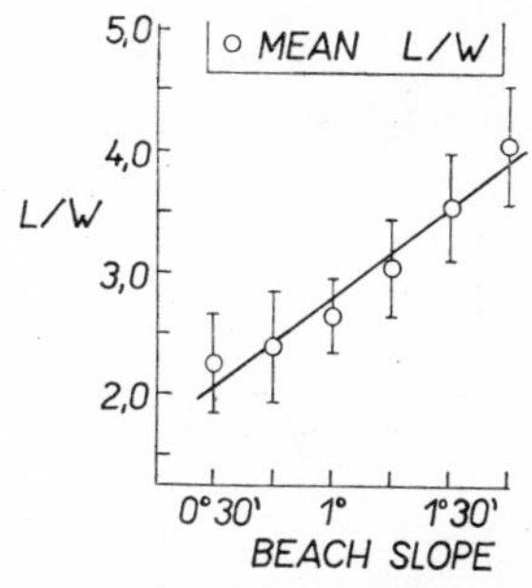

Fig. 134. Graph showing relation on length-width ratio of rhomboid ripple-marks to beach slope. Correlation coefficient equals 0·71. Beach material 60—90% fine sand. L — Length, W — Width. After J. H. Hoyt - V. J. Henry (1963).

The wave length of aqueous and eolian ripple-marks depends on the grain-size and the intensity of the current. The coarser the sediment, the greater must have been the intensity of current or the orbital velocity of the waves for the ripples to develop. The direct relation between the grain-size and the length of wave has been proved many times both in Recent and fossil deposits.

The occurrence of ripple-marks cannot be bathymetrically limited. On submarine photographs, they have been described from the seafloor at 1,000 m deep, where they are produced by the oscillation of water particles induced by currents. The conception that deep-water ripples are more liable to be preserved than shallow-water ones is probably incorrect, although it has been several times put forward and discussed. Forces which generate the ripple-marks are roughly of the same intensity as those disturbing them, which implies that in deep waters there are certainly ripple-disturbing forces when the ripple-forming agencies exist there. Moreover, it has been directly observed that some ripple fields were preserved after their burying by clay.

The occurrences of ripple marks in other than sandy sediments are remarkable. In Recent deposits, they are often developed in shallow-water calcarenites; they are also known from silty clays of the North German wades.

The position and direction of the shore line can be determined from the trend of the crests of ripples only in some cases (see p. 219, 221), although they mostly run parallel to each other. It is because the predominant trend of crests depends primarily on the position of beaches and sand flats in relation to the direction of currents.

Mud cracks

Mud cracks are quite frequent both in Recent and ancient sediments. They are generally interpreted as proving the deposition above the water level. In fact, they occur in Recent deposits most frequently in clays of ephemeral lakes, playas, alluvial flats, flood-plains, swamps, and in the upper part of deltas and tidal flats. Areas

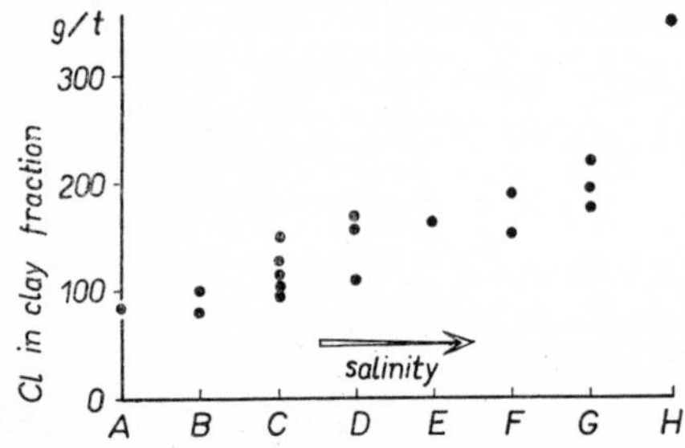

Fig. 135. Relationship between total Cl content and salinity of environment. A. Guadelope River, B. Guadelope Delta, C. San Antonio Bay (upper), D. San Antonio Bay (central), E. San Antonio Bay (lower), F. Aransas Bay, G. Mesquite Bay, H. Gulf of Mexico. After W. D. Johns (1963).

attaining a size of up to several tens of kilometres square can be covered by them. Recent investigation, however, has shown that their origin is not limited to the surface above water level, that mud cracks indistinguishable from the former ones can develop also under water in the following ways:

1. By the effect of synaeresis, i. e. rapid dewatering of clay sediments composed of colloidal particles after sedimentation.
2. By tremors of the bottom.

Thus, sporadic finds of mud cracks need not prove the origin of sediment above the water level. Their mass occurrence in large areas indicates, however, their development in some of the above mentioned sedimentary environments.

Structures originating as a result of the escape of air

These structures occur in finer-grained sediments of the beaches of lagoons or tidal flats. The escape of air manifests itself on the surface of sediments by the presence of bubbles which burst and leave behind pits encircled by minute ridges. The resulting pit and mound structure can be mistaken for some sort of bioglyph. Remarkable structures in the vertical sections through several beds generated by the escape of air were observed also in fine beach sediments of lagoons.

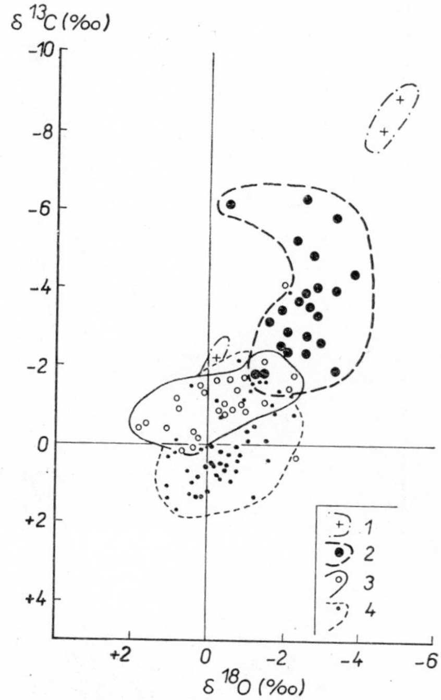

Fig. 136. Carbon and oxygen isotopic composition of mollusc shells from various sedimentary environments. Salinity increasing from 1 to 4. 1. Estuary, Mobile Bay, 2. Mississippi Delta an Sound, 3. Open Gulf shell and littoral areas, 4. Marginal lagoons. After M. L. Keith - R. H. Parker (1965).

Rill marks

Rill marks are indicative of an intermittent sedimentation above the water level, both at backshore and foreshore (see p. 221,276). They occur in sands and silts. Being formed probably by the water flowing down the beach during the low tide, they appear as landward-ramified rills. The occurrence of these structures in ancient sediments is rather scarce. According to some modern views it is doubtful whether rill marks can appear in ancient sediments.

Reconstruction of the environment from the orientation of clastic components

The orientation of particles can serve as a reliable indicator of sedimentary environment only in isolated cases; it is most convenient to use it combined with other criteria. One of this is the imbrication of pebbles, developed both under marine and fluviatile conditions. According to A. Cailleux (1945) the boundary between the

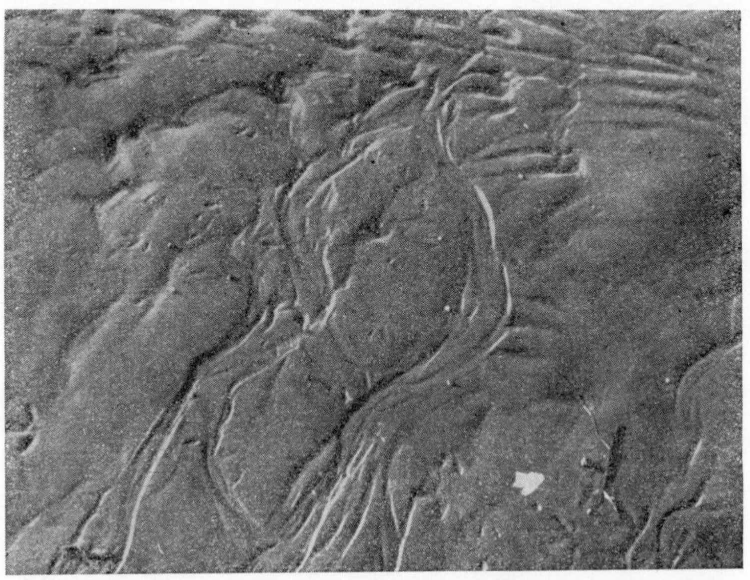

Photo 21. Two examples of rill marks. Upper photo: Rill marks in the beach sand. After J. M. Trefethen *et al.* (1960). Lower photo: Rill marks in the waden mud ("Schlick"). After F. Trusheim (1936.)

marine and fluviatile environment is indicated by 18—20° dip of pebbles. Those of fluviatile gravels are inclined at a higher angle and those of marine gravels at a lower angle. Pebbles dip upstream and the marine ones seawards. The origin of preferred orientation of rod-shaped (elongated) pebbles is rather questionable. Only in deposits of turbidity currents, as has also been corroborated by experiments, the longer axes of disc- and rod-shaped pebbles run parallel to the direction of currents.

The study of the orientation of sand grains is more difficult than that of pebbles, but nevertheless, several exact methods have been elaborated which give promising results.

First of all, it has been found that a minimum of 65% of longer grain-axes of the marine sandy sediments show a preferred orientation, whereas the grains of eolian sediments display a wider scatter of the axes.

J. R. Curray (1956) has proved that sand grains are oriented with the longest axis perpendicular to the shore line even in those places where a strong longshore current exists. He has also ascertained a close relationship between the orientation of grains and the shape of sand bodies. Table 228 presents an approach to the determination of an unknown parameter with the help of the recognized parameters.

Table 228

Relationship between the orientation of grains and the shape of sand bodies (after J. R. Curray 1956)

Grain orientation	Form of sandstone body and its relation to grain orientation	Origin of sediment
Strongly preferential	Linear sandstone bodies, preferred grain orientation parallel to their elongation	River sands
Strongly preferential	Linear sandstone bodies, preferred grain orientation perpendicular to their orientation	Beach and barrier sands
Strongly preferential	Tabular sand bodies, without relation to grain orientation	Various shallow-water sands
Slightly preferential	Various	Eolian sands

Bioglyphs

Organic structures exist in all environments, but differ in amount and kind. The quantity of bioglyphs in various environments is fairly well known, better than their qualitative composition and dependence on the environment. Thanks to A. SEILACHER (1953, 1962, 1963, 1964, 1967), who differentiated five kinds of bioglyphs, we can, to a certain extent, use them for the reconstruction of fossil environments. They are as follows:

1. Repichnia: Trails or burrows left by vagile benthos during directed locomotion.

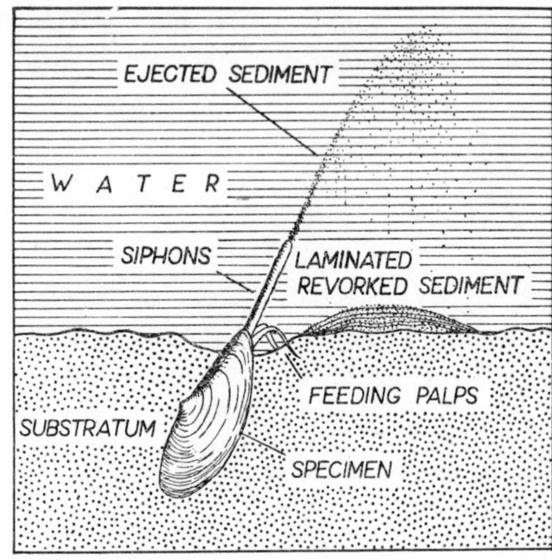

Fig. 137. Feeding and reworking orientation of Yoldia limatula. After D. C. Rhoads (1963).

2. Pascichnia: Winding trails or burrows of vagile mud easters which reflect a "grazing" search for food by covering a given surface more or less efficiently and avoiding double coverage.

3. Fodichnia: Burrows made by hemisessile deposit feeders. They reflect the search for food and at the same time fit the requirements for a permanent shelter.

4. Domichnia: Permanent shelters dug by vagile or hemisessile animals procuring food from outside the sediment as predators, scavengers, or suspension feeders.

5. Cubichnia: Shallow resting tracks, left by vagile animals hiding temporarily in the sediment, usually sand, and obtaining their food as scavengers or suspension feeders.

The shallow-water origin can be indicated only by the so-called *Cubichnia* which occur solely in photic zone, roughly to a depth of 100 m. *Cubichnia* of starfish, found at a depth of a few thousand metres, are the only exception. It must be noted, however, that in relation to other kinds of bioglyphs, the *Cubichnia* are extremely rare.

Otherwise, A. SEILACHER (1964, 1967) divided bioglyphs into several associations (ichnofacies), which can be characteristic of the respective sedimentary environments (Fig. 138).

Bioglyphs are most frequent in shallow-water, slowly deposited sediments. The succession of environments based on the quantity of bioglyphs is as follows:

1. Fine-grained sands and silts of beaches, tidal flats and places of slow sedimentation on shelves.
2. Fine-grained sediments of lagoons, silts and sands of deltas.
3. Medium-grained sands of beaches and sand bars, coarser-grained sediments of shelves and sediments of continental slopes.
4. Other sediments.

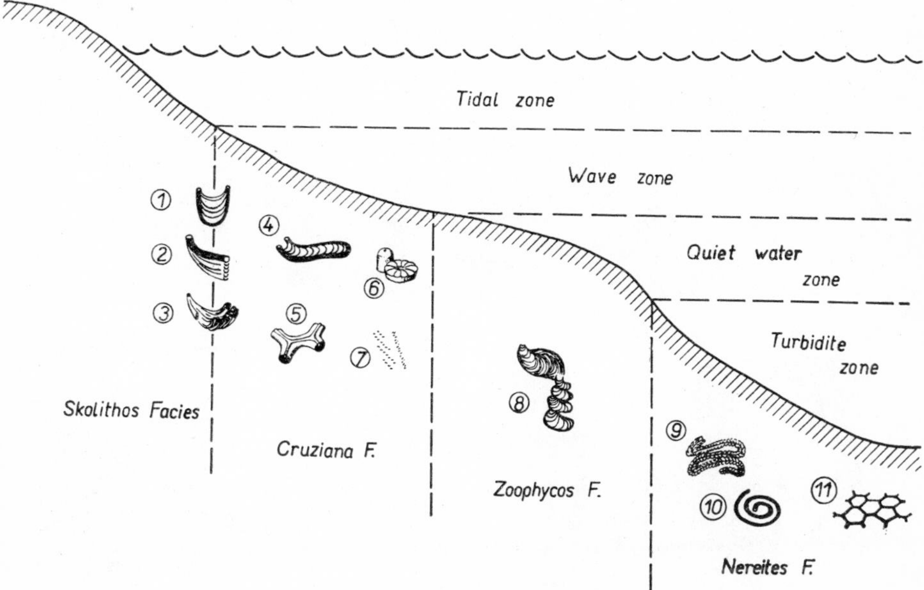

Fig. 138. All communities of trace fossils, regardless of geologic age, can be assigned to one of four major types of ichnofacies which show parallel differences in lithology and inorganic sedimentary structures. 1. *Corophioides*, 2. *Teichichnus*, 3. *Phycodes*, 4. *Rhizocorallium*, 5. *Thalaösincides*, 6. *Bergaueria* and *Solicyclus*, 7. Tracks of trilobites, 8. *Zoophycos*, 9. *Nereites*, 10. *Cerasophycus*, 11. *Paleodictyon*. After A. Seilacher (1964).

Palaeontological criteria for the reconstruction of sedimentary environments

The reconstruction of sedimentary environments from palaeontological criteria is the subject of specialized study and it can be mentioned here only in broad outlines. In discussing the significance of individual groups or species of organisms, due regard must be paid to their ecology. Planktonic organisms are most sensitive to the temperature of water, whereas benthonic organisms are to the character of sediments. Consequently, the former can be used as indicative of conditions in the upper water layer, the latter for the estimation of the depth of the basin.

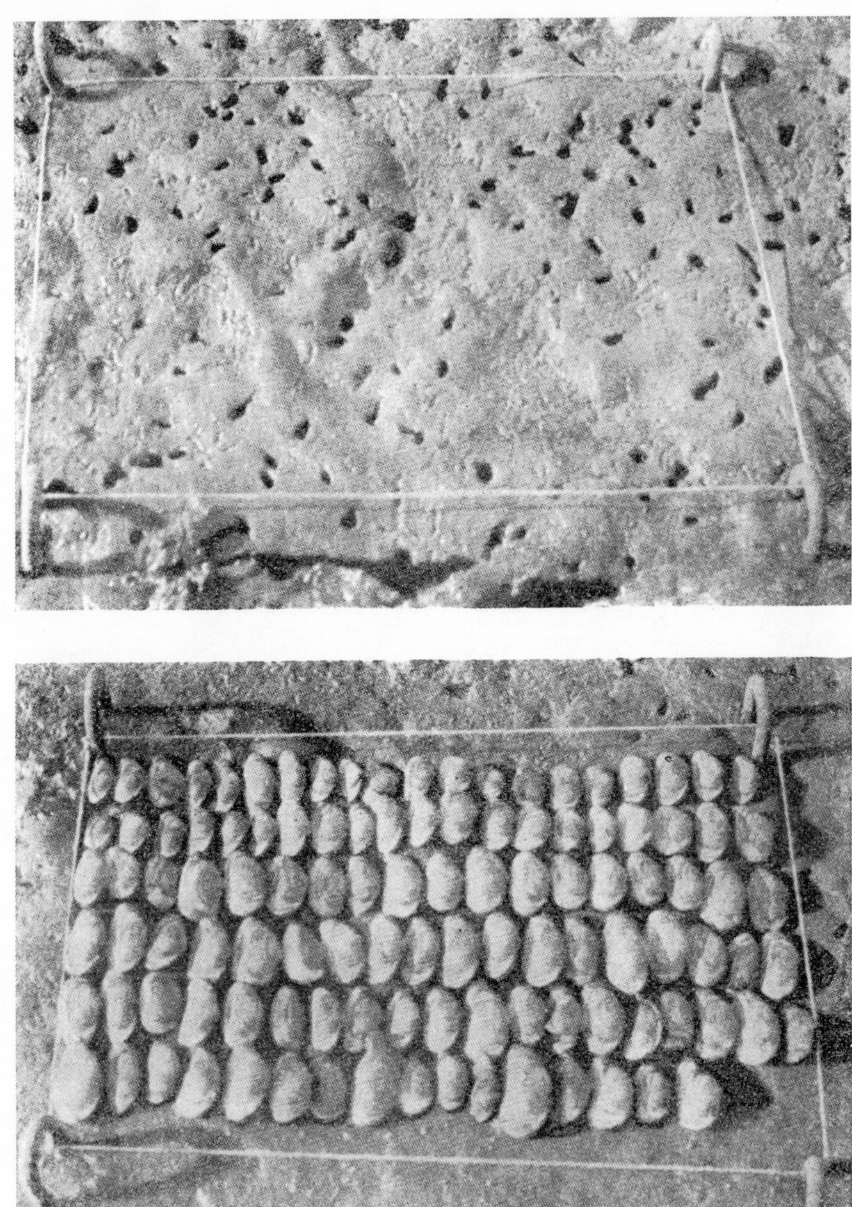

Photo 22. Enormous quantity of pelecypod *Mya arenaria*, living on the sand bottom (encircled part of the bottom is only 1/10 m^2). After F. Kühl (1951) and F. Gessner (1957). North German wades.

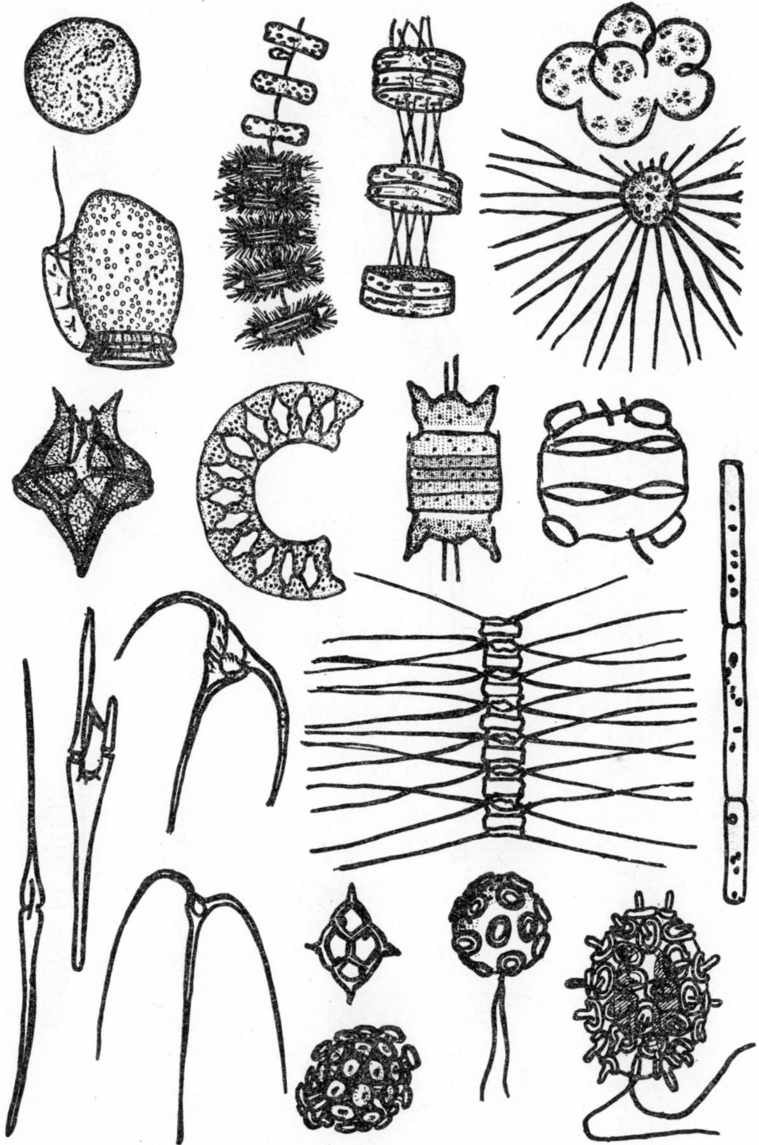

Fig. 139. Phytoplankton of the North Sea waters. After F. Gessner (1957).

In some regions, characteristic associations of of invertebrates for various environments have been elaborated (e. g. in the Gulf of Mexico — R. H. PARKER 1956, in Paria Bay — TJ. VAN ANDEL - H. POSTMA 1954, in the Northeastern Pacific — J. R. NAYUDU - B. J. ENBYSK 1964, etc.). Besides, in many places, associations of planktonic and benthonic foraminifers characteristic of waters of different salinity have been

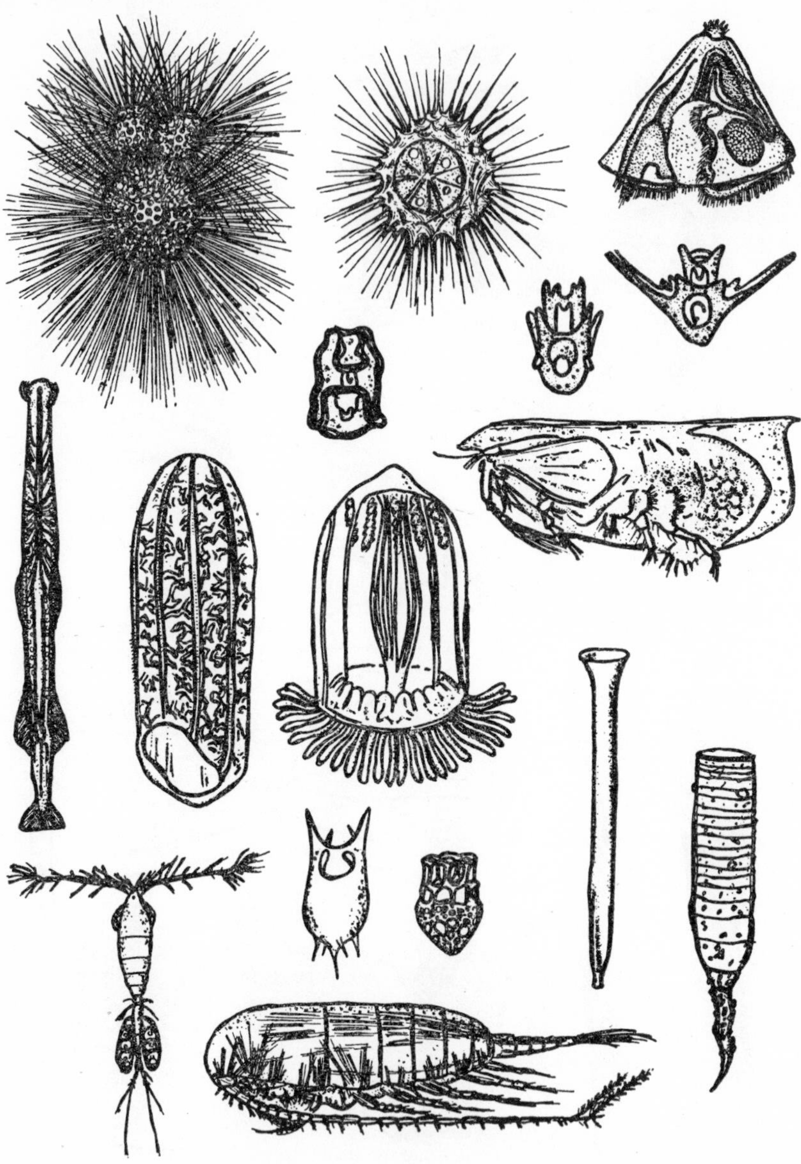

Fig. 140. Zooplankton of the North Sea waters. After F. Gessner (1957).

established. With a few exceptions, it remains true that many groups of organisms are restricted to one or the other environment — for example in marine environments are radiolaria, reef-building corals, brachiopods (although *Lingula* seems to enter brackish waters), cephalopods (but *Logilidae* are found in the Bay of Kiel at a salinity of 17 to 18 per mille), sessile echinoderms, most *Rhodophyceae* and *Phaeophyceae.*

On the contrary in fresh waters, are nearly all the aquatic cormophytic cryptograms, and most flowering plants. Not only systematic but even ecological groups may be confined to one salinity environment: sinupalliate lamellibranchiata (with the exception of some *Limnocardiidae* and *Adacnidae*), animals boring in rock, and most aquatic animals boring in wood, rhizosessile forms of the macrofauna, etc. are marine.

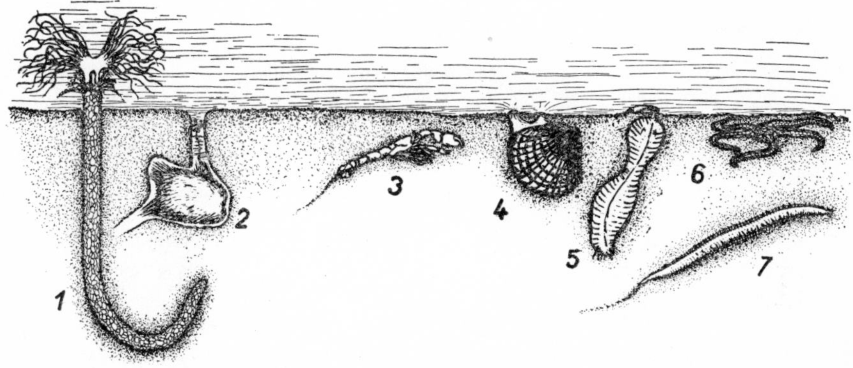

Fig. 141. The most abundant benthos in North Sea sediments. 1. *Lanice conchilega*, 2. *Echinocardium cordatum*, 3. *Callianassa*, 4. *Cardium*, 5. *Echiurus*, 6. *Amphiura*, 7. *Nepthtys*. After F. Gessner (1957).

Representatives of certain groups of organisms inhabiting diverse environments frequently show distinctive characteristics. Brackish forms are usually smaller and less calcareous than marine forms, but *Mactra* reaches its greatest size in brackish water. Foraminifera in brackish waters frequently have organic or siliceous, instead of calcareous, tests. Many organisms grow to their maximum size under certain conditions of salinity and temperature, but the conditions that allow a species to reach maximum size are frequently not identical with those governing greatest abundance. The importance of foraminifera and radiolaria as depth indicators in the marine environment was discussed by B. B. Fumell (1967). Well-known eye-reduction in some Devonian and Carboniferous trilobites can be also used as depth indicator (E. N. K. Clarkson 1967). The dominance of blind and small-eyed trilobites may reflect dark or dim environment. The presence of some of those species in shallow-water environments could result from their migration from the troughs.

The present-day conditions, however, cannot be applied to earlier conditions without some corrections. As is well-known, the surface and abyssal oceanic waters were warmer in the Tertiary than they are today and, therefore, a number of foraminiferal species could reach to greater depths. Analogously, organic assemblages characteristic of greater depths can exist in environments where quiet water, rich in suspension, is photic to a small depth only. Clay bottoms of these environments would be inhabited by the benthos which, for instance, on open shelves occur at substantially greater depths.

The total amount of organic remains can be a suitable diagnostic feature of the environment. They can be accumulated in two ways. *In situ*, i. e. at their habitat where they were accumulated as a result of slow inorganic sedimentation, or by their aggradation in certain places in the zone of active shallow-water sedimentation. The differences between these two modes of occurrence are summed up in Table 229.

Table 229

Differences between two modes of occurrence of organic remains

Shell ridges	*In situ* accumulations of organic remains
Very well sorted according to size and form. Several organic species form individualized accumulations	Shells not sorted according to size and form, Associations of various species
Preferrential orientation of shells, mostly with convex side downward. Orientation of shells is influenced by their mutual contacts	Random orientation of shells. Their orientation responsive to growth position
Features of transport and abrasion	Organic remains in growth position or only slightly displaced

The prospects for a correct solution of environmental condition on the basis of organic remains decrease with geometric progression from the later to the earlier geological periods. In the Late Tertiary, foraminiferal associations allow for fairly correct conclusions on the bathymetrical conditions (with an accuracy of ± 100 m).

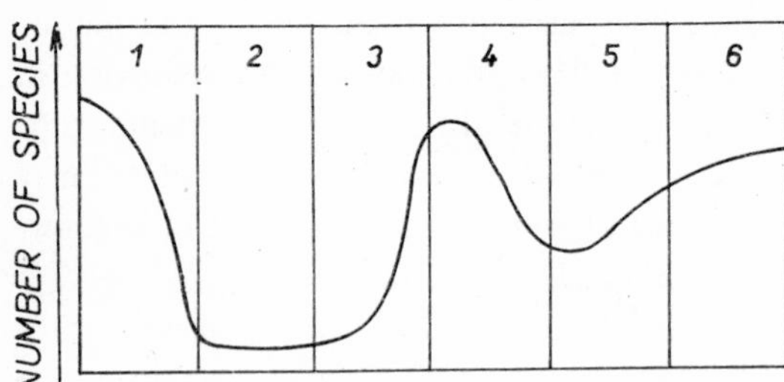

Fig. 142. Relationship between number of animal species with respect to substrate in shallow water environments. 1. Rock, 2. Gravel, 3. Sand, 4. Silt, 5. Clay, 6. Calcareous muds. After J. M. Weller (1960).

Corals, algae and foraminifers serve as suitable bathymetric indicators also for the Mesozoic. All we know with certainty about the Paleozoic, is that water plants had to live in a photic zone, i. e. to a depth of several tens of metres; the reef-building corals and some specific genera, such as *Lingula*, are used as indicators of the environments of that period.

On the whole, the present biological and palaeontological knowledge is mostly insufficient for a successful reconstruction of sedimentary environments. Therefore, these data have to be used only jointly with other criteria.

References

ADAMS T. D. - HAYNES J. R. - WALKER Ch. T. (1965): Boron in Holocene illites in the Dovey estuary, Wales, and its relationship to paleosalinity in cyclotherms. Sedimentology, vol. 4: 189—195, Amsterdam.

ALLEN J. R. L. (1968): Current ripples. P. 1—433, Amsterdam.

BEHNE W. (1953): Untersuchungen zur Geochemie des Chlors and Broms. Geochim. Cosmochim. Acta, vol. 3: 86—114, London.

BOTVINKINA L. N. (1962): Op. cit. on page 106.

BROMLEY R. G. (1967): Marine phosphorites as depth indicators. Marine Geology, vol. 5: 503—509, Amsterdam.

CADIGAN R. A. (1961): Geologic interpretation of grain-size distribution measurements of Colorado Plateau in sedimentary rocks. Jour. Geology, vol. 69: 121—144, Chicago.

CAILLEUX A. (1943): Distinction des sables marins et fluviatiles. Bull. Géol. Soc. France. N. S., vol. 3: 125—138, Paris.

— (1945): Distinction des galets marins et fluviatiles. Bull. Géol. Soc. France. N. S., vol. 5: 375—404, Paris.

CAROLL D. - STARKEY H. C. (1960): Effects of sea water on clay minerals. Proc. of 5th Conf. on Clay and Clay Min., p. 80—101, Washington.

CHESTER R. (1965): Geochemical data for differentiating reef from non-reef facies in carbonate rocks. Bull. Am. Assoc. Petrol. Geol., vol. 49: 258—276, Tulsa.

CHILINGAR G. V. (1963): Ca/Mg and Sr/Ca ratio of calcareous sediments as a function of depth and distance from shore. Jour. Sedimentary Petrology, vol. 33: 256, Menasha.

CLARKSON E. N. K. (1967): Environmental significance of eye-reduction in trilobites and recent arthropods. Marine Geology, vol. 5: 367—375, Amsterdam.

CLAYTON R. N. - DEGENS E. T. (1959): Use of carbon isotope analysis of carbonates for differentiating fresh-water and marine sediments. Bull. Am. Assoc. Petrol. Geol., vol. 43: 890—897, Tulsa.

CURRAY J. R. (1956): Dimensional grain orientation studies of recent coastal sands. Bull. Am. Assoc. Petrol. Geol., vol. 40: 2440—2456, Tulsa.

DAL CIN R. (1968): Climatic significance of roundness and percentage of quartz in conglomerates. Jour. Sedimentary Petrology, vol. 38: 1093—1099, Menasha.

DEGENS E. T. - WILLIAMS E. G. - KEITH M. L. (1957): Environmental studies of carboniferous sediments. P. I. Bull. Am. Assoc. Petrol. Geol., vol. 41: 2427—2455; P. II., vol. 42: 981—997, Tulsa.

DICKINSON W. R. (1968): Singatoga dune sands, Viti Levu (Fiji). Sedim. Geology, vol. 2: 115—124, Amsterdam.

DOEGLAS D. J. (1950): De interpretatie van Korrelgrootanalysen. I—V. Verh. Nederl. Geol. Mijnb. Genootschap. geol. ser. 15; 236—328, Amsterdam.

— (1959): Sedimentology of Recent and old sediments. A comparison. Geologie en Mijnbouw, 21 Jg., p. 228—230, Amsterdam.

DOYLE L. J. - CLEARY W. J. - PILKEY D. U. (1968): Mica: its use in determining shelf-depositional regime. Marine Geology, vol. 6: 381—389, Amsterdam.

EMERY K. O. (1955): Grain-size of California Basin Sediments. Jour. Geology, vol. 58: 503—512, Chicago.

— (1968): Position of empty pelecypod valves on the continental shelf. Jour. Sedimentary Petrology, vol. 38: 1264—1269, Menasha.

ERNST W. - WERNER H. (1960): Die Bestimmung der Salinintätsfazies mit Hilfe der Bor-Methods. Glückauf, vol. 97: 1064-1070, Dortmund.

FRIEDMAN G. M. (1961): Distinction between dune, beach, and river sands from their textural characteristics. Jour. Sedimentary Petrology, vol. 31: 514—529, Menasha.

— (1967): Trace elements as possible environmental indicators in carbonate rocks. Bull. Am. Assoc. Petrol. Geol., vol. 51: 464, Tulsa.

FÜCHTBAUER H. (1963): Zum Einfluss des Ablagerungmilieus auf die Farbe von Biotiten und Turmalinen. Fortschr. d. Geol. Rheinl. u. Westf., Bd. 10: 331—336, Krefeld.

FUMELL B. B. (1967): Foraminifera and radiolaria as depth indicators in the marine environment. Marine Geology, vol. 5: 333—347, Amsterdam.

GESSNER F. (1957): Meer und Strand. P. 1—426, Berlin.

GOLDBERG E. D. - SEMAYAJULU B. L. K. - GALLOWAY J. - KAPLAN I. R. - FAWRE G. (1969): Differences between barites of marine and continental origins. Geochim. Cosmochim. Acta, vol. 33: 287—289, London.

GRIM R. E. - DIETZ R. S. - BRADLEY W. F. (1942): Clay mineral composition of some sediments of the California coast and the Gulf of California. Bull. Geol. Soc. Am., vol. 60: 1785—1808, Baltimore.

GRIM R. E. - KULBICKI G. - CAROZZI A. V. (1960): Clay mineralogy of the Great Salt Lake, Utah. Bull. Geol. Soc. Am., vol. 71: 515—520, Baltimore.

HARDER H. (1963): Inwieweit ist das Bor ein marines Leitelement. Fortschr. Geol. Rheinld. u. Westf., vol. 10: 239—252, Krefeld.

HJULSTRÖM F. J. (1935): Studies of the morphological activity of rivers as illustrated by the river Fyris. Bull. Geol. Inst. Uppsala, vol. 25: 221—528, Uppsala.

HOYT J. H. - HENRY V. J. (1963): Rhomboid ripple marks, indication of current direction and environment. Jour. Sedimentary Petrology, vol. 33: 604—608, Menasha.

INMAN D. L. - CHAMBERLAIN T. K. (1956): Particle-size distribution in nearshore sediments. Finding Ancient Shorelines. Soc. Econ. Palaeontolog. and Mineralog. Spec. Publ., p. 106—127, Tulsa.

JOHNS W. D. (1963): Die Verteilung von Chlor in rezenten marinen und nichtmarinen Sedimenten. Fortschr. Geol. Rheinl. u. Westf., vol. 10: 215—230, Krefeld.

JOHNS W. D. - GRIM R. E. (1958): Clay mineral composition of Recent sediments from Mississippi River delta. Jour. Sedimentary Petrology, vol. 28: 186—199, Menasha.

KATCHENKOV S. M. (1959): Trace elements in sedimentary rocks, and petroleum (in Russian). Trudy Vses. Naučn. Isl. Geol. Razv. Inst., vol. 143: 1—271, Leningrad.

KEITH M. L. - DEGENS E. T. (1959): Geochemical indicators of marine and fresh water sediments. Researches in Geochemistry, p. 38—61, New York.

KEITH M. L. - PARKER R. H. (1965): Local variation of ^{13}C and ^{18}O content of mollusk shells and the relatively minor temperature effect in marginal marine environment. Marine Geology, vol. 3: 115—129, Amsterdam.

KELLER W. D. (1956): Clay minerals as influenced by environments of their formation. Bull. Am. Assoc. Petrol. Geol., vol. 40: 2689—2710, Tulsa.

KLEIN DE VRIES G. (1968): Paleocurrent analysis in relation to modern marine sediment dispersal pattern. Bull. Am. Assoc. Petrol. Geol., vol. 51: 366—382, Tulsa.

KOCZY F. F. - ANTAL P. S. - JOENSUU O. (1963): Die natürlichen radioaktiven Elemente in Sedimenten. Geochim. Cosmochim. Acta, p. 201—214, London.

KORITNIG S. (1963): Zur Geochemie des Fluors in den Sedimenten. Fortschr. d. Geol. Rheinl. Westf., vol. 10: 231—238, Krefeld.

KREJCI-GRAF K. (1964): Geochemical diagnosis of facies. Proc. Geol. Soc. Yorkshire, vol. 34, No 23: 469—521, York.

KREJCI-GRAF K. - KLEIN K. - KREHER A. - ROSSUROM H. - WENZEL G. (1965): Versuche zur geochemischen Fazies-Diagnostik. Chemie d. Erde, vol. 24: 115—146, Jena.

KUENEN PH. H. (1959): Experimental abrasion. F. Fluviatile action on sand. Am. Jour. Sci., vol. 257: 172—190, New Haven.

— (1964): Experimental abrasion. 6. Surf action. Sedimentology, vol. 3: 21—43, Amsterdam.

KÜHNEL R. (1960): Geochemistry of the sediments of Ostrava-Karviná Basin. Unpublished, Ostrava.

LADD H. S. (1957): Treatise on marine ecology and paleoecology. Geol. Soc. Am. Mem., 67, vol. I: 1—1297; vol. II: 1—1077, Baltimore.

LANDERGREN S. - MANHEIM F. T. (1963): Über die Abhängigkeit der Verteilung von Schwermetalen von der Fazies. Fortschr. Geol. Rheinl. u. Westf., vol. 10: 173—192, Krefeld.

MABECONE J. M. (1963): Depositional environment and provenance of the sediments in the Guadelete Estuary (inv. Spain). Proc. 6th Internat. Sediment. Congr., Amsterdam.

MABERRY J. O. (1969): Paleoecologic aspects of trace fossils. Bull. Am. Assoc. Petrol. Geol., vol. 53: 730, Tulsa.

MCKEE E. D. (1964): Inorganic marine structures. In "Approaches to paleoecology". P. 275—295, New York.

MISHRA S. K. (1968): Granulometric studies of Recent sediments in the Firth of Tay region (Scotland). Sediment. Geology, vol. 2: 191—200, Amsterdam.

MÜLLER G. (1964): Die Korngrössenverteilung in den rezenten Sedimenten des Golfes von Neapel. Deltaic and shallow marine sediments, p.293—300, Amsterdam.

MUN A. I. - BAZILEVICH Z. A. (1962): Distribution of boron in clayey sediments of continental basins (in Russian). Geochimija, vol. 2: 175—180, Moscow.

MURRAY H. H. - SAYYAB A. S. (1955): Clay mineral studies of some Recent marine sediments off the North Carolina coast. Proc. 3rd Nat. Conf. Clays and Clay Min., p. 430—441, Washington.

NEVIN CH. (1946): Competency of moving water to transport debris. Bull. Geol. Soc. Am., vol. 57: 713—732, Baltimore.

NICHOLLS C. D. (1967): Trace elements in sediments: an assessment of their possible utility as depth indicators. Marine Geology, vol. 5: 539—555, Amsterdam.

PARKER R. H. (1956): Macro-invertebrate assemblages as indicators of sedimentary environments in East Mississippi delta region. Bull. Am. Assoc. Petrol. Geol., vol. 40: 235—276, Tulsa.

PASSEGA R. (1957): Texture as characteristics of clastic deposition. Bull. Am. Assoc. Petrol. Geol., vol. 41: 1952—1984, Tulsa.

— (1962): Problem of comparing ancient with Recent sedimentary deposits. Bull. Am. Assoc. Petrol. Geol., vol. 46: 114—124, Tulsa.

PETTIJOHN F. J. (1962): Paleocurrents and paleogeography. Bull. Am. Assoc. Petrol. Geol., vol. 46: 1468—1493, Tulsa.

POPOV V. I. - BABADAGLY V. A. (1963): Harmonic series of ripple marks. (In Russian.) Deltaic and shallow-water littoral marine deposits, p. 97—101, Moscow.

PORRENGA D. H. (1963): Bor in Sedimenten als Index für den Salinitätsgrad. Fortschr. Geol. Rheinld. Westf., p. 267—271, Krefeld.

— (1967): Glauconite and chamosite as depth indicators in the marine environment. Marine Geology, vol. 5: 495—501, Amsterdam.

POTTER P. E. - SHIMP N. F. - WITTERS J. (1963): Trace elements in marine and fresh-water argillaceous sediments. Geoch. Cosmoch. Acta, p. 669—694, London.

POWERS M. C. (1957): Adjustment of land derived clays to the marine environment. Jour. Sedimentary Petrology, vol. 27: 68—80, Menasha.

REMANE A. (1963): Biologische Kriterien zur Unterscheidung von Süss- und Salzwassersedimente. Fortschr. Geol. Rheinl. u. Westf., vol. 10: 9—34, Krefeld.

REYNOLDS J. H. - VERHOOGEN J. (1953): Natural variations in the isotopic constitution of silicon. Geochim. Cosmochim. Acta, vol. 3: 224—234, London.

RHOADS D. C. (1963): Rates of sediment reworking by Yoldia limatula in Buzzard Bay, Massachusetts, and Long Island Sound. Jour. Sedimentary Petrology, vol. 33: 723—727, Menasha.

ROYSE Ch. F. (1968): Recognition of fluvial environments by particle-size characteristics. Jour Sedimentary Petrology, vol. 38: 1171—1178, Menasha.

RONOV A. B. - KHLEBNIKOVA Z. V. (1957): Chemical composition of the most important genetic types of clays (in Russian). Geokhimia, No 6: 449—469, Moscow.

RUKHIN L. B. (1947): Granulometry and genesis of sands (in Russian). Izd. Leningr. Univ., p. 1—111, Leningrad.

RUSSELL R. J. (1968): Where most grains of very coarse sand and fine gravel are deposited. Sedimentology, vol. 11: 31—38, Amsterdam.

SAADALLAH A. - KUKAL Z. (1969): Grain size and carbonate content in coastal sediments of Iraq. Jour. Iraq. Geol. Society, vol. 2: 3—10, Baghdad.

SEILACHER A. (1953): Über die Methoden der Palichnologie. N. Jhrb. Geol. Pal. Abh., vol. 96: 426—452, Stuttgart.

— (1962): Paleontological studies on turbidite sedimentation and erosion. Jour. Geology, vol. 70: 227—234, Chicago.

— (1963): Lebensspuren als Salinitätsfazies. Fortschr. Geol. Rheinl. u. Westf., vol. 10, 81—94, Krefeld.

— (1964): Biogenic sedimentary structures. In "Approaches to paleoecology". P. 296—316, New York.

— (1967): Bathymetry of trace fossils. Marine Geology, vol. 5: 413—428, Amsterdam.

SHEPARD F. P. (1964): Criteria in modern sediments useful in recognizing ancient sedimentary environment. Deltaic and shallow marine deposits, p. 1—25, Amsterdam.

STEWART J. H. (1958): Sedimentary reflection of depositional environment in San Miguel Lagoon, Baja California, Mexico. Bull. Am. Assoc. Petrol. Geol., vol. 42: 737—788, Tulsa.

TANNER W. F. (1967): Ripple mark indices and their uses. Sedimentology, vol. 9: 89—104, Amsterdam.

TOMŠÍK J. (1960): Conductivity as the indicator of paleosalinity (in Czech). Sborník I. konf. ostravsko-karvínského revíru, p. 57—61, Ostrava.

WALKER Ch. T. (1968): Evaluation of boron as a paleosalinity indicator and its application to offshore prospects. Bull. Am. Assoc. Petrol. Geol., vol. 52: 751—766, Tulsa.

WALKER CH. T. - PRICE N. B. (1963): Departure curves for computing paleosalinity from boron in illites and shales. Bull. Am. Assoc. Petrol. Geol., vol. 47: 833—841, Tulsa.

WARME J. E. (1967): Graded bedding in the Recent sediments of Mugu Lagoon, California. Jour. Sedimentary Petrology, vol. 37: 540—547, Menasha.

WEAVER CH. (1959): The clay petrology of sediments. Proc. of the 6th Conf. on Clays and Clay Minerals, p. 154—187, Washington.

WEBER J. N. (1965): Chemical composition of siderite nodules in the environmental classification of shales. Bull. Am. Assoc. Petrol. Geol., vol. 49: 362, Tulsa.

Conclusion

Analytical studies of Recent sediments have supplied a wide basis for a synthesis and a general picture of contemporary sedimentary processes.

Further investigation is directed to approach as much as possible the methods of investigation of ancient sediments. At present, the entire complexes of shallow-water environments, through the whole thickness of sediments of the latest geological periods, are being studied. It remains to carry out the correlations of cores of shallow-water sediments with a greater precision and to study in detail the causes of changes in sedimentation in the course of the thousands and tens of thousands of years. The study of Recent, particularly deep-sea sediments will give us a deep insight into the latest geotectonic history of the Earth's crust, the age of oceans, the origin of continents and oceans and other fundamental problems.

The developing knowledge of Recent deposits poses greater and greater problems for the geologists. The demands on the accuracy of the determination of depositional environments and conditions of ancient deposits are steadily increasing. This question can be satisfactorily solved only when the laws controlling Recent sedimentation are perfectly known. In using the data obtained we must avoid the mechanical application, we must pay due regard to the pecularities of the present geological period and take account of the diagenetic and epigenetic processes in rocks. A concise outline of useful information which will aid the correlation of ancient and Recent sediments is presented in this book. Those interested in certain special questions can study the literature referred to in the preceding pages.

Author index

Adams, J. E., 373
Adams, T. D., 422, 423, 459
Ahrens, L. H., 332, 339
Ahrock, R. R., 181
Akimcev, V. V., 18, 24
Alderman, A. R., 181, 373
Aleksina, I. A., 135
Alexander, A. E., 120, 135
Al-Habeeb, K. M., 37, 40, 90, 91, 106
Allen, J. R. L., 72, 106, 193, 459
Altschaeffl, A. G., 262
Andel, Tj. van, 16, 47, 52, 106, 193, 254, 261, 263, 404, 455
Andrée, K., 9, 322
Antal, P. S., 461
Arnal, R. E., 181
Arrhenius, G., 54, 120, 136, 322, 335, 337, 340
Avdusin, P. P., 106
A-Zen, E., 313, 340

Baas-Becking, L.G. M., 12, 16
Babadagly, V. A., 447, 461
Bagnold, A., 123, 135
Baier, C. R., 59
Baker, J. L., 55
Banu, A. C., 193
Barghoorn, E. S., 375
Basumallick, S., 87, 108
Bathurst, R. G. C., 348, 373
Baturin, V. P., 9, 84
Bazilevich, Z. A., 461
Beal, M. A., 135
Beales, E. V., 350, 374
Becke, F., 118, 119, 135
Behne, W., 459
Behre, C. H., 78, 106
Bernard. H. A., 88, 106, 193
Berner, R. A., 375, 390, 396
Berrit, C. R., 340
Berry, L., 24
Berryhill, M. L., 223
Beverage, J. P., 82, 86, 108
Bezrukov, P. L., 315, 320, 340
Biedermann, E. W., 127, 136
Bien, G. S., 374, 375
Bigarella, J. J., 136
Biggs, D. L., 397
Birnstein, J. A., 206, 208
Bischoff, J. L., 289
Bissell, H. J., 374
Black, M., 62, 69, 374
Blacktin, S. C., 136
Blackwerder, B. W., 370, 375
Blissehbach, E., 113, 114
Bluck, B. J., 114, 223
Bock, W. D., 69
Bogoslovskij, B. B., 181
Bohrmann, F. H., 24
Bonatti, E., 120, 136, 331, 340
Borch, C. von, 149, 156
Bordovskiy, O. K., 390, 391, 396
Borreswara, R. C., 86, 97, 108
Botvinkina, L. N., 102, 106, 150, 151, 155, 223, 435, 459
Bourcart, J., 340
Boyd, D. W., 374
Bradley, J., 136
Bradley, W. F., 460
Bray, E. E., 52, 397
Bregger, I. A., 69
Brockmann, Ch., 272, 279
Bromley, R. G., 459
Bruns, E., 9
Bruun, A. F., 340
Bruun, P., 106
Buch, K., 208
Bull, W. B., 110, 111, 113, 114
Burri, C., 80, 84, 106
Burst, J. F., 396
Butcher, H. S., 261
Byrne, J. V., 106, 252, 261

Cadigan, R. A., 403, 459
Cailleux, A., 136, 412, 413, 449, 459
Caine, N., 115
Calvert, S. E., 252, 261
Campbell, D. H., 127, 136
Carlson, O. R., 52

Carroll, D., 459
Carozzi, A. V., 70, 460
Castaing, R., 55
Chamberlain, T. K., 262, 403, 460
Chave, K., 57, 59, 60, 61, 70, 340, 343, 374
Chayes, K. E., 342
Chester, R., 432, 459
Chilingar, G. V., 9, 374, 397, 425, 459
Chow, T. J., 340
Chronic, J., 375
Chumley, J., 316, 341
Clarke, F. W., 59, 70, 332, 333, 340
Clarkson, E. N. K., 457, 459
Clayton, R. N., 70, 459
Cleary, W. J., 460
Cloud, P. E., 356, 357, 366, 367, 374
Coakley, J. P., 181
Colby, B. R., 77, 78, 106
Coleman, J. M., 74, 87, 91, 106, 193, 194
Collet, L. W., 35, 40
Conolly, J. R., 340
Conover, J. T., 70
Corbel, J., 28, 29, 30, 31, 40
Correns, C. W., 313, 340
Cronan, D. S., 332, 334, 340
Crosby, E. J., 16, 108
Crowell, J. C., 155
Cultberson, J. K., 80, 82, 108
Curray, J. R., 216, 223, 232, 451, 459
Curtis, C. D., 70

Dal Cin, R., 459
Dapples, E. C., 106, 394, 396
Davies, D. K., 340
Davis, S. N., 106, 146, 155
Davis, W. M., 25, 40
Debyser, J., 289
Degens, E. T., 70, 289, 421, 423, 426, 459, 460
Delany, A. C., 120, 136
Denny, Ch. S., 115
Dickinson, K. A., 223
Dickinson, W. R., 404, 459
Dietz, R. S., 331, 332, 340, 460
Dill, R. F., 340, 342
Dodd, J. R., 59, 70
Doeglas, D. J., 126, 132, 136, 261, 404, 459
Donner, J. J., 155
Dow, R. L., 224
Doyle, L. J., 416, 460
Dreimanis, A., 148, 155
Droste, J. B., 155
Duboul-Razavet, Ch., 194
Dygas, J. A., 342

Eardley, A. J., 158, 165, 181, 374, 396
Eckis, R., 115
Ehrhardt, H., 265, 266, 267, 279, 436
Eichler, R., 70
El Wakeel, S. K., 316, 317, 340
Emery K. O., 78, 106, 153, 223, 235, 245, 247, 249, 252, 261, 262, 321, 326, 361, 362, 365, 372, 375, 409, 410, 426, 460
Enbysk, B. J., 53, 262, 455
Engelhardt, W., 107, 417
Epstein, M., 150
Epstein, S., 348, 374
Ericson, D. B., 340, 382
Ericson, K. G., 52
Ernst, W., 423, 460
Evans, G., 48, 52, 268, 269, 271, 273, 276, 279
Ewing, G. C., 252, 253, 262
Ewing, J., 340
Ewing, M., 340, 341

Fairbridge, R. W., 9, 374
Fawre, G., 460
Fay, R. C., 52
Ferm, J. C., 193
Fessenden, F. W., 24
Filoux, J., 262
Firbas, E., 45, 52
Fischer, A. G., 396
Fisk, H. H., 105, 181, 191, 194, 261, 441
Fleming, R. H., 10, 342
Flemming, N. C., 223
Flint, R. L., 138, 151, 155
Folk, R. L., 86, 98, 100, 107, 109, 124, 136, 368, 374, 417
Förstner, U., 40
Frakes, L. A., 155
Frazier, D. E., 107
Frederickson, K., 55
Freyberg, B. v., 260, 261
Friedman, G. M., 136, 350, 370, 374, 392, 393, 396, 417, 430, 460
Frye, J. C., 137
Füchtbauer, H., 261, 345, 352, 374, 416, 460
Fumell, B. B., 457, 460
Fütterer, D., 261

Gagliano, S. M., 193, 194
Galloway, J. 460
Game, P. M., 118, 119, 120, 121, 136
Gardner, D. E., 223
Garrels, R., 397
Gerasimov, I. P., 42
Gershanovich, D. E., 261
Gessner, F., 223, 454, 455, 456, 457, 460
Gilberg, M., 155
Gilbert, G. K., 10, 72, 107
Giles, A. W., 53
Ginsburg, R. N., 70, 353, 374, 375
Glass, B. P., 54, 55
Glen, J. W., 155
Goldberg, E. D., 50, 53, 136, 137, 340, 341, 426, 460
Goodell, H. G., 61, 71, 261
Goreckij, G. I., 107
Gorham, E., 182
Gorshkova, T. I., 290, 396
Gorsline, D. S., 223, 233, 248, 261, 262, 263
Gotthard, R., 223
Gould, H. R., 261
Grady, J. R., 52
Graf, D. L., 374
Granelli, N. C., 341
Greenman, N. N., 261
Griffin, J. J., 136, 341
Griffiths, J. C., 107
Grim, R. E., 194, 418, 460
Gripp, E., 155, 261
Grippenberg, S., 290, 384, 388
Gross, M. G., 261
Gudelis, V., 136
Gvosdetsky, V., 396

Hain-Yi-Ling, 261
Hamilton, E. L., 374
Hamilton, W. B., 115
Hansen, K., 177, 178, 182
Harder, H., 460
Harris, R. C., 70, 136
Harris, S. A., 122, 123
Harrison, P. W., 155
Harrison, W., 262
Hartman, M., 290
Harvey, H. W., 208
Hathaway, J. C., 396
Hayes, M. O., 231, 262
Haynes, J. R., 459
Hedgpeth, J. W., 16
Heezen, B. C., 301, 338, 340, 341
Henry, V. J., 223, 447, 460
Hentschel, E., 208
High, L. R., 182
Hill, M. N., 10, 54, 208, 209, 223, 326, 338 425
Hill, N., 108
Hjulström, F. J., 37, 40, 72, 73, 107, 400, 460
Hobbs, W. H., 130, 136
Hodge, P. W., 55
Hoffmeister, I. E., 53
Hollister, C., 341
Holmes, C. C., 155
Holmes, C. D., 136, 149
Holmes, Ch. W., 223, 224
Hocke, R. L. B., 115
Hoskin, Ch. M., 370, 374
Houbolt, J. H. C., 344, 345, 374
Hough J. L., 182
Howard, C. S., 34, 40
Hower, J., 71, 349, 375, 392, 397
Hoyt, J. H., 223, 262, 460
Hubbell, D. W., 107
Hubbs, L. V., 374
Hubert, J. F. F., 326, 341
Hülsemann, J., 261
Hume, J. D., 78, 107
Hummel, H., 279
Hunzicker, A. A., 181
Hutchinson, G. E., 182

Illing, H., 348, 374
Imbrie, J., 374
Ingerson, E., 70
Ingle, J. C., 223
Inman, D. L., 21, 22, 24, 136, 214, 215, 223, 262, 403, 460

Jäckli, H., 25, 40
Jahn, A., 136
Jell, J, S., 374
Jenny, H., 17, 24
Joensuu, O., 461
Johansson, C. E., 93, 107, 155
Johns, W. D., 194, 448, 460
Johnson, J. W., 223
Johnson, M. W., 10, 342
Johnson, N. M., 18, 24

Kagami, H., 262

Kalle, K., 9, 10, 195, 203, 204, 208
Kalterherberg, J., 107
Kamel, A. M., 223
Kaplan, I. R., 16, 396, 460
Karcz, I., 88, 107
Karczewski, A., 155
Kartashov, I. P., 81, 107
Katchenkov, S. M., 460
Keith, M. L., 70, 421, 423, 425, 426, 435, 449, 459, 460
Keller, W. D., 155, 223, 460
Kellog, C. E., 24
Kelly, W. C., 24
Kharkar, D. P., 53
Khlebnikova, Z. V., 421, 462
King, C. A. M., 208, 209, 223
King, C. I., 447
Kinsman, B., 263
Kinsman, D. J. J., 70, 251, 262
Kitazaki, W., 316, 342
Kleim, K., 461
Klein deVries, G., 279, 444, 460
Klenova, M. V., 10, 34, 36, 40, 262, 283, 285, 288, 290, 385, 388
Klimova, L. T., 108
Knauff, W., 61, 70
Koczy, F. F., 461
Kodymová, A., 99, 107
Kofoed, J. W., 262
Koide, M., 53, 340
Koldewijn, B. W., 107
Kopstein, F. P. H. W., 53
Koritnig, S., 461
Kornicker, L. S., 70, 355, 374
Koshechkin, B. I., 383, 388
Krasinceva, V. V., 341
Kreher, A., 461
Krejci-Graf, K., 461
Krinsley, D., 70
Krinsley, D. H., 136, 155
Krumbein, W. C., 15, 16, 84, 107, 396
Kuenen, Ph. H., 9, 10, 125, 136, 272, 259, 296, 298, 303, 304, 327, 341, 410, 413, 461
Kühl, F., 454
Kuhlman, H., 136
Kühnel, R., 427, 461
Kukal, Z., 53, 93, 107, 136, 194, 279, 341, 430, 462
Kulbicki, G., 460
Kullenberg, B., 397
Kürsten, M., 93, 107

Laagaij, R., 53, 194
Ladd, G. E., 107
Ladd, H. S., 374, 461
Lalou, C., 70
Land, L. L., 129, 130, 132, 136
Landergren, S., 422, 424, 461
Landim, P. B., 155
Lane, D. W., 107
Lankford, R. R., 194
Laporte, E. F., 10, 16
Laprade, K. E., 117, 119, 120, 137
Larsen, G., 397
Lattman, L. H., 107
LeBlanc, R. J., 261
Leontiev, O. K., 223
Leopold, E. B., 367, 369, 375
Leopold, L. B., 18, 19, 24, 27, 40, 72, 79, 89, 103, 107
Lerman, A., 70
Leuchs, K., 130, 136
Leutwein, F., 58, 59, 70
Likens, G. E., 24
Lisicyn, A. P., 53, 153, 156, 262, 292, 293, 340, 341
Lloyd, R. M., 70, 375
Logan, W. B., 63, 70
Logvinenko, N. V., 215, 216, 219, 223
Lombard, A., 10
Long, J. F., 137
Lopatin, G. V., 28, 29, 30, 35, 40, 79, 80
Love, L. G., 397
Lowestam, H. A., 56, 57, 70, 348, 353, 374
Lucia, F. J., 249, 262
Lucke, F. J., 262
Lüders, K., 268, 274, 279
Ludwig, G., 108
Lukashev, K. I., 137
Lundbeck, J., 173, 174, 182
Lynch, M. P., 262

Mabecone, J. M., 194, 419, 420, 461
Maberry, J. O., 461
MacCarthy, G. R., 223
Mackenzie, E. T., 340
Major, C. F., 88, 106
Makedonov, A. V., 397
Manheim, F. F., 422, 424, 461
Mann, J. F., 182

Margolis, S. V., 223
Martens, J. H. C., 217, 223
Maslov, V. P., 70
Matejka, D. Q., 107
Matthews, W. H., 194
Mattox, R. B., 137
Maxwell, W. G. H., 353, 374
Mayer, E. G., 290
McCarter R. S., 397
McClelland, B., 261
McEwen, M. C., 20, 24
McKee, E. D., 90, 108, 129, 137, 218, 219, 224, 361, 367, 369, 370, 374, 440, 442, 461
McKeller, R. C., 374
McKelvey, V. E., 108, 182
McManus, D. A., 261
McMaster, R. L., 70
Meader, R. W., 182
Mellis, O., 341
Menard, H. W., 36, 40, 72, 108, 319, 321, 332, 341
Menzies, R. J., 341
Mero, J. L., 341
Miller, D. N., 375
Miller, J. P., 24, 40, 107
Miller, J. I., 224
Minder, L., 182, 378
Minkevichius, V., 136
Mishra, S. K., 461
Mohr, E. C. J., 18, 19, 22, 24
Moldvay, L., 117, 137
Moore, D., 16
Moore, D. G., 16, 256, 257, 262
Moore, D. J., 341
Moore, H. B., 65, 70
Moore, J. E., 164, 182
Moore, J. R., 240, 262
Morisawa, M., 72, 108
Mortimer, C. H., 182
Müller, G., 22, 24, 40, 246, 247, 262, 404, 408, 461
Multer, H. G., 52
Mun, A. I., 461
Münchenhausen, L., 24
Muravenskij, C. D., 181
Murdmaa, I. O., 228, 262, 341
Murray, H. H., 417, 461
Murray, J., 302, 305, 306, 312, 313, 314, 315, 316, 341
Murray, R. C., 352, 375
Naidu, A. S., 86, 97, 108
Nalivkin D. V., 10, 15, 16
Naumann, E., 182
Nayadu, J. R., 53, 262, 331, 340, 341, 455
Neev, D., 223
Nesteroff, W. D., 341
Nevin, Ch., 72, 108, 400, 461
Newell, N. D., 346, 347, 354, 375
Nicholls, C. D., 461
Nielsen, H., 290
Niino, H., 235, 245, 253, 262
Nikolaieva, V. K., 36, 40
Nordin, C. F., 80, 82, 86, 108
Norin, E., 341
Nota, D. J. G., 194, 262, 282, 286, 404

Okko, V., 40, 146, 156
Olausson, E., 388
Olsson, I. U., 388
Oostdam, B. L., 262
Oosthuizen, C. O., 339
Oppenheimer, C. H., 71
Osanik, A., 107
Ostroumov, E. A., 281, 282, 290, 319, 341
Otvos, E. G., 224

Panin, N., 224
Pannekoek, A. J., 115
Pardé, M., 40
Parker, R. H., 70, 193, 194, 341, 435, 449, 455, 460, 461
Parrin, D. W., 136
Passega, R., 404, 461
Patterson, C. C., 340
Petterson, M. N., 342, 374
Pettijohn, F. J., 94, 108, 442, 461
Péwé, T. L., 44, 53, 137
Phleger, F. B., 244, 252, 253, 262, 263
Pia, J., 171, 172, 182
Picard, K., 223
Picard, M. D., 182
Pierce, R. S., 24
Pilkey, D. U., 460
Pilkey, O. H., 61, 71, 370, 371, 375
Pipkin, B. W., 182
Plumley, W. J., 95, 96, 100, 108
Pollack, J. M., 84, 96, 108
Poole, F. G., 137
Popov, V. I., 84, 446, 447, 461
Porrenga, D. H., 434, 461

Porter, J., 137
Postma, H., 16, 47, 52, 254, 261, 404, 455
Potter, P. E., 108, 422, 462
Power, W. R., 115
Powers, M. C., 263, 419, 462
Pratt, R. N., 342
Pratt, W. L., 263
Prattje, O., 263
Pray, I. C., 397
Pray, L. C., 350, 375
Price, N. B., 462
Price, W. A., 224
Purdy, E. G., 70, 374

Rad, U. von, 342
Radczewski, O. E., 130, 131, 135, 137
Rankama K., 208
Rao, C. B., 224
Rao, M. S., 227, 229, 230, 263
Raupach, F., 290
Reavely, G. H., 155
Reesman, A. L., 155
Reeves, C. C., 157, 165, 166, 167, 182
Reid, J. R., 182, 405
Reimann, B. E. F., 136
Reineck, H. E., 52, 53, 264, 279
Remane, A., 462
Remizov, I. N., 215, 216, 219
Renard, A. F., 302, 305, 306, 312, 313, 314, 315, 341
Reuter, J. H., 52
Revelle, R., 208, 302, 311, 312, 328, 342, 375
Rex, R. W., 137
Reynolds, J. H., 462
Rezak, R., 70
Rhoads, D. C., 277, 279, 439, 452, 462
Rhodes, M. L., 373
Richards, A. F., 40, 320, 342
Richardson, S. H.., 52, 397
Richter, R., 277, 278, 279
Richter, V. G., 397
Rigby, J. K., 375
Riley, J. P., 316, 317, 340
Rittenberg, S. C., 396
Rittenhouse, G., 108
Riviére, A., 404
Robertson, E. C., 396
Robinson, G. W., 22, 23, 24
Robles, R., 374
Rodolfo, K. S., 342
Regers, J. J. W., 24
Rominger, J. F., 106
Ronov, A. B., 421, 462
Rosenfeld, M. A., 107
Ross, D., 289
Rossurom, H., 461
Royse, Ch. F., 462
Rukhin, L. B., 82, 84, 95, 108, 403, 462
Rukhina, E. V., 140, 141, 142, 143, 144, 145, 146, 147, 156
Rusnak, G. A., 108, 214, 215, 223, 247
Russell, R. D., 97, 108
Russell, R. J., 406, 462
Rust, B. R., 109, 181
Ruxton, E. P., 20, 24

Saadallah, A., 93, 107, 136, 430, 462
Sackett, W. M., 35
Sahama, P., 208
Samoilov, I. V., 28, 40
Sanders, J. E., 279, 350, 374
Sapoznikov, D. G., 182
Sarkar, S. K., 87, 108
Sarkisjan, S. G., 108
Saunders, G. W., 182
Sayyab, A. S., 417, 461
Schäfer, W., 208, 275, 279, 326
Schlanger, S. O., 374
Schlee, J., 108, 137
Schneider, E. D., 341
Scholl, D. W., 182, 250, 263
Schott, G., 295, 328, 341
Schumm, S. A., 85, 88, 89, 90, 108, 109
Schwarzbach, M., 156
Scott, M. R., 53
Scruton, P. C., 187, 189, 194, 262
Sears, M., 342
Seibold, E., 256, 263, 375
Seilacher, A., 452, 453, 462
Sekyra, J., 123, 127, 128, 137
Semayajulu, B. L. K., 460
Semenovich, N. N., 182
Shancer, E. V., 103, 109
Sharp, R. P., 115, 137
Shepard, F. P., 9, 10, 16, 135, 137, 185, 190, 194, 208, 209, 220, 224, 258, 261, 263, 301, 302, 326, 342, 414, 462
Shepps, V. C., 156
Shilov, V. M., 319, 341
Shimp, N. F., 374, 462

Shinn, E., 375
Shishkina, O. V., 341, 394, 395, 397
Shor, G. G., 261
Siegel, F. R., 61, 71, 375
Siever, R., 397
Sindowski, K. H., 263, 404, 406
Singh, I. B., 279
Sioli, H., 194
Sitler, R. F., 156
Skinner, H. C. W., 182, 375
Skolnick, H., 55
Skopincev, B. A., 64, 71
Skorniakova, N. S., 340
Slánská, J., 182
Sloss, L. L., 15, 16, 394, 396
Smalley, I. J., 130, 137
Smith, D. D., 262
Smith, P. V., 208
Smith, S. V., 329, 342
Smolík, L., 22, 23
Sneed, E. D., 98, 109
Soliman, S. M., 194
Stanley, D. J., 156
Starkey, H. C., 459
Stehli, F. G., 349, 375, 392, 397
Stetson, H. C., 263
Stewart, J. H., 224, 251, 257, 263, 402, 403, 462
Stewart, R. A., 263
Stockman, K. W., 348, 375
Stokes, W. L., 137
Storr, J. F., 375
Straaten, L. M. J. U. van, 194, 203, 269, 272, 279, 342, 443
Strachov, N. M., 9, 10, 32, 34, 40, 287, 381, 396, 397
Stringham, B., 181
Suess, H. E., 374
Sundborg, A., 72, 109
Suzuki, K., 316, 342
Sverdrup, H. U., 9, 10, 198, 208, 304, 305, 316, 318, 342
Swain, F. M., 182
Sweeting, M. M., 31, 40
Swineford, A., 137

Taft, W. H., 182, 375
Takahashi, T., 136
Tanner, W. F., 224, 263, 447, 462
Taylor, J. C. M., 374
Taylor, R. E., 97, 108
Teichert, C., 358, 359, 360, 364, 375
Teichmüller, R., 182
Termier, G., 9, 10
Termier, H., 9, 10
Terry, R. D., 261
Teruggi, M. E., 137
Thomas, L. A., 397
Thompson, W. O., 224
Tibbitts, G. C., 129, 137
Tomšík, J., 427, 462
Tongiorgi, E., 258, 263
Tooms, J. S., 332, 340
Trask, P. D., 10, 68, 71, 232, 329, 342
Trefethen, J. M., 224, 450
Trevisan, L., 263
Tricart, J., 194
Trumbull, J. V. A., 137
Trusheim, F., 279, 450
Turekian, K. K., 53, 208
Twenhofel, W. H., 182

Uchupi, A., 137, 261
Usdowski, H. A., 182
Utech, K., 55

Valentin, H., 279
Van Baren, P. A., 18, 19, 22, 24
Vatter, A. E., 155
Vejcher, A. A., 109
Verhoogen, J., 462
Vigneaux, M., 263
Viselkuna, M. A., 182
Vita-Finzi, G., 130, 137
Vollbrecht, K., 108
Volkov, I. I., 281, 282, 290

Wadell, H., 109
Walker, Ch. T., 422, 459, 462
Walker, R. G., 298, 342
Walther, J., 9, 10
Ward, W. C., 107
Warme, J. E., 444, 462
Waskowiak, H., 58, 59, 70
Wasmund, E., 53, 182
Wattenberg, H., 208
Weaver, Ch. E., 370, 362, 375
Webb, J. E., 194
Weber, J. N., 431, 462
Wedepohl, K. H., 334, 342
Wegman, E., 42

Wegner, Th., 279
Weller, J. M., 297, 428, 458
Wells, A. J., 374, 375
Wentworth, Ch. K., 148, 149, 156
Wenzel, G., 461
Werner, F., 53, 263
Werner, H., 423, 460
West, R. G., 155
Weyl, F., 279
Wheeler, W. C., 59, 70
White, G. H., 155
Whitehouse, I. G., 397
Wildt, R., 55
Williams, E. G., 459
Williams, G., 137
Williams, M., 375
Williams, P. F., 109
Willis, J. P., 339
Winterer, E. L., 149, 156
Witters, J., 462
Wolf, K. H., 78, 109
Wollin, G., 340
Wolman, M. G., 24, 40, 89, 107
Wong, W. H., 35, 37, 40, 109
Wüst, G., 208

Yakovleva, S. V., 139, 156
Yañez, A., 241, 263
Young, R., 137

Zenkevich, L. A., 206, 208
Zenkovich, V. P., 224
Zingg, T., 156
ZoBell, C. E., 68, 69, 71
Zumberge, J. H., 24

Subject index

Abachis avara semiplicata, 255
Abra liocia, 255
Abrasion,
—, pebbles, 84, 97, 98, 414
—, shore, 211
Abyssal ocean,
— —, mechanism of sedimentation, 15, 402, 409
— —, sedimentary structures, 438
Abyssal plains,
— —, graded bedding, 445
Acanthopora, 254
Acropora, 360, 361, 366, 372
— *cervicornis*, 355
— *palifera*, 372
— *palmata*, 355
— *rambleri*, 361
— *rayneri*, 361
— *reticulata*, 361
— *vaughani*, 361
Acropora zone, 361
Adacnidae, 457
Adratic Sea, 251, 256
Aegropropila, 167
Aequipecten irradians amplicostatus, 255
Afja, 176, 177
Africa,
—, coast, 227
—, deep-sea currents, 294, 295
Agullas Bank, 321
Alacran Reef, 355, 370, 371
Alcohols, 65
Algae,
—, calcareous, 51
—, coral reefs, 355, 356, 357
—, lakes, 167, 168, 169
Algal balls, 167, 168
Algal biscuits, 168
Algal mats, 248, 249, 348
Algal mud, 348
Algal ridge, 360, 361, 362
Alkaloids, 65
Allopora, 359
Allsa plantago, 178
Alluvial fans, 15, 104, 110—115
— —, sedimentary structures, 437
Alluvial lakes, 103
Alluvial plains, 14, 74, 101—103
— —, graded bedding, 445
— —, preservation of sediments, 386, 387
— —, sedimentary structures, 437
Alm, 168
Almatinka River, 75
Alophane, 316
Alpine Rhine, 33
Alps,
—, denudation, 25, 26, 29
—, lakes, 170, 174, 176
—, rivers, 100, 187
Alternathera philoxeroides, 179
Alum Shale, 424
Amazon River, 184, 185, 187, 307
America, shelf, 229
Amnicola limosa, 171, 172
Amphistegina lessoni, 367
— *madagascariensis*, 367
Amphistegina zone, 370
Amphiura, 457
Amu Darya River, 33, 88, 90, 91, 102, 104, 106, 121, 185, 187
Ancylus, 171, 172
Anhydrite, 289
Anodonta, 172
Antarctic Ocean,
—, continental slope, 381
—, deep-sea currents, 294, 295
—, shelf, 227
Anthozoa, 253
Antidune transport, 77
Apex, 110
Aphanocapsa, 168
Aphotic layer, 204
Aptian corals, 373
Aragonite,
—, coral reefs, 166, 167, 367, 368, 369, 371
—, lakes, 166, 167
—, old sediments, 430
—, palaeogeographic indicator, 430

Aragonite, shells, 56, 57, 59
—, stability, 392, 393
Aragonitic mud, 345, 346, 348
— —, compaction, 390
— —, needles, 346, 347, 348
Aral Sea, 161, 287
Aransas Bay, 448
Arctic Sea,
— —, nitrates, 203
Arenicola, 227, 278
Arroyo Hondo, 110
Ash falls, 320
Asterias, 277
Astreopora, 372
— *tabulata*, 355
Atlantic Ocean,
— —, brown clay, 310
— —, carbonate in water, 200, 201
— —, carbonatic sediments, 328
— —, coral reefs, 371
— —, deep-sea currents, 294, 295
— —, deep-sea sands, 325
— —, diatom ooze, 314
— —, dolomite, 351
— —, dust falls, 130, 131, 135
— —, foraminiferal ooze, 311, 312, 313
— —, grain-size of sedinents, 404
— —, nitrates in water, 203
— —, oceanic currents, 294
— —, pyroclastic sedinents, 320
— —, shelf, 227, 233, 234
— —, silica in waters, 197
— —, stratigraphy, 379—382
— —, suspension in waters, 293
Atolls, 353—363
Australia,
—, lagoons, 351
—, lakes, 165
Autotropic bacteria, 207
Avicena oficialis, 260
Azores, 320
Azov Sea, 242, 243
— —, beaches, 215, 219

Backshore, 218
Backswamp, 14
Backswash, 210
Bacteria, 67—69, 204, 207, 208
—, aerobic, 68
—, anaerobic, 68
Bacteria, calcareous, 69
—, denitrifying, 68
—, lake waters, 168, 177
—, nitrifying, 68
—, sulphate reducing, 68
Bahamas, 62, 300, 345—351
—, diagenesis of carbonates, 392
Bahamites, 350
Baikal Lake, 285
Baja California,
— —, lagoons, 253
Balkhash Lake, 158, 287
— —, dolomite. 351
Baltic Sea, 41, 384
— —, beaches, 212, 237, 287, 288, 289
— —, clays, 424, 431
— — suspension in waters, 291
Bank deposits, 88, 89, 91, 106
Banks, submarine, 409
Barakar River, 87
Barataria Bay, 404
Barbados Island, 120
Barchans, 123
Barentz Sea, 231, 385, 431
Barite,
—, deep-sea sediments, 321, 322, 334—335
—, Mn nodules, 332
—, palaeosalinity indicator, 424, 426
Barium,
—, carbonate rocks, 392
—, coral reefs, 370
—, palaeosalinity indicator, 424
Barrier, 216
Barrier reefs, 353
Bathyal environment, 303
Bathymetric distribution of sediments, 407, 408, 415, 428
Batillaria minima, 67
Battle Creek, 95, 96
Bay of Bengal, 237
— —, suspension in waters, 293
Bay of Kiel, 240, 241, 456
Bay of Naples, 239, 240, 246, 247
Bays, 14, 15, 46, 47, 236—263
—, biological sediments, 251—253
—, carbonate, 246—248
—, clay minerals, 250
—, fauna, 254—255
—, graded bedding, 445
—, grain-size of sediments, 243—245

Bays, organic matter, 249
—, sedimentary structures, 255—257, 438
—, thickness of deposits, 259
Beach, 14—15, 209—224
—, carbonate sediments, 343—344
—, chemical components in sediments, 217
—, definition, 209
—, fauna, 222
—, grain-size of sediments, 212—215, 403, 409
—, gravels, 212—213
—, mineralogy of sediments, 216—217
—, morphology, 210—211
—, preservation, 386—387
—, roundness of grains, 216
—, sands, 122—123, 213—215
—, slopes, 211—212
—, thickness of sediments, 222
Beach cusps, 210
Beachrock, 343—345, 364
Bear Butte Creek, 95, 96
Bed load transport, 35, 76
Beidellite, 316
Benthos, 206, 207, 439
Bergaueria, 453
Bering Sea, 390, 391
— —, glacial sediments, 154
Bermuda, 300
—, carbonate sediments, 395
—, coral reefs, 374
Bikini atoll, 361, 365, 371, 372
Bimini Island, 66
Bimodal sediments, 83—84, 96, 97, 405—406
Biogenic factors, 11, 13
Bioglyphs, 451—453
—, beaches, 221
—, tidal flats, 275—279
Bioherms, 10, 353
Biological components of sediments, 56
Biological sediments, 49
— —, bays, 251—255
Biostromes, 353
Biotite, 416
Bituminous matter, 65, 429—430
Black clays, 427—429
— —, shallow and deep-water, 427—429
Black muds, 302
— — composition, 302
— —, organic matter, 329
Black Sea,
— —, beaches, 212
Black Sea, carbonate, 281
— —, coastal abrasion, 211
— —, pore waters, 394—396
— —, sediments, 280—282
— —, sulphides, 281—282
Blue muds, 305
— —, composition, 306
— —, Mn nodules, 331
— —, org. C., 329
Bohemian Massif, 98, 99
Borneo, coral reefs, 371
Boron,
—, deep-sea sediments, 335
—, palaeosalinity indicator, 422—423
—, sea water, 195
Brachiopoda, 253, 456
—, coral reefs, 357, 359
Brahmaputra River, 74, 91, 185
Brandwine Creek, 89
Brazil,
—, lakes, 170
—, mangrove swamps, 259—260
Brazos River, 88
Breaks in sedimentation, 385
Bromine, palaeosalinity indicator, 424
Brown clay, 302—304
— —, chemical composition, 333
— —, composition, 315—319
— —, definition, 315
— —, distribution, 317—318
— —, Mn nodules, 331
— —, org. C., 329
— —, origin, 315—316
Bruchschill, 272
Bruguiera criopetala, 260
— *gymnorhiza*, 260
Bryozoa, 253, 356, 357, 359
Bubble impressions, 276
Buffalo, 120
Buried soils, 134
Burton Lake, 405
Buzzard Bay, 240, 241, 242
Bythinia, 173

$CaCO_3$ saturation, 13
Cadmium, deep-sea sediments, 336
Calcarenites, 345, 347
Calcareous mud, 305
— —, compaction, 389—390
Calcareous ooze, see Foraminiferal ooze

Calcareous sand, 331
Calcarina spengleri, 372
Calcite in shells, 56—57, 59
Calcium, deep-sea sediments, 334
California,
—, alluvial fans, 112, 113
—, bays, 244, 246
—, beaches, 214, 215
—, deep-sea currents, 295, 296, 409
—, shelf, 244, 246
Callianasa, 457
Calliergum giganteum, 178
Ca/Mg ratio, 425
Campeche Bank, reefs, 371
Campeloma, 172
Canada, tidal flats, 264
Canary Islands, 62
Canthares cancellarius, 255
Caraico Trench, 305
Carbohydrates, 65, 66
Carbonate,
—, Baltic Sea, 281
—, bays, 246—249
—, beaches, 217
—, Black Sea, 281
—, deep-sea sediments, 327—329
—, delta, 188—189
—, environmental indicator, 429—433
—, foraminiferal ooze, 310—314
—, glacial sediments, 147—148
—, lakes, 166—169
—, mangrove swamps, 260
—, migration, 33
—, minerals, 165
—, precipitation, 69
—, sea-water, 200—201
—, shelf sediments, 231—232
—, tidal flat sediments, 27, 271
—, trenches, 308—309
Carbonate sediments,
— —, diagenesis, 392, 393, 395
— —, littoral, 343—345
— —, shallow basins, 345—349
Carboniferous sediments,
— —, fresh-water, 420
— —, marine, 421
— —, claystones, 425
Cardium, 272, 457
— *edule*, 276, 278, 284
Carex gracilis, 178
Caribbean Sea, 379
Caspian Sea,
— —, bimodal sediments, 405—406
— —, carbonates, 284
— —, dustfalls, 135, 159, 161
— —, oolites, 348
— —, sediments, 283—285
— — shells, 284
— —, stratification, 284
— —, suspension in waters, 291
Catastrophic changes, 384—385
Celtis laevigata, 179
Cave sediments, 409
Cellulose, 65, 66
Cenomanian corals, 373
Central Asia, dust falls, 130
Cephalopods, 456
Cerasophycus, 453
Ceratophyllum demersum, 178
Cerithium muscarum, 255
— *variable*, 255
Chalk, lacustrine, 168—169
Chamosite, 434
Channel sediments, 91, 103, 106
Chara, 178
Chesapeake Bay, 419
Chile,
—, coasts, 321
—, tidal flats, 264
Chitons, 355, 357
Chlorine, palaeosalinity indicator, 425, 448
Chlorinity, 197, 198
Chlorite,
—, brown clay, 316, 317
—, delta sediments, 187
—, diagenetic changes, 418—420
—, hemipelagic sediments, 305
Chlorophyll, 305
Chlorosity, 198
Chondrites, 54
Chromium, deep-sea sediments, 335, 336
Chroococcus, 168
Circulation, nearshore, 209
Clarophora, 179
Classification, sedimentary environments, 13
Clay balls, 130, 275, 276
Clay,
—, chemical composition, 419—420
—, delta sediments, 187
—, diagenesis, 390, 395

Clay, mechanical compaction, 389
Clay minerals,
— —, bays, 250, 351
— —, brown clay, 316—318
— —, delta, 187
— —, diagenetic changes, 416—420
— —, fresh-water, 416—420
— —, glacial sediments, 319
— —, loess, 133
— —, sea-water, 416—420
Claystones,
—, fresh-water, 420—421
—, marine, 420—421
Clay pebbles, 88
Cl'/HCO_3', ratio, 427
Climatic changes, 375—377
Climatic effects, 378—382
Clyde Bay, 66
Clymenella, 277
CM diagram, 404
Coastal dunes, 128
Coastal plains, 14
Coatings, Fe-hydroxides, 127, 135
Cobalt, deep-sea sediments, 336
Coccolith ooze, 311
Coccolithus fragilis, 206, 311
Coefficient of chemical-mechanical weathering, 36
Coefficient of grain-size changes, 95
Coefficient of sorting, 21
Collenia, 62
Collophanite, 321
Colorado River, 33, 34, 77, 95, 98, 100, 159, 164, 185, 187
Columbia River, 300
Concentrates, heavy minerals, 217, 219
Concretions, palaeosalinity indicator, 430—431
Conductivity method, 426—427
Congo submarine canyon, 300, 302
Connecticut River, 89
Constanze Lake, 167
Constrative phase, 81
Continental environment, 14
Continental shelf, see shelf
Continental slope, see slope
Convection currents, 203
Convolute structures, 91
Copper, deep-sea sediments, 336
Coprolites, 67
Coralline algae, 355, 356
Coral mud, 302
— —, Mn nodules, 331
Coral reefs, 14—16
— —, borings, 371—372
— —, classification, 353—354
— —, definition, 353
— —, origin, 372—373
— —, sediments, 356, 363, 371
— —, thickness of deposits, 371—373
— —, types, 353—356
Coral sand, 302, 366
Corals,
—, ahermatypic, 358—359, 364
—, hermatypic, 358—359, 364
—, growth, 49
—, reef building, 355—358, 368
—, staghorn, 368
Corophium, 276—278
— *volutator*, 277
Corophioides, 453
Cosmic dust, 44, 54—55
Cosmic material, 54—55, 310
Cosmic spherules, 54
Crassinella marticensis, 255
Creeks, 14
Crescent ridges, 210
Cr/Ni ratio, 425
Cross-bedding, 442—444
— —, beach sediments, 219—221
— —, eolian sediments, 129
— —, river sediments, 87
Crustacean, 357—359
Cruziana Facies, 453
Cr/V ratio, 425
Cryptozoon, 62
Cuba, mangrove swamps, 259
Cubichnia, 452
Cultivated land, 13
Current stratification, 442—444
Currents,
—, deep-water, 293—296
—, ocean, 293—294
—, sea, 226—227
—, tidal, 226
—, velocity, 72, 226—227, 401
Cyclopypeus, 365
Czechoslovakia, eolian sands, 127

Dahlite, 321

Danmark,
—, lacustrine sediments, 178
—, tidal flats, 264—270
Danube River, 87, 185
Darwin's subsidence theory, 371
Decay products, 63—66
Deep-sea delta fans, 302
Deep-sea sediments, 291—342
— — —, absolute age, 379
— — —, carbonates, 327—329
— — —, chemical composition, 332—337
— — —, classification, 302—303
— — —, coarse admixtures, 323—326
— — —, distribution, 304—305
— — —, gravels, 323—327
— — —, iron and manganese, 330—332
— — —, sands, 323—327
— — —, stratigraphy, 379
— — —, trace elements, 424
Deep Spring Lake, 165
— — —, dolomite, 351
Delta, 15, 16, 47, 74, 105, 183—194
—, classification, 186
—, fauna, 193
—, gravels, 187
—, mechanism of sedimentation, 402
—, morphology, 184—186
—, preservation, 386—387
—, processes, 184—187
—, sedimentary structures, 186—188, 438
—, sequences, 192—193
Denudation,
—, chemical, 27, 31
—, glacial, 29
—, limestones, 30, 31
—, mechanical, 26, 27, 31
—, rate, 25—30, 38, 42
—, silicites, 31
—, types, 25
Deposition-Transport-Erosion diagram, 73
Depth of weathering, 17
Desert, 12, 14, 16
Desert quartz, 135
Diagenesis, 389—396
—, beach sediments, 222
Diatomaceous sediments,
— —, Baikal Lake, 285
— —, bays, 252, 256, 257
— —, lakes, 175—176
— —, vertical changes, 376
Diatom ooze, 302, 303, 304
— —, chemical composition, 333, 337, 338, 339
— —, composition, 313—314
— —, definition, 313
— —, distribution
— —, Mn nodules, 331
— —, org. C, 329
Diatoms,
—, Atlantic Ocean, 205
—, coral reefs, 369
—, ice, 153
—, lagoons, 251—252
—, lakes, 168, 204
—, Pacific Ocean, 205
—, tidal flats, 270—272
Didacna, 284
Diplanteria wrighti, 254
Diploporia, 61
Dissimmetry coefficient, 413
Dnieper River, 185, 187
Dolomite,
—, Balkhash Lake, 287
—, environmental indicator, 430—431
—, foraminiferal ooze, 312, 313
—, lakes, 155, 156
—, recent, 350, 352
—, shells, 59
Domichnia, 452
Don River, 38, 48
Donax, 272
Dovey Estuary, 422
Downstream changes, rivers, 94—101
Dreissensia, 284
— *polymorpha*, 173
Drumlins, 151
Dull surface of grains, 126—127, 411—414
Dune-ripple transport, 77
Dune sands, 121—125, 215, 409
— —, heavy minerals, 414
— —, shells, 414
Dust falls, 117, 119, 130, 131, 409
— —, rate of sedimentation, 134
Dy, 176—181
Dysphotic layer, 204

East Asian shelf, 234—235
East China Sea, 245
East Indian Archipelago, 303, 305, 320
Echinocardium cordatum, 279, 457

Echinodermata, 253, 456
—, coral reefs, 357, 359
Echinoidea, 355, 357
Echinus miliaris, 278
Echiurus, 278, 457
Egypt, 348
—, pyramids, weathering, 19
Eh, 12
Eh-pH characteristics, 12
Einkippungsregel, 277
Elements, migration, 31, 32
Energy of environment, 12
England, beaches, 212
English Channel, 197
Eniwetok Atoll, 351, 371
Enrichment factor, 204
Environment of sedimentation, 13
Eolian processes, 116—118
Eolian sands, 121—127
— —, bimodality, 124
— —, grain size, 123—127, 403
— —, mineralogy, 127—128
— —, roundness of grains, 124
— —, sedimentary structures, 128
Eolian sediments, 116—137
— —, preservation, 387
— —, source, 117
— —, sedimentary structures, 437
Ephemeral streams, 85, 88, 89, 90
Epilimnion, 158
Equatorial Pacific current, 293
Equisetum heleochantis, 178
Erosion, 25
Eskers, 150
Etocha Basin, 352
Eupelagic sediments, 302
— —, composition, 309
— —, grain size, 309—310
Euphotic layer, 204
Euphrates River, 37, 91, 102, 185
Eustatic movements, 41
Evaporites, 251, 434
Excrements of organisms, 66
Eyre Lake, 167

Faecal pellets, 66
— —, Bahamas, 345, 347, 350
— —, bathyal sediments, 306
— —, glaciomarine sediments, 319
— —, shelf sediments, 232, 233
Fats, 65
Faunal associations,
— —, bays, 254, 255
— —, beaches, 222, 223
Fe-hydroxides, lakes, 170
Feldspar,
—, beach sediments, 216
—, deep-sea sediments, 316, 318
Fennoscandia, rate of tectonic movements, 41
Fishes, 355, 357
Fjords,
—, Norwegian, 307
—, postglacial changes, 383
—, sedimentary structures, 438
Flagellates, 206
Flame structure, 77
Flatness coefficient, 413
Flood-plain, 74, 76, 81, 88, 89, 103, 106
— —, deposits, 88—91
— —, graded bedding, 435
— —, mechanism of deposition, 402
— —, sedimentary structures, 437
Floods, 37, 75, 76, 89, 90, 102
Florida, 220
—, carbonates, 349, 392, 393
—, coral reefs, 353, 356, 371
—, mangrove swamps, 259, 260
—, oolites, 348
Florida Bay, 49, 62, 68, 250
— —, dolomite, 351
Flotation, 75, 78
Flowage structure, 276
Fluor-apatite, 321
Fluorine, palaeosalinity indicator, 426
Fodichnia, 452
Foraminifera, 206
—, bays, 252, 253, 310—313, 314
—, coral reefs, 356, 357, 358
—, depth indicators, 457. 458
—, tests, 328
Foraminiferal number, 310
Foraminiferal ooze, 310—314
— —, composition, 311—313
— —, definition, 310
— —, distribution, 310, 311
— —, Mn nodules, 331
— —, org. C, 329
Foreshore, 218
Forests, 13
Förna, 176, 177

Francolite, 321
Fresh water, 12
Fringing reefs, 353
Fyris River, 37
Funafuti Atoll, 371, 373

Galapagos Islands, 300
Gallium, palaeosalinity indicator, 423
Ganges River, 185
Garnet, 414
Gases, sea water, 199
Gastropoda, 253
—, coral reefs, 357, 368
Gaylussite, 166
Geosynclines, 309
Gezeitenschichtung, 364
Glacial-Postglacial transition, 380—382
Glacial processes, 138—139
Glacial sediments, 15, 44, 138—156
— —, chemical composition, 147
— —, composition, 139—153
— —, fluviatile reworking, 141—143
— —, grain-size, 140—144
— —, mineralogy, 144—147
— —, sedimentary structures, 150—152
— —, thickness, 152
Glacial transport, 138, 152
Glaciers, Alpine, 139
— —, rate of abrasion, 139
Glaciofluvial sediments, 150, 152
Glaciomarine sediments, 319
Glauconite,
—, hiatuses, 386
—, palaeosalinity indicator, 433, 434
—, shelf sediments, 232, 233
Gleocapsa, 168
Gleotheca, 168
Globigerina bulloides, 380
— *deutertrei*, 380
— *dubia*, 380
— *inflata*, 380
— ooze, see Foraminiferal ooze
— *pachyderma*, 380
Globigerinoides conglobata, 380
— *rubra*, 380
— *sacculifera*, 380
Globorotalia canariensis, 380
— *crassula*, 380
— *flexuosa*, 380
— *hirsuta*, 380
Globorotalia menardii, 380
— *punctulata*, 380
— *scitula*, 380
— *tumida*, 380
Goethite, 289
—, Mn nodules, 332
Goniastrites, 360
Gorgoniae, 368
Graded bedding, 444, 445
— —, rivers, 91, 111
— —, trenches, 308
— —, turbidites, 297
Grain-size analyses
— — —, alluvial channels, 111, 112, 403
— — —, beach sediments, 212—216, 403
— — —, cave deposits, 125
— — —, coral reefs, 366—367
— — —, eolian suspension, 118—120
— — —, glacial sediments, 403
— — —, loess, 126, 131, 132
— — —, mud-flow deposits, 403
— — —, open shelf sediments, 229—230
— — —, pyroclastic sediments, 125
— — —, river deposits, 80, 403
— — —, sea-water suspension, 292
— — —, tidal flat sediments, 267—269
— — —, turbidites, 403
Grain-size, bimodal, 83, 84, 96, 97
Grain-size curves, 403—405
Grain-size parameters, 20
Grand Banks, 300
Granite weathering, 20
Grapestone, 347
Gravels,
—, abrasion, 163, 164
—, bathymetric conditions, 407—408, 415
—, beach, 212—213
—, deltas, 187
—, grain-size, 409, 410
—, river-borne, 163, 164
—, shape, 412—414
—, shelf, 229
Great Barrier Reef, 371
Great Laba River, 95
Great Lakes, longshore currents, 210
Great Plöner Lake, 173, 174
Great Salt Lake, 62, 158, 159, 165, 166, 384
— — —, dolomite, 351
Greenland,
—, loess deposits, 130

Greenland, river denudation, 29
Green muds, 302
— —, composition, 306, 307
— —, Mn nodules, 331, 379
Grey muds, 307
— —, org. C, 321
Guadelete River, 420
Guadelupe River, 187, 448
Guayana Shelf, 236, 286
Guatemala, tidal flats, 264
Gulf of California, 242, 252, 256, 257, 390, 419
Gulf of Mexico, 187, 206, 448
— —, beaches, 215
— —, grain-size of sediments, 404, 419
— —, longshore currents, 210
— —, shelf, 234, 236
Gulf of Naples, 406, 408
Gulf of Paria, 47, 236, 237, 239, 254
— —, associations of invertebrates, 455
Gulf Stream, 293
Guyots, 373
Gypsum, lakes, 167, 168, 176, 352
Gyttja, 176—181, 288, 376
—, trace elements, 426

Halimeda, 300, 346, 355, 357, 366, 368, 372
— *macroloba*, 367
Halmyrolisis, 316—319
Hard-ground, 386
Hawaian islands, 335
Hawaian trench, 309
Heavy minerals, 414, 415, 416
— —, beaches, 217, 218
— —, glacial sediments, 147
— —, loess, 132
— —, river sediments, 94
Hecla, 320
Heliopora, 360
—, zone, 360
Helium, sea water, 199
Hemicellulose, 65
Hemipelagic — abyssal environment, 303
Hemipelagic sediments, 49, 302
— —, chemical composition, 332—333, 337 to 339
— —, composition, 306
— —, distribution, 303
— —, grain-size, 304
Hermes Reef, 356
Heteromastus filiformis, 278
Heteromastus sp., 276
Heteropoda, 311
Heterotrophic bacteria, 207
Hiatuses, 385, 386
Hidden Glacier, 29
Himalayas, denudation, 38
Holocene transgression, 381—384
Homothermy, 158
Honolulu, 356
Hudson submarine canyon, 300, 302
Humic matter, 65, 428—429
Humus, 66, 178
Hwang-Ho River, 36, 37, 102, 103, 185, 187, 239
Hydrobia, 278
— *ulvae*, 276
Hydrocarbons, production, 205
Hydrogene sulphide, sea water, 199
Hypnea, 254
Hypolimnion, 158

Icebergs, 153, 154
Illite,
—, bay, 251
—, brown clay, 316—318
—, delta, 187
—, diagenetic changes, 418—420
—, hemipelagic sediments, 305
India, shelf, 229, 230
Indian Ocean,
— —, carbonate, 329
— —, foraminiferal ooze, 312, 313
— —, nitrates in water, 203
— —, radiolarian ooze, 315
— —, stratigraphy, 379
— —, suspension in waters, 292, 293, 304
Indus River, 185
Inert gases, sea water, 199
Initial movement, 76
Inland seas, 12, 14, 45, 46
— —, characteristics, 280—281
— —, mechanism of deposition, 402
— —, preservation of sediments, 386
— —, waters, 280
Instrative phase, 81
Intercontinental seas, 15
Intermountainous troughs, 14, 16
Iodine, palaeosalinity indicator, 425, 426
Iraq, 93
Iron,

Iron, deep-sea sediments, 318—319, 330—332
—, diagenesis, 390—391, 395, 396
—, horizons, 385—386
—, sea-water, 195, 197
Iron-montmorillonite, 289
Isla Perez, coral reefs, 368
Island arcs, 307, 308
Isotopes, 380, 381
—, carbon, 435
—, oxygen, 344, 349, 434
—, sulphur, 435
Issyk-kul Lake, 163, 407

Jade Bay, 265, 268, 272
Japan, continental slope, 321
Java, 320
Juncus roemerianus, 179

Kamchatka, 320
Kames, 150
Kansas River, 89
Kaolinite,
—, bay sediments, 251
—, brown clay, 316, 317
—, diagenetic changes, 418—420
Kapingaramangi Atoll, 361, 367, 369, 370
Kara-Bougas-Gol Bay, 167, 384
Karelia, 152
Katmai, 320
Kelp, 324
Kemka River, 74
Krakatoa Island, 17, 19, 320
Krivan Lake, 351
Knudsen's formula, 198
Kish Chai River, 75, 76
Kita-Daito-Zima, 371, 373
Kolyma River, 185
Kosi, River, 29
Kura River, 33, 38, 284
Kuriles, 320
Kuril-Kamchatka trench, 206, 207, 307, 308, 309, 407
Kuro-shio current, 293
Kyrkfjörden fjord, 424

Lacustrine sediments, 44, 157—182
— —, biochemical, 165—167
— —, chemical, 165—167
— —, classification, 161
— —, clastic, 161—164
Lacustrine sediments, development, 180—181
— —, distribution, 180—181
— —, grain-size, 162—164
— —, mineralogy, 165—168
— —, organic, 171—179
— —, preservation, 387
— —, sedimentary structures, 437
Lagoon of Terminos, 241
Lagoons, 14, 15, 46, 47
—, atolls, 362, 373
—, postglacial development, 383, 384
—, tropical and subtropical, 248, 249
La Jolla submarine canyon, 300, 302
Lake Mead, 166
Lakes, 10, 14
—, classification, 157—159
—, dystrophic, 159—161
—, eutrophic, 159—161, 175
—, oligotrophic, 159—161, 175
—, Swedish, 424
Lamellibranchiata, 457
Laminary flow, 72
Lamination, 87—89, 91, 189
—, bay sediments. 252, 256, 257
—, Black Sea sediments, 282
—, diatom ooze, 314
Lanice, 278
— *conchilega*, 276, 277, 278, 457
Lanis coheni, 278
Last Glacial, 377, 378, 379, 380, 381, 382
Lateral accretion, 88, 103, 437
Lena River, 89, 185
Leptoseris incrustans, 361
Levee, 14
Libya, dunes, 129
Lignin, 65, 66
Limestone denudation, 30
Limnaea, 172, 173
— *stagnalis*, 457
Limnocardium, 457
Liggula, 456, 458
Lipoids, 66
Lithium, 424
Lithological criteria, 435—453
Lithophyllum, 346
— *incrustans*, 51
Lithothamnion lenormandi, 51
Lithothamnium ridge, 360, 361, 362
Lithothamnium zone, 361
Lithuania, glacial sediments, 152

Littoridina sphinctostoma, 255
Littorina, 276
— *littorea*, 278
Load,
—, dissolved, 27
—, suspended, 27, 38
Loess, 130—134
—, composition, 132—133
—, grain-size, 126, 131—132
—, preservation, 387
—, sedimentary structures, 134
Logilidae, 456
Longshore currents, 209, 210
Lower Godawari River, 86, 95, 97, 100
Luga River, 74
Lutraria elliptica, 276
Luzern Lake, 176

Macoma baltica, 276
— *mitchelli*, 255
— *tagelliformis*, 253
Mactra, 272, 457
Madrepora, 358
— *cribripora*, 358
— *ramea*, 359
Magdalena River, 300
Magnesium,
—, coral reefs, 370—371
—, deep-sea sediments, 334
—, palaeosalinity indicator, 425
Magnetic spherules, 54
Magnetite, 414
Manganese,
—, Baltic Sea sediments, 289
—, carbonate sediments, 392
—, deep-sea sediments, 318—319, 330—332
—, lakes, 171
Manganese nodules, 331—332
— —, chemical composition, 333—334
— —, composition, 331—332
— —, origin, 331—332
— —, trace elements, 334
Manganite, 289, 332
Manganosiderite, 259
Mangrove swamps, 14, 249, 260
Mariana Trench, 307
Marginal barrier, 14
Marginal estuary, 14
Marginopora, 372
— *vertebralis*, 367
Marine environment, 14
Marl, lacustrine, 168—169
Marsh,
—, brackish, 179
—, fresh-water, 179, 409
—, salt-water, 179
—, sedimentary structures, 437
—, sediments, 106
Md-So diagram, 402—403
Mean, grain-size, 22
Mechanical factors, 11, 12
Median, grain-size, 21
Mediterranean Sea, 205, 212, 303, 305, 311
— —, coral reefs, 359
— —, grain-size of sediments, 404, 407
Megaripples, 87
Mendota Lake, 172
Mercenaria compochiensis texana, 255
Merimac River, 125
Mesopotamia, 43
Mesozoic, bathymetric conditions, 458
Mesquite Bay, 448
Metalimnion, 158
Methan, sea water, 199
Mg-calcite, 57—59
—, coral-reefs, 367, 368, 369, 371
—, lakes, 165, 167
—, shells, 59, 61
Mica, shelf sediments, 416
Michigan Lake, 163, 164, 169, 407
Microcystis, 168
Micro-foraminifera, 369
Microtectites, 54
Mid-Atlantic Ridge, 311
Midia clay, 280
Mid-Pacific Ridge, 373
Millepora, 355
Miriophyllum, 178
Mississippi River, 33, 47, 87, 95, 97, 100, 102, 105, 106
— —, delta, 184, 185, 190, 191, 193, 449
— —, delta swamps, 179
Mobile Bay, 449
Mohenjo Daro, 43
Molasse Tertiary, Alps, 416
Molluscs, shells, 58, 59
Monodacna caspia, 284
— *edentula*, 284
Mono Lake, 167
Montastrea annularis, 49

Montipora marshallensis, 361
Montmorillonite,
—, bay sediments, 251
—, brown clay, 316, 317
—, delta sediments, 187
—, diagenetic changes, 418—420
Moraines, 138
—, ablation, 138, 141, 151
—, basal, 138, 139, 141, 144, 146, 151
—, end, 138
Mosses, 169
Mottled structure, 89, 91, 256
Mountain ranges, 14, 16
Mud cracks, 88, 92, 275, 448, 449
Mudflow processes, 111, 112
— sediments, 111, 112, 149, 297, 409
Mulinia lateralis, 255
Murray Lake, 356
Mya arenaria, 278, 454
Mytilus, 272, 278
— *californianus*, 56
— *edulis*, 276

Nassa reticulata, 278
Nassarius acutus, 278
Natural concentrates, 416
— —, beaches, 217—218
— —, rivers, 94
Natural levee, 80
Nephtys, 276, 457
Nereis, 277, 278
— *diversicolor*, 276
Nereites, 453
—, facies, 453
Neritina, 173
— *reclinata*, 255
Netherlands, 17, 48
—, swamps, 179
—, tidal flats, 264, 269, 272
Neusiedler Lake, dolomite, 352
Nevada, alluvial fans, 113
New Hampshire, 18
New Zealand, dust falls, 130
Nickel,
—, deep-sea sediments, 336
—, lacustrine sediments, 171
Niger River, 185
Nile River, 38, 48, 90, 185, 187
Nitrates, lake water, 176
Nitrogen cycle, 69
Nitrogen, sea water, 119, 203
Non-sedimentation, 51
Nontronite, Mn nodules, 332
North America, glacial sediments, 143, 148
North Germany,
— —, lakes, 164, 167, 168, 407
— —, tidal flats, 264, 265, 266, 267, 268, 269, 272, 274, 275, 277, 278
Norway,
—, continental shelf and slope, 417
—, coral reefs, 359
Nostoc, 168
Nuculana eborea, 255
Nuphar luteum, 178
Nutrients, lake water, 176
Nymphaea alba, 178

Oceans, 49
Ohio River, 78
Ohio Valley, 89
Oligocene,
—, reefs, 372
—, unconformity, 372
Olivella mutica, 255
Ooids,
—, Bahamas, 345, 346, 347
—, Caspian Sea, 284
—, lakes, 166
Oolites, 348, 349
Opal, 252
—, deep-sea sediments, 318
—, Mn nodules, 332
Open shelf, 14
Ophiuroidea, 355, 357
Orbiculina universa, 379
Ores, lacustrine, 170—171
Organic acids, 65
Organic matter,
— —, bay sediments, 249
— —, black clays, 427, 428, 429
— —, Black Sea sediments, 282
— —, composition, 66
— —, deep-sea sediments, 329—330
— —, diagenesis, 390, 391, 395
— —, mangrove swamps, 260
— —, palaeosalinity indicator, 425—429
— —, production, 13, 64
— —, shelf sediments, 232
— —, tidal flats, 271
— —, trenches, 308

Organic substances, migration, 32
Org. C/org. N ratio, 65
— — —, Baltic Sea sediments, 289
— — —, bay sediments, 249
— — —, lakes, 178, 179
— — —, mangrove swamps, 260
— — —, shelf sediments, 232
Orientation of clastic components,
— — —, beach sediments, 216
— — —, environmental indicators, 449—451
— — —, eolian sediments, 130
— — —, river sediments, 92—93
Orinoco River, 48
Orléansville earthquake, 300
Ostrava-Karviná coal basin, 426
Overbank deposits, 89, 91
Oxidation-reduction potential, 12
Oxygen, 198, 199
— minimum layer, 199
Oyster bioherms, 252

Pacific Ocean, 68, 304
— —, atolls, 354, 355, 360, 361, 362, 363
— —, brown clay, 316, 317
— —, carbonate in sediments, 328—329
— —, carbonate in waters, 200
— —, deep-sea sands, 325
— —, diatomaceous ooze, 316
— —, diatoms, 205
— —, foraminiferal ooze, 311, 322
— —, Mn nodules, 332
— —, nitrates in water, 203
— —, organic matter, 197
— —, phillipsite, 323
— —, pyroclastic sediments, 320—321
— —, radiolarian ooze, 315
— —, silica in water, 197
Palaeontological criteria, 453
Palaeosalinity, 421—422, 423—428
Palaeozoic, depth indicators, 458
Palagonite, 318, 321
Paleodictyon, 453
Paludina, 173
Panicum repens, 179
Paramshir Isle, 228
Pascichnia, 452
Patelina corrugata, 61
Patch reefs, 356
Peat, 105, 178, 179
Peat bogs, 45, 178, 179
Pebbles,
Pebbles, glacial sediments, 144—146, 150
—, orientation, 92, 93, 111, 150
—, river sediments, 92, 93, 111
—, shape, 148, 149
—, stability in transport, 99, 100
—, surface textures, 149, 150, 413, 414
Pectinaria, 277
Pelagic-abyssal environment, 303
Pelagic ocean, 12, 14
Pelagic sediments, 302, 303
Pelecypoda,
—, coral reefs, 355, 356, 357, 368
—, lakes, 168, 170, 172, 173, 253
Peloglea, 66, 348
Percussion marks, 127
Persian Gulf, 134, 245, 249, 344, 345, 348, 349
Perstrative phase, 81
Peru—Chile Trench, 313, 320, 351
Petrographic methods, 400—405
pH, 12, 68, 69, 427
—, sea water, 199, 200
Phaeophycae, 456
Phaseolina clay, 280
Phialidium, 277
Phi Deviation Measure, 21
Phillipsite, 317, 318, 322, 323
Phi scale, 21
Phi Skewness Measure, 21
Pholas candida, 276
Phosphates, 44
—, lake waters, 176
—, sea waters, 202
—, shelf sediments, 233, 386, 434
Phosphorite, 321
Photosynthesis, 67, 68, 205
Phragmites communis, 178, 179
Phycodes, 453
Physa, 171, 172
— *ancillaris*, 171, 172
— *gyrina*, 171, 172
— *heterostropha*, 172
Physico-chemical factors, 11. 12
Phytoplankton, 65, 67, 68, 204, 205, 455
Pigments, 65
Pinnus-Järvi Lake, 170
Pisidium, 171, 172
— *abyssorum*, 172
— *solitudum*, 172
Planar cross-stratification, 442, 444

Planorbis, 172
Planorbis, bicarinatus, 171, 172
— *companatus*, 172
— *parvus*, 172
Plant associations,
— —, marshes, 178
— —, swamps, 178, 179
Playas, 409
Pleistocene,
—, beaches, 222
—, carbonate sediments, 392, 393
—, sands and gravels, 398, 399
—, shelf sediments, 228, 229, 243
Pleurocera elevatum, 171, 172
Point bars, 14, 87
Pocillopora, 361
Poland, dust falls, 130
Polished grains, 410—412
Pollen, 369
Polysacharids, 65
Populus deltoides, 179
Pore waters, 394, 395, 396
Porifera, 253
Porites, 360, 361
— *limosus*, 358
— *porites*, 355
— *zone*, 360, 361
Post-Glacial eustatic changes, 383, 400
Potamogeton foliosus, 179
Potomac River, 89
Pozzuoli Bay, 407
Preservation of sediments, 386, 387, 388
Prielen, 264, 268
Proteins, 65, 66
Protodolomite, 165, 166
Pteropod ooze, 303, 304, 311
— —, chemical composition, 333
— —, Mn nodules, 331
Pteropods, 311
Puerto Rico Trench, 307, 309
Pulleniatina obliquiloculata, 380
Pygospis, 276
Pyrite,
—, grains, 414
—, palaeosalinity indicator, 426
—, sediments, 250
Pyroclastic sediments, 319—321
— —, grain-size, 125

Quartz/feldspar ratio, 100

Quercus virginiana, 179
Radioactive isotopes, 195, 197
Radioactivity, deep-sea sediments, 323
Radiolaria, 204, 206
—, depth indicator, 457
Radiolarian ooze, 302—304
— —, chemical composition, 333
— —, composition, 313, 314, 315
— —, definition, 313
— —, distribution, 313
— —, organic matter, 329
— —, trace elements, 333
Ragunda Lake, 17
Raindrop impressions, 276
Rangia cuneata, 255
— *flexuosa*, 255
Rapid Creek, 95, 96
Rare earths, deep-sea sediments, 335
Rate of denudation, 25
Ratio light/heavy minerals, 414
Recent sediments, characteristics, 398—399
Red clay, see brown clay
Red muds, 302
— —, composition, 302
— —, Mn nodules, 331
— —, organic matter, 329
Red Sea, 264, 289, 345
— —, grain-size of sediments, 404
— —, oolites, 348
Reef flat, 360, 362
— front grooves, 366
— knolls, 370
— limestones, trace elements, 432
— slope, 362
— talus, 362
— terraces, 362
Relic sediments, 228, 385
Repichnia, 452
Resins, 65, 66
Rhine River, 100
Rhizocorallium, 453
Rhizopora, 260
— *conjugata*, 260
— *mucronata*, 260
Rhodophyceae, 456
Rhône River, 408
— —, delta, 106, 187, 189, 192, 193
Rill marks, 221. 276, 449, 450
Rio Grande, 82, 86
Rip currents, 209, 210

Ripple bedding, 87, 89
Ripple-drift lamination, 91, 275
Ripple fields, 447
Ripple marks, 444—448
— —, beaches, 219, 221
— —, eolian, 116
— —, genetic classification, 441
— —, rhomboid, 221, 446, 447
— —, river, 77, 78, 87, 91
— —, tidal flats, 275—276
Riukiu Trench, 307
River channels, 80
— —, deposits, 81
— —, graded bedding, 445
— —, mechanism of deposition, 402
— —, sedimentary structures, 437
Rivers, 15
—, Caucasian, 83
—, discharge, 83
—, ice, 152
—, lowland, 82
—, mountainous, 82
—, sediments, 403
—, suspension, 33—40
—, valleys, 14, 16
Rivularia, 168
Romanche Deep, 325
Rongelap Atoll, 362
Roundness,
—, beach sand, 216
—, changes downstream, 100—101
—, grains, 410, 411, 412, 413, 414
Ruppia maritima, 179
Russian Plateau,
— —, glacial sediments, 140, 143, 152
— —, rivers, 35, 36
Rutile, 332

Sabellaria, 277, 278, 279
Sacramento River, 39
Sage Creek, 88, 90
Sagittaria, 178
Sahara,
—, dust stroms, 119
—, formations, 412
Saipan Island, 356, 357, 360, 371
Salinity, 197, 198
Salix migra, 179
Saltation, 76
—, eolian, 117
Salt Flat Graben, 165
Salton Lake, 159, 164
Salts, diagenesis, 395
San Antonio Bay, 448
Sand banks, 94
Sand bars, 86, 217, 218
Sands, 163
—, bathymetric conditions, 407—408, 415
—, beaches, 213—218
—, delta, 187
—, lakes, 163
Sand-silt-clay ratio, 405, 406
Sand waves, 72, 86, 87
San Miguel Lagoon, 251, 402
San Salvador, tidal flats, 264
Santa Barbara Basin, 407
Sapelo Island, 132
Sapropel, 176, 181, 376
—, chemical composition, 426
—, trace elements, 426
Santorini, 320
Sargasso Sea, 203, 205
Sargassum, 355
Scandinavia,
—, denudation, 29
—, lakes, 170
Scaphopoda, 253
Scattered transport, 76
Schill, 272
Schizotrix fasciculata, 167
— *lateritia*, 167
Schlick, 269, 270, 272
Schönau Lake, 169
Scirpus americanus, 179
— *olneyi*, 179
Scoloplogs armiger, 276
Scour and fill, 77, 78
Scrobicularia plana, 276, 277, 278
Sea ice, 152, 153
Sea mammals, 326
Sea of Japan, 381
Sea water, 12
— —, biology, 201—208
— —, carbonate system, 200—201
— —, chemical composition, 195—201
— —, gases, 198, 199
— —, nitrogene cyclus, 202, 203
— —, pH, 199, 200
— —, phosphates, 202
Sebastian Viscaino Bay, 247, 248, 249

Sedimentary environments, 11
Sedimentation rate, 42, 43, 53
Seekreide, 168—169
Seif dunes, 129
Seine River, 185
Selective transport, 89, 97, 98
Seneca Creek, 89
Seriatopora, 360
Shallow-marine sediments,
— — —, classification, 225
— — —, definition, 225
Shatt-al-Arab, 38
Shark Bay, 63
Sheetfloods, 25, 26, 72
Shelf, 15
—, mechanism of sedimentation, 402
—, sediments, 303, 399, 400
Shelf, open,
— —, chemical composition of sediments, 233
— —, glauconite, 232, 233
— —, grain-size, 229—231
— —, morphology, 227
— —, phosphates, 233
— —, preservation of sediments, 366, 387
— —, sedimentary structures, 234, 348
— —, sediments, 228—234
— —, thickness of sediments 234
Shelf, sheltered, 235, 263
— —, preservation of sediments, 386—387
Shell,
— banks, 284
— ridges, 430
— zone in lakes, 173—175
Shells,
—, composition, 56—60
—, trace elements, 430
Siderite, 431
Silica,
—, amorphous, 314
—, foraminiferal ooze, 312
—, lake water, 176
—, sea water, 252
Siliceous ooze, 304, 313—315
Siliceous sediments, diagenesis, 393, 394
Silt, deltaic sediments, 187
Shoeshoe River, 103
Skeletal material, 57
Skolithos Facies, 453
Slide deposits, 303
Slope, continental,
Slope, continental, mechanism of sedimentation, 402, 409
— —, sedimentary structures, 438
Smooth traction, 77
Smooth transport, 76
Soft bodies of organisms, 63
Soils, 12
—, arctic, 22, 23
—, coastal, 409
—, formation, 17, 19
—, tropical and subtropical, 22, 23
Solicyclus, 453
Solution load, 35
Sonora Desert, 120
South America,
— —, coast, 227
— —, deep-sea currents, 294
South Bonaire, 249
South Canadian River, 96, 100
South China Sea, 245
Southern California, 238, 394
Spargentum natans, 178
Spartina alterniflora, 179
— *patens*, 179
Sphaerium, 173
— *occidentale*, 171, 172
Sphaerodinella defiscens, 380
Sphagnum marsh, 178
Sphericity of grains, 411, 412
— —, changes downstream 100, 101
Spits, 216
Sponges, 59
—, coral reefs, 355, 357, 359
Sponge spicules, 311, 313, 369
Spongia ottoi, 277
Spülsäume, see tidal ridges
Starches, 65
Steppe land, 13
Sternberg law, 98
Straits, 409
Stratification, 440—444
—, bays, lagoons, 255—259
—, beach, 218—221
—, causes, 376—379, 437—438
—, deep-sea sands, 327
—, delta, 189, 192
—, Recent-Pleistocene, 376
—, shelf, 234
—, tidal flats, 274—275
Streaky bedding, 91

Streamfloods, 25, 26, 72
Stromatolites, 49, 62—63, 67
Strontium,
—, carbonate rocks, 392
—, coral reefs, 370, 371
— deep-sea sediments, 334
—, palaeosalinity indicator, 424, 431—433
—, shells, 61, 62
Structures, sedimentary, 437—438
— —, bays, lagoons, 255—259
— —, beaches, 218, 219—259
— —, delta, 189, 192
— —, eolian sands, 128—130
— —, primary, 439—440
— —, secondary, 439—440
— —, shelf, 234
— —, tidal flats, 274—277
Stump Pass, 220
Stylaster, 359
Stylopora, 372
Subaerial delta, 14
Subenvironments, 15
Submarine canyons, 227, 301—302
— —, sedimentary structures, 301—302
— —, sediments, 301—302
Succinea, 172
Sugars, 65
Sulphates, diagenesis, 391
Sulphides, 272
—, Baltic Sea, 288
—, Black Sea, 281, 282
—, diagenesis, 390, 391, 395
—, Red Sea, 289
Sulphur cyclus, 79
— —, diagenesis, 390—391
Sumatra, mangrove swamps, 259
Surf, 210, 211
Surface features of grains, 126, 127
Suspended matter, sea water, 291, 292, 293
Suspension,
—, eolian, 117
—, grain-size, 117—119
—, mineralogy, 120—121
Suspension load, 34—35, 39
— —, chemical composition, 35, 36
Suva, Fiji, 300
Swamps, 45, 103, 168, 178, 179
Swash, 210
Szeged, 119
Table reefs, 353
Tahiti, coral reefs, 356
Tajo River, 300
Talus, sediments, 114
Taman Bay, 383
Tampa Bay, 242
Tannin, 65
Tectites, 54
Tectonic movements, 41
— —, rate, 41, 42
Teichichnus, 453
Terrigenous sediments, see hemipelagic sediments
Tertiary,
—, deep-sea sediments, 337
—, ecological conditions, 457
—, foraminifera, 458
Tessin River, 95
Texas Bay, 232, 238, 252, 255, 257, 258
Thalaosincides, 453
Thalassia, 355, 357
Thalassogenic elements, 195
Thalassophilous elements, 195
Thermal zoning of water, 158, 159
Thoracophelia mucronata, 222
Th/U ratio, 425
Tidal creeks, 264, 266, 273
Tidal currents, 264
Tidal flats, 12, 48
— —, carbonate, 271
— —, definition, 264
— —, fauna, 277—279
— —, grain-size, 268—270
— —, mineralogy, 269, 270
— —, organic matter, 271, 272
— —, sedimentary structures, 274, 277, 438, 443, 444
— —, sulphides, 272
Tidal ridges, 266, 267, 268
Tidal stratification, 264, 274, 275
Tides, 210
Tigris River, 102
Till, 138
Titanium, deep-water sediments, 335
Tonga Trench, 309
Tonkin Bay, 235, 245
Tourmaline, 414, 416
Trace elements,
— —, deep-sea deposits, 333—334
— —, Mn nodules, 332

Trace elements, palaeosalinity indicators, 422—427
— —, shells, 58
Transitional environments, 14
Transport in rivers, 75—79
— — —, bed load, 76
— — —, flotation, 78
— — —, saltation, 76
Transport in suspension, 75
Trenches, deep-water, 14, 15
— — —, graded bedding, 445
— — —, mechanism of deposition, 402
— — —, morphology, 307—309
— — —, sedimentary structures, 438
— — —, sediments, 307—309
Trenches, Pacific Ocean, 301, 307, 308, 309
Trilobites, 457
—, depth indicators, 457
—, Devonian, Carboniferous, 457
—, tracks, 453
Trona, lakes, 166
Tropical weathering, 17
Trough cross-stratification, 442, 444
Turbidites, 303
Turbidity currents, 296—301
— —, definition and characteristics, 296—297
— —, origin, 299
— —, recent, 300, 301
— —, submarine canyons, 301, 302
— —, trenches, 308, 309, 338
— —, types, 298, 299
— —, velocity, 298
Turbo, 372
Turbonilla, 255
Turbulent flow, 72
Typha latifelia, 179
Tyrphopel, 177, 178
Tyrrhenian Sea, 320, 422

Ulva, 254
Unio, 173
Ural River, 187
Uranium, Baltic Sea, 289
Urionic acid, 65

Valvata tricarinata, 171, 172
Vanadium, deep-sea sediments, 335
Vaterite, 57
Vaucheria, 179
Velocity of currents, 12
Venus, 272
Vesuvius, 320
Visla River, 38
Volcanism, submarine, 314, 321
Volcanic glass, 318
Volcanic material,
— —, loess, 132
— —, trench sediments, 308
Volcanic mud, 331
Volga River, 33, 74, 87, 90, 95, 185, 187

Waddens, see tidal flats
Wades, see tidal flats
Wash, 48, 264, 268, 269, 270, 272, 273
Wattenpapier, 272
Watts Branch, 89
Wave action, 211, 216
Wave energy, 12, 401
Waxes, 65, 66
Weathering, 17
—, chemical 37
—, rate, 17, 42
Wiesenkreide, 168
Wilcox Playa, 55
Worms, 427
—, coral reefs, 357, 359

Xenophora, 204

Yangtze Kiang River, 38
Yellow Sea, 235, 239
Yoldia limatula, 439, 452
Yuba River, 39
Yucatan, tidal flats, 264

Zambezi River, 185
Zizianopsis miliacea, 179
Zircon, 414
Zirconium, deep-sea sediments, 335
Zoophycus, 453
Zoophycus Facies, 453
Zooplankton, 65, 205, 206, 207, 456
Zooxanthellae, 358
Zostera, 300
Zurich Lake, 299, 378